ELEMENTARY COLLEGE MATHEMATICS

Sol Weiss

West Chester State College
West Chester, Pennsylvania

PRINDLE, WEBER & SCHMIDT, INC.
BOSTON, MASSACHUSETTS

To Joshua

Third printing: July 1978

Library of Congress Cataloging in Publication Data

Weiss, Sol.
Elementary college mathematics.

Includes index.
1. Mathematics—1961- I. Title.
QA39.2.W443 510 76-54344

ISBN 87150-217-8

Printed in the United States of America.

PREFACE

Elementary College Mathematics is intended as a text for prospective teachers of elementary school mathematics. It can also be used as a reference and sourcebook for those who are already teaching. The book assumes a minimal background in mathematics on the part of the student.

Every chapter, except Chapter 1, begins with a statement of *objectives* and *prerequisites*, which is followed by three main parts: *introduction*, *content*, and *enrichment material*. The introductions, written very informally, motivate the content of the chapters. The content, which focuses on important mathematical ideas and their interrelationships, includes a variety of *comments* that provide a broader view or deeper insight into the concept under discussion. Enrichment material, called Extended Study, gives interested students a higher-level development of the content.

In addition to the regular exercises throughout the chapters, each chapter has a section called Basics Revisited, which has computational exercises designed to refresh the student's memory. Each chapter also includes several mathematical puzzles related to the chapter content. Chapter 10 has five main sections dealing with different topics in geometry. If all sections are used, there is enough material for a course in geometry for elementary teachers.

Many years of experience in the training of teachers at both the pre-service and in-service levels and in the teaching of introductory mathematics courses to liberal arts students have contributed to the development of this book.

A companion book to *Elementary College Mathematics* is available to teachers and teachers in training. The book, *Elementary Mathematics: Teaching Suggestions and Strategies*, puts at the disposal of the teacher a wealth of materials, ideas, suggestions, and strategies for teaching the content in *Elementary College Mathematics.* While each book can be used independently of the other, or in combination with other books, these two books, together, offer a unique combination of mathematical content and teaching strategies for the teacher of elementary mathematics.

I would like to thank Professors Bruce Allen (University of Maine, Gorham), C. Michael Lohr (Virginia Commonwealth University), Harold L. Schoen (University of Iowa), David O'Neil (Georgia State University), J. W. McGhee (University of California, Los Angeles), Gary Musser (Oregon State University), and Roberta Flexner (University of Colorado) for their valuable suggestions. I would also like to thank the staff at Prindle, Weber and Schmidt for their generous help, especially Eileen Katin for her conscientious production work and Anne Schick for her competence in carrying through her editorial responsibilities.

Sol Weiss

CONTENTS

Chapter 1

WHAT IS MATHEMATICS?

INTRODUCTION

At the National Radio Observatory in Green Bank, West Virginia, stands a giant saucer-shaped antenna pointed at two stars named Tau Ceti and Epsilon Eridani. This antenna is poised to receive any message that may be sent to earth from intelligent beings on the planet of another star.

Should such a message ever be received, what will it be like and how will it be decoded? Scientists believe that the one kind of message most likely to make sense to any intelligent form of life anywhere would be a mathematical one, and that the first messages might consist of a simply coded bit of arithmetic such as a series of beeps that might mean: "Counting, one, two, three." "Dot-dash-dot"—pause—"dot—dot" might mean: "One plus one equals two." Once these signals have been acknowledged, more complex bits could then follow to establish a basic vocabulary for further communication.

These conjectures attest to the universality of mathematics, a quality that has made mathematics a vital part of not only all human endeaver but probably essential to any intelligent form of life that may exist on other planets.

What is mathematics? Mathematicians and philosophers have been trying for centuries, and without much success, to give a simple definition of mathematics. Some think of mathematics as "the science of numbers" or "the science of quantity." The Pythagoreans, for example, understood mathematics to be the answer to the two questions, "How many?" and "How much?" But to think of mathematics only as "the science of quantity" is to overlook the fact that mathematics includes geometry, algebra, calculus, and more than 80 other subjects, much of which has nothing to do with quantity.

Perhaps a more fruitful way of thinking of mathematics is to consider its *method* rather than the subjects it embraces. Is there a characteristic method used in mathematics no matter what its subject matter? Yes, there is and it is characterized by a *special way of thinking* about problems. When we apply this way of thinking to space, we end up with a subject called geometry. When we apply this way of thinking to number, we end up with a subject called arithmetic. When we apply it to the study of rate of change, we end up with a subject called calculus. This special way of thinking underlies all of mathematics no matter what its subject matter.

1.1 Mathematical Thinking

DEDUCTIVE REASONING

What is this special way of thinking that characterizes all of mathematics? Suppose you are told that John is taller than Frank. What can you say about Frank in relation to John? Of course, you can say that Frank is shorter than John.

Notice that to obtain your answer you did not need to do any computing with numbers, nor to count, measure, or draw figures. The only thing you had to do was to *reason*. You said that if a certain thing is true, then a certain conclusion has to follow. That is, *if* John is taller than Frank, *then* Frank must be shorter than John.

This kind of reasoning, where we say that if something is true then something else must *necessarily* be true, is called *deductive reasoning*. The *if* part is called the *hypothesis*, and the *then* part is called the *conclusion*. Deductive reasoning is at the heart of that special way of thinking that characterizes all of mathematics. Deductive reasoning plays such a crucial role in mathematics that mathematics has been defined by at least one mathematician as "the science which draws necessary conclusions."

INDUCTIVE REASONING

Now suppose you are given the set of numbers 2, 4, 6, 8, 10, 12, 14, . . . and are asked to name the next number in the set. Most people would say 16. The reason most people would say 16 is that it *appears* that the set of numbers in question consists of even numbers arranged in consecutive order. But *must* this be so? Is there anything in the statement of the question which makes this assumed pattern necessarily true? Can we prove that the next number is 16? The answer is that the next number is *not necessarily* 16. It can be any other number. There is, however, a *greater probability* that the next number is 16 because our conjecture works in seven instances. Our confidence in a conjecture will be greater or smaller depending on the number of instances where it works. So that in these three cases,

	Set	*Next Number in Set*
1	2, 4, 6, 8, 10, 12, 14, . . .	16
2	2, 4, 6, 8, . . .	10
3	2, 4, 6, 8, 10, 12, 14, 16, 18, 20, 22, 24, 26, 28, 30, . . .	32

we have the greatest confidence in the answer to *3* and the least confidence in the answer to *2*. In each instance, however, we have drawn only a *probable* conclusion.

Reasoning of this kind, where you draw a conclusion from a limited number of specific examples, is called *inductive reasoning*. Inductive reasoning is another special way of thinking used in mathematics. It is the way of thinking used by the scientist in his experiments and by the mathematician in discovering new ideas. Deductive reasoning is used to *prove* conclusions, while inductive reasoning is used to discover ideas, make conjectures, and suggest possibilities.

After a scientist tries out a new vaccine on a limited number of patients, how does he proceed to prove (or demonstrate) whether the vaccine is the solution to his problem? He tries it out on more and more patients. If the vaccine works on a certain percentage of the cases, it is considered a solution. A mathematician, too, may believe that he has arrived at a correct conclusion by having seen it work in a number of specific examples. But, unlike the scientist, he does not accept corroboration by more and more cases as *proof* that his conclusion is valid. Ultimately, the only way he can *prove* his conclusion is by deductive reasoning.

The important difference, then, between deductive and inductive reasoning is that in deductive reasoning our conclusions are *inescapable*, while in inductive reasoning they are only *probable*. The fact that deductive reasoning leads to inescapable conclusions implies that all such conclusions are contained implicitly in the hypothesis. In this sense, deductive reasoning is a roundabout way of saying that A is A. However, by using deductive methods we make explicit the implications of the hypothesis.

1.2 *Russell's Definition of Mathematics*

Let us go back to the statement that "If John is taller than Frank, then Frank is shorter than John." Note that the validity of our conclusion does not depend on the truth of the hypothesis. Even if it were not true, in fact, that John is taller than Frank, our conclusion would still be valid because it is based on the *supposition* that John is taller than Frank. That is, we assert that Frank is shorter than John *if* John is taller than Frank. We say nothing about whether John is actually taller than Frank, and for purposes of logical reasoning we don't care whether he is or he is not taller than Frank. We are concerned here only with what is true in the *hypothetical* situation that John is taller than Frank.

Furthermore, we don't care whether the statement refers to John and Frank, Joe and Don, or to any other two people or things. Our only concern is with the conclusion that if *anything* is taller than something else, then the second thing must be shorter than the first thing. The validity of our conclusion does not depend on the *meaning* of the hypothesis, but only on its *form*.

Since the meaning of the hypothesis does not affect the validity of the conclusion, we prefer to express our argument in a more generalized form: if x is taller than y, then y is shorter than x. This conclusion is independent of any particular subject matter and is valid regardless of any concrete meanings we give to x and y and regardless of whether x is actually taller than y. It is this extraordinary use of abstract symbols that gives mathematics so much of its power and universality.

This is the meaning of Bertrand Russell's witty definition of mathematics as "the subject in which we never know what we are talking about nor whether what we are saying is true."

To say that the process of reaching valid conclusions is independent of physical truth is not to say that mathematics is independent of the material world. The history of mathematics shows that most basic mathematical concepts have their roots in people's physical needs. For example, the basic concept of counting is behind the more abstract concept of number, and Sir Isaac Newton developed the calculus in order to describe the behavior of moving objects. So even if the method of mathematics involves abstract reasoning, the vitality of mathematics arises from the fact that its concepts are rooted in the physical world, and its results find application in our daily lives.

1.3 Conclusion

Mathematics is not a static body of eternal truths that was discovered by a few mathematicians centuries ago. On the contrary, mathematics is improving, changing, and growing every day to such an extent that the twentieth century is already regarded as the "golden age" of mathematics. On its growth depends progress in every facet of human activity from the most prosaic daily activities to the most fundamental investigations into the secrets of the universe.

One need not be a mathematician to enjoy mathematics. Mathematics is fascinating to many people not only because of its utility but also because of its opportunities for discovery, because it offers insights into everyday affairs, and because it can be fun. Mathematics is a subject where with only a few basic symbols at our command, we come out with the most amazing formulas and conclusions. Some of these have completely changed the face of the earth.

What is mathematics? Perhaps a good way to answer the question is by actual experience with mathematics itself.

PUZZLES: JUST FOR FUN

1 A barber is required to shave all those people, and only those people, who do not shave themselves. Does the barber shave himself?

2 "This statement is false." Is the statement at the left true or false?

3 You have eight boxes of candy. Seven of them weigh the same, but the eighth is underweight. In two weighings on a balance scale, how can you determine which of the eight boxes is underweight?

4 There are 15 people at a party. How can you prove by deductive reasoning that at least two of them have their birthdays in the same month?

5 A number of people applied for a position requiring a high ability to reason logically. After a series of eliminations, three applicants remained. To select the one who can best reason logically and quickly, the employer decided on the following method:

The three candidates will be blindfolded and either an ink mark or a water mark will be placed on their foreheads. (The ink mark will show, while the water mark will remain invisible.)

The blindfolds will then be removed and these directions will be followed: (a) If anyone sees an ink spot on somebody else's forehead he is to knock on his desk; (b) If anyone concludes that he himself has an ink spot on his forehead, he is to stand up.

The three applicants were blindfolded and marks were put on their foreheads. As soon as the blindfolds were removed, all three knocked. Shortly thereafter one of them stood up. The one who stood up was adjudged to be the fastest, logical thinker and got the job.

By the use of deductive reasoning, how did the man who stood up conclude that he had an ink mark on his own forehead?

BASICS REVISITED

[The Basics Revisited exercises for Chapters 1–12 are designed to refresh and strengthen basic computational skills. For a review of percent and rounding off numbers see Appendix B.]

1 In the numeral 498,063, which digit names
a) tens? b) thousands?

2 What is the place value of the digit 7 in each of the following numerals:
a) 73 b) 297 c) 750

3 Round off 398,573 to the nearest thousand.

4 What number is 6 more than 7?

5 Locate $\frac{5}{3}$ on the number line.

6 If a heart normally beats about 4320 times every hour, how many times does it beat every minute?

7 Add: 1.3 + .24 + 0.009.

8 Subtract 0.04 from 1.

9 Divide 3.205 by 6. (Round to 3 places.)

10 Multiply 5.007 by 0.3.

11 If an inch of precipitation in the form of rain produces ten inches of snow, how many inches of snow correspond to 17.53 inches of rain?

12 Find the sum of $\frac{1}{3}$, $\frac{3}{4}$, and $\frac{5}{8}$.

13 Find the difference between $\frac{7}{8}$ and $\frac{2}{5}$.

14 Find the product of $\frac{2}{3}$ and $\frac{6}{25}$.

15 Find the quotient of $\frac{2}{3}$ and $\frac{6}{25}$.

16 Find the missing number: $3\frac{3}{4} = 2\frac{\quad}{24}$.

17 Which is smaller: $\frac{5}{7}$ or $\frac{2}{3}$?

18 Which is larger: .48 or .481?

19 Write eleven hundredths as a percent, a decimal, and a fraction.

20 Round off \$7.4849 to the nearest cent.

Solve for x *and check your answer in Exercises 21–23.*

21 $.20(75) = x$ 22 $8 = \frac{x}{100} \cdot 16$ 23 $.6x = 90$

24 A man died in 1935 at the age of 79. When was he born?

25 Max bought a car for \$3,280. If he paid \$250 in cash and arranged to pay off the balance in eight equal monthly payments, what is the amount of each payment?

Chapter 2

SETS, RELATIONS, AND FUNCTIONS

Objectives

At the end of this chapter, you should be able to:

1. Recognize well-defined sets.
2. Describe a set, in *set notation*, by listing its elements.
3. Describe a set by translating from set notation to a verbal description, and from a verbal description to set notation.
4. Decide whether an element is a member of a given set.
5. Translate between set notation and verbal description of sets; translate between set builder notation and verbal description of sets.

6 Define the *empty set* and write it in standard notation.
7 Distinguish between *finite* and *infinite* sets.
8 Define and identify *subset*; *proper subset*.
9 List all the subsets of a given set.
10 Define the *universal set* and be able to recognize it when it is only implied rather than explicitly stated.
11 Define and find the *complement* of a given set.
12 Define and recognize *equal sets*.
13 Set up a *one-to-one correspondence* between the elements of two sets when possible, and recognize when this is impossible.
14 Define and recognize *equivalent sets*.
15 Distinguish betweeen equality of sets and equivalence of sets.
16 Picture set relations by Venn diagrams.
17 Draw conclusions from given statements involving sets and subsets, such as: If $A \subseteq B$ and $B \subseteq A$, what must also be true about A and B?
18 Define and perform the operations of *union*, *intersection*, and *Cartesian product*.
19 Picture the union and intersection of sets by Venn diagrams.
20 Define a *relation*.
21 Define an *equivalence relation*.
22 Determine whether a given relation is an equivalence relation.
23 Define a *function*.
24 Determine whether a given relation is a function.

Prerequisites None.

INTRODUCTION

A fascinating aspect of mathematics is that we do not know today what may turn out to be useful tomorrow. For thousands of years people have been thinking and talking about collections or sets of things—herds of cattle, tribes, nations, families, teams. But it was not until about one hundred years ago that this ancient and obvious idea, the idea of a *set* of things, was recognized as useful and important.

The concept of a set has become important because by using it we can express ideas in a more clear, exact, and concise manner. The language of sets learned in connection with numbers and arithmetic can be applied to geometry and algebra as well as any other branch of mathematics.

By using the ideas of sets, we can solve problems that might be very difficult or even impossible to solve without them. For example, one of the fascinating and unanswered puzzles for a long time was the question of whether, say, a 6-inch segment contains more points than, say, a 1-inch segment. It was not until the mathematician Georg Cantor (1845–1918) developed certain ideas about sets that we finally got a clear and definite answer to this question.

Today we use sets to define whole numbers; the operations of addition, subtraction, multiplication, and division with the set of whole numbers; and the properties of the set of whole numbers. When we count, all that we are really doing is attaching a number to a set. The study of arithmetic is based on the study of sets, and arithmetic is the beginning of mathematics. Today we also use sets to give us clearer, more exact, and more concise definitions in geometry. We think of a curve as a set of points and of space as the set of all points.

Sets are useful not only in mathematics but are used frequently in other fields as well. For instance, the mathematics of life insurance divides the population into *sets of people* of similar age. Physicists do not think of an atom as a single object but as a member of a *set of atoms* of similar type.

In this chapter you will be introduced to the language of sets, and you will see how we use sets and how we operate on them. You will need this background for the rest of the book.

2.1 Set Notation

A *set* is any collection of objects. The objects in the set are called its *elements* or *members*.

Comments

1. *The* objects *of a set need not be tangible. They can be abstract ideas, meaningless symbols, or anything at all.*

2. *The set concept is useful in that it enables us to think of a collection of objects as a* single entity. *A set, as a single entity, has characteristics not possessed by its members. For example, the set of people who make up the U.S. Senate,* as a whole, *has the authority to pass laws. This characteristic is not possessed by any single member.*

In dealing with sets, we adopt, for convenience, the following symbolism:

Capital letters, such as A, B, or C are the names for sets.

The elements of the set are listed within braces. For example, $A = \{1, 5, 13\}$. This is read "A is the set whose elements are 1, 5, and 13."

The statement "5 is an element of set A" may be abbreviated $5 \in A$. To show that 3 is not an element of set A, we write $3 \notin A$.

A set must be *well defined*, which means that there is no doubt whether or not any given element belongs to the given set. A set may be defined:

1. By *listing* the elements of the set. For example, $A = \{2, 3, 5\}$.
2. By giving a *rule* or description by which the elements of the set may be found. For example, $B = \{$all odd whole numbers less than 7$\}$; or

$$C = \{x | x \text{ is a counting number greater than } 5\}, \quad (1)$$

which is read "C is the set whose elements are all x such that each x is a counting number greater than 5."

Comments

1. *A set is distinguished from another set only by its members, not by the* order *in which its members are listed. The set denoted by* $\{2, 3, 5\}$ *and the set denoted by* $\{3, 5, 2\}$ *are exactly the same set.*

2. *The expression "*$x|x$ *is a counting number greater than 5" is called a* set builder *because it selects from some larger set (the set of counting numbers) a set consisting only of certain members of the set.*

3. *Set* C (1) *can be expressed as* $C = \{6, 7, 8, 9, \ldots\}$, *where the three dots mean "and so on." We assume that the words "and so on" make perfectly clear which numbers are and which numbers are not members of* C.

4. *The* rule *method for defining a set is useful when we wish to discuss a set containing a larger or unlimited number of elements. For example,*

   a) *If we are talking about the set of 35 students in Mr. Cole's class in Central High School, it would be more convenient to write* $S = \{$*all students in Mr. Cole's class*$\}$ *than to list all 35 names within braces.*

   b) *If we are talking about the set of rational numbers, there is no simple way to represent this set by the* listing *method because any number in the set that is listed has no "next larger" number in the set.*

5. *Certainty as to whether a given element is or is not a member of a given set is the most important purpose in defining a set. If membership in a set depends on* opinion *(the set of interesting people in a room), then the set is not clearly defined. In a well-defined set, there must be complete agreement on its membership.*

Although it is natural to assume that a collection has two or more elements in it, we find it useful to refer to collections containing just *one* element, or even *no* elements, as sets. The set that has *no* members is called the *empty set* or the *null set.* We denote this set by the symbol ϕ or { } (braces without any elements inside). When a set has a limited number of elements, it is called a *finite set.* When a set has an unlimited number of elements, it is called an *infinite set.*

Comments

1 *We talk about* the *empty set rather than* an *empty set because there is only one empty set, the set that contains no elements.*

2 *The symbol ϕ is an adaptation of a letter of the Danish alphabet.*

3 *Students sometimes represent the empty set by $\{\phi\}$, which is incorrect. What is being represented by $\{\phi\}$ is a set containing one element, ϕ. For a similar reason $\{0\}$ is not the empty set.*

4 *The statements about* finite *and* infinite *sets are not definitions, only intuitive descriptions.*

5 *The empty set is a finite set.*

6 *It is incorrect to say that a set is finite simply because it has a largest element. The set {all fractional numbers from* 1 *to* 2} *has a largest element,* 2, *but is not a finite set.*

2.2 Relations Between Sets

SUBSETS

DEFINITION 1 If every element of set A is also an element of set B, then A is a *subset* of B. We denote this relation by $A \subseteq B$.

Example 1

a) If $A = \{1, 2\}$ and $B = \{1, 2, 3\}$, then $A \subseteq B$.

b) If $A = \{1, 2, 3\}$ and $B = \{1, 2, 3\}$, then $A \subseteq B$.

Comments

1 *Since the definition of* subset *does not specify that A must contain* fewer *elements than B, A may have exactly the same elements as B (as in Example 1b).*

2 *The definition of a subset may be expressed in several other ways:*

a) $A \subseteq B$ if and only if $x \in A$ implies $x \in B$.

b) Set A is a subset of set B if A contains no element that is not in B.

c) A is not a subset of B if and only if A contains an element not in B.

If A is a subset of B and B has at least one element that is not an element of A (for example, $A = \{1, 2\}$, $B = \{1, 2, 3\}$) then A is said to be a *proper subset* of B, and we write $A \subset B$.

DEFINITION 2 If A is a subset of B but B is not a subset of A, then we say that A is a *proper subset* of B.

Comments

1 *When the subset A does not contain all the elements of set B, we call A a proper subset of B. When the subset A contains exactly the same elements as set B, A is called an* improper subset *of B.*

2 *Just as* $2 < 3$ *means the same thing as* $3 > 2$, *so the expression* $A \subset B$ *means the same thing as* $B \supset A$, *read "B is a* superset *of A."*

PROPERTIES OF SUBSETS

Every subset of B, other than B itself, is a *proper* subset of B.

Every set is an *improper* subset of itself.

The empty set is a proper subset of every set except itself. Since the empty set contains no elements, all its elements belong to *every* set.

For every set of n elements there are exactly 2^n subsets.

Comments

1 *The concept that the empty set is a subset of every set is difficult for some students to grasp. Sometimes, posing the following question helps convince them: Can you name an element in the empty set that is not contained in any given set? (See Comment 2b on page 11.)*

2 *To see why every set of n elements has exactly* 2^n *subsets, form all possible subsets for sets containing one, two, three, and four elements. Tabulate the results and then look for a pattern.*

Number of Elements	*Number of Subsets*
1	2
2	4
3	8
4	16

3 *To form* all *possible subsets of any set, we consider all possible subsets containing* no *elements,* one *element,* two *elements,* three *elements,* . . . , all *the elements, and then find the sum of all these possibilities.*

When all the sets in a given discussion are subsets of a given set, we call the given set the *universal set*, or the *universe*. We shall usually denote the universal set by U.

DEFINITION 3 If $A \subseteq U$, then the elements of U that are *not* in A form another subset of U called the *complement of* A, denoted by A'.

Example 2 If $U = \{1, 3, 5, 7\}$ and $A = \{5\}$, then $A' = \{1, 3, 7\}$.

It is possible to have a set whose elements are sets themselves. For example, the set of all the subsets of $A = \{1, 2, 3\}$ is

$$S = \{\{1, 2, 3\}, \{\ \ \}, \{1\}, \{2\}, \{3\}, \{1, 2\}, \{1, 3\}, \{2, 3\}\}.$$

DEFINITION 4 A set whose elements are all the subsets of a given set A is called the *power set* of A.

Since the number of subsets in a set S containing n elements is 2^n, the number of elements in the power set of S is 2^n.

VENN DIAGRAMS

A way of picturing relations between sets is to use Venn diagrams. In these diagrams we let a region, such as a rectangular region, represent the universal set U (Figure 2.1), and let subregions, such as circular regions completely contained in U, represent subsets of U (Figure 2.2).

FIGURE 2.1

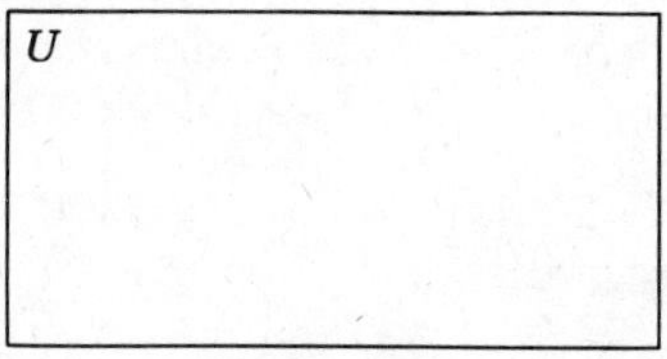

U

FIGURE 2.2

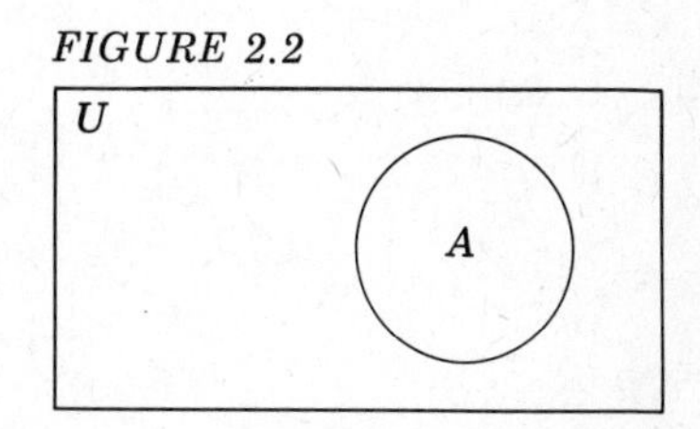

Figure 2.3 shows $A \subseteq U$, $B \subseteq U$, A and B having no common elements. If two sets have no common elements, they are said to be *disjoint sets.*

Figure 2.4 shows C' represented by the shaded region.

FIGURE 2.3

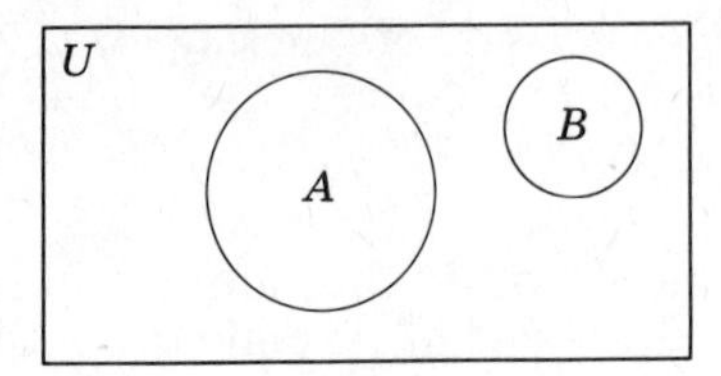

FIGURE 2.4

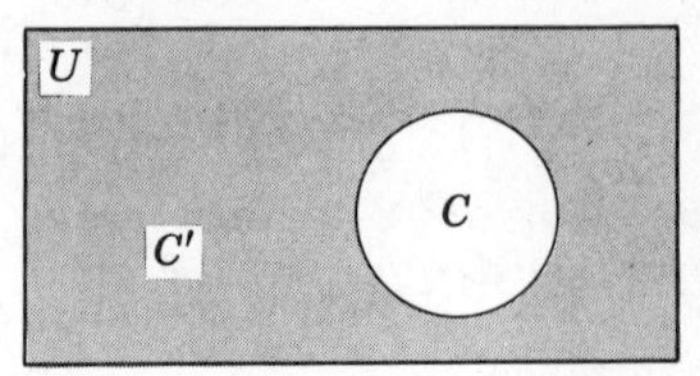

Comments

1 *Although we are using rectangles and circles to construct Venn diagrams, any simple closed curves would do. What is important is the* relationship *between the interiors of the curves used rather than their shapes.*

2 *Venn diagrams are so named in honor of the English mathematician John Venn (1834–1923) who popularized their use in the study of logic.*

EQUAL SETS

DEFINITION 5 Two set are *equal* if each is a subset of the other. That is, $A = B$ if $A \subseteq B$ and $B \subseteq A$.

Example 3 If $A = \{2, 3, 7\}$ and $B = \{2, 3, 7\}$ then $A = B$ since A is a subset of B because every element of A is in B, and B is a subset of A because every element of B is in A.

Comments

1 *It follows from this definition that two sets are equal if they contain the same elements. The sentence "A and B contain the same elements" can be broken down into "every element of A is an element of B, and every element of B is an element of A." This is another way of saying that* $A \subseteq B$ *and* $B \subseteq A$.

2 *It also follows that two sets are equal if they contain the same elements* regardless of the order *in which they are listed.*

3 *Equal sets A and B are pictured by a Venn diagram in Figure 2.5.*

FIGURE 2.5

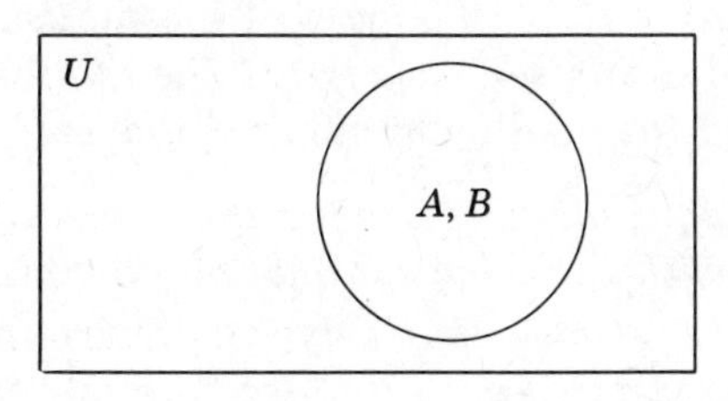

EQUIVALENT SETS

A pairing that assigns to each member of each set one and only one member of the other set is called a *one-to-one correspondence.*

DEFINITION 6 Two sets are *equivalent* if there exists a one-to-one correspondence between their elements. If set A is equivalent to set B, we write $A \sim B$.

Example 4 If $A = \{a, b, c\}$ and $B = \{\bigcirc, \square, \triangle\}$, then $A \sim B$ since there exists a one-to-one correspondence between the elements of the two sets; e.g.,

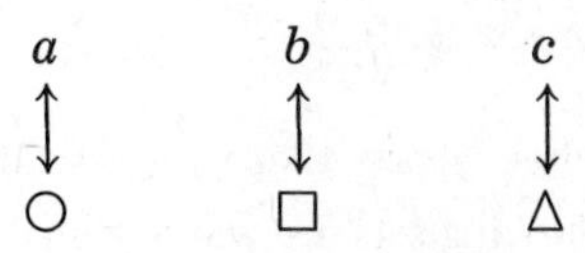

Comments

1 *The way in which the elements of two sets are matched has no bearing on the equivalence of the sets as long as:*

To every element in the first set there corresponds some element in the second set, and to every element in the second set there corresponds some element in the first set.

For example, if $A = \{a, b, c\}$ and $B = \{1, 2, 3\}$, there are six *different ways in which the elements can be matched:*

(1)	(2)	(3)	(4)	(5)	(6)
$a \longleftrightarrow 1$	$a \longleftrightarrow 1$	$a \longleftrightarrow 2$	$a \longleftrightarrow 2$	$a \longleftrightarrow 3$	$a \longleftrightarrow 3$
$b \longleftrightarrow 2$	$b \longleftrightarrow 3$	$b \longleftrightarrow 1$	$b \longleftrightarrow 3$	$b \longleftrightarrow 1$	$b \longleftrightarrow 2$
$c \longleftrightarrow 3$	$c \longleftrightarrow 2$	$c \longleftrightarrow 3$	$c \longleftrightarrow 1$	$c \longleftrightarrow 2$	$c \longleftrightarrow 1$

2 *Finite sets whose elements can be paired in a one-to-one correspondence contain the same* number *of elements. This number is the same regardless of the method by which the one-to-one correspondence is established. (For a discussion of* infinite *sets* *see page 101.)* *The number of elements in a set is called the* cardinal number *of the set.*

3 *All equivalent sets have the same cardinal number, and all sets with the same cardinal number are equivalent.*

4 *We now have two ways of determining the equivalence of finite sets: First, by determining whether there exists a one-to-one correspondence between the elements of the two sets; second, by counting the elements in each set. By counting, we not only determine whether the sets are equivalent, but we also find out how many elements each set contains. When we set up a one-to-one correspondence, we determine only whether or not the sets are equivalent.*

5 *The difference between equal and equivalent sets must be seen clearly. Equal sets contain the* same elements; *equivalent sets contain only the* same number *of elements. All equal sets are equivalent, but all equivalent sets are not necessarily equal.*

6 *The usefulness of the equivalence property depends on the fact that it does not matter what kind of elements the sets contain. Equivalence is concerned only with the fact that we can pair the elements in such a way as to produce a one-to-one correspondence.*

EXERCISE SET 2.1, 2.2

1 What is meant when we say that a set must be "well-defined"?

2 Which of the following sets are well defined?
a) All positive integers greater than 2 and less than 19.
b) All great Presidents of the U.S.
c) Four consecutive small numbers.
d) All grains of sand on earth.

3 Express in set notation what is specified in each of the following statements.
a) The set of all elements x such that x is a whole number less than 9 and greater than 0.
b) The set of all odd numbers.
c) All A's are B's but not all B's are A's.
d) Every set S is a subset of itself.
e) The empty set is a subset of every set G.
f) a is an element of the set whose only element is a.

4 Translate the following into verbal statements:
a) $\{a | a < 4\}$. b) $\{x | x \text{ is a prime}\}$. c) $\{y | 4 \leq y < 12\}$.

5 Are the following statements true?
a) $\{\phi\}$ is a set. b) $2 \in \{\{1\}, \{2\}, \{3\}\}$.
c) If $A = \{x | x$ is an odd number less than $12\}$, then $1 \in A$.

6 Use one of the symbols $\subset$, $\supset$, $\in$, $\subseteq$ to indicate the relationship between the given expressions:
a) ϕ, set A. b) Set B, set B. c) $2, \{1, 2, 3\}$. d) $\{a, b, c\}, \{a, c\}$.

7 Which of the following pairs of sets are equal, equivalent, equal and equivalent, or neither equal nor equivalent?
a) $\{1, 4, 9\}, \{1, 9, 9, 4\}$ b) $\{2, 4, 6, 8\}, \{2, 4, 6\}$
c) $\{\triangle, \bigcirc, \square\}, \{\bigcirc, \square, 7\}$

8 If U is the set of the counting numbers and A is the set of the even numbers, what is A'?

9 If the universal set is the children in a given class, what is the complement of the set of all the boys in the class?

10 List the elements of the power set of $A = \{2, 5, r\}$.

11 What is the difference between the concept *is an element of* and the concept *is a subset of*?

12 What numbers are associated with the following sets?
a) $\{\triangle, 2, b\}$ b) $\{1, 2, 3\}$ c) {whole numbers less than zero}
d) {zero}

13 Let U, the universal set, be the numbers 15 to 32 inclusive. Let D be the numbers in U greater than 17 and less than 29.
a) List the elements in D. b) Draw a Venn diagram of U and D.

14 How can you tell whether there are more men than women at a dance, without counting?

2.3 Operations on Sets

Just as numbers may be combined by addition and multiplication to form other numbers, so may sets be combined by certain operations to form other sets. The operations we shall be concerned with here are *union*, *intersection*, and *Cartesian product*.

UNION

DEFINITION 1 The *union* of two sets is the set of all elements belonging to either of the two sets or to both of them. The union of sets A and B is symbolized by $A \cup B$, read "the union of A and B."

Example 1 If $A = \{1, 3, 5\}$ and $B = \{2, 3, 4\}$, then $A \cup B = \{1, 2, 3, 4, 5\}$.

Comments 1 *Note that the elements contained in both A and B are written only once.*

2 *Note that the* union *of two sets is another* set.

3 *Since the union of two sets contains all the elements of each set, each of the original sets is a subset of the union set. That is, for all sets A and B,* $A \subseteq (A \cup B)$ *and* $B \subseteq (A \cup B)$.

4 $x \in (A \cup B)$ *is equivalent to* $x \in A$ or $x \in B$.

5 *The union of sets is closely related to the meaning of the word "or." The test of membership in* $A \cup B$ *is whether the element in question is* or *is not a member of set A* or *set B* (or *of both A and B).*

We can show the union of two sets with a Venn diagram (Figure 2.6).

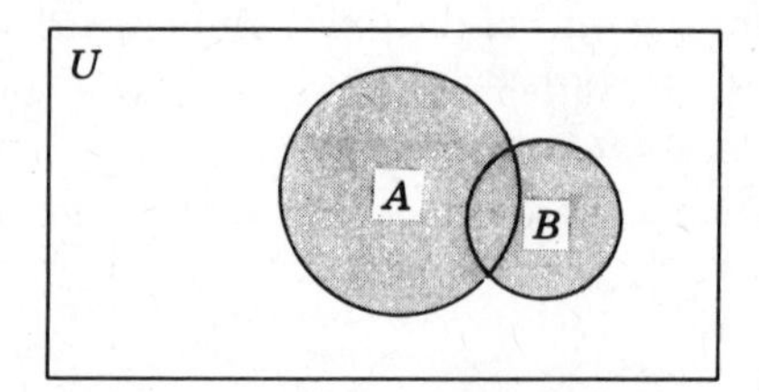

FIGURE 2.6 $A \cup B$ *is represented by the shaded region.*

PROPERTIES OF UNION

Commutative Property The *order* in which we form the union of two sets has no effect on the result. For example, if $A = \{1, 2\}$ and $B = \{0, 2, 4\}$, then

$$A \cup B = \{1, 2, 0, 4\} = \{0, 1, 2, 4\}$$

and $$B \cup A = \{0, 2, 4, 1\} = \{0, 1, 2, 4\}.$$

That is, $A \cup B = B \cup A$. Thus we say that *union is commutative.*

Associative Property The way in which we *group* the sets when we take their union has no effect on the result. For example, if $A = \{1, 2\}$, $B = \{0, 2, 4\}$, and $C = \{1, 3, 5\}$, then

$$(A \cup B) \cup C = \{1, 2, 0, 4\} \cup \{1, 3, 5\} = \{1, 2, 0, 4, 3, 5\}$$
$$= \{0, 1, 2, 3, 4, 5\}$$

and $$A \cup (B \cup C) = \{1, 2\} \cup \{0, 2, 4, 1, 3, 5\} = \{1, 2, 0, 4, 3, 5\}$$
$$= \{0, 1, 2, 3, 4, 5\}$$

That is, $(A \cup B) \cup C = A \cup (B \cup C)$. Thus we say that *union is associative.*

INTERSECTION

DEFINITION 2 The *intersection* of two sets is the set of elements belonging to *both* sets. The intersection of sets A and B is symbolized by $A \cap B$, read "the intersection of A and B."

Example 2 If $A = \{1, 3, 5\}$ and $B = \{2, 3, 4, 5\}$, then $A \cap B = \{3, 5\}$.

Comments

1. *Note that the* intersection *of two sets is another set.*
2. *Since the elements of the intersection set belong to both sets, $A \cap B$ is a subset of set A and of set B. That is, for all sets A and B, $A \cap B \subseteq A$ and $A \cap B \subseteq B$.*
3. *The statement "$x \in (A \cap B)$" is equivalent to* $x \in A$ and $x \in B$.
4. *The words "intersect" and "intersection" do not necessarily imply the same thing. The statement "A and B intersect" implies that A and B have at least one element in common. But the phrase "the intersection of A and B" can also refer to the empty set.*
5. Disjoint *sets may now be defined as sets whose intersection is the empty set. That is, A and B are disjoint if $A \cap B = \phi$.*
6. *The intersection of sets is closely related to the meaning of the word "and." The test of membership in $A \cap B$ is whether the element in question is or is not a member of both A* and *B.*

We can show the intersection of two sets with a Venn diagram (Figure 2.7).

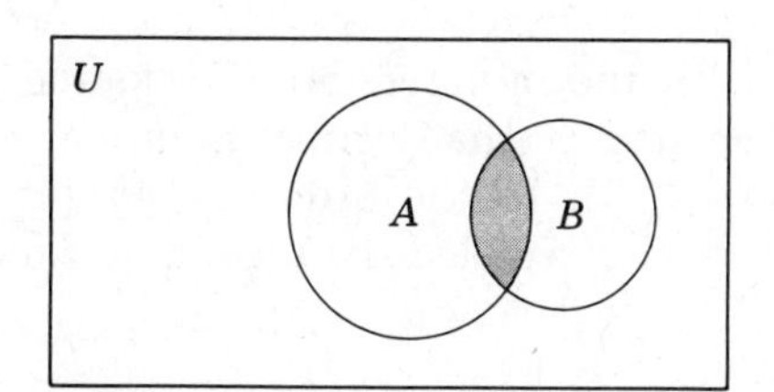

FIGURE 2.7 $A \cap B$ is represented by the shaded region.

PROPERTIES OF INTERSECTION

Commutative Property That is, $A \cap B = B \cap A$. For example, if $A = \{1, 2\}$ and $B = \{0, 2, 4\}$, then $A \cap B = \{2\}$ and $B \cap A = \{2\}$.

Associative Property That is, $(A \cap B) \cap C = A \cap (B \cap C)$. For example, if $A = \{1, 2\}$, $B = \{0, 2, 4\}$, and $C = \{1, 2, 5\}$, then

$$(A \cap B) \cap C = \{2\} \cap \{1, 2, 5\} = \{2\}.$$

and $$A \cap (B \cap C) = \{1, 2\} \cap \{2\} = \{2\}.$$

Distributive Properties

$$A \cap (B \cup C) = (A \cap B) \cup (A \cap C), \qquad (1)$$

and $$A \cup (B \cap C) = (A \cup B) \cap (A \cup C). \qquad (2)$$

Note that each operation is *distributive* over the other.

Comments 1 *Illustrate both parts of the distributive property by letting A, B, and C represent any three sets.*

2 *The distributive property can be verified very neatly with a Venn diagram. Let A, B, and C represent the following sets in Figure 2.8:*

$A = \{1, 2, 3, 4\}$, $B = \{2, 3, 5, 6\}$, $C = \{3, 4, 6, 7\}$.

a) To establish Equation 1, show that $A \cap (B \cup C)$ *and* $(A \cap B) \cup (A \cap C)$ *represent the same region in the diagram.*

b) To establish Equation 2, show that $A \cup (B \cap C)$ *and* $(A \cup B) \cap (A \cup C)$ *represent the same region in the diagram.*

FIGURE 2.8

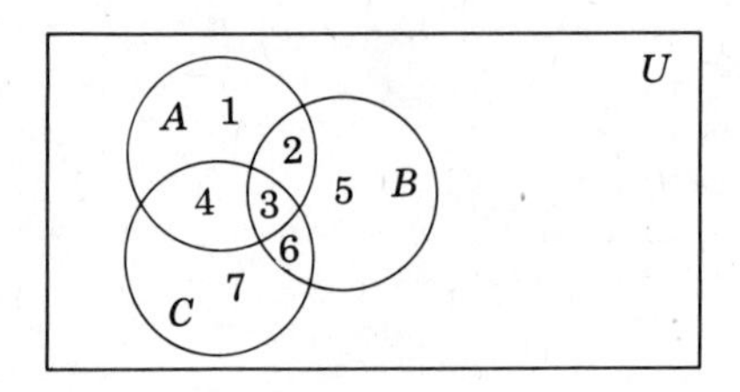

CARTESIAN PRODUCT

Before talking about Cartesian products, we shall define an *ordered pair.* An *ordered pair* is a pair of numbers (or objects) where the *order* in which the numbers are considered is important. If 2 is the first number and 3 the second number, then the ordered pair is denoted by (2, 3). In the ordered pair (2, 3), 2 is called the *first component*, and 3 is called the *second component.*

Comments 1 *Do not refer to the* 2 *and the* 3 *as the first and second* elements *since we have reserved the word element to mean* member of a set, *and* (2, 3) *is not a set in the sense in which we have used the term.*

2 *Ordered pairs appear throughout mathematics, although sometimes in disguised form. Consider the ordered pair* (6, 2). *In the operation of addition it is expressed as* $6 + 2$, *in subtraction as* $6 - 2$, *in multiplication as* 6×2, *and in division as* $6 \div 2$. *All these operations involve ordered pairs of numbers because the order in which we consider the* 6 *and the* 2 *is important. For example,* $6 \div 2$ *is not the same number as* $2 \div 6$.

3 *Although an ordered pair is not a set, we can have a* set of

ordered pairs, *that is, a set whose elements are ordered pairs. For example,*

$J = \{(1, 65), (2, 82), (5, 62)\}$

is a set of ordered pairs.

4 *Two ordered pairs are equal if and only if they have identical first components and identical second components. That is, $(a, b) = (c, d)$ if and only if $a = c$, and $b = d$.*

Let $A = \{1, 2\}$ and $B = \{a, b, c\}$. We shall now form *all possible* ordered pairs such that the first component is an element of A and the second component is an element of B. We obtain the following set of ordered pairs:

$$A \times B = \{(1, a), (1, b), (1, c), (2, a), (2, b), (2, c)\}.$$

We call this set the *Cartesian product of A and B* and is denoted by $A \times B$, read "A cross B."

DEFINITION 3 If A and B are sets, then the set $A \times B$, called the *Cartesian product of A and B*, is the set of all possible ordered pairs such that the first component of an ordered pair is an element of A and the second component is an element of B.

Comments

1 Cartesian product *takes its name from René Descartes (1596–1650) who related ordered pairs of numbers to points in a plane and thereby conceived the ideas of analytic geometry.*

2 *Unlike union and intersection, the formation of a Cartesian product produces a set whose elements are* not *in the universal set.*

PROPERTIES OF THE CARTESIAN PRODUCT

1 $A \times B \neq B \times A$. That is, the formation of Cartesian products of sets is *not* commutative.

2 $A \times \phi = \phi$. Since ϕ has no elements, we cannot have ordered pairs because there are no second components. Therefore, $A \times \phi$ has no elements.

3 $\phi \times A = \phi$. Why?

4 $\phi \times \phi = \phi$. Why?

5 $A \times B \sim B \times A$. That is, the number of elements in each of the Cartesian products $A \times B$ and $B \times A$ will be the same since to each element (a, b) in $A \times B$ there corresponds the element (b, a) in $B \times A$ and to each element (b, a) in $B \times A$ there corresponds the element (a, b) in $A \times B$.

EXERCISE SET 2.3

1. Let $A = \{2, 4, 6, 8\}$, $B = \{1, 3, 5\}$, $C = \{0, 2, 4, 6, \ldots\}$, $D = \{1, 3, 5, 7, \ldots\}$. Find:
   a) $A \cup B$ b) $B \cup C$ c) $B \cup D$ d) $A \cap C$ e) $B \cap D$
   f) $B \cap C$ g) $A \cup (B \cap C)$ h) $C \cap D \cap A$
2. If U is the set of counting numbers, $A = \{1, 2, 3, 4\}$, and $B = \{5, 6\}$, find
   a) A' b) B' c) $B \times B$
3. Use a Venn diagram to show that $A' \cap B' = (A \cup B)'$.
4. Complete the following statements if neither set E nor set K is the empty set:
   a) If $E \subseteq K$, then $E \cup K =$ ____.
   b) If $E \subset (C \cup K)$, then $E \neq$ ____.
   c) If $E \subseteq K$, then $E \cap K =$ ____.
5. In a group of 19 people, 11 like pizza, 12 drive a car, and the rest like pizza and also drive a car. How many people like pizza and also drive a car?

2.4 Relations

Think of the students in class Math 101-5. There are many ways in which these students can be related to each other. Some may be brothers, some may have been born in the same city, some are taller than others, some are the same age, etc. Consider the relation *is taller than*, and let a and b be any two students in the class. When we say that *a is related to b*, written $a \mathcal{R} b$, we mean that *a is taller than b*; $x \mathcal{R} y$ means that *x is taller than y*. The relation *is taller than* would thus represent the set of all pairs of students in class Math 101-5 where the first student in the pair *is taller than* the second student in the pair. A *relation R*, then, is a set of ordered pairs where the first component is related to the second component in the sense in which we choose to define *related*—in the present example, *related* means *is taller than.* If $a \mathcal{R} b$, then the ordered pair (a, b) is an element in the set R; if $x \mathcal{R} y$, then the ordered pair (x, y) is an element in the set R.

We shall now consider another example of a relation. Let

$$A = \{\text{Chicago, Munich, London}\}, B = \{\text{U.S., France, Germany}\}.$$

The Cartesian product of sets A and B is as follows:

$$A \times B = \{(\text{Chicago, U.S.}), (\text{Chicago, France}), (\text{Chicago, Germany}), (\text{Munich, U.S.}), (\text{Munich, France}), (\text{Munich, Germany}), (\text{London, U.S.}), (\text{London, France}) (\text{London, Germany})\}.$$

Let x be any element in set A and let y be any element in set B. Suppose that we are interested not in the entire Cartesian product but only in that part of it where x is a city in y. We are thus seeking only those ordered pairs in $A \times B$ where the relation *is a city in* is true. This relation will yield the subset R of $A \times B$ where

$$R = \{(\text{Chicago, U.S.}), (\text{Munich, Germany})\}.$$

R is called a *relation* from set A to set B.

DEFINITIONS

1. Given two sets A and B, a *relation* from set A to set B is any subset of the Cartesian product $A \times B$.
2. If R is a subset of $A \times A$, then R is said to be a *relation on set A*.
3. The set of first components of R is called the *domain* of the relation R. The *domain* in $R = \{(\text{Chicago, U.S.}), (\text{Munich, Germany})\}$ is $\{\text{Chicago, Munich}\}$.
4. The set of second components of R is called the *range* of the relation. The *range* in R is $\{\text{U.S., Germany}\}$.

Let us consider several other examples of relations.

Example 1 $A = \{1, 2, 3\}$ and $B = \{1, 2\}$. We wish to find all ordered pairs that belong to the relation R in $A \times B$ such that if $x \in A$ and $y \in B$ then $x > y$. Also, find the *domain* and *range* of R.

Solution $A \times B = \{(1, 1), (1, 2), (2, 1), (2, 2), (3, 1), (3, 2)\}$. Since the relation requires that x be *greater* than y, we select from $A \times B$ only those ordered pairs that satisfy this requirement.

$$R = \{(2, 1), (3, 1), (3, 2)\}.$$

The domain of R is $\{2, 3\}$, and the range is $\{1, 2\}$.

Example 2 Find R in $A \times B$ as given in Example 1 if the relation is $y = 2x$.

Solution Here we select from the Cartesian product $A \times B$ only those ordered pairs where the second component is twice the first component. The relation consists of only one element, $R = \{(1, 2)\}$. The domain is $\{1\}$ and the range is $\{2\}$.

Example 3 If $A = \{1, 2, 3\}$, find the relation R on set A such that

$$R = \{(x, y) | y < x^2\}.$$

Solution

$$A \times A = \{(1, 1), (1, 2), (1, 3), (2, 1), (2, 2), (2, 3), (3, 1), (3, 2), (3, 3)\}.$$

$$R = \{(2, 1), (2, 2), (2, 3), (3, 1), (3, 2), (3, 3)\}.$$

The domain of R is $\{2, 3\}$ and the range is $\{1, 2, 3\}$.

EXERCISE SET 2.4a

1. If $A = \{1, 2\}$ and $B = \{3, 4, 5\}$, and $x \in A$, $y \in B$, find the following relations, and indicate the domain and range for each relation.
   a) $\{(x, y)|x = y\}$ b) $\{(x, y)|x + y < 5\}$ c) $\{(x, y)|y = x^2\}$
2. Find all the ordered pairs that belong to the following relations. Then state the domain and range of each relation.
   a) $\{(x, y)|x < y$; x and y are whole numbers between 3 and 6$\}$
   b) $\{(x, y)|y = x + 2$; x and y are whole number between 1 and 4$\}$
   c) $\{(x, y)|x - y < 3$; x is a whole number less than 5, y is a whole number less than 7, and $x - y$ is a whole number$\}$
3. If set A contains two elements and set B contains three elements, determine the number of possible relations that there are from set A to set B.
4. Let $A = \{2, 3, 4\}$ and let R be a relation on set A.
   a) Find R where the relation is x *is more than* $y + 1$.
   b) Find R where the relation is $x - y = 4$.
   c) Describe in words the relation $R = \{(3, 2), (4, 3)\}$.
   d) Describe in words the relation $R = \{(2, 2), (3, 3), (4, 4)\}$.

EQUIVALENCE RELATIONS

In considering any relation we are particularly interested in three properties that the relation may possess. Let $S = \{$all students in Math 101-5$\}$; let x, y, and z be any three students in the class; and let the relation be *is taller than.* We shall now ask three questions about this relation.

1. Is $x \mathcal{R} x$? That is, is every student taller than himself? Since the answer is no, we say that this relation is *not reflexive.*
2. If $x \mathcal{R} y$, then is $y \mathcal{R} x$? That is, if student x is taller than student y, then is student y taller than student x? Again the answer is no so we say that this relation is *not symmetric.*
3. If $x \mathcal{R} y$ and $y \mathcal{R} z$, then is $x \mathcal{R} z$? That is, if x is taller than y and y is taller than z, then is x taller than z? Since the answer is yes, we say that this relation is *transitive.*

So we conclude that the relation *is taller than* is not reflexive, not symmetric, but is *transitive.*

Now consider the same set S, but let the relation be *is the same sex as.* We shall now test this relation to see whether it is reflexive, symmetric, or transitive.

1. *Reflexive* Is $x \mathcal{R} x$? That is, is every student the same sex as himself? Since the answer is yes, the relation is reflexive.

2 *Symmetric* If $x \mathcal{R} y$, then is $y \mathcal{R} x$? That is, if x is the same sex as y, then is y the same sex as x? Since the answer is yes, the relation is symmetric.

3 *Transitive* If $x \mathcal{R} y$ and $y \mathcal{R} z$, then is $x \mathcal{R} z$? That is, if x is the same sex as y and y is the same sex as z, then is x the same sex as z? Since the answer is yes, the relation is transitive.

Any relation that possesses these three properties, that is, any relation that is *reflexive, symmetric, and transitive*, is called an *equivalence relation.*

Comment *From the definitions of the three properties of an equivalence relation, it follows that:*

1 *Under the* reflexive property, *if x is an element of R, then $(x,x) \in R$.*

2 *Under the* symmetric property, *if* $(x, y) \in R$, *then* $(y, x) \in R$.

3 *Under the* transitive property, *if* $(x, y) \in R$, *and* $(y, z) \in R$, *then* $(x, z) \in R$.

DEFINITION 1 Let R be a relation defined over the set S and let x, y, and z represent arbitrary elements in S. The relation R is said to be an *equivalence relation* if the following three properties are satisfied:

Reflexive $x \mathcal{R} x$ for all elements x in S.

Symmetric If $x \mathcal{R} y$, then $y \mathcal{R} x$ for all x and y in S.

Transitive If $x \mathcal{R} y$ and $y \mathcal{R} z$, then $x \mathcal{R} z$ for all x, y, and z in S.

Let us take a closer look at the effect of an equivalence relation on the set over which it is defined. The relation *is the same sex as*, defined over the set of students in Math 101-05, breaks up the class into two sets of students:

S_1 = {all male students in Math 101-05},

S_2 = {all female students in Math 101-05}.

Every student in the class belongs to either S_1 or S_2 but not to both. We call the two disjoint subsets S_1 and S_2 *equivalence classes.* All students in S_1 are equivalent in the sense of our relation; that is, all of them belong to the same sex. Similarly, all students in S_2 are equivalent. If x and y are elements in the same equivalence class then $x \mathcal{R} y$.

We therefore describe the effect of an equivalence relation R on the set S over which it is defined by saying that R *partitions* S into disjoint subsets called *equivalence classes.*

THE INVERSE OF A RELATION

If $R = \{(1, 3), (2, 5), (3, 7)\}$, we can form another relation from R by merely interchanging the first and second components in each ordered pair:

$$R^{-1} = \{(3, 1), (5, 2), (7, 3)\}.$$

R^{-1} is called the *inverse of R*. Note that if R is a relation from set A to set B, then R^{-1} is a relation from set B to set A.

DEFINITION 2 If R is a relation from set A to set B and $x \in A$ and $y \in B$, the *inverse of R*, written R^{-1}, is defined by

$$R^{-1} = \{(y, x) | (x, y) \in R\}.$$

Comments

1 *From the definition of an inverse relation, it follows that R^{-1} is a subset of $B \times A$; the domain of R^{-1} is the range of R; and the range of R^{-1} is the domain of R.*

2 *If R is a symmetric relation, then it satisfies the property that if $x \mathcal{R} y$, then $y \mathcal{R} x$. This means that if $(x, y) \in R$, then $(y, x) \in R$. From this and the definition of the inverse relation it follows that if R is a symmetric relation, then $R = R^{-1}$.*

EXERCISE SET 2.4b

1 Determine whether each of the following relations is reflexive, symmetric, or transitive.

Relation R	*Set S*
a) Is less than	numbers
b) Has the same name as	people
c) Is a subset of	sets
d) Is congruent to	geometric figures
e) Is the father of	people
f) Is equal to	numbers
g) Is perpendicular to	lines in a plane
h) Is a proper subset of	sets
i) Weighs within 10 pounds of	people
j) Is older than	houses
k) Is parallel to	lines in a plane
l) Is at least as old as	people

2 Give an example of a relation that is:

a) Transitive, but neither reflexive nor symmetric.
b) Reflexive and transitive, but not symmetric.
c) Neither reflexive nor symmetric nor transitive.

3 Let $S = \{$Ann, John, Ben, Laura, Eve$\}$, and let R be defined to be *is the same sex as.* Partition S into equivalence classes.

4 Which of the following are equivalence relations? For those that are,

describe the equivalence classes into which S is partitioned.
a) Is the brother of.
b) Congruence of triangles.
c) $\{(3, 3), (4, 4), (3, 5), (4, 5)\}$. $S = \{3,4,5,6,7\}$
d) Is the same age as.
e) Was born in the same city as.

5 If $A = \{1, 2, 3, 4, 5\}$, give an example of a.
a) Reflexive relation on set A.
b) Symmetric relation on set A.
c) Transitive relation on set A.
d) Equivalence relation on set A.
e) Transitive, but neither reflexive nor symmetric relation on set A.

6 Given the relation $T = \{(a, 1), (a, 3), (b, 2), (b, 5)\}$, what are the elements of T^{-1}?

7 Write the inverse of each relation in Exercise Set 2.4a, Exercises 1 and 2.

2.5 Functions

Let S be a relation that matches students with their grades on a test:

$$S = \{(\text{Joe}, 70), (\text{Frank}, 85), (\text{Tom}, 75), (\text{George}, 90), (\text{Carl}, 85)\}.$$

Note that for ever student there is only one grade; that is, there are no two ordered pairs with the same first component and different second components. A relation, such as S, in which the first component is associated with only one second component is called a *function.*

The relation

$$T = \{(1, 3), (2, 6), (3, 9), (4, 12)\}$$

is a *function* since for each first component there is a unique second component. However, the relation

$$R = \{(2, 3), (5, 3), (2, 4)\}$$

is not a *function* because the first component, 2, is associated with more than one second component; 2 is associated with 3 and with 4.

DEFINITION 1 A *function* is a relation in which each element of the domain is paired with a single element of the range.

In other words, a relation R is a function if whenever $(a, b) \in R$ and $(a, c) \in R$, $b = c$. There are examples of functions all around us:

Example 1 If A is the set of Americans with social security and B is the set of counting numbers, then the set of ordered pairs which associates each person with his social security number is a *function* from A to B.

Example 2 If A is the set of students in a class and B is the set of counting numbers, then the set that associates each student with the number representing his height is a *function*.

Example 3 The cost of gasoline is a function of the price per gallon; your weight is a function of how much you eat; the cost of mailing a package is a function of its weight; the area of a square is a function of the length of its side.

Comments

1 *Figures 2.9 and 2.10 illustrate the difference between a relation and a function.*

FIGURE 2.9

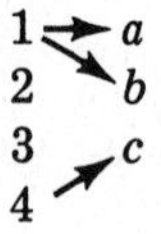

1 a
2 b
3 c
4

FIGURE 2.10

1 a
2 b
3 c
4

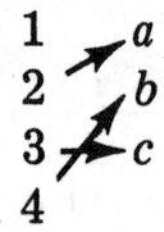

If $A = \{1, 2, 3, 4\}$ and $B = \{a, b, c\}$, then Figure 2.9 suggests a relation from A to B that is not a function *because two ordered pairs, $(1, a)$ and $(1, b)$ contain the same first component but different second components. Figure 2.10 suggests a relation that is a function because every first component is matched with one and only one second component.*

2 *In the ordered pair (x, y) of a function F, y is sometimes called the* image *of x under the function F.*

3 *A function is sometimes called a* single-valued relation *because each element x in the domain is matched with exactly one element in the range. If we allow the same value of x to be matched with more than one value of y, we have a* many-valued relation.

4 *The* inverse of a function *is defined and obtained the same way as the inverse of a relation: interchange the components of each of the ordered pairs in the function. For example, if*

$F = \{(2, 3), (3, 4), (4, 5)\}$,

then the inverse of F is

$F^{-1} = \{(3, 2), (4, 3), (5, 4)\}$.

Note that both F and F^{-1} are functions. If, however,

$F = \{(2, 3), (3, 4), (4, 3)\}$,

then its inverse, $\{(3, 2), (4, 3), (3, 4)\}$,

is a relation but not a function. We use the symbol F^{-1} when the inverse of a function is a function.

5 *A function from set A to set B is sometimes called a* mapping *from A to B.*

a) A mapping of A onto *B is a correspondence where each element of A is mapped to a unique element of B, and each element of B is the image of at least one element of A. For example, if A = {1, 2, 3} and B = {a, b}, then the mapping {(1, a), (2, b), (3, a)} is an* onto *mapping from A to B.*

b) A mapping from A to B is one-to-one *if each element of A corresponds to a unique element of B, and each element of B corresponds to a unique element of A. It follows from this definition that if the inverse of a mapping F is also a mapping, then F is a* one-to-one *mapping. Also, every one-to-one mapping is an onto mapping.*

The function $\{(x, y)|y = 3x; x = 1, 2, 3, 4\}$ is

$$F = \{(1, 3), (2, 6), (3, 9), (4, 12)\}.$$

Since x and y take on different values, both are called *variables.* x is called the *independent variable* because its values are chosen first, independently of the y. y is called the *dependent variable* because each of its values depends on the corresponding value of x.

Since the values of y depend on the values of x, we speak of y as *a function of* x and express this relationship by writing $y = f(x)$, read "y is a function of x." The expression $f(1) = 3$ means that when $x = 1$, $y = 3$. Similarly, $f(2) = 6$ means that when $x = 2$, $y = 6$.

EXERCISE SET 2.5

1 **Which of the following relations are functions?**
- **a)** $R = \{(1, 2), (2, 3), (3, 4)\}$
- **b)** $S = \{(1, 1), (2, 2), (3, 3), (4, 4)\}$
- **c)** $A = \{(1, 3), (1, 4), (2, 5), (3, 7)\}$
- **d)** $C = \{(1, 5), (2, 7), (3, 5), (4, 9)\}$

2 **Which of the following relations are functions?**
- **a)** $T = \{(a, b)|b = 2a; a = 2, 3, 4\}$
- **b)** $R = \{(a, b)|b = a + 3; a = 1, 2, 3, \ldots\}$
- **c)** $B = \{(a, b)|b = a^2; a = 1, 2, 3, 4\}$
- **d)** $D = \{(a, b)|b > a + 1; a = 2, 4, 6, 8\}$

3
- **a) Write the inverse relation for each of the relations given in Exercise 1.**
- **b) Which of the inverse relations are functions?**

4
- **a) Write the inverse relation for each of the relations given in Exercise 2.**
- **b) Which of the inverse relations are functions?**

5 *If* $R = \{(x, y) | y = 2x + 1; x = 1, 2, 3, \ldots, 10\}$,

a) What is the rule for determining the function?

b) Write the table of values for this function.

c) Express this function as a set of ordered pairs.

6 Write a formula for the function represented by each of the following tables.

a)

x	1	2	3	4	5
y	4	8	12	16	20

b)

x	0	1	2	3	4
y	7	8	9	10	11

c)

x	4	5	6	7	8
y	1	2	3	4	5

d)

x	1	2	3	4	5
y	1	3	5	7	9

7 Express each function in Exercise 6 as a set of ordered pairs.

8 If $A = \{9, 11, 13\}$ and $B = \{a, b, c, d\}$, give:

a) Two examples of relations from A to B that are not functions.

b) Two examples of relations from B to A that are functions.

c) The inverses of the functions in part b.

9 The distance x, measured in feet, through which a body falls in t seconds is given by the formula $s = 16t^2$.

a) Make a table of values for this function for $t = 0$ to $t = 4$.

b) Write this function as a set of ordered pairs.

c) If an object, dropped from a cliff, reaches the ground in 2.5 seconds, how high is the cliff?

d) If a cliff is 150 feet high, in how many seconds will an object reach the ground?

PUZZLES

1 At a certain hotel there are 60 guests who had traveled to London, 45 who had traveled to Paris, and 40 who had traveled to Rome. Seventeen of these guests had been to both London and Paris, 15 had been to both Paris and Rome, 13 had been to both Rome and London, and 5 visited all three cities. How many guests are there altogether at the hotel? (Use a Venn diagram to solve this problem.)

2 Three of the characteristics possessed by the inhabitants of a certain island are onx, bonx, and conx. Two of the inhabitants are onx and bonx, 2 are bonx and conx, 2 are conx and onx, and 2 are at the same time onx, bonx, and conx. Altogether there are 15 inhabitants on the island. Why are the conditions in this problem contradictory?

BASICS REVISITED

1 Find the sum of 382, 816, 69, and 5007.

2 Write in standard notation: 400 + 600 + 40 + 9.

Perform the indicated operations in Exercise 3–14.

3 $17 \overline{)31759}$ (Round to 2 places.)

4	$0.08 + 0.5 + 0.0043$	5	$2 - .023$
6	$\$15.08 - \$.59$		
7	$.6\overline{)43.07}$ (Round to 2 places.)		
8	$5\frac{3}{4} - 3\frac{1}{8}$	9	$\frac{4}{17} \div \frac{4}{17}$
10	$\frac{2}{3} \div 100$	11	$1\frac{2}{3} \times 3\frac{5}{8}$
12	$\frac{5}{6} + \frac{1}{2} + \frac{3}{4}$	13	$12 \times 6 + 11$
14	$(13 \times 5)^0$		

15 Write .045 as a fraction and as a percent.

16 Find the average of 8, 5, 6, 9, and 7 by inspection.

17 Express in exponential form:

$17 \times 17 \times 17 \times 17 \times 17 \times 17$

18 Express $12\frac{1}{2}$% as a common fraction.

In Exercises 19–21, solve for x and check your answers.

19	$36 = \frac{x}{5} \cdot 9$	20	$\frac{3}{8} \cdot 92 = x$
21	$\frac{2}{7}x = .8$		

22 Express 512 as a product of its prime factors.

23 How many 25-kilogram bags of flour can you get out of a load of 4573 kilograms?

24 A student pays \$73 a month rent, \$52 a month for food, and \$85 a month for other expenses during nine months at college. What are his total expenditures during these nine months?

25 If 80% of a boy's money amounts to \$3.60, how much money does he have?

EXTENDED STUDY

2.6 Set Theory

The subject of set theory was developed by Georg Cantor (1845–1918) toward the end of the nineteenth century. As a result of his work, it has become possible to answer such questions as (1) Are there more natural numbers than positive even numbers, and (2) Does a 5-inch segment contain more points than a 2-inch segment?

We have already seen that the equivalence of two finite sets can be established either by counting or by a one-to-one correspondence. Although we cannot count the number of elements in an infinite set, it is possible to devise techniques to ascertain whether there exists a one-to-one correspondence between the elements of two infinite sets. If such a correspondence can be shown to exist, then the two sets must be equivalent.

To answer the question of whether there are more natural numbers than positive even numbers, let us denote the set of natural numbers by N and the set of positive even numbers by G and pair their elements this way:

$$\begin{array}{l} N = \{1,\ 2,\ 3,\ 4,\ 5,\ \ \ldots,\ n,\ \ \ldots\}, \\ \qquad \updownarrow\ \updownarrow\ \updownarrow\ \updownarrow\ \updownarrow \qquad\ \ \updownarrow \\ G = \{2,\ 4,\ 6,\ 8,\ 10,\ \ldots,\ 2n,\ \ldots\}. \end{array}$$

This pairing establishes a one-to-one correspondence between N and G since for every natural number there exists a positive even number and for every positive even number there exists a natural number. Therefore, G is equivalent to N, and the answer to question (1) is that there are as many positive even numbers as there are natural numbers!

Although G is equivalent to N, N also contains the positive odd numbers. Since G is a subset of N but N contains elements that are not in G, G is a proper subset of N.

We are now confronted with the curious paradox that the set of positive even numbers is a proper subset of the set of natural numbers and *at the same time* equivalent to it! In other words, the set N is no greater than a piece of itself. There are many other such pieces of N that are as great as N; for example, the set of square numbers, the set of positive cube numbers, and the set of positive multiples of 3. This paradox involves a property that we use to define finite and infinite sets.

DEFINITION

A set is an *infinite* set if it is equivalent to a proper subset of itself. If a set is *not* an infinite set, then it is a *finite* set.

The famous Euclidean axiom that "the whole is greater than any of its parts" is not valid for infinite sets.

2.7 Proving Some Properties of Sets

We have seen that union and intersection are commutative. We shall now prove these properties for any sets A and B. In these proofs we shall rely on the definition of equality of sets; that is, two sets are equal if each is a subset of the other.

THEOREM 1 Union is commutative. That is, $A \cup B = B \cup A$.

Proof

1. Each element of $A \cup B$ is an element of $B \cup A$, and each element of $B \cup A$ is an element of $A \cup B$.
2. Therefore, $A \cup B \subseteq B \cup A$ and $B \cup A \subseteq A \cup B$.
3. Therefore, $A \cup B = B \cup A$, since each is a subset of the other.

THEOREM 2 Intersection is commutative. That is, $A \cap B = B \cap A$.

Proof We need to show that (a) $A \cap B \subseteq B \cap A$, (b) $B \cap A \subseteq A \cap B$.

1. To prove statement (a), we must show that any element x of $A \cap B$ is also an element of $B \cap A$.
   a) $x \in A \cap B$ means $x \in A$ and $x \in B$;
      $x \in A$ and $x \in B$ means $x \in B$ and $x \in A$.
   b) $x \in B \cap A$.
   c) Therefore, $A \cap B \subseteq B \cap A$.
2. To prove statement (b), we show that any element y of $\mathrm{B} \cap \mathrm{A}$ is also an element of $\mathrm{A} \cap \mathrm{B}$. Follow the same procedure as for statement (a).
3. Therefore, $\mathrm{A} \cap \mathrm{B} = \mathrm{B} \cap \mathrm{A}$, since each is a subset of the other.

2.8 Partitioning a Set

Many problems related to sets can best be described in terms of *partitions* of a set. A *partition* is a subdivision of a set into mutually disjoint subsets:

If A, B, and F are sets such that $A \cup B = F$ and $A \cap B = \phi$, then we say that A and B *partition* F. For example, the set of positive even numbers E and the set of positive odd numbers O partition the set of natural numbers N because

1. Every natural number is either even or odd; that is, $E \cup O = N$.
2. No natural number can be both even and odd; that is, $E \cap O = \phi$.

A set may be partitioned into more than two sets. For example, A, B, and C partition F if and only if

1. $A \cup B \cup C = F$, and
2. $A \cap B = \phi$, $A \cap C = \phi$, and $B \cap C = \phi$.

If A and B partition F, then we call A the *complement of B with respect to F*, and B is the *complement of A with respect to F*.

2.9 Functions and their Graphs

Graphs provide a quick way to determine whether a relation is a function. Since, by the definition of a function, two different values of y cannot exist for the same value of x, any vertical line meets the graph of a function in at most one point; a vertical line meets the graph of a relation that is not a function in more than one point.

Graphs of relations that are functions are shown in Figure 2.11. Notice that no vertical line intersects the graph in more than one point.

FIGURE 2.11

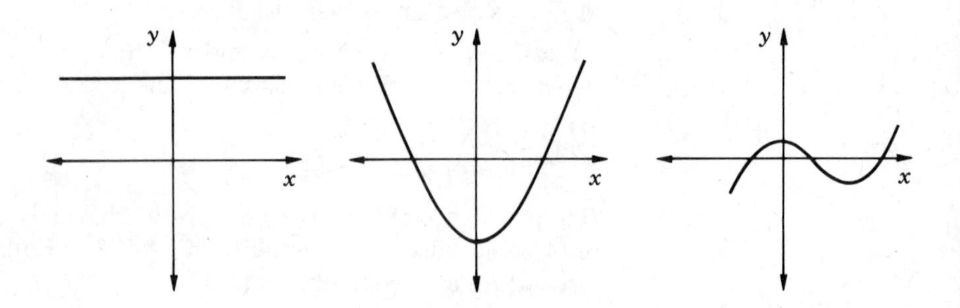

Graphs of relations that are not functions are shown in Figure 2.12. In each of these cases, there exists at least one vertical line that intersects the graph in two or more points.

FIGURE 2.12

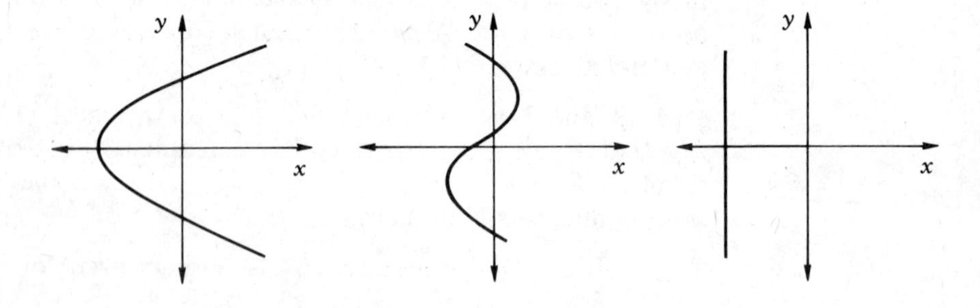

2.10 Useful Functions

Among the most commonly used functions in mathematics and science are the *linear*, the *quadratic*, and the *exponential* functions.

THE LINEAR FUNCTION

A function of the form $\{(x, y) | y = mx + k\}$, where m ($m \neq 0$) and k are real numbers, is called a *linear function*. For example,

$$\{(x, y) | y = 2x + 3\}$$

is a linear function. We call such a function *linear* because its graph is always a straight line. In such a function both the dependent variable and the independent variable appear only to the first power. The graph of the linear function

$$\{(x, y) | y = 2x + 3\}$$

is shown in Figure 2.13.

FIGURE 2.13

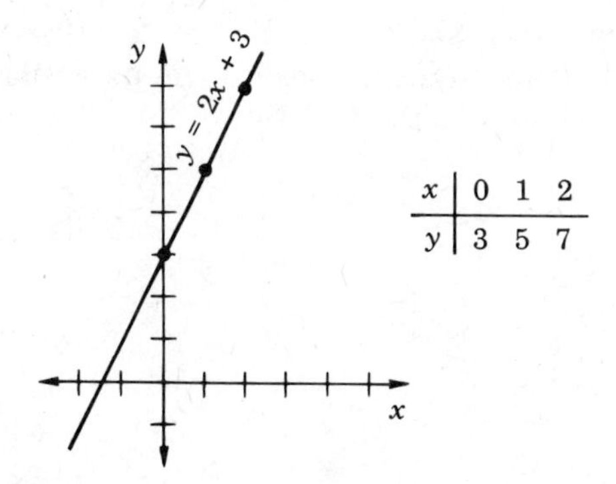

x	0	1	2
y	3	5	7

The graph of a linear function cannot cross the x-axis more than once. Any ordered pair which belongs to the function represents a point on the line, and any point on the line is represented by an ordered pair that belongs to the function.

Another important characteristic of linear functions is that the ratio of a change in y to the corresponding change in x is constant. This means that if y_1 corresponds to x_1 and y_2 corresponds to x_2, then

$$\frac{y_2 - y_1}{x_2 - x_1}$$

is constant, that is, will always equal the same number no matter what values we assign to x_1 and x_2. In the linear function $\{(x, y) | y = 2x + 3\}$, this ratio is always 2 no matter which two points on the line we may select. For example, if we choose the two ordered pairs (1, 5) and (0, 3) that satisfy the equation $y = 2x + 3$, we get

$$\frac{y_2 - y_1}{x_2 - x_1} = \frac{5 - 3}{1 - 0} = \frac{2}{1} = 2.$$

If we choose (4, 11) and (2, 7), we again obtain the ratio of 2:

$$\frac{y_2 - y_1}{x_2 - x_1} = \frac{11 - 7}{4 - 2} = \frac{4}{2} = 2.$$

THE QUADRATIC FUNCTION

A function of the form

$$\{(x, y) | y = ax^2 + bx + c\},$$

where a ($a \neq 0$), b, and c are real numbers, is called a *quadratic function.* In this function the independent variable appears to the second power. The word *quadratic* is derived from the Latin word *quadratus* which means *squared.*

The graph of the quadratic function is a parabola whose axis is perpendicular to the x-axis and cannot cross the x-axis more than twice. The graph of the quadratic function

$$\{(x, y) | y = x^2 - 2x + 1\}$$

is the parabola shown in Figure 2.14. Any ordered pair which belongs to this function lies on the parabola, and any point on the parabola belongs to the function.

FIGURE 2.14

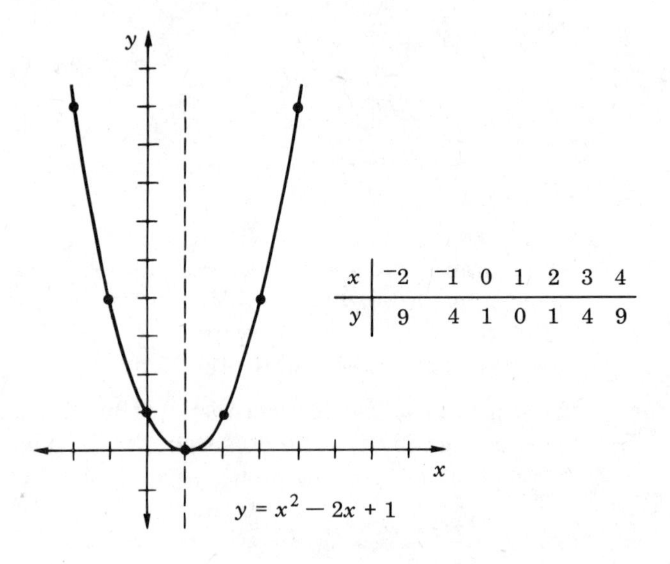

x	$^-2$	$^-1$	0	1	2	3	4
y	9	4	1	0	1	4	9

A parabola is a very useful curve. A projectile, fired into space, follows a path that is approximately a parabola. Cables of a suspension bridge hang in parabolic arcs if the weight of the bridge is uniformly distributed. A reflecting surface in the shape of a parabola reflects light from the focus in the direction of the axis. This idea is put to use in automobile headlights, in telescopes, and in radar. The same principle, applied to sound waves, is used in listening and broadcasting devices. Parabolic reflectors with a microphone at the focus catch distant sounds and "focus" them on the microphone.

The Greeks discovered how the parabola can be used to control light. Archimedes was said to have put this discovery to use by constructing a huge parabolic mirror which focused the sun's rays on the Roman ships besieging Syracuse and so set them on fire.

THE EXPONENTIAL FUNCTION

Consider the growth pattern of a germ that grows to a certain size every hour and then splits into two distinct germs. At the end of the first hour there will be two germs; at the end of the second hour these two germs will split into four germs; at the end of the third hour there will be eight germs; etc. Table 2.1 illustrates this growth pattern.

TABLE 2.1

x (number of hours)	0	1	2	3	4	5	6
y (number of germs)	1	2	4	8	16	32	64

Since at the end of x hours there will be 2^x germs, the equation of this function is $y = 2^x$. Such a function is called an *exponential function.* The graph of the exponential function

$$\{(x, y) | y = 2^x\}$$

is shown in Figure 2.15.

FIGURE 2.15

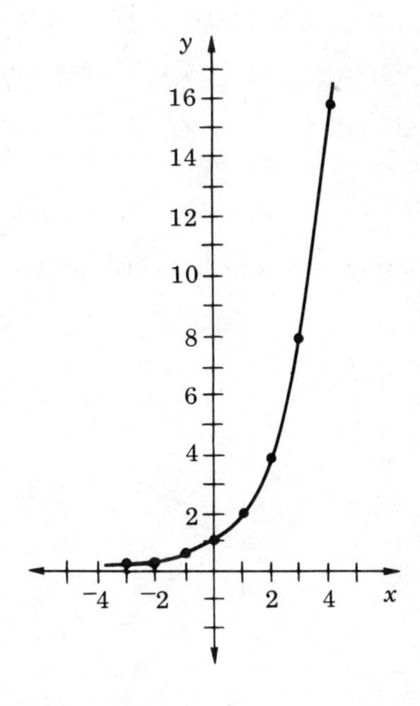

In the case of the germs, x (the number of hours) must have a positive value; but in a nonphysical situation, x can have either positive or negative values such as those shown in Table 2.2. y, however, always assumes only positive values.

TABLE 2.2

x	$^-3$	$^-2$	$^-1$	0	1	2	3
y	$\frac{1}{8}$	$\frac{1}{4}$	$\frac{1}{2}$	1	2	4	8

The graph in Figure 2.15 reflects the negative as well as the positive values of x for the function $\{(x, y)|y = 2^x\}$. From Figure 2.15 we note that the function y increases very slowly as x increases toward zero, and then increases more and more rapidly as x continues to increase.

If the growth pattern of the germ were such that a germ splits into *three* distinct germs every hour, then this pattern could be described by the exponential function $\{(x, y)|y = 3^x\}$. More generally, if the germ splits into b germs every hour, then its growth pattern is described by the exponential function $\{(x, y)|y = b^x\}$. If we start with, say, 7 germs instead of one germ, there will be 7 times as many germs at the end of x hours, and the equation becomes $y = 7(b^x)$. If we start with a germs, the equation becomes $y = a(b^x)$.

An *exponential function*, then, is a function of the form

$$\{(x, y)|y = b^x\}, \qquad \text{where } b > 1.$$

A more general form of the function is

$$\{(x, y)|y = ab^x\}.$$

The inverse of the exponential function $\{(x, y)|y = b^x\}$ is

$$\{(x, y)|b^y = x\} \qquad \text{or} \qquad \{(x, y)|y = \log_b x\}$$

which is a *logarithmic function.* If the exponential function is $\{(x, y)|y = 2^x\}$, then its inverse is the logarithmic function

$$\{(x, y)|2^y = x\} \qquad \text{or} \qquad \{(x, y)|y = \log_2 x\}.$$

The graphs of these two functions are shown in Figure 2.16.

FIGURE 2.16

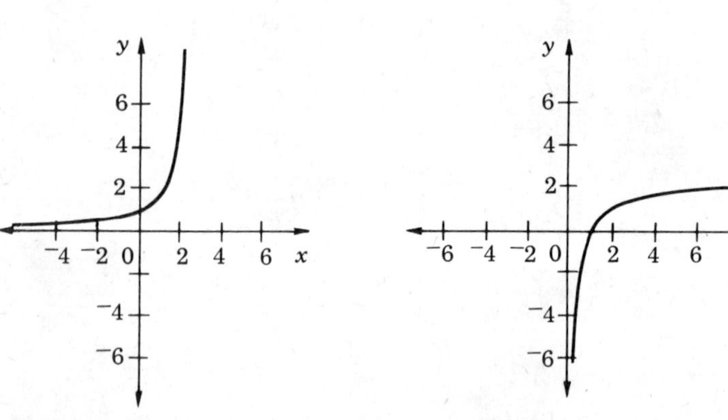

Exponential Function $\{(x, y)|y = 2^x\}$

Logarithmic Function $\{(x, y)|2^y = x\}$ or $\{(x, y)|y = \log_2 x\}$

Note that in the *logarithmic* function, x assumes only positive values, and y always increases as x increases. This increase is very rapid for $x < 1$, but after this point y increases slower and slower.

APPLICATIONS OF THE EXPONENTIAL FUNCTION

The exponential function derives its importance from its wide applicability to many real-life situations such as the growth of bacteria, the growth of populations, the growth of capital and interest, and the rate of decay of radioactive substances. Each of these situations has the property that *the rate of its change is proportional to the rate of change of its variable.* That is, the quantity present at the end of one of the equal intervals of time is proportional to the quantity present at the beginning of the interval. In the case of the germs, the number of germs present at the end of any hour is *twice* the number of germs present at the beginning of the hour. Thus, the ratio of any term to the preceding term in Table 2.1 is 2. We sometimes describe such a sequence of numbers where the ratio of any two successive values is the same as a *geometric progression.*

An interesting example of the use of functional relations to predict population growth was provided by Malthus in 1789 when he concluded that drastic means, such as wars, are needed to limit the growth of the world population if we are to have enough food to survive. According to Malthus, the functional relation between time and world population can be expressed by the exponential function, $y = ab^x$, and the functional relation between time and food supply can be expressed by the linear function, $y = mx + k$. Figure 2.17 shows the graphs of these functions.

FIGURE 2.17

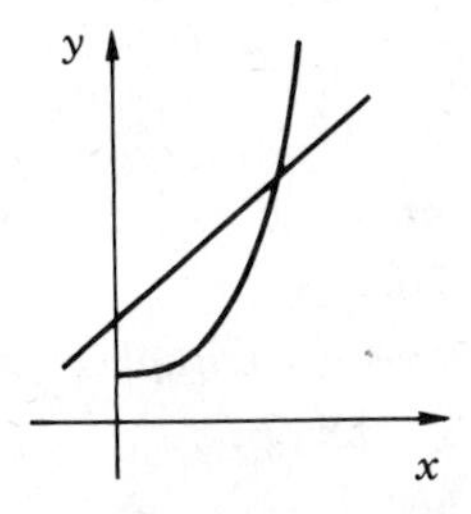

As long as the straight line remains above the exponential curve, the food supply is adequate. The point of intersection of the two graphs represents a time when there is an exact balance between the amount of food needed and the amount available. When the straight line is below the curve, the food supply is inadequate.

More recent studies, however, have shown that the graph in Figure 2.18 more nearly approximates population growth than the graph in Figure 2.17. The graph in Figure 2.18 suggests a built-in restraint on the growth of the population not considered by Malthus.

FIGURE 2.18

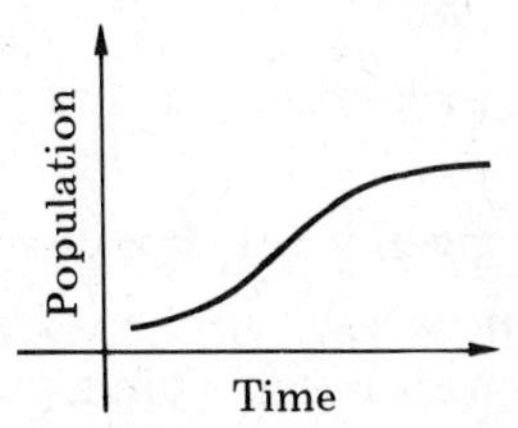

Chapter 3

SYSTEMS OF NUMERATION

Objectives At the completion of this chapter, you should be able to:

1 Distinguish between a *number* and a *numeral.*

2 List the characteristics of ancient systems of numeration, including the Egyptian, Roman, Babylonian, and Mayan systems.

3 Identify the basic features of a *place-value numeration system.*

4 Describe the properties of the Hindu-Arabic numeration system.

5 Represent any given whole number in different bases.

6 Use *standard notation*, *expanded notation*, and *exponential notation* to express numbers in a given base.

7 Count in different bases.

8 Convert from base x to base ten, from base ten to base x, and from base x to base y.

9 Construct addition and multiplication tables for various bases.

10 Add, subtract, multiply, and divide in various bases.

Note Objectives 5 through 10 refer to base two through base twelve.

Prerequisites

You should know that:

1 Two sets are *equivalent* if there exists a one-to-one correspondence between their elements.

2 Equivalent sets contain the same number of elements.

INTRODUCTION

How long did it take you to learn to count? How long did it take you to learn to *write* the numbers you counted? Isn't it amazing that what you learned so easily and quickly in childhood has taken mankind more than 20,000 years to develop?

About 25,000 years ago man was slowly learning to count. But it wasn't until about 2,000 years ago that a truly efficient system for *writing* numbers was developed. This system, the Hindu-Arabic system that we use today, is simple enough to make everyday calculations easy as well as refined enough to express with ease the weight of an atom.

Many systems of writing numbers were developed by earlier civilizations before the Hindu-Arabic system came into being. But those who devised the earlier systems did not learn the secret of how to use only a few basic symbols to represent easily infinitely many numbers no matter how large or small. In this chapter we shall see what this secret is and why it has been called one of the great inventions of the human mind.

We shall also see that although the system we use today is very efficient, it is not the only system in use. One other system, in particular, is of great importance today. We shall see how using this other system of writing numbers brought into the world the modern electronic computer. This computer made possible the exploration of interplanetary space as well as other astounding achievements brought about through the computer's ability to make calculations with lightning speed, to remember enormous amounts of information, and to make decisions on the basis of this information.

Studying different ways of using symbols to represent numbers is not only fascinating in itself but helps us see more clearly why and how the system we have been using all our lives really works.

3.1 Ancient Numeration Systems

Before discussing numeration systems, we wish to distinguish between a *number* and a *numeral.* A *number* is an idea associated with a set, such as the set $A = \{\triangle, \bigcirc, \square\}$. The same number is associated with all other sets equivalent to set A, such as $B = \{*, ?, \#\}$, $C = \{\text{apple, star, glass}\}$, or $D = \{1, 2, 3\}$. If we could collect *all* the sets which contain three elements, this collection would completely describe the concept of *three.* The property common to all these sets is called the *number three. Numbers*, then, are ideas in our head, mathematical abstractions associated with sets. Numbers cannot be seen, touched, heard, smelled, or tasted. They cannot be erased or moved.

A *numeral* is the symbol or name we use to *represent* a number. A number may be represented by many different numerals. For example, the *number three* may be written 3, III, 5 − 2, and in many other ways. A numeral can be seen, touched, moved, or erased.

A *numeration system* is a way of representing numbers. It consists of a set of basic symbols, and rules for combining these symbols to represent various numbers.

Comment

A numeration system *should not be confused with a* number system. *A numeration system is concerned with a way of writing numerals to represent numbers. It is not directly concerned with the properties of these numbers.*

A number system *is concerned with classifying numbers, such as natural numbers, integers, rational numbers, or real numbers. The properties of these number systems remain the same irrespective of the numeration system used to denote the numbers. For example, two is less than five, and* $3 \times 5 = 5 \times 3$ *regardless of the numeration system used to write these numbers.*

The earliest numeration systems used knots in a rope, piles of pebbles, or tally marks to represent numbers. These tallies might have been in the form of notches in a stick

or scratches on a cave wall.

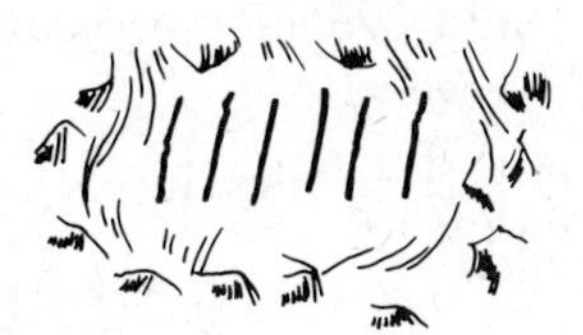

Such a numeration system becomes inadequate when we wish to write large numbers. Imagine using tally marks to represent the population of the United States!

EGYPTIAN SYSTEM

The early Egyptians avoided the disadvantage of tally numerals by using a *single numeral* to represent a *group of tallies.* For example, they used the numeral ∩ (heel bone) to represent the group of ten tallies ||||||||||.

Other basic numerals introduced by the Egyptians are shown in Figure 3.1.

FIGURE 3.1

1	10	100	1,000	10,000	100,000	1,000,000
Stroke	Heel Bone	Coiled Rope	Lotus Flower	Pointed Finger	Fish	Astonished Man

Characteristics of the Egyptian System

1 The number ten was used as the basis for grouping tallies.

2 Different symbols were used to represent groups of different size.

3 A numeral represented a number which was the *sum* of the numbers represented by the individual numerals; that is,

∩∩||| stands for 10 + 10 + 1 + 1 + 1 or 23.

4 It was not a *place-value* system; that is, the *location* of a symbol in a numeral did not affect its meaning. For example, "twenty-three" might have been written in several different ways, such as

∩∩||| or |∩|∩| or |||∩∩

This is not true of our decimal system where 23 does not represent the same number as 32.

5 It did not use a zero.

6 A disadvantage of the Egyptian numeration system was that computation could become very cumbersome. For example, to add 28 and 39 we get

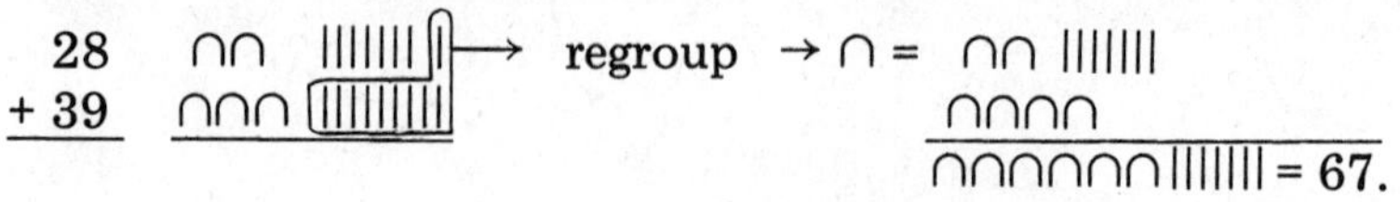

```
  28    ∩∩   ||||||||  ─→  regroup → ∩ =  ∩∩ |||||||
+ 39   ∩∩∩  |||||||||                     ∩∩∩∩
                                         ∩∩∩∩∩∩ ||||||| = 67.
```

ROMAN SYSTEM

The Roman numeration system uses these basic symbols:

1	5	10	50	100	500	1000
I	V	X	L	C	D	M

To represent various numbers, the basic symbols are combined according to these rules:

1 When the basic symbols in a numeral appear in the order M, D, C, L, X, V, I, the number represented is the *sum* of the numbers represented by the individual numerals. For example,

XVI = X + V + I = 10 + 5 + 1 = 16

2 If a smaller number precedes a larger number, the smaller number is subtracted from the larger number. For example, a) IV = V − I = 5 − 1 = 4   b) XL = L − X = 50 − 10 = 40.

3 To represent large numbers, bars are drawn above the numeral, or portions of the numeral, and then the multiplication principle is put to use. For example, $\overline{\text{V}}$ means 5 × 1000, or 5000; and $\overline{\text{XX}}$VII means (20 × 1000) + 7, or 20,007.

Characteristics of the Roman System

1 It uses addition and subtraction of the individual numbers in a numeral to represent various numbers.

2 It uses grouping. For example, five I's are grouped as V; two V's are grouped as X.

3 The basic symbols X, L, C, D, and M represent multiples of ten.

4 It does not use a zero.

5 It is not a place-value system.

BABYLONIAN SYSTEM

The next big step in the improvement of numeration systems was the introduction of the idea of *place value.* This idea permits us to attach many different values to the same symbol by making its value depend on the place that it occupies in a numeral. The Babylonians, the ancient Chinese, and the Mayans devised such systems. In the Babylonian system, for example, the symbol ▼ could stand for one, or sixty, or sixty times sixty, or any power of sixty, depending on its *position* in the numeral. Although the Babylonian system used the place-value principle, it did not do so consistently. In this system, ▼ and ◄ stood for one and ten, respectively.

The number twenty-three would be written ◄◄ ▼ ▼ ▼ . Here the symbols are repeated in the same way that the Egyptians and Romans had done. It was only in writing numbers *larger* than fifty-nine that the Babylonians used the place-value idea. For example,

a) ◄ ▼ ▼ ▼ meant

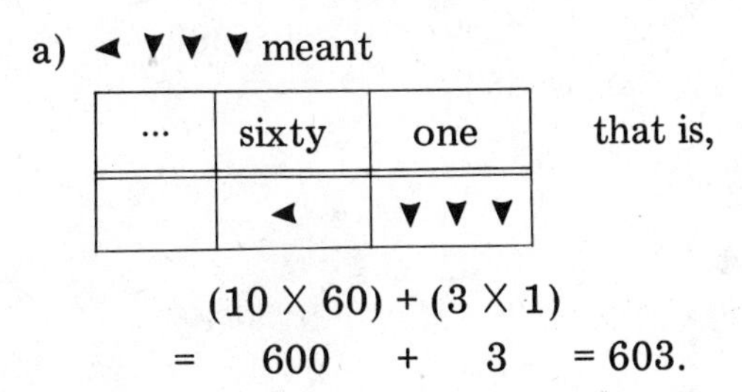

...	sixty	one
	◄	▼ ▼ ▼

that is,

$$(10 \times 60) + (3 \times 1)$$
$$= \quad 600 \quad + \quad 3 \quad = 603.$$

b) ▼ ▼ ◄ ▼ meant

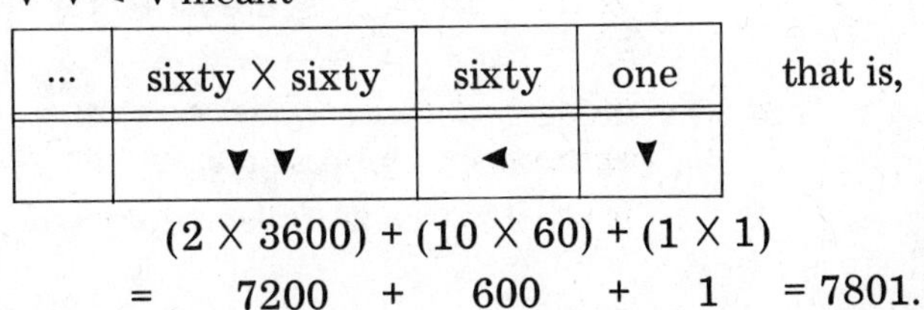

...	sixty × sixty	sixty	one
	▼ ▼	◄	▼

that is,

$$(2 \times 3600) + (10 \times 60) + (1 \times 1)$$
$$= \quad 7200 \quad + \quad 600 \quad + \quad 1 \quad = 7801.$$

Because the early Babylonians did not have a symbol for zero, they sometimes left empty spaces where we write zeros. The absence of such a symbol led to problems in interpretation. For example, does "▼   ▼" mean "(1 × 3600) + (1 × 1)" = 3600 + 1 = 3601, or does it mean "(1 × 60) + (1 × 1)" = 60 + 1 = 61?

Characteristics of the Babylonian System

1. It was the first system to use the concept of place value. Place value was used to represent numbers larger than 59.
2. Only two symbols, ▼ and ◄, were used to write all numerals.
3. The early Babylonians did not have a symbol for zero.
4. It used the additive and multiplicative principles to represent a number.
5. It was a base-sixty, or a *sexagesimal*, system.

MAYAN SYSTEM

This numeration system, dating back to about 300 A.D., was developed by the Mayan Indians of Guatemala and Honduras. The Mayans used dots and horizontal lines to represent numbers, a dot representing one and a line representing 5. The number six would be represented by $\dot{-}$, and the number nineteen by $\overset{\cdots\cdot}{\equiv}$. Numbers larger than 19 were written in terms of base twenty, with one exception: instead of $(20)^2$ they used $18(20)$; instead of $(20)^3$ they used $18(20)^2$. That is, their higher place values were of the form $18(20)^n$. They used zero as a place holder and its symbol was ◎. Like the Chinese and Japanese, the Mayans wrote their numerals in vertical form. For example,

a) · → 1(20) = 20
◎ → 0(1) = 0
20

b) $\dot{\dot{-}}$ → 7(18)(20) = 2520
◎ → 0(20) = 0
··· → 3(1) = 3
2523

c) ···· → $4(18)(20)^2$ = 28800
— → 5(18)(20) = 1800
◎ → 0(20) = 0
$\dot{\dot{-}}$ → 7(1) = 7
30607

Characteristics of the Mayan System

1. The Mayan system used the ideas of repetition and addition of its two basic symbols to represent all numbers from 1 to 19.
2. It used a place value system, essentially in base twenty, to represent numbers greater than 19.
3. It used a zero as a place holder.

CHINESE-JAPANESE SYSTEM

The traditional Chinese system of numeration, dating back to the third century B.C., is a base ten system which uses nine symbols for the numbers one to nine and additional symbols for the powers of ten. Some of these symbols, later adopted by the Japanese, are shown in Figure 3.2.

FIGURE 3.2

1	2	3	4	5	6	7	8	9	10	100	1000
一	二	三	四	五	六	七	八	九	十	百	千

Numerals were written from the top downward or from left to right. When numerals for numbers larger than ten are written, the place-value positions are indicated by using symbols for 10, 100, 1000 next to the symbols for one through nine. For example, the numeral

三
百
七
十
八

represents $3 \times 100 + 7 \times 10 + 8$ or 378.

Characteristics of the Chinese-Japanese Systems

1 It is a decimal system.

2 It uses place value.

3 It uses multiplication and addition for writing numerals.

4 A symbol for zero, such as ours, was first introduced about 800 years ago.

3.2 The Decimal System

The Hindu-Arabic system, called the *decimal system*, is a vastly more efficient numeration system than all the earlier systems because:

1 The place-value principle is used with complete consistency;

2 The need for a zero was recognized, and a symbol was chosen.

Characteristics of the Decimal System

1 In this system, many objects are counted in *groups of ten.* The same principle applies to our use of a dime to represent a group of ten pennies and a dollar to represent a group of ten dimes.

2 The decimal system uses *ten basic symbols*, {0, 1, 2, 3, 4, 5, 6, 7, 8, 9}, each symbol representing a different number. By arranging these symbols in various positions, any number, or quantity, can easily be represented. These ten symbols are called *digits.* The word *digit* comes from the Latin word meaning *finger.*

3 It is a *place-value numeration system.* By this we mean that the value of any digit in a numeral depends on the position or place that it occupies in the numeral. For example, the numeral 235 denotes these place-values:

...	100's	10's	1's
	2	3	5

and represents 2 groups of 100's, 3 groups of 10's, and 5 groups of 1's.

4 Note that each place in a numeral represents a group ten times the value of the next place to its right. From right to left, the first position represents groups of 1; the second position represents groups of 10, or (10×1); the third position represents groups of 100, or (10×10); etc. So we see that in the decimal system

a) There is a value assigned to each position in the numeral. This is called the *place value* of the position:

...	1000	100	10	1

b) Each digit in a numeral represents the product of the number, or quantity, it represents and the place value assigned to its position.

100	10	1
3	7	8

In the numeral 378,
the 3 represents $3 \times 100 = 300$
the 7 represents $7 \times 10 = 70$
the 8 represents $8 \times 1 = 8$
378.

c) The quantity represented by the numeral is the sum of the above products.

5 The decimal system uses 0 as the symbol for zero. Zero is needed to fill places which would otherwise be empty and lead to misunderstandings. For example, the zero in the numeral 20 tells us that the number contains 2 groups of 10 and *no* groups of 1. Without zeros we would not know whether the 2 represents groups of 1's, 10's, or 100's, etc.

6 Since we group by *tens* in the decimal system, we say that it is a *base-ten* system.

7 By agreement, $2 \times 2 \times 2$ may be written 2^3, and $10 \times 10 \times 10 \times 10$ may be written 10^4. In the expression 2^3, the numeral 2 is called the *base* and 3 is called the *exponent.* A number written in this way is said to be written in *exponential form.* It is further agreed that any number with a

zero exponent is 1. Thus, $5^0 = 1$, $14^0 = 1$, and $10^0 = 1$.

The base-ten place values may, therefore, be expressed in exponential form this way:

$\cdots \quad 10^3 \quad 10^2 \quad 10^1 \quad 10^0.$

By this notation, 2539 may be written as

$(2 \times 10^3) + (5 \times 10^2) + (3 \times 10^1) + (9 \times 10^0)$, or

$(2 \times 10^3) + (5 \times 10^2) + (3 \times 10) + 9$

which is called the *expanded form* of the number 2539.

In general, the *decimal expansion* of a whole number is an expression of the form

$$\cdots + (a_3 \times 10^3) + (a_2 \times 10^2) + (a_1 \times 10) + a_0$$

where a_3, a_2, a_1, and a_0 are whole numbers.

Comments

1 *A place value is, therefore, a number which is a power of the base.*

2 *Exponents are introduced here to make our notation more concise, and to establish an easily recognizable pattern for other positional numeration systems that follow.*

3 *Where and when the idea of zero originated is not certain, but the Hindus are known to have used a symbol for zero about 600* A.D. *Eight hundred years earlier the Babylonians introduced the symbol ⋞ to denote the absence of a figure, but they did not use it in computation the way we use our 0. The Mayans, about 300 A.D., used a zero as a place holder.*

4 *The way we write numerals in the decimal system was developed by the Hindus and brought to Europe by the Arabs. Interestingly enough, most Arabs never actually used these symbols.*

5 *The great superiority of the decimal system over all the previous systems is due not only to the greater ease with which numbers can be represented, but to the far greater efficiency with which computations can be performed.*

6 *Because this system uses groups of ten, it is called a* decimal system. *The word* decimal *comes from the Latin word* decem *which means* ten.

7 *According to one story, the Bank of London used tally sticks as late as 1790 to record transactions. Horizontal tally marks were made across the stick, and the stick was then split vertically. Half of the stick was given to the investor and half was kept by the bank. When the bank decided to modernize its accounting system, the occasion was celebrated by setting the tally sticks on fire. But the fire got out of control and the bank itself was destroyed.*

EXERCISE SET 3.1, 3.2

1. What numbers are represented by the following Egyptian symbols?
   a) ||| b) ∩∩|||| c) 𓍢𓍢∩||

2. True or false:
   a) The Egyptians used a single numeral to represent a group of tallies.
   b) The Egypitans used a zero.
   c) In the Egyptian system of numeration, the numerals ||∩ and ∩|| represent the same number.
   d) The Egyptian numeration system was a place-value system.
   e) The Egyptian numeration system was an additive system in which a numeral represented the number that is the sum of the numbers represented by the individual numerals.

3. What numbers are represented by the following Roman numerals?
   a) XVII   b) XLIX   c) CXX   d) CLIV   e) DCLXXVIII

4. What numbers are represented by the following Babylonian numerals?
   a) ▼ ▼   b) ◀◀   c) ◀ ▼ ▼   d) ▼ ◀ ▼ ◀ ◀

5. True or false:
   a) The Babylonians were the first to use the concept of place value.
   b) The Babylonian system was a base-sixty system.
   c) Neither the Romans nor the Babylonians used a zero.

6. What numbers are represented by the following Mayan symbols.
   a) ≐   b) ⁚≡   c) ∷ over —   d) ⁚ over — / 𓁹 / ⁝ over ═   e) · over ≡ / 𓁹 / ⁚

7. True or false:
   a) The Mayans used zero as a place holder.
   b) The Mayans used a place-value system to represent numbers greater than 19.
   c) The traditional Chinese system used a decimal system, a place-value system, but not a symbol for zero.

8. What is the difference between a numeration system and a number system?

9. What are the major differences between the ancient numeration systems and the Hindu-Arabic numeration system?

10. Describe the characteristics of the decimal system.

11. Express the following numbers in Roman numerals.
    a) 27   b) 109   c) 369   d) 68   e) 750   f) 1976

12. Express the following numbers in Egyptian, Babylonian, Mayan, and Chinese numerals.
    a) 5   b) 175   c) 350   d) 989

13. Write the base-ten numeral for each of the following.
    a) 8 hundreds, 9 tens, 6 ones   b) $3 \times 1000 + 2 \times 100 + 4 \times 1$
    c) $5 \times 10^4 + 8 \times 10^3 + 0 \times 10^2 + 1 \times 10 + 0$

14. What does the digit 2 mean in each of the following?
    a) 729   b) 972   c) 279

15. The sum of the digits of a two-digit number is 11. If 45 is added to the number, the result is the number with its digits interchanged. Find the original number.

3.3 *Other Numeration Systems*

Although ten basic symbols are used in our decimal system, we can create other numeration systems based on a different number of basic symbols. We can build numeration systems on as many basic symbols as we wish, except 0 and 1. If we form a numeration system that employs seven basic symbols, we call it a *base-seven* system. If the system uses n basic symbols, then it is a *base-n* system. Conversely, if the base of a system is the whole number r, then the system uses r basic symbols.

Since an essential component of any place-value numeration system is a zero, we always use 0 as one of the basic symbols. Thus a *base-three* system would use the basic symbols 0, 1, 2; a *base-five* system would use the basic symbols 0, 1, 2, 3, 4; and a *base-two* system would use the symbols 0, 1.

BASE FIVE

In a *base-five* numeration system

1 We use *five* basic symbols: 0, 1, 2, 3, 4.

2 We group in multiples of *five.*

3 The place values are in powers of five:

···	5^4	5^3	5^2	5^1	5^0

or

···	625	125	25	5	1

4 We count as follows:

Base Ten	1	2	3	4	5	6	7	8	9	10	11	12	···
Base Five	1	2	3	4	10	11	12	13	14	20	21	22	···

Comments

1 *Note that the symbol* 5 *has no meaning in base five since the only symbols used in this base are* 0, 1, 2, 3, 4. *In item 3 above, we are merely* translating *the base-five place values into our own language, i.e., into our everyday base-ten system.*

2 *Note that each place value in base five is* five times *that of the next place to its right.*

3 *This numeration system is a further illustration of the essential components of any positional numeration system: a base which determines the number of basic symbols used in the system; a zero; ordered symbols; and place values that use successive powers of the base.*

4 *Note that the base-five place values follow the same pattern as the base-ten place values:*

$$\cdots \quad 10^4 \quad 10^3 \quad 10^2 \quad 10^1 \quad 10^0$$
$$\cdots \quad 5^4 \quad 5^3 \quad 5^2 \quad 5^1 \quad 5^0$$

5 *Since numbers may be represented by numerals written in various numeration systems, we must identify the numeration system being used. The base of the system is usually indicated by a subscript to the right of the numeral, like* 21_{five} or 578_{nine}.

Many people prefer to indicate the base of the system by a word rather than by a numeral, e.g., 23_{five} rather than 23_5 *because the symbol* 5 *does not occur in a base-five numeration system. Using the word* five *rather than the numeral* 5 *emphasizes this fact.*

Whenever we write a numeral like 73 *without any subscript, the numeral is considered to be in the base-ten system.*

6 *A base-ten numeral and its corresponding base-five numeral are* different names for the same number. *For example, if we wish to represent the number of dots in Figure 3.3 in base ten we write* 7_{ten}; *in base five we write* 12_{five}. *Although* 7_{ten} *and* 12_{five} *are different symbols,* $7_{\text{ten}} = 12_{\text{five}}$ *because both represent the same number of dots.*

FIGURE 3.3

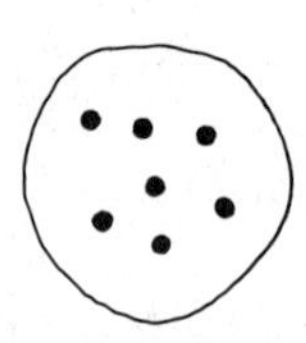

7 *Note that* 12_{five} *is not read "twelve base five," but "one-two base five";* 314_{five} *is not read "three-hundred-fourteen base five," but "three-one-four base-five."*

Conversion from One Base to the Other To convert a *base-five* numeral *to* its *base-ten* equivalent, we use the base-five place values. For example,

1 24_{five} means

5	1
2	4

, that is,

$$\begin{aligned} 2 \times 5 &= 10 \\ \text{and } 4 \times 1 &= \underline{\ \ 4} \\ & \quad 14_{\text{ten}} \end{aligned}$$

Therefore, $24_{\text{five}} = 14_{\text{ten}}$.

2 423_{five} means

25	5	1
4	2	3

, that is,

$$\begin{aligned} 4 \times 25 &= 100 \\ 2 \times \ 5 & \quad \ \ 10 \\ 3 \times \ 1 & \quad \ \ \underline{\ \ 3} \\ & \quad 113_{\text{ten}} \end{aligned}$$

Therefore, $423_{five} = 113_{ten}$.

To convert a *base-ten* numeral *to* its *base-five* equivalent, we again use the base-five place values. For example,

1 $6_{ten} = ?_{five}$ asks us, in effect, to determine the number of ..., 25's, 5's, and 1's that are contained in 6_{ten}. The highest of these place values contained in 6_{ten} is 5. It also contains one 1:

6_{ten} =

5	1
1	1

, that is,

$$\begin{array}{lr} 1 \times 5 = & 5 \\ 1 \times 1 & 1 \\ \hline & 6_{ten} \end{array}$$

Therefore, $6_{ten} = 11_{five}$.

2 $37_{ten} = ?_{five}$ asks us to determine the number of ..., 25's, 5's, and 1's that are contained in 37_{ten}. The highest of these place values contained in 37_{ten} is 25. So,

37_{ten} =

25	5	1
1	2	2

, that is,

$$\begin{array}{lr} 1 \times 25 = & 25 \\ 2 \times 5 & 10 \\ 2 \times 1 & 2 \\ \hline & 37_{ten} \end{array}$$

Therefore, $37_{ten} = 122_{five}$

Comments *1* *Students generally find it harder to change from base ten to base five than the reverse. Converting from base-ten to base-five can be better understood by the use of a set of discs or other objects. For example,* $17_{ten} = ?_{five}$ *can be illustrated by a set of discs grouped in 10s that we are asked to rearrange in groups of 5s; that is, we start with*

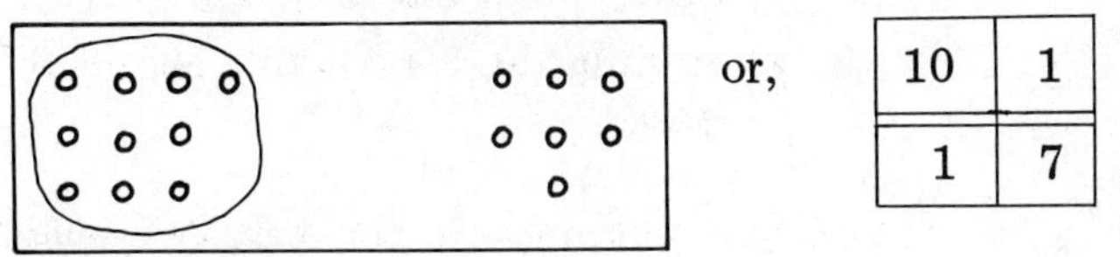

or,

10	1
1	7

.

Rearranging these discs in groups of five, we get

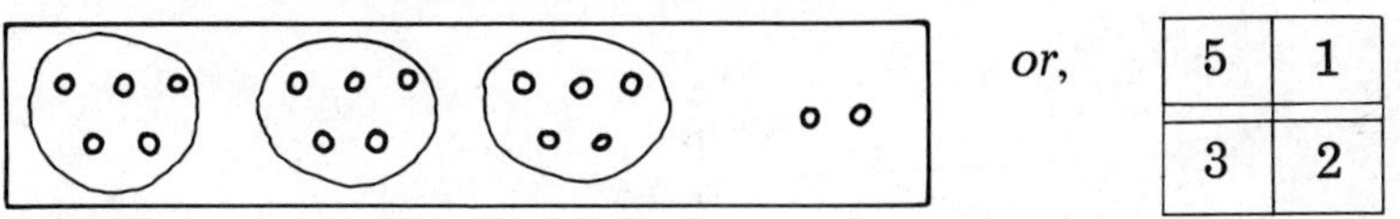

or,

5	1
3	2

Therefore, $17_{ten} = 32_{five}$.

2 *Another way that may be helpful in converting a base-ten numeral to its base-five equivalent is to imagine a currency system that contains \$125 bills, \$25 bills, \$5 bills, and \$1 bills.*

$37_{ten} = ?_{five}$ *illustrates the problem of how to pay off a debt of \$37 using the least number of bills.*

Total debt	$37
Pay one $25 bill	25
Unpaid balance	12
Pay two $5 bills	10
Unpaid balance	2
Pay two $1 bills	2
Unpaid balance	0

...	$125	$25	$5	$1
		1	2	2

Since $37 can be paid off with one $25 bill, two $5 bills, and two $1 bills, then $37_{\text{ten}} = 122_{\text{five}}$.

Addition and Multiplication in Base Five The *addition* and *multiplication* tables for base five are:

+	0	1	2	3	4
0	0	1	2	3	4
1	1	2	3	4	10
2	2	3	4	10	11
3	3	4	10	11	12
4	4	10	11	12	13

×	0	1	2	3	4
0	0	0	0	0	0
1	0	1	2	3	4
2	0	2	4	11	13
3	0	3	11	14	22
4	0	4	13	22	31

To *add* numbers in the base-five system, we follow the same procedure used to add in the base-ten system, but we use the base-five addition table.

Example 1

$$\begin{array}{r} 43_{\text{five}} \\ +\ 24_{\text{five}} \\ \hline 122_{\text{five}} \end{array}$$

a) Adding the 1's column, we get 4 + 3 = 12. So we put down the 2 and carry the 1 to the 5's column.

b) Adding the 5's column, we get 1 (that we carried) + 4 + 2 = 12.

To *multiply* numbers in the base-five system, we follow the same procedure used to multiply in the base-ten system, but we use the base-five multiplication table.

Example 2

$$\begin{array}{r} 34_{\text{five}} \\ \times\ 12_{\text{five}} \\ \hline 123 \\ 34 \\ \hline 1013_{\text{five}} \end{array}$$

Comment

A good way to check our computation in base five is to translate the problem to base ten and then compare the answers. For example,

	Base Five		Base Ten
1	43	⟷	23
	+ 24	⟷	+ 14
	122	⟷	37
2	34	⟷	19
	× 12	⟷	× 7
	123		
	34		
	1013	⟷	133

Subtraction and Division in Base-Five Subtraction and division in base five are defined and done the same way as in base ten. *Subtraction* is defined in terms of addition.

Example 3 $34_{\text{five}} - 13_{\text{five}}$ means "what number must be added to 13_{five} to get 34_{five}?" Using the base-five addition table, we get

$$\begin{array}{r} 34_{\text{five}} \\ -\ 13_{\text{five}} \\ \hline 21_{\text{five}} \end{array}$$

To subtract 23_{five} from 42_{five}, we must use the idea of *regrouping*:

$$\begin{array}{rcl} 42_{\text{five}} & = 4 \text{ fives} + 2 \text{ ones} = & 3 \text{ fives} + 12 \text{ ones} \\ -\ 23_{\text{five}} & = 2 \text{ fives} + 3 \text{ ones} = & 2 \text{ fives} + \ \ 3 \text{ ones} \\ \hline & & 1 \text{ five} \ + \ \ 4 \text{ ones} \end{array} \quad \text{or} \quad \begin{array}{r} 42_{\text{five}} \\ -\ 23_{\text{five}} \\ \hline 14_{\text{five}} \end{array}$$

Note that the "12 ones" means "1 five and 2 ones."

Division is defined in terms of multiplication.

Example 4 $4_{\text{five}}\overline{)202_{\text{five}}}$ means "what number multiplied by 4_{five} equals 202_{five}?" Using the base-five multiplication table, we get

$$\begin{array}{r} 23_{\text{five}} \\ 4_{\text{five}}\overline{)202_{\text{five}}} \\ \underline{13}\phantom{_{\text{five}}} \\ 22\phantom{_{\text{five}}} \\ \underline{22}\phantom{_{\text{five}}} \end{array}$$

BASE TWO

In a numeration system to *base two*

1 We use *two* basic symbols: 0 and 1.

2 We group in multiples of *two*.

3 The place-values are in powers of *two*:

···	2^4	2^3	2^2	2^1	2^0

or

···	16	8	4	2	1

4 We count as follows:

Base Ten	1	2	3	4	5	6	7	8	9	10
Base Two	1	10	11	100	101	110	111	1000	1001	1010

Comments

1 *The base-two system is also called the* binary *system.*

2 *Each place value in base two is* two times *that of the next place to its right.*

3 *Note that the base-two place values follow the same pattern as the base-ten and base-five place-values:*

$$\begin{array}{cccccc} \cdots & 10^4 & 10^3 & 10^2 & 10^1 & 10^0 \\ \cdots & 5^4 & 5^3 & 5^2 & 5^1 & 5^0 \\ \cdots & 2^4 & 2^3 & 2^2 & 2^1 & 2^0 \end{array}$$

To convert a *base-two numeral to its base-ten equivalent*, we use the base-two place-values.

Example 5 11011_{two} means

16	8	4	2	1
1	1	0	1	1

that is,

$$\begin{array}{r} 1 \times 16 \\ 1 \times 8 \\ 1 \times 2 \\ 1 \times 1 \\ \hline 27_{ten} \end{array}$$

or, $11011_{two} = 27_{ten}$.

To convert a *base-ten numeral to its base-two equivalent* we again use the base-two place values.

Example 6 $15_{ten} = ?_{two}$ asks us to determine the number of ..., 16's, 8's, 4's, 2's, and 1's that are contained in 15_{ten}. The highest of these place values contained in 15_{ten} is 8.

15_{ten} =

8	4	2	1
1	1	1	1

or $15_{ten} = 1111_{two}$.

Addition and Multiplication in Base Two The *addition* and *multiplication* tables for base two are:

+	0	1
0	0	1
1	1	10

×	0	1
0	0	0
1	0	1

To *add* or *multiply* in base two, we follow the same procedure used to add or multiply in base ten or in base five, but we use the base-two tables.

Example 7

a)
$$\begin{array}{r} 110101_{two} \\ +\ 100111_{two} \\ \hline 1011100_{two} \end{array}$$

b)
$$\begin{array}{r} 1011_{two} \\ \times\ \ 101_{two} \\ \hline 1011 \\ 10110 \\ \hline 110111_{two} \end{array}$$

Subtraction and Division in Base-Two

Example 8

a)
$$\begin{array}{r} 1010_{two} \\ -\ \ 101_{two} \\ \hline 101_{two} \end{array}$$

b)
$$1011_{two}\ \overline{)\,110111_{two}}$$ quotient 101
$$\begin{array}{r} 1011 \\ \hline 1011 \\ 1011 \\ \hline \end{array}$$

Comments 1 *Numbers written in base two are of special interest because of their application in electronic computers. In an electric circuit there are only two conditions:* on *or* off. *In a panel of light bulbs, each light can be either* on *or* off. *In the binary system only two symbols are needed to express any number:* 0 *or* 1.

Let us agree that if a light is on *it stands for the symbol* 1; *and if the light is* off *it stands for the symbol* 0. *This agreement gives us an electronic way of representing any number in the binary system.*

For example, Figure 3.4 shows a panel of six lights. Let each position in the panel represent a place value in the binary system, and let each light represent a binary numeral: 1 *if the light is* on, *and* 0 *if the light is* off. *Then the panel shown represents the binary numeral* 101101_{two} *or* 45_{ten}.

FIGURE 3.4

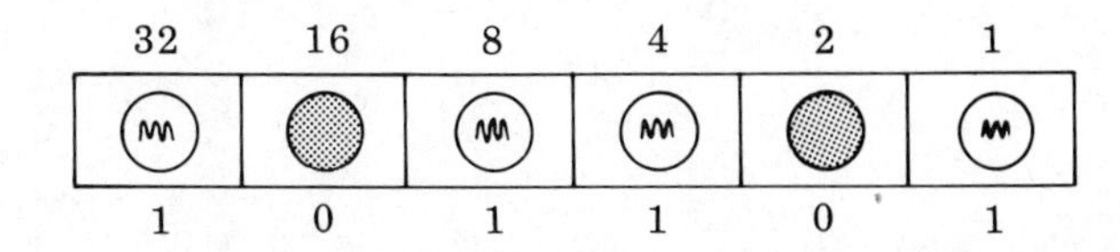

Since electrical energy moves with the speed of light, we can understand why electronic computers solve problems with such fantastic speeds.

2 *The binary system is simple because it is based on only two basic symbols and on only four addition and four multiplication facts. But is has the disadvantage of requiring many more digits to express numbers than are required in the decimal system. For example, in the two numerals* 1000_{ten} *and* $1{,}111{,}101{,}000_{two}$, *which represent equal quantities, the base-ten numeral requires only four digits, while the binary numeral requires ten digits.*

BASE TWELVE

In a *base-twelve* numeration system:

1 We need *twelve* basic symbols. We use the same ten symbols that are used in the decimal system and then add two *new* basic symbols. The symbols *t* and *e* are often used to represent the numbers *ten* and *eleven* in this system. So the twelve basic symbols used in the base-twelve numeration system are: 0, 1, 2, 3, 4, 5, 6, 7, 8, 9, *t*, *e*.

2 We group in multiples of *twelve*.

3 The place values are in powers of *twelve*:

···	12^4	12^3	12^2	12^1	12^0

or

···	20736	1728	144	12	1

4 We count as follows:

Base Ten	1	2	3	4	5	6	7	8	9	10	11	12	13	14	15
Base Twelve	1	2	3	4	5	6	7	8	9	*t*	*e*	10	11	12	13

Comments

1 *The base-twelve system is also called the* duodecimal system.

2 *Note that each place value in this system is* twelve times *that of the next place to its right.*

3 *Note that the base-twelve place values and, more generally, the place values of any system to base B follow the same pattern as seen before:*

$$\begin{array}{cccccc} \cdots & 10^4 & 10^3 & 10^2 & 10^1 & 10^0 \\ \cdots & 5^4 & 5^3 & 5^2 & 5^1 & 5^0 \\ \cdots & 2^4 & 2^3 & 2^2 & 2^1 & 2^0 \\ \cdots & 12^4 & 12^3 & 12^2 & 12^1 & 12^0 \\ \cdots & B^4 & B^3 & B^2 & B^1 & B^0 \end{array}$$

4 *Some people believe that the duodecimal system would be preferable to the decimal system for the following reasons.*

a) Twelve *has more divisors than* ten.

10: 1, 2, 5, 10.
12: 1, 2, 3, 4, 6, 12.

Base twelve would, therefore, simplify work with common fractions.

b) Twelve *is related to many of our units of measure, such as* 12 *inches in* 1 *foot,* 12 *hours on the face of a clock,* 12 *units in a dozen,* 60 *minutes in an hour,* 144 *units in a gross, and* 360 *degrees in a circle.*

For these reasons an organization called the Duodecimal Society of America is urging a change from the decimal system to the duodecimal system of notation.

Conversion from base twelve to base ten and from base ten to base twelve, and the operations of addition, multiplication, subtraction, and division are performed as before. But in the duodecimal system the operations are based on an addition table containing 144 basic addition facts and on a multiplication table containing 144 basic multiplication facts.

EXERCISE SET 3.3

1 Express the number of x's shown in the figure in:
a) Base five. b) Base two. c) Base seven. d) Base twelve.

x x x x x x x x
x x x x x x x x
x x x x x x x x
x x x

2 Convert each numeral to a base-ten numeral.
a) 468_{nine} b) 4023_{five} c) 1011011_{two} d) 100_{twelve}

3 Convert from one numeration system to the other.
a) 121_{three} = ___ten
b) 75_{ten} = ___three
c) 29_{ten} = ___two
d) 150_{ten} = ___five
e) 145_{ten} = ___twelve
f) 34_{five} = ___six
g) 11011_{two} = ___five
h) 123_{four} = ___seven
i) 56_{eight} = ___two
j) 1010111_{two} = ___eight

4 Which number is larger and by how much: 120_{nine} or 235_{six}?

5 In what base a) is $5 \times 5 = 31$, b) is $5 \times 5 = 41$?

6 The place value of c in the numeral $123c0_{\text{eight}}$ is ___.

7 Name a base in which a) 13 is an even number, b) 32 is an odd number.

8 Write the first twenty-five counting numbers in base nine.

9 a) $246_{\text{eight}} + 75_{\text{eight}} =$
b) $403_{\text{five}} \times 34_{\text{five}} =$
c) $524_{\text{six}} - 235_{\text{six}} =$
d) $11000_{\text{two}} \div 110_{\text{two}} =$

10 What number is represented by c?
a) $c_{six} + 5_{six} = 34_{six}$ b) $3_c + 24_c = 30_c$.

11 Write the addition and multiplication tables for base four.

12 Write in expanded notation using exponents:
a) 4001_{six} b) $e2t_{twelve}$.

13 Write the addition and multiplication tables for
a) Base-three, b) Base-eight, c) Base-twelve.

14 Write one of the symbols =, < , or > in the blank space to make each statement true.
a) 200_{three} ___ 102_{four} b) 43_{six} ___ 43_{five} c) 24_{five} ___ 24_{eight}

15 A numeration system uses the basic symbols 0, *, ?, and [to represent the numbers from 0 to three.
a) Write the numerals for numbers from zero to twenty-five in this system.
b) What number does the numeral ?*[00? represent?

16 State a rule for determining when a number written in base six is divisible by six.

17 A number is written as 45 in base six. In what base would the same number be written as 35?

18 Two numbers are written in the A base as 26 and 35. The same two numbers are written in the B base as 40 and 52. Find A and B.

19 Write the following numeral in expanded notation using exponents: .
$\rightarrow * ? >_{pati}$.

20 In a base-b numeration system:
a) What are the basic symbols used to represent any number?
b) Express the number rst in expanded form if r, s, and t are basic symbols in the system.

21 When a number is multiplied by x, how is its numeral in base x changed?

22 a) How many different base-five numerals can you write that do not have more than three digits?
b) How many such numerals can you write in base eight?

23 Write the following binary numerals as octal numerals (see Extended Study, Section 3.4).
a) a) 11011_{two} b) 11111111_{two} c) 10000111_{two}

24 Write the following octal numerals as binary numerals.
a) 37_{eight} b) 200_{eight} c) 561_{eight}

25 a) A set of weights to be used in a balance scale consists of 1-ounce, 2-ounce, 4-ounce, 8-ounce, 16-ounce, etc. measures. How many weights of each measure would be required in order to be able to weigh any whole number of ounces?
b) How many weights of each measure would be required if the set consisted of 1-ounce, 5-ounce, 25-ounce, 125-ounce, etc. measures?
c) How many weights of each measure would be required if the set consisted of 1-ounce, b-ounce, b^2-ounce, b^3-ounce, etc. measures?

In Exercises 26–36, state true or false and give a reason for each answer.

26 The 4 in 437_{eight} stands for *four hundred.*

27 10^3 means "10 + 10 + 10".

28 The numeral 7 represents the same number in the decimal system as in the duodecimal system.

29 The smaller the base, the more basic combinations there are in the addition table.

30 We can make a numeral mean what we wish.

31 A numeration system must be based on twelve or fewer basic symbols.

32 Any number may be represented by more than one numeral.

33 In the statement $1_a + 1_a = 2_a$, a may represent two, three, or four.

34 In the ancient Egyptian numeration system, ∩|| and ||∩ represented the same number.

35 The larger the base of a numeration system, the fewer the digits needed in a numeral to represent a number.

36 The numeral immediately before 500_{six} is 455_{six}.

37 The natives on Watuchi Island count in the following way:

|, ∧, △, □, ⬠, ⬡, |*, ||, |∧, |△, |□, |⬠, |⬡, ∧*, ∧|, ∧∧, etc.

a) What is the base of this numeration system?
b) Count in this sytem from ⬠ to □⬠.

38 Which of the following numbers is the largest: 54_{six}, 54_{seven}, 54_{eight}, or 54_{nine}?

39 Find the answer in base-seven: $42_{eight} - 42_{five}$.

40 Make up your own numeration system containing five basic symbols different from the ones we used.
a) Write the numerals for 1 to 15 in your system.
b) Write your numeral for 48; 79.

PUZZLES

1 What is the smallest number of weights needed for a balance scale that can weigh any amount from 1 pound to 15 pounds, and what is the weight of each?

2 Here is an addition example in base five:

```
   L A T E
+  A H E L
----------
 H L T A T
```

Each letter stands for a base-five digit. What digit does each letter represent?

3 Translate this story into the decimal system:

I live at 1256 Grant Ave. My friend who is 32 years old lives three houses from my home, at 1265 Grant Avenue. Last summer we both got jobs. I saved $3041; my friend saved $4115, which was $1044 more than I saved. At the end of the summer we took a 41-day trip and traveled 4313 miles altogether.

BASICS REVISITED

1 Write the numeral for: three million, two hundred sixty-five thousand, four.

2 Express 32.4 in expanded notation.

3 Divide 100 by $\frac{1}{4}$.

4 Two-thirds of what number is 34?

5 What part of 40 is 16?

6 Multiply 256 by $12\frac{1}{2}$ in two different ways.

7 $\frac{3}{4}$ of what number is 57?

8 What is a numeral?

9 List all pairs of factors of 72.

10 What are the multiples of 8?

11 What are the least common multiples of (a) 12 and 57, (b) 36 and 92?

12 Seawater contains 1.4 milligrams of fluorine per kilogram of seawater, and 0.00003 milligrams of mercury per kilogram of seawater. How much more fluorine than mercury is there in 5 kilograms of seawater?

13 Between each pair of fractions insert $>$, $<$, or $=$ to form a true statement.

a) $\frac{2}{3} \quad \frac{3}{4}$

b) $\frac{9}{5} \quad \frac{11}{7}$

Perform the indicated operations in Exercises 14–19.

14 $\frac{1}{4} + .57\frac{1}{2}$

15 $\dfrac{\frac{3}{5}}{\frac{8}{9}}$

16 $\left(3\frac{1}{2} \times 2\frac{1}{3}\right) \div 1\frac{1}{4}$

17 $3 + \dfrac{2}{1 + \frac{3}{4}}$

18 $\dfrac{\frac{3}{4} + \frac{2}{3}}{1\frac{1}{2} - \frac{3}{5}}$

19 $5 - \left(\frac{2}{7} \div \frac{9}{4}\right)$

Translate Exercises 20–22 into equations. Then solve and check.

20 What is 35% of $120?

21 15 is what percent of 75?

22 75% of what number is 60?

23 What are the digits in the decimal system?

24 How many cards measuring 3 inches by $1\frac{3}{8}$ inches can be cut out of a cardboard sheet measuring 3 inches by 24 inches?

25 A man travels 112.7 kilometers in $1\frac{3}{4}$ hours. What is his average speed?

EXTENDED STUDY

3.4 Binary and Octal Numerals

We have seen that the large number of digits required to write binary numerals makes the system cumbersome. Fortunately, binary numerals are easily converted to octal (base-eight) numerals which require fewer digits. For example, to convert 1011111_{two} to 137_{eight}:

1 Break up the binary numeral in groups of three digits, *starting at the right*:

$$1011111_{\text{two}} = \underbrace{001,}_{\substack{\text{group}\\3}}\ \underbrace{011,}_{\substack{\text{group}\\2}}\ \underbrace{111}_{\substack{\text{group}\\1}}{}_{\text{two}}.$$

2 The sum of the place values of the digits in each group results in the octal numeral (why?). From the Table of Numerals we see that:

Group 3: $001_{\text{two}} = 1_{\text{eight}}$
Group 2: $011_{\text{two}} = 3_{\text{eight}}$
Group 1: $111_{\text{two}} = 7_{\text{eight}}$

Therefore,

$$1, 011, 111_{\text{two}} = 137_{\text{eight}}.$$

Similarly, 10101110_{two} breaks up into

$$010, 101, 110_{\text{two}} = 256_{\text{eight}}.$$

TABLE 3.1 Table of Numerals

Binary	*Octal*
000	0
001	1
010	2
011	3
100	4
101	5
110	6
111	7

To convert an octal numeral to binary notation, we reverse the process by expressing each digit in the octal numeral as a binary numeral. For example, to convert 367_{eight} to a binary numeral, we note that

$3_{\text{eight}} = 011_{\text{two}}$
$6_{\text{eight}} = 110_{\text{two}}$
$7_{\text{eight}} = 111_{\text{two}}$

Therefore, $367_{\text{eight}} = 11, 110, 111_{\text{two}}$.

3.5 A Number-Guessing Trick

A "mind-reading" trick based on binary notation can be used to guess a number between 1 and 15 that your friend may think of.

Prepare a set of cards A, B, C, and D as shown below:

A		
8	9	10
11	12	13
14	15	

B		
4	5	6
7	12	13
14	15	

C		
2	3	6
7	10	11
14	15	

D		
1	3	5
7	9	11
13	15	

Ask your friend to think of a number between 1 and 15. Show him the set of cards, and ask him to tell you the letters of the cards on which his number appears. To guess his number merely add all the *first* numbers on the cards that contain the number. For example, if the number appears on cards A, C, and D, then add 8, 2, and 1, and the number is 11. If the number appears only on card C, then the number is 2.

The trick is based on the use of binary notation to prepare the cards. Each card represents a place value in a four-digit binary numeral:

A	B	C	D
8	4	2	1

To get the decimal equivalent of a binary numeral, we merely add the place values of the columns in which a 1 appears.

In Table 3.2, the numbers 1 to 15 are expressed in binary notation. Note that the number 3, for example, will appear on cards C and D whose place values are 2 and 1; the number 13 will appear on cards A, B, and D whose place values are 8, 4, and 1. By adding the first numbers on the cards containing a given number, we are merely adding the place values whose sum represents the number desired. By looking down each column in Table 3.2, you can see which numbers must appear on each card.

TABLE 3.2

A (8)	B (4)	C (2)	D (1)	Decimal Equivalent
			1	1
		1	0	2
		1	1	3
	1	0	0	4
	1	0	1	5
	1	1	0	6
	1	1	1	7
1	0	0	0	8
1	0	0	1	9
1	0	1	0	10
1	0	1	1	11
1	1	0	0	12
1	1	0	1	13
1	1	1	0	14
1	1	1	1	15

The trick can be extended to include as many numbers as we wish, provided we prepare enough cards. For instance, to include numbers 1 to 31, we need five cards to represent the place values of a 5-digit binary numeral. To include numbers 1 to 63, we need six cards to represent the place values of a 6-digit binary numeral.

3.6 Binary Notation and Russian Multiplication

Binary notation explains why a certain method of multiplication once used by Russian peasants extensively works.

The method operates as follows: To multiply 15 by 28, form two columns (Table 3.3), one headed by 15 and the other by 28. Successively multiply one column by 2 and divide the other by 2. When an odd number is divided by 2, drop the remainder.

TABLE 3.3

Divide	*Multiply*
15	28
7	56
3	112
1	224

Add all the numbers in the *multiply* column that lie opposite an *odd* number in the *divide* column. The sum is the product of 15 and 28, or 420 (Table 3.4).

TABLE 3.4

Divide	*Multiply*	*Numbers Opposite Odd Numbers*
15	28	$28 = 2^0 \times 28$
7	56	$56 = 2^1 \times 28$
3	112	$112 = 2^2 \times 28$
1	224	$224 = 2^3 \times 28$
		$420 = 15 \times 28$

This method works because if we express 15 in binary notation we get

2^3	2^2	2^1	2^0
1	1	1	1

or, $2^3 + 2^2 + 2^1 + 2^0$.

Therefore,

$$\begin{aligned} 15 \times 28 &= (2^3 + 2^2 + 2^1 + 2^0) \times 28 \\ &= (2^3 \times 28) + (2^2 \times 28) + (2^1 \times 28) + (2^0 \times 28) \\ &= (2^0 \times 28) + (2^1 \times 28) + (2^2 \times 28) + (2^3 \times 28). \end{aligned}$$

This is precisely what we did as we successively multiplied the 28 by 2 and then added. It makes no difference which column is divided and which is multiplied. If we reverse our columns, we obtain Table 3.5.

TABLE 3.5

Divide	*Multiply*	*Numbers Opposite Odd Numbers*
28	15	$= (2^0 \times 15)$
14	30	$= (2^1 \times 15)$
7	60	$60 = (2^2 \times 15)$
3	120	$120 = (2^3 \times 15)$
1	240	$240 = (2^4 \times 15)$
		$420 = 28 \times 15$

In this case we think of 28 as

2^4	2^3	2^2	2^1	2^0
1	1	1	0	0

or, $2^4 + 2^3 + 2^2$,

and

$$\begin{aligned} 28 \times 15 &= (2^4 + 2^3 + 2^2) \times 15 \\ &= 2^4 \times 15 + 2^3 \times 15 + 2^2 \times 15. \end{aligned}$$

This again is what we did when we successively multiplied the 15 by 2 but added only the numbers opposite the odd numbers.

3.7 *More on the Binary System*

The binary system, though often regarded as an example of the "new" mathematics, was actually mentioned in a Chinese book about 5,000 years ago. Many centuries later it was rediscovered by Leibniz (1646–1716) who not only liked this system for its simplicity but also read into it religious significance. He regarded 1 (unity) as representing God, and 0 as representing nothingness. Since God created everything out of nothingness, Leibniz said, 0 and 1 are an expression of the entire universe. Several hundred years later, the binary system helped revolutionize our world by making possible the introduction of the modern electronic computer.

Apart from its fundamental importance in electronic computers, the binary system can also be used to explain or solve several interesting games and problems. It can also lead to the construction of some interesting devices.

NIM

An ancient mathematical game, possibly Chinese in origin, in which the binary system can be put to use is called *Nim.* Translating the moves into binary notation reveals the requirements of a winning strategy. In this game, two players use a number of counters such as pennies that are placed in several piles. The players take turns picking up as many of the pennies as they wish (but at least one). The player who picks up the last penny wins.

In a simple version of the game, 12 pennies are arranged in three rows as shown in Figure 3.5. The players alternate in removing one or more pennies from the same row. Whoever takes the last penny wins. (The game can also be played so that whoever takes the last penny *loses.*)

FIGURE 3.5

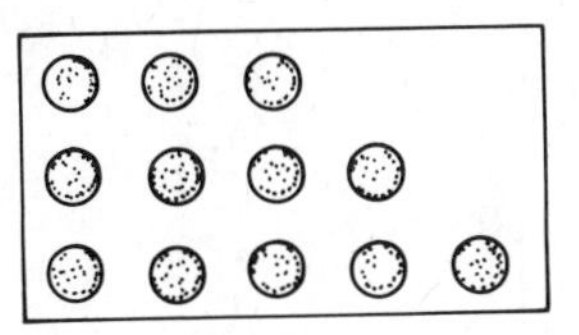

A player can always win, that is, will pick up the last penny, if:

1 His move leaves one penny in one row, two pennies in a second row, and three pennies in a third row; or

2 One of his moves leaves two rows with more than one penny in a row and the same number in each.

The first player can always win if, on his first move, he takes two pennies from the top row and, thereafter, plays according to the analysis given below.

To see why these strategies win, we make use of the binary system of notation. First express the number of pennies in each row in the binary system. If each column adds up to zero or to an even number, then the position is a winning one. In Table 3.6, the numbers of pennies in each row of Figure 3.5 are expressed in the binary system. Since the middle column adds up to 1, an *odd* number, this position is not a winning one. However, by removing two pennies from the top row, the player can convert this position into a winning one since the new position becomes one where every column is either 0 or an even number (Figure 3.6).

TABLE 3.6

Decimal	*Binary*
3	11
4	100
5	101
	212 (Total in *decimal* system.)

FIGURE 3.6

Decimal	*Binary*
1	1
4	100
5	101
	202

Note that in adding each column: (1) the *sums* are expressed in the decimal system; and (2) we do not carry over (as in normal addition) from one column to the next. Only the sums of the individual columns are considered.

Strategy *1* on page 67 leads to a winning position since each column adds up to an even number:

Decimal	*Binary*
1	1
2	10
3	11
	22

Show that strategy *2* on page 67 leads to a winning position.

The game can be played with an arbitrary number of counters distributed into three rows in an arbitrary way. The winning strategy is still the same. Express the number of counters in each row in the binary system. Then the position is a winning one if and only if the sum of the digits in each column is zero or an even number. For a *proof* of this statement see *Mathematical Recreations* by Maurice Kraitchik, Dover Publications, Inc., New York, 1953, page 87.

HOME-MADE COMPUTERS

A simple computer to illustrate binary numerals can be made out of a cardboard box and some Christmas tree bulbs. A bulb that's on represents 1; a bulb that's off represents 0 (Figure 3.7).

FIGURE 3.7

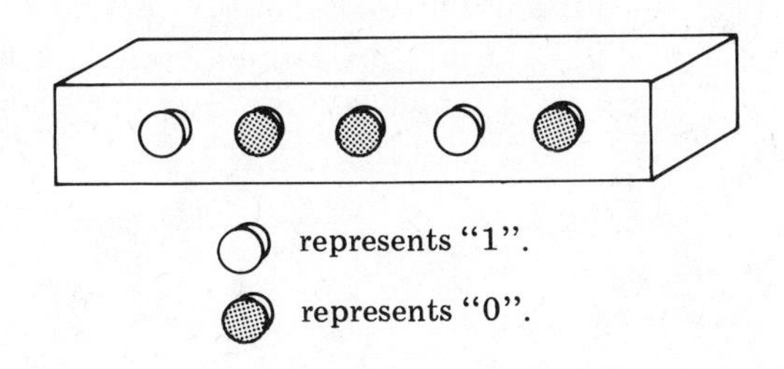

A peg board and some pegs can be used to perform computations in the binary system by letting a peg inserted into a hole in the board represent 1, while a hole without a peg represents 0. Set up two numbers, one below the other, and add them. Do the same with subtraction. Do the same with multiplication and division, regarding multiplication as repeated addition and division as repeated subtraction.

CARD SORTER

Binary notation can be used for sorting a set of cards. For example, to sort a set of 32 cards numbered 0 to 31,

1 Punch 5 holes in each card (Figure 3.8). (Snip off the upper right-hand corner of the card. This helps determine whether the card is right side up.)

FIGURE 3.8

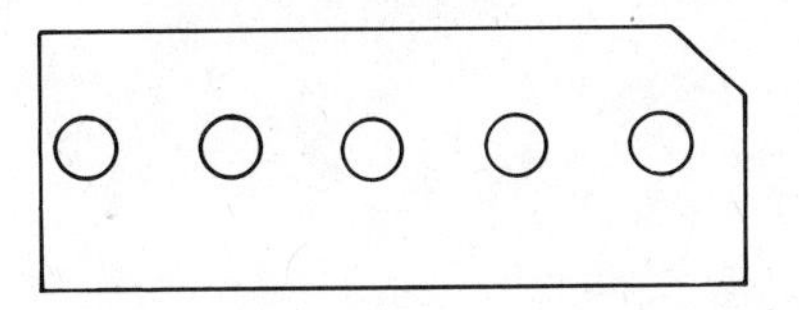

2 Number the cards 0 to 31 in binary notation. Let a punched-out hole represent 0; a cut-out U represent 1. Figure 3.9 shows the card that represents the number 22.

FIGURE 3.9

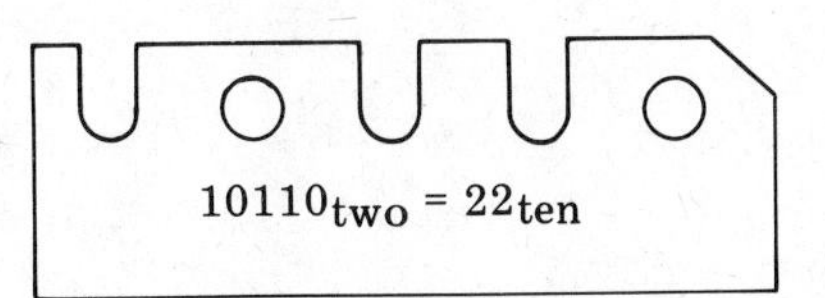

3 Shuffle the cards thoroughly. You can now arrange the 32 cards in numerical order 0 to 31 by repeating a simple operation 5 times:

a) Stick a pencil through the first hole at the right and lift all the cards that "hang on." Only the cards whose binary numerals end with 0 will "hang on" and be lifted.
b) Place the lifted cards in front of the other cards, and then repeat the same operation for the remaining four holes, moving from right to left.

When you have finished, the cards should be in numerical order from 0 to 31.

Since this sorting process is based on the binary place values, one additional "lift-up" will sort twice as many cards. Therefore, 6 "lift-ups" will sort 64 cards; 7 "lift-ups" will sort 128 cards; and 20 "lift-ups" will sort more than 500,000 cards.

Chapter 4

THE WHOLE NUMERS

Objectives

At the completion of this chapter, you should be able to:

1. Define the set of *natural numbers* and the set of *whole numbers*.
2. Recognize that the set of whole numbers is an extension of the set of natural numbers.
3. Represent whole numbers on the number line.
4. Define the properties of *closure*, *commutativity*, and *associativity* for addition and multiplication.
5. Define addition in terms of the union of two disjoint sets.
6. Define multiplication in terms of Cartesian product.

7 Show that multiplication can be regarded as repeated addition.

8 Define subtraction in terms of addition.

9 Define division in terms of multiplication.

10 Show that division can be regarded as repeated subtraction.

11 Show that a given property of the whole numbers is the consequence of a particular property of sets.

12 Construct an addition table.

13 Perform subtraction using an addition table.

14 Construct a multiplication table.

15 Perform division using a multiplication table.

16 Explain why division by zero is undefined.

17 Use the inverse relationship between addition and subtraction to express addition sentences as subtraction sentences.

18 Use the inverse relationship between multiplication and division to express multiplication sentences as division sentences.

19 Demonstrate basic multiplication and division facts by using arrays.

20 Identify examples of the various properties of the whole numbers.

21 Add, subtract, multiply, and divide whole numbers on the number line.

22 State the Cancellation Law for addition and multiplication.

23 State the distributive property of multiplication over addition.

24 Apply the distributive property to remove parentheses and to remove common factors.

25 Evaluate expressions such as $12 + 6 \times 3 - 15 \div 5 + 2$.

26 Define the inequalities $a < b$, and $a > b$.

27 Determine by definition which is the greater of two whole numbers.

Prerequisites

1 You should have knowledge of sets: the empty set; disjoint sets; set equivalence; union, intersection, Cartesian product, ordered pair; properties of union and intersection.

2 You should know what a number line is and how to construct it.

INTRODUCTION

Most people appreciate that inventions like television, computers, and space ships are creations of the human mind. But how many of us realize that the number system we use in everyday arithmetic is also the creation of the human mind? And can we see that without the invention of such a number system the other inventions might not have been possible?

No great invention comes about all at once. Its roots can be found in earlier inventions and ideas, and, as time goes on, it is enlarged and refined. The invention of the number system is no exception. It was developed by many people over a period of thousands of years. As with all other inventions, number systems were invented to help meet some human need. When a number system became inadequate to meet new needs, it was extended to include new numbers that would meet these new needs. This happened over and over again, and, as a result, today we have not one number system but many number systems. What we often call *our* number system is, in reality, a nest of number systems, one within the other. It started with the counting numbers and then expanded to the integers, the rational numbers, and the real numbers. Later, a need was found for expanding the real number system to the complex number system.

The first numbers we use are the *counting numbers*, often called the *natural numbers*: the numbers 1, 2, 3, 4, ... These numbers were invented to help us answer the question "how many?" Since, in order to count, we must have a set of objects that are to be counted, the properties of sets are closely tied to the counting numbers. We can not only count objects with these numbers, but we can add them and multiply them and the result is always another counting number. For a long time these counting numbers were the only mathematical tool that people had. Later, the number *zero* was invented. We can think of zero as a special kind of counting number that tells us that there are no objects in the set which we are counting. When we join *zero* to the other counting numbers, we obtain the set of *whole numbers*.

The whole numbers are adequate for counting and telling how many objects there are in a set, but how can we use these numbers to distinguish between a \$10 *profit* and a \$10 *loss*, between 17° *above* zero and 17° *below* zero, between 2 blocks *east* and 2 blocks *west*? To meet this need, a larger number system had to be created. This larger system includes the counting numbers, zero, and the numbers $-1, -2, -3, \ldots$. This larger system is called the set of *integers*.

While the integers are suitable for telling how many people there are in a room or for distinguishing between a profit and a loss, they are unsuitable for other purposes such as measuring. How can we use integers to measure *part* of a mile, of an inch, or of a pie? To meet this need, we again had to enlarge our number system to include, besides integers, other numbers like $\frac{1}{2}, \frac{1}{4}, -\frac{2}{3}$. This larger system is called the *rational number system*.

Now let us suppose that we wish to represent the length of the hypotenuse of an isosceles right triangle whose leg is one inch in length. By the Pythagorean theorem, the length of the hypotenuse

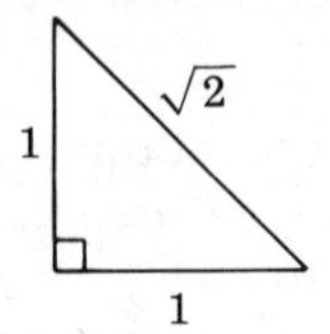

comes out to be $\sqrt{2}$, but $\sqrt{2}$ is *not* a rational number. In fact, no rational number exists that can represent the length of the hypotenuse of such a triangle. To correct this inadequacy we once again had to enlarge our number system to include not only rational numbers but also numbers like $\sqrt{2}$, $\sqrt{3}$, $\sqrt{17}$, called *irrational numbers*. This expanded system, made up of the rational numbers as well as the irrational numbers, is called the *real number system*. Later, inadequacies in the real number system led to its expansion into the *complex number system*. So we see that our number system is not *one* number system but a *nest* of number systems, each an outgrowth of the system that preceded it.

While the geometrical concepts of *point* and *line* served as the basis of Greek mathematics, the concept of *number* is the basis of modern mathematics. Since all number systems derive from the whole numbers, a good understanding of the basic properties of the whole number system is essential to an understanding of all other number systems and of the structure of elementary mathematics as well.

We shall now begin with the system of whole numbers and examine the assumptions and properties that make this system behave the way it does. We shall then move to each succeeding system, examine *its* assumptions and properties, and show how these relate to the previous systems.

4.1 Number

We have seen that a number is an abstract idea, that it isn't anything we can see or touch. The number three, for instance, has no reference to the individual characteristics of the objects that make up a set of three. Instead, it is an abstraction from *all* sets containing three objects. At first, we become acquainted with the concept of number by counting sets of objects. Later, we can more easily think of number as a purely abstract idea without any reference to counting.

A fundamental property of the concept of number is that the number of objects in a set, called its *cardinal number*, is the same regardless of the order in which you count them.

DEFINITIONS

The set N of *natural numbers* or *counting numbers* is

$N = \{1, 2, 3, 4, ...\}$.

The set W of *whole numbers* is the set $N \cup \{0\}$. That is,

$W = \{0, 1, 2, 3, ...\}$.

We shall find it useful to represent the whole numbers as points on the number line:

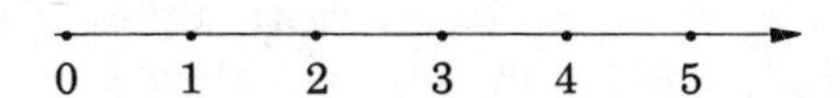

Comments

1 *On the number line, a whole number tells us how many steps of equal length are needed to reach a given point to the right of the point* 0.

2 *Note that the number line need not be placed horizontally.*

3 *Not everyone defines the set of* counting numbers *the same way. Some texts treat* zero *as a counting number.*

THE EQUALITY RELATION

The *relation of equality*, denoted by =, between any two numbers has the following important properties. For all numbers a, b, and c, the equality relation is

1 *Reflexive* For all a, $a = a$. That is, any number is equal to itself. Sometimes this property is called the *law of identity.*

2 *Symmetric* For all a and b, if $a = b$, then $b = a$. That is, we may read an equality from left to right or from right to left.

3 *Transitive* For all a, b, and c, if $a = b$ and $b = c$, then $a = c$. That is, if each of two numbers is equal to the same third number, then the two numbers are equal to each other.

Any relation that is reflexive, symmetric, and transitive is called an *equivalence relation.* Therefore, the *relation of equality* between any two numbers is an *equivalence relation.*

Comment

Not all relations between numbers are equivalence relations. Consider, for example, the relation $<$ *(is less than). Are these statements true:* (1) $a < a$; (2) *if* $a < b$, *then* $b < a$; (3) *if* $a < b$ *and* $b < c$, *then* $a < c$?

Obviously, the relation $<$ *is not reflexive since it is not true that every number is less than itself; it is not symmetric; but it is transitive. The relation* $<$ *is not, therefore, an equivalence relation.*

The *equality relation* has the following additional properties:

4 *Addition Property* For all a, b, and c, if $a = b$, then

$a + c = b + c$ and $c + a = c + b$.

That is, if we add the same number to both sides of an equality, the result is an equality.

5 *Multiplication Property* For all a, b, and c, if $a = b$, then

$ac = bc$ and $ca = cb$.

That is, if we multiply both sides of an equality by the same number, the result is an equality.

6 *Substitution Property* For all a and b, if $a = b$, then in any statement involving a, b may be substituted for a without changing the truth (or falsity) of the statement. For example, if $a = b$ and $a + 9 = 15$, then $b + 9 = 15$.

4.2 Order Relation on the Set of Whole Numbers

Whole numbers come in a definite order: 0, 1, 2, 3, ... This is the order of counting. When a whole number a precedes a whole number b in the order of counting, we say that *a is less than b* and write $a < b$. Another way of saying a is less than b is to say *b is greater than a*, written $b > a$. Therefore, the statements $a < b$ and $b > a$ mean exactly the same thing. More formally, we define the relation $<$ as follows:

DEFINITION If a and b are any whole numbers, then $a < b$ if and only if there exists a counting number c such that $a + c = b$. If $a < b$, then $b > a$.

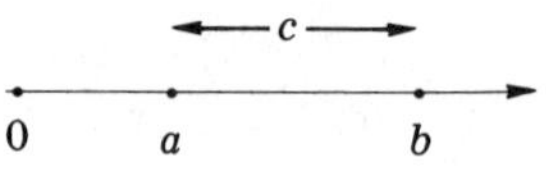

←c→

0 a b

$a + c = b$

Example $3 < 5$ because there exists a counting number 2 such that $3 + 2 = 5$. If $3 < 5$, then $5 > 3$.

Comments

1 *Note that c in the definition must be a counting number and cannot be 0. If c could be 0, then $a + 0 = a$ would mean, by the definition, that $a < a$.*

2 *On the number line, $a < b$ means that a is to the* left *of b.*

3 *We say that the relation $<$ describes an* order relation *for the whole numbers.*

PROPERTIES OF THE ORDER RELATION

1. *Trichotomy Law* Any two whole numbers a and b can be related to each other in one and only one of three ways:

   $a < b, \quad a = b, \quad a > b.$

2. *Transitive Property* For all whole numbers a, b, and c,

   if $a < b$ *and* $b < c$, then $a < c$.

3. *Addition Property* For all whole numbers a, b, and c,

   if $a < b$, *then* $a + c < b + c$ *and* $c + a < c + b$.

4. *Multiplication Property* For all whole numbers a, b, and c,

   if $a < b$ *and* $c \neq 0$, *then* $ac < bc$ *and* $ca < cb$.

Comment *These properties are also true when the symbol* $<$ *is replaced by the symbol* $>$.

4.3 Operations on Whole Numbers

A *binary operation* associates an ordered pair of numbers with a unique third number. For example, the operation of addition associates one and only one third number, 8, with the ordered pair (6, 2). The operation of multiplication associates one and only one number, 12, with the ordered pair (6, 2). We call addition and multiplication *binary operations* because they are performed on *two* numbers at a time. More formally and more precisely,

DEFINITION 1 A *binary operation*, denoted by $*$, on a set S associates with each ordered pair (a, b) of elements of S a unique element $a * b$ of S.

Comments

1. *Even when we add three or more numbers, we are performing a* binary *operation since the sum of these numbers will be obtained by adding only two numbers at a time.*

2. *Note that the definition requires that the operation be defined for* every *ordered pair (a, b), and that the result $a * b$ must be an element of S. Therefore, if S is the set of whole numbers, subtraction is* not *a binary operation on S since $(a - b)$ will not result in a whole number if $a < b$. For a similar reason division is not a binary operation on the set of whole numbers.*

3. *In a binary operation,* every *ordered pair* in the set corresponds to a unique single element in the set; i.e. $(a, b) \to c$. If *every* single *element in the set corresponds, under the*

operation, to a unique element in the set, the operation is called a unary operation. *For example, the operation of squaring a number is a unary operation since* $1 \to 1$, $2 \to 4$, $3 \to 9$, $4 \to 16$, *etc.*

ADDITION OF WHOLE NUMBERS

Addition of whole numbers may be defined in terms of disjoint sets, recalling that a whole number is the cardinal number of a set. We shall denote the cardinal number of set A by $n(A)$. For example, if $A = \{2, 5, 9\}$, then $n(A) = 3$; if $B = \{a, c\}$, then $n(B) = 2$.

If $A = \{2, 9\}$ and $B = \{a, b, c\}$, then

$$A \cup B = \{2, 9, a, b, c\}.$$

The cardinal numbers of these sets are: $n(A) = 2$, $n(B) = 3$, and $n(A \cup B) = 5$. If $A = \{5, 7, 8\}$ and $B = \{a, e, i, o, u\}$, then

$$A \cup B = \{5, 7, 8, a, e, i, o, u\};$$

the cardinal numbers of these sets are: $n(A) = 3$, $n(B) = 5$, $n(A \cup B) = 8$. Similarly, if $n(A) = r$ and $n(B) = s$, then

$$n(A \cup B) = r + s \text{ (see Table 4.1).}$$

TABLE 4.1

$n(A)$	$n(B)$	$n(A \cup B)$
2	3	5
3	5	8
r	s	$r + s$

The definition of the sum of two whole numbers follows the above approach.

DEFINITION 2 If A and B are disjoint sets, if r is the cardinal number of A, and if s is the cardinal number of B, then $r + s$ is the cardinal number of the union of A and B.

That is, if $n(A) = r$, $n(B) = s$, and $A \cap B = \phi$, then

$$n(A) + n(B) = n(A \cup B)$$

or

$$r + s = n(A \cup B).$$

The whole numbers r and s are called *addends;* the whole number $(r + s)$ is called their *sum.*

Comments 1 *Note that sets A and B in Definition 2 must be disjoint. Why?*

2 *In Definition 2 we are making several assumptions:*

a) That for any two whole numbers r and s, there exist disjoint sets A and B such that $n(A) = r$ and $n(B) = s$.

b) That $r + s$ is unique. That is, $r + s$ is independent of the sets we choose as long as the two sets are disjoint, the cardinal number of one of them is r, and the cardinal number of the other is s.

c) That if r is a whole number and s is a whole number, then $r + s$ is a whole number.

3 *On the number line, addition of whole numbers may be seen as* counting to the right. (*See Figure 4.1.*)

FIGURE 4.1

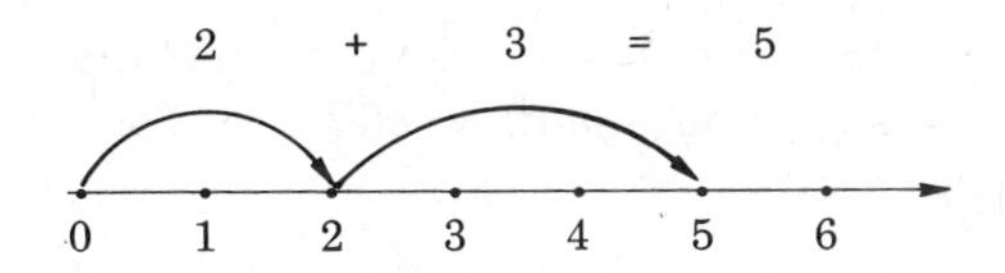

4 *Note that we take the union of* sets, *but find a sum of* numbers. *We do not add sets and do not take the union of numbers.*

5 *By our definition of addition,*

$$n(A) + n(B) = n(A \cup B)$$

will be true only if A and B are disjoint. Can we adjust this formula to make it work if A and B are not disjoint*? Yes, we can generalize the formula to make it applicable to both situations by writing it as:*

$$n(A) + n(B) = n(A \cup B) + n(A \cap B).$$

a) When A and B are disjoint, the formula becomes

$$n(A) + n(B) = n(A \cup B),$$

since $n(A \cap B) = 0$.

b) The following example shows what happens when A and B are not *disjoint. Let*

$A = \{1, 2, 3, 4,\}$ *and* $B = \{3, 5, 7\}$.

Then $A \cup B = \{1, 2, 3, 4, 5, 7\}$, $A \cap B = \{3\}$, $n(A) = 4$, $n(B) = 3$, $n(A \cup B) = 6$, and $n(A \cap B) = 1$.

If we substitute these values in the new formula, we get

$$n(A) + n(B) = n(A \cup B) + n(A \cap B)$$
$$n(A) + n(B) = 6 + 1$$
$$n(A) + n(B) = 7.$$

PROPERTIES OF ADDITION

1. *Closure* If a and b are any two whole numbers, then their sum, $a + b$, is a unique whole number. Thus addition is a binary operation.
2. *Commutative* $a + b = b + a$. That is, the sum of any two whole numbers is the same regardless of the order in which the numbers are added.
3. *Associative* If a, b, and c are any whole numbers, then

   $(a + b) + c = a + (b + c)$.

   That is, the sum of any three whole numbers is the same regardless of the way in which the numbers are grouped.
4. *Identity* $a + 0 = 0 + a = a$. That is, when zero is added to any whole number the sum is that number. The number zero is called the *identity element for addition* or the *additive identity*.
5. *Equality* If a, b, and c are any whole numbers, and if $a = b$, then $a + c = b + c$. That is, if we add the same number to both sides of an equality, the result is an equality.
6. *Cancellation* If a, b, and c are any whole numbers, and if $a + c = b + c$, then $a = b$. That is, if we subtract the same number from both sides of an equality, the result is an equality.

Comments

1. *By using the number line, we can see the commutative, associative, and identity properties in action. For example,*

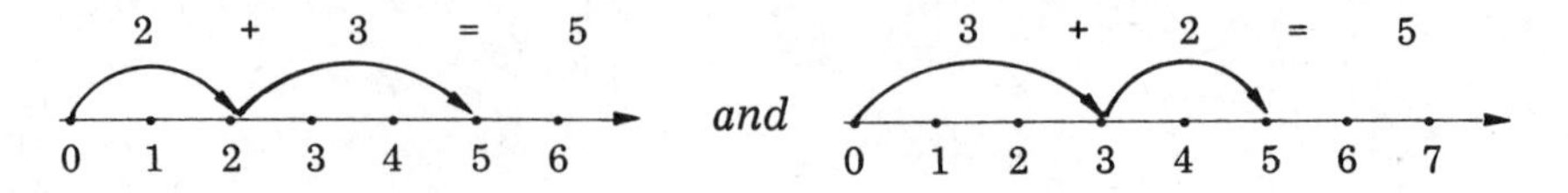

yield the same result.

2. *Since we defined addition in terms of the union of sets, we can see intuitively that the properties of addition are based upon corresponding properties of the union of sets. We can, however, prove these properties more formally. For example, we can prove the commutative property in the following way:*

Given *Sets A and B such that* $a = n(A)$, $b = n(B)$, $A \cap B = \phi$.
Prove $a + b = b + a$.

Proof

a)	$a = n(A), b = n(B)$	(*Given*)
b)	$a + b = n(A \cup B)$	(*Definition of addition*)
c)	$b + a = n(B \cup A)$	(*Definition of addition*)
d)	$n(A \cup B) = n(B \cup A)$	(*Since* $A \cup B = B \cup A$)
e)	$b + a = n(A \cup B)$	(*Transitivity:* c *and* d)
f)	$a + b = b + a$	(*Transitivity:* b *and* e)

3 *With these properties, we can now prove the properties of the order relation on page 77. For example, let us prove the addition property.*

Given $a < b$ *Prove* $a + c < b + c$

Proof

a)	*If* $a < b$, *then* $a + n = b$ (*where* n *is a counting number*)	(*Definition of* $<$)
b)	$a + n + c = b + c$	(*Addition property for* =)
c)	$a + (n + c) = b + c$	(*Associative*)
d)	$a + (c + n) = b + c$	(*Commutative*)
e)	$(a + c) + n = b + c$	(*Associative*)
f)	$a + c < b + c$	(*Definition of* $<$)

4 *The commutative and associative properties for addition are used to check the addition of a column of numbers by adding either from bottom to top or top to bottom.*

SUBTRACTION OF WHOLE NUMBERS

We define *subtraction* in terms of addition. For example, $5 - 2 = ?$ asks us to find a whole number which, when added to 2, gives 5. We say that $5 - 2 = 3$ because $2 + 3 = 5$.

DEFINITION 3 If a and b are whole numbers such that $a \geqslant b$, then $a - b$ is defined to be that whole number c such that

$$a = b + c.$$

In the expression $a - b$, the number a is called the *minuend*, the number b is called the *subtrahend*, and $a - b$ is called the *difference*. (The symbol $\geqslant$ means "is greater than or equal to.")

Comments

1 *Note that the definition states that a must be greater than or equal to b. If a is less than b, the difference is not a whole number.*

2 *With every addition sentence there are associated two subtraction sentences. For example, associated with the addition sentence* $2 + 3 = 5$ *are the subtraction sentences*

$$5 - 2 = 3 \quad and \quad 5 - 3 = 2.$$

All three sentences say the same thing in a different way, so if any one of them is true they are all true. This is why the addition table can be used as a subtraction table.

3 *On the number line, subtraction may be seen as* counting to the left. (*See Figure 4.2.*) *For example,* $5 - 2$ *means*

a) Starting at 0, *we count* 5 *units to the right, taking us to the point* 5.

b) Then, from point 5 *we count* 2 *units to the* left, *taking us to point* 3.

FIGURE 4.2

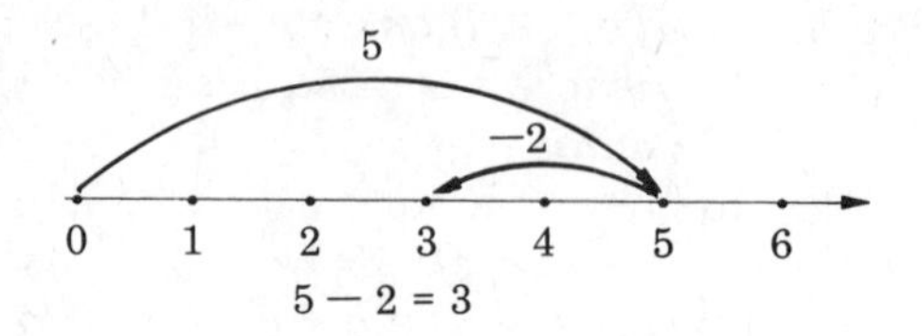

4 *Subtraction of whole numbers may also be defined in terms of sets. We shall show this for the example* $5 - 2 = ?$ *Choose sets A and B so that* $n(A) = 5$ *and* $n(B) = 2$. *We shall now look for a set C such that*

a) B and C are disjoint, and

b) $B \cup C$ *is equivalent to A.*

We get $A = \{x\ x\ x\ x\ x\}$, $B = \{x\ x\}$, $C = \{x\ x\ x\}$. $n(C)$ *must be* 3.

5 *We can also view subtraction in terms of* subsets, *using the example* $5 - 2 = ?$ *Let* $B \subset A$, *and* $n(A) = 5$, $n(B) = 2$. *When we remove the subset B from set A, the* remainder set *is C, where* $n(C) = 3$.

$$A = \{\overbrace{xx}^{B}\overbrace{xxx}^{C}\}$$

6 *If we think of addition as finding the sum, given two addends, we can think of subtraction as finding a missing addend if one addend and the sum are known. For example,*

Addition		*Subtraction*	
4	addend	7	sum
+ 3	addend	− 3	addend
7	sum	4	missing addend

PROPERTIES OF SUBTRACTION

1 *Not closed* since $a - b$ is not a whole number when $b > a$. For example, $2 - 7$ is not a whole number. Thus subtraction is not a binary operation.

2 *Not commutative* since $a - b \neq b - a$, except when $a = b$. For example, $5 - 3 \neq 3 - 5$.

3 *Not associative* since $(a - b) - c \neq a - (b - c)$, except when $c = 0$. For example, $(7 - 5) - 1 = 1$, while $7 - (5 - 1) = 3$.

4 *No identity element;* although $a - 0 = a$, $0 - a \neq a$. For example, $5 - 0 = 5$, but $0 - 5 \neq 5$.

Comment

Inverse operations *are operations that counteract each other. For instance, opening and closing a door are inverse operations as well as walking two blocks east and then walking two blocks west. Not every operation or activity has an inverse. For example, eating a sandwich or singing a song have no inverses.*

The inverse of adding 5 *to a number is subtracting* 5; *the inverse of subtracting* 2 *is adding* 2. *We, therefore, say that* addition and subtraction are inverse operations. *This relationship becomes evident when we add and subtract on the number line.*

MULTIPLICATION OF WHOLE NUMBERS

We shall define multiplication of whole numbers in terms of the Cartesian product of two sets. For example, we can think of 3×4 in terms of two sets A and B such that $n(A) = 3$ and $n(B) = 4$. Let us choose $A = \{a, b, c,\}$ and $B = \{1, 2, 3, 4\}$. Then

$$A \times B = \{(a, 1), (a, 2), (a, 3), (a, 4), (b, 1), (b, 2), (b, 3), (b, 4), (c, 1), (c, 2), (c, 3), (c, 4)\},$$

and $n(A \times B) = 12$. The product of the cardinal numbers of sets A and B, 3×4, is the same as the cardinal number of the Cartesian product of A and B, 12.

In other words, each of the two whole numbers whose product we wish to find is the cardinal number of some set. The product of these numbers is the same as the cardinal number of the Cartesian product of the two sets.

DEFINITION 4 The *product* of any two whole numbers a and b is defined to be:

$$a \times b = n(A \times B)$$

where A and B are any sets such that $n(A) = a$ and $n(B) = b$. That is,

if a is the cardinal number of set A,
and b is the cardinal number of set B,
then $a \times b$ is the cardinal number of set $(A \times B)$.

$a \times b$, or ab, is called the *product* of a and b; a and b are called *factors;* and the operation is called *multiplication.* We also say that ab is a *multiple* of both a and b.

Example 1 In the product $3 \times 5 = 15$, 15 is the *product*, 3 and 5 are *factors*, and 15 is a *multiple* of both 3 and 5.

Comments

1 *In the definition of multiplication, it is not required that* $A \cap B = \phi$, *as in the definition of addition, since the number of elements in* $A \times B$ *is not affected by whether* A *and* B *are disjoint sets.*

2 *A schematic way to represent the Cartesian product* $A \times B$ *if* $A = \{a, b, c\}$ *and* $B = \{1, 2, 3, 4\}$ *is shown in Figure 4.3. We call such a representation an array.*

FIGURE 4.3

	1	2	3	4
a	$(a, 1)$	$(a, 2)$	$(a, 3)$	$(a, 4)$
b	$(b, 1)$	$(b, 2)$	$(b, 3)$	$(b, 4)$
c	$(c, 1)$	$(c, 2)$	$(c, 3)$	$(c, 4)$

More generally, we can represent the product 3×4 *by a rectangular array of dots (Figure 4.4). The first factor,* 3, *gives the number of rows in the array. The second factor,* 4, *gives the number of columns in the array. The product,* 12, *gives the total number of dots in the array.*

FIGURE 4.4

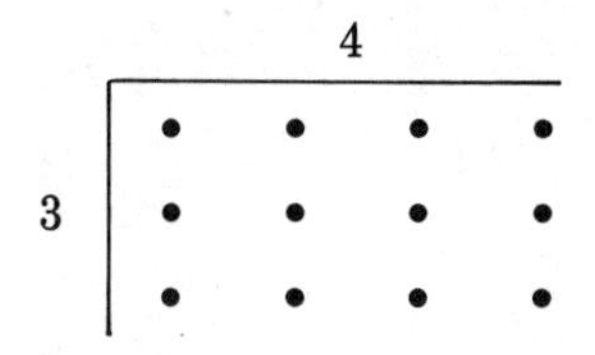

3 *To multiply* 3 *and* 4, *we can put pegs into a pegboard so that we have three rows of pegs, each row containing four pegs. To find the product,* 3×4, *we count the total number of pegs in the board and find that it is* 12. *A similar procedure can be followed with a geoboard by using rubberbands to indicate the number of rows and the number of columns and then counting the total number of nails in the array.*

4 *A closer look at the array of dots in Figure 4.4 suggests that* 3×4 *may be regarded as the union of three disjoint sets each having four elements. In other words,* 3×4 *is the sum of* 3 *fours:*

$$3 \times 4 = 4 + 4 + 4 = 12$$

Similarly, 2×3 *is the sum of* 2 *threes, and* 13×5 *is the sum of* 13 *fives.*

The repeated addend *approach is an easy and meaningful way to understand multiplication but presents a problem when we wish to find* 0×5, *for instance. What does* "0 fives" *mean? A way to get around this difficulty is to use the commutative property and say* $0 \times 5 = 5 \times 0$. *The product* 5×0 *means*

$$0 + 0 + 0 + 0 + 0 = 0.$$

5 *On the number line, multiplication of whole numbers may be seen as* counting to the right in groups. *For example,* 3 × 4 *means counting* 3 *groups of* 4 *to the right, beginning at* 0 (*Figure 4.5*).

FIGURE 4.5

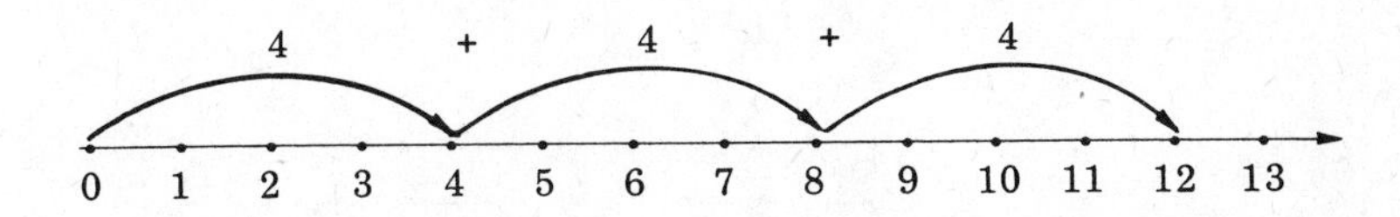

PROPERTIES OF MULTIPLICATION

1 *Closure* If a and b are any two whole numbers, then their product ab is a unique whole number. Thus multiplication is a binary operation.

Comment

Since the product of two whole numbers is defined as the cardinal number of the Cartesian product of two sets, and since cardinal numbers are whole numbers, the product of two whole numbers is a whole number. Our intuition tells us that this whole number is unique since it is independent of our choice of sets as long as the cardinal number of the first set is a and the cardinal number of the second set is b.

2 *Commutative* If a and b are any two whole numbers, then $ab = ba$.

Comments

1 *We justify the commutative property on the basis that ab and ba are the cardinal numbers of two equivalent sets. Let us choose sets A and B so that* $n(A) = a$ *and* $n(B) = b$. *Then* $ab = n(A \times B)$ *and* $ba = n(B \times A)$. *Although* $A \times B$ *is not equal to* $B \times A$, *the two sets can be shown to be* equivalent *because we can always set up a one-to-one correspondence between the elements of both sets.*

For example, if $A = \{1, 2, 3\}$ *and* $B = \{a, b\}$, *then*

$$A \times B = \{(1, a), (1, b), (2, a), (2, b), (3, a), (3, b)\}, \text{ and}$$
$$B \times A = \{(a, 1), (b, 1), (a, 2), (b, 2), (a, 3), (b, 3)\}.$$

It is evident that each element in $A \times B$ *has an obvious corresponding element in* $B \times A$ *and vice versa:*

$$(1, a) \leftrightarrow (a, 1), \quad (1, b) \leftrightarrow (b, 1), \quad (2, a) \leftrightarrow (a, 2)$$
$$(2, b) \leftrightarrow (b, 2), \quad (3, a) \leftrightarrow (a, 3), \quad (3, b) \leftrightarrow (b, 3).$$

More generally, for any two finite sets A and B, to each ordered pair (a, b) *in* $A \times B$ *there corresponds the unique ordered pair* (b, a) *in* $B \times A$ *and vice versa and, therefore,* $A \times B$ *is equivalent to* $B \times A$.

2 *The commutative property for multiplication enables us to learn the entire multiplication table by learning only half the*

basic multiplication facts because if we know that, say, 3 × 5 = 15 *we also know that* 5 × 3 = 15.

3 *An interesting structural pattern exists in tables of operations that are commutative. For example, Figures 4.6 and 4.7 are the basic addition and multiplication tables for the counting numbers.*

FIGURE 4.6

+	1	2	3	4	5	6	7	8	9
1	2	3	4	5	6	7	8	9	10
2	3	4	5	6	7	8	9	10	11
3	4	5	6	7	8	9	10	11	12
4	5	6	7	8	9	10	11	12	13
5	6	7	8	9	10	11	12	13	14
6	7	8	9	10	11	12	13	14	15
7	8	9	10	11	12	13	14	15	16
8	9	10	11	12	13	14	15	16	17
9	10	11	12	13	14	15	16	17	18

FIGURE 4.7

×	1	2	3	4	5	6	7	8	9
1	1	2	3	4	5	6	7	8	9
2	2	4	6	8	10	12	14	16	18
3	3	6	9	12	15	18	21	24	27
4	4	8	12	16	20	24	28	32	36
5	5	10	15	20	25	30	35	40	45
6	6	12	18	24	30	36	42	48	54
7	7	14	21	28	35	42	49	56	63
8	8	16	24	32	40	48	56	64	72
9	9	18	27	36	45	54	63	72	81

The outlined diagonal line of numerals in Figures 4.6 and 4.7 is called the main diagonal. *If we locate the answers for, say,* 3 × 5 *and* 5 × 3, *notice that*

a) both answers are the same, 15, *and*

b) both answers are "balanced" about the main diagonal.

This "balance" about the main diagonal is described by saying that 3 × 5 *and* 5 × 3 are symmetric with respect to the main diagonal.

Investigate other combinations in both tables to see whether the same structural pattern holds.

A quick way, therefore, to determine whether an operation is commutative is to study its table for symmetry about the main diagonal.

4 *The number line may be used to demonstrate the commutative property for multiplication. For example,* $2 \times 3 = 3 \times 2$ *can be seen on the number line as illustrated in Figure 4.8.*

FIGURE 4.8

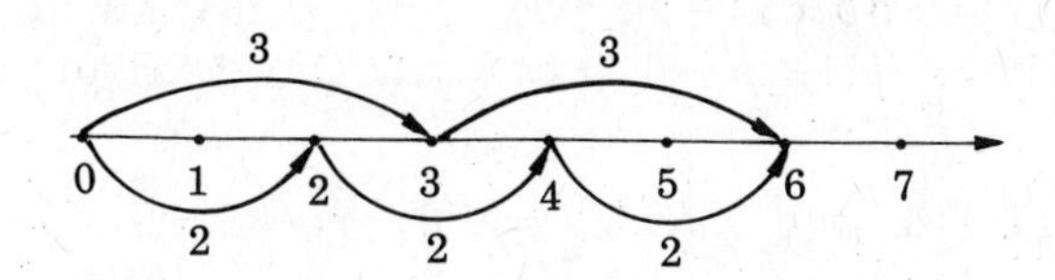

5 *An array may be used to demonstrate the commutative property for multiplication. For example, to show that* $3 \times 4 = 4 \times 3$ *set up a* 3×4 *array and a* 4×3 *array as shown in Figure 4.9. Note that the* 4×3 *array is the* 3×4 *array turned on its side, or rotated* $90°$.

FIGURE 4.9

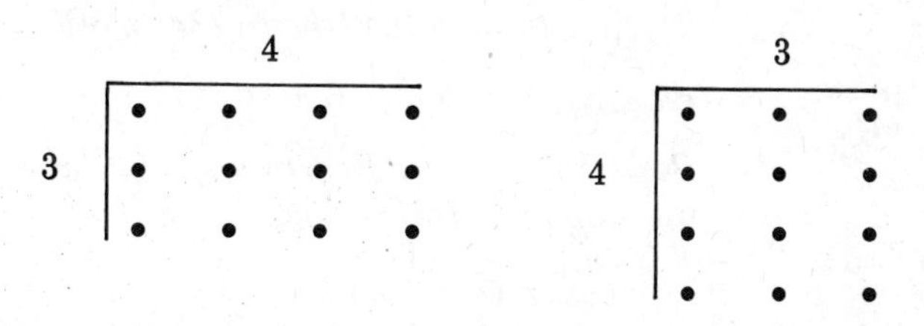

3 *Associative* If a, b, and c are any whole numbers, then

$(ab)c = a(bc)$.

Comment

The associative property for multiplication is justified by the fact that when we form the Cartesian product of sets A, B, and C, we obtain the same number of elements when we form the product $(A \times B) \times C$ *as when we form the product* $A \times (B \times C)$. *That is,* $n[A \times (B \times C)] = n[(A \times B) \times C]$.

4 *Identity* If a is any whole number, then

$a \times 1 = 1 \times a = a$.

The number 1 (one) is called the *identity element for multiplication* or the *multiplicative identity*.

Comments

The number 1 *has other interesting properties:*

1 *It is the smallest counting number.*

2 *For any whole number we get the next larger number by adding* 1.

3 *If we start with* 1 *and keep adding* 1s, *we can build any whole number.*

5 *Cancellation* For any whole numbers a and b and any whole number c ($c \neq 0$), if $ac = bc$, then $a = b$.

6 *Distributive* If a, b, and c are whole numbers, then

$a(b + c) = ab + ac$ (*Left* distributive property)

and $(b + c)a = ba + ca$ (*Right* distributive property)

Example 2 $4(7 + 3) = (4 \times 7) + (4 \times 3)$ and $(5 + 9)\,17 = (5 \times 17) + (9 \times 17)$.

Comments

1 *To find the perimeter of the rectangle as shown in Figure 4.10, we can:*

a) Either add the length and width and then multiply the sum by 2; i.e.,

$2(3 + 5) = 2 \times 8 = 16$; or

b) Add the two widths to the two lengths; i.e.,

$(2 \times 3) + (2 \times 5) = 6 + 10 = 16$.

Both methods will produce the same result. This choice of methods illustrates the distributive property; i.e.,

$2(3 + 5) = (2 \times 3) + (2 \times 5)$.

FIGURE 4.10

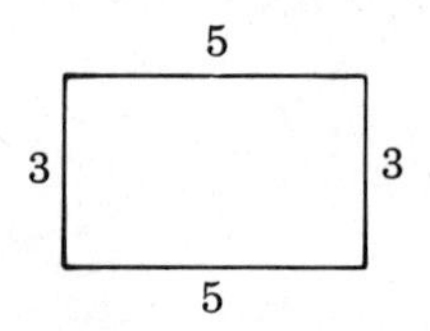

2 *In the expression $a(b + c)$, there are two factors, a and $(b + c)$. The first factor is distributed over the numbers being added in the second factor $(b + c)$. This is why we say that the whole numbers are* distributive for multiplication over addition. *Note that the distributive property is the only property so far which involves* two operations.

Although whole numbers are distributive for multiplication over addition, they are not *distributive for addition over multiplication. For example,*

$2 + (3 \times 5) \neq (2 + 3) \times (2 + 5)$

since $2 + (3 \times 5) = 2 + 15 = 17$,

while $(2 + 3) \times (2 + 5) = 5 \times 7 = 35$.

3 *To prove the distributive property of whole numbers we rely upon the fact that the Cartesian product operation is distributive over union; that is,*

$A \times (B \cup C) = (A \times B) \cup (A \times C)$.

4 *The distributive property must be seen from* both *directions:*

$a(b + c) = ab + ac$ and $ab + ac = a(b + c)$,
$(b + c)a = ba + ca$ and $ba + ca = (b + c)a$.

5 *By using arrays, we can see that the distributive property enables us to multiply by separating numbers into two or more parts. For example,* 2 × 7 *can be seen as*

```
          7
   ┌────────────────
2  │•  •  •  •  •  •  •
   │•  •  •  •  •  •  •
```

But 2 × 7 *can also be seen as*

```
      3        4
   ┌────────  ───────
2  │•  •  •  •  •  •  •      or      (2 × 3) + (2 × 4).
   │•  •  •  •  •  •  •
```

This idea can be seen even more easily if a set of movable discs is used to form the array. Lay out 2 × 3 *and* 2 × 4 *arrays side by side:*

```
      3                 4
   ┌────────        ┌───────────
2  │•  •  •      2  │•  •  •  •
   │•  •  •         │•  •  •  •  .
```

Pushing the two arrays together we get

```
          7
   ┌────────────────
2  │•  •  •  •  •  •  •
   │•  •  •  •  •  •  •
```

showing that 2 × 7 = (2 × 3) + (2 × 4).

6 *The distributive property can be extended in the following ways:*

a) $a(b_1 + b_2 + \cdots + b_n) = ab_1 + ab_2 + \cdots + ab_n$.

where a number is distributed over the sum of more than two addends.

b) $(a + b)(c + d) = (a + b)c + (a + b)d = ac + bc + ad + bd$,

where a sum is distributed over another sum.

c) $a(b - c) = ab - ac$,

where a number is distributed over a difference.

7 *More generally, the distributive property may be defined as follows: Let* $\circ$ *and* $*$ *be two binary operations on a set S. Then* $\circ$ *is distributive over* $*$ *if and only if for all elements a, b, and c of S,*

$a \circ (b * c) = (a \circ b) * (a \circ c)$.

8 *The distributive property is of the greatest importance in mathematics because it justifies the common procedures we use for multiplying whole numbers, as we shall see later, and for factoring in algebra.*

DIVISION OF WHOLE NUMBERS

We define division in terms of multiplication. For example, $6 \div 2 = ?$ asks us to find a whole number which when multiplied by 2 is equal to 6. We say that $6 \div 2 = 3$ because $2 \times 3 = 6$.

DEFINITION 5 Let a, b, and c be whole numbers, $b \neq 0$. Then

$$a \div b = c \text{ if and only if } a = bc.$$

The number a is called the *dividend*, b the *divisor*, and c the *quotient*.

Example 4 In the division $175 \div 7 = 25$, 175 is the *dividend*, 7 is the *divisor*, and 25 is the *quotient*.

Comments

1 *The definition of division requires that $b \neq 0$. The reason for this restriction will be seen on page 91.*

2 *Division and multiplication are* inverse *operations because one operation counteracts the other. We counteract the effect of multiplying* 7 *by* 5 *by dividing the result by* 5.

3 *The three statements* $2 \times 3 = 6$, $6 \div 2 = 3$, *and* $6 \div 3 = 2$ *are three ways of saying the same thing. If any one of these statements is true, they are all true.*

4 *In multiplication we are given two factors and asked to find their product; in division we are given the product and one of the factors and asked to find the other factor.*

5 *On the number line, division of whole numbers may be seen as* counting to the left in groups. *For example,* $6 \div 2$ *means counting to the left in groups of* 2, *beginning at* 6 *and ending at* 0.

FIGURE 4.11

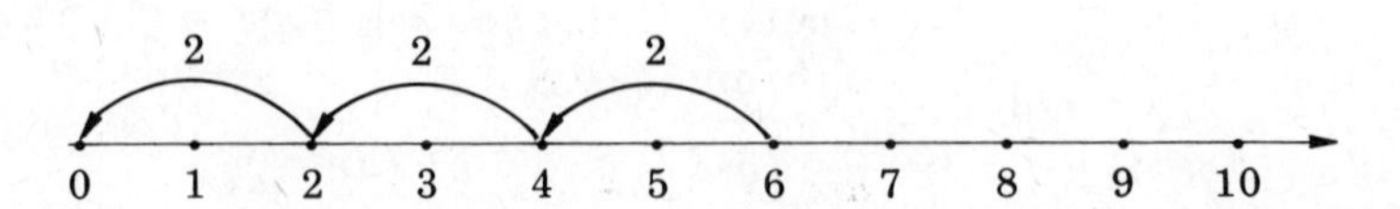

6 *We can think of division as* repeated subtraction. $6 \div 2 = ?$ *asks: If we have a set of* 6 *objects, how many sets of* 2 *objects can we remove? Obviously,* 3 *groups of* 2 *can be subtracted from* 6. (*Figure 4.12*).

FIGURE 4.12

7 *We can also think of division in terms of an array. If we set up the array on a pegboard, we carry out division by reversing the procedure we used in multiplication. To divide* 6 *by* 2 *we start with* 6 *pegs and arrange them in rows of* 2. *We find that there are* 3 *rows. Therefore,* $6 \div 2 = 3$. *Or we can arrange the* 6 *pegs in columns of* 2 *and then find that there are* 3 *columns.*

PROPERTIES OF DIVISION

1 Since division of whole numbers is not always defined (e.g. $3 \div 2$ is not a whole number), division is *not closed.* Thus division is not a binary operation.

2 Since $6 \div 2 \neq 2 \div 6$, division of whole numbers is *not commutative.*

3 Since $(12 \div 4) \div 2 = 3 \div 2$ which is not defined, while $12 \div (4 \div 2) = 12 \div 2 = 6$, division of whole numbers is *not associative.*

4 Since $5 \div 1 \neq 1 \div 5$, *1 is not an identity element* for division.

5 The *cancellation property holds* for division under certain conditions. For any whole numbers a, b, and c *such that* $a \div c$ *and* $b \div c$ *are defined:* If $a \div c = b \div c$, then $a = b$.

6 Division by zero

a) When zero is divided by any counting number a the result is zero; i.e., $\frac{0}{a} = 0$ because $a \times 0 = 0$.

b) We cannot divide a counting number a by zero; i.e., $\frac{a}{0}$ is *undefined* because *no* counting number multiplied by $0 = a$.

c) We cannot divide 0 by 0; i.e., $\frac{0}{0}$ is *undefined* since *any* counting number multiplied by $0 = 0$. Such a result is unsatisfactory because we want a *unique* solution.

Therefore, we say that *division by zero is not defined.*

EXERCISE SET 4.1–4.3

1. What is the least whole number? What is the largest whole number?

2. A binary operation $*$ on the set of whole numbers is defined as "adding three to the first number and multiplying the result by the second number." For example,

   $2 * 5 = (2 + 3) \times 5 = 5 \times 5 = 25.$

   Find the value of each of the following.
   a) $3 * 7$ b) $4 * 9$ c) $9 * 4$

3. Is the operation $*$ as defined in Exercise 2 (a) closed, (b) commutative?

4. Which properties of the equality relation are illustrated by each of the following?
   a) $m = m$. b) If $y + 2 = 7$, then $y + 2 - 2 = 7 - 2$.
   c) If $x = 2$, then $5x = 10$.

5. Which properties of the order relation are illustrated by each of the following?
   a) If $5 < 8$ and $8 < 9$, then $5 < 9$.
   b) If $c < d$, then $c + t < d + t$.
   c) If $e > b$, then $ab < ae$.

6. If x and y are whole numbers, write statements which are equivalent to:
   a) $x \neq y$. b) $y \not> x$. c) $x \neq y$ and $x \not< y$. d) $x \geqslant y$.

7. Which properties of whole numbers are illustrated by each of the following?
   a) $0 \times 17 = 0$ b) $4 + (7 + 2) = (2 + 7) + 4$ c) $7 \times 8 \times 9 = 9 \times 8 \times 7$
   d) $21 + 78$ is a whole number.

8. Which of the following sets are closed?
   a) The even numbers with respect to multiplication.
   b) The odd numbers with respect to multiplication.
   c) The squares of the counting numbers with respect to multiplication.

9. If you display six discs arranged this way:

   ○○○ ○○○,

   and then this way:

   ○○ ○○ ○○,

   which property are you demonstrating?

10. If $*$ is a binary operation defined over the set $R = \{\#, \$, \%\}$, which properties are illustrated by each of the following?
    a) $\# * \$ = \$ * \#$ b) $\# * \% = \% * \# = \%$ c) $(\% * \$) * \# = \% * (\$ * \#)$

11. Which of the following statements are true for all whole numbers r and t?
    a) $r + t = t$ b) $r + t = r + t + 0$ c) $r + 6 + t = r + t + 6$

12. Use the distributive property to write the following products as the sum of two addends.
    a) $9\,(13 + 25)$ b) $(9 + 17)\,8$

13. Compute the following expressions in two different ways.
    a) $(5 + 9)\,12$ b) $17\,(3 + 8)$

14. Fill in the blanks so as to demonstrate the distributive property.
    a) ____ $= 4 \times 9 + 4 \times 6$ b) $5 \times 7 + 9 \times 5 =$ ____

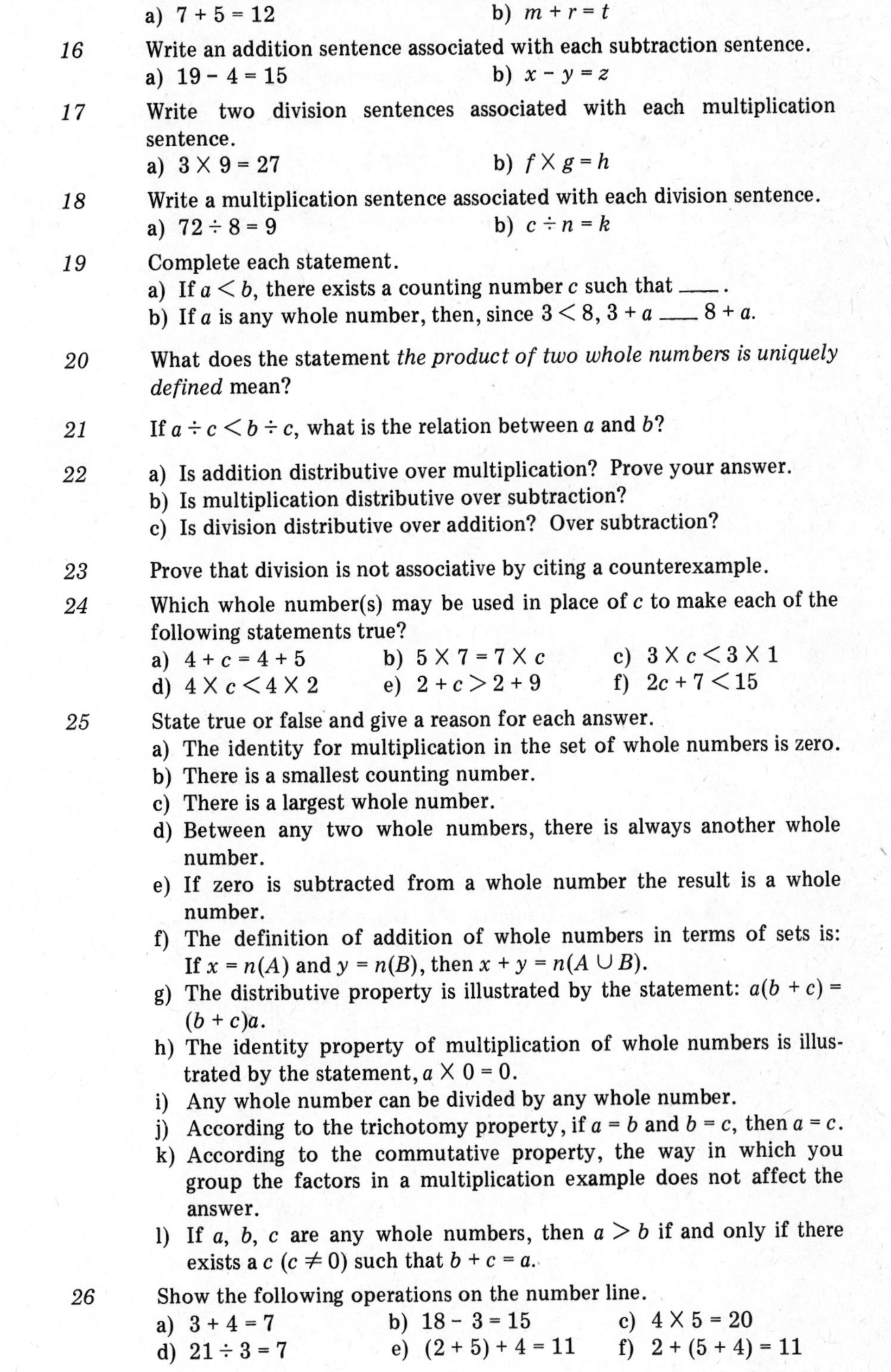

15 Write two subtraction sentences associated with each addition sentence.
a) $7 + 5 = 12$ b) $m + r = t$

16 Write an addition sentence associated with each subtraction sentence.
a) $19 - 4 = 15$ b) $x - y = z$

17 Write two division sentences associated with each multiplication sentence.
a) $3 \times 9 = 27$ b) $f \times g = h$

18 Write a multiplication sentence associated with each division sentence.
a) $72 \div 8 = 9$ b) $c \div n = k$

19 Complete each statement.
a) If $a < b$, there exists a counting number c such that ____.
b) If a is any whole number, then, since $3 < 8$, $3 + a$ ____ $8 + a$.

20 What does the statement *the product of two whole numbers is uniquely defined* mean?

21 If $a \div c < b \div c$, what is the relation between a and b?

22 a) Is addition distributive over multiplication? Prove your answer.
b) Is multiplication distributive over subtraction?
c) Is division distributive over addition? Over subtraction?

23 Prove that division is not associative by citing a counterexample.

24 Which whole number(s) may be used in place of c to make each of the following statements true?
a) $4 + c = 4 + 5$ b) $5 \times 7 = 7 \times c$ c) $3 \times c < 3 \times 1$
d) $4 \times c < 4 \times 2$ e) $2 + c > 2 + 9$ f) $2c + 7 < 15$

25 State true or false and give a reason for each answer.
a) The identity for multiplication in the set of whole numbers is zero.
b) There is a smallest counting number.
c) There is a largest whole number.
d) Between any two whole numbers, there is always another whole number.
e) If zero is subtracted from a whole number the result is a whole number.
f) The definition of addition of whole numbers in terms of sets is: If $x = n(A)$ and $y = n(B)$, then $x + y = n(A \cup B)$.
g) The distributive property is illustrated by the statement: $a(b + c) = (b + c)a$.
h) The identity property of multiplication of whole numbers is illustrated by the statement, $a \times 0 = 0$.
i) Any whole number can be divided by any whole number.
j) According to the trichotomy property, if $a = b$ and $b = c$, then $a = c$.
k) According to the commutative property, the way in which you group the factors in a multiplication example does not affect the answer.
l) If a, b, c are any whole numbers, then $a > b$ if and only if there exists a c $(c \neq 0)$ such that $b + c = a$.

26 Show the following operations on the number line.
a) $3 + 4 = 7$ b) $18 - 3 = 15$ c) $4 \times 5 = 20$
d) $21 \div 3 = 7$ e) $(2 + 5) + 4 = 11$ f) $2 + (5 + 4) = 11$

PUZZLES

1. We can use four 4's, together with the basic operations, to represent many different counting numbers. For example,

   $1 = (4 - 4) + (4 \div 4), \quad 2 = (4 \div 4) + (4 \div 4), \quad 3 = (4 + 4 + 4) \div 4.$

   How would you represent the numbers 4, 5, 6, 7, 8, and 9?

2. a) Arrange the numbers 1, 2, 3, 4, 5, 6, 7, 8, 9 in the square so that each row, column, and diagonal adds up to the same total (15).

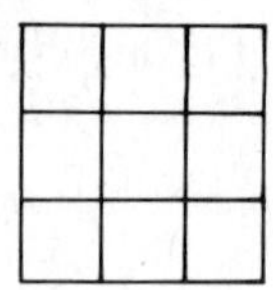

   b) Arrange the numbers 1, 2, 3, 4, 5, 6, 7, 8, 9, 10, 11, 12, 13, 14, 15, 16 in the square so that each row, column, and diagonal adds up to the same total (34).

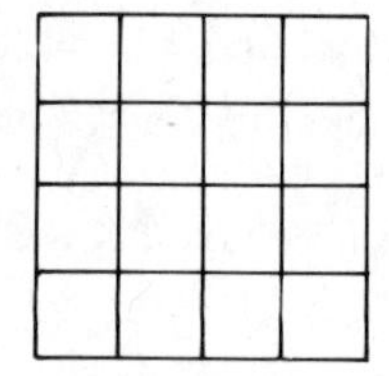

3. A customer bought $5 worth of groceries and paid the clerk with a $20 bill. The clerk had no change so he took the $20 bill to a neighbor and got $20 in change. He then gave the customer the groceries and $15 change.

   The customer left and shortly thereafter the neighbor came and told the clerk that the $20 bill he got from the customer was counterfeit. The clerk gave him a good bill in its place. How much did the grocer lose?

4. Given ten discs numbered from 0 to 9:

   a) Can you arrange the discs in the shape of a triangle in such a way that the sum along each side of the triangle is 17?

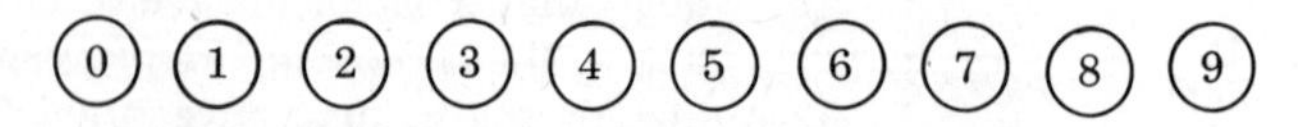

   b) Can you arrange the discs so that the sum along each side of the triangle is 18?

BASICS REVISITED

1 Round off 385,179,863 to the nearest million.

Perform the indicated operations in Exercises 2-7.

2 $\frac{1}{3} + \frac{2}{5} + \frac{5}{6}$

3 $\frac{7}{5} - \frac{3}{4}$

4 $\frac{3}{4} \times \frac{4}{5} \times \frac{2}{7}$

5 $\frac{2}{3} \div \frac{3}{8} \times 5$

6 .09 - .0023

7 $.3\overline{)43.551}$

8 After spending $15 for slacks and $7 for a shirt, John has $13 left. How much did he have to begin with?

9 Change $\frac{2}{7}$ to 28ths.

10 Find the missing numbers: $5\frac{2}{3} = 5\frac{}{24} = 4\frac{}{24}$

11 $\frac{1}{6}$ of what number is 14?

12 Express as a decimal: two and sixteen thousandths.

13 Which is smaller: .53 or .529?

14 Use the short method to find:
a) 3.4695 × 1000. b) 17.56 ÷ 1000.

15 Solve for y and check.
a) $3(y + 2) = 15$ b) $18 = 2(y - 7)$

16 Find $22\frac{1}{2}$% of $80.

17 60% of what number is 96?

18 12 is what percent of 72?

19 What do we mean by *place value*?

20 Find the value of t if $100^3 = 10^t$.

21 For its first 60 miles, the Earth's atmosphere consists of 78.088% nitrogen and 0.0001 methane. How many times the amount of methane is the amount of nitrogen?

22 What part of his $15,600 a year income does a man spend on housing if his monthly rent is $275?

23 $\frac{5}{8}$ of a high school graduating class are college bound. If 200 students will enter college, how many students are there in the graduating class?

24 A student receives a grade of 85% on a test containing 25 questions. How many problems did he get wrong?

25 Two-thirds of the human body is water. If a man's body contains 150 pounds of water, how much does the man weigh?

EXTENDED STUDY

4.4 Other Methods of Multiplication

THE LATTICE METHOD

To multiply 27 × 53 by this method, we follow these steps:

1 Form a *lattice* of rows and columns so that there is a row and a column for each digit in each number to be multiplied (Figure 4.13).

FIGURE 4.13

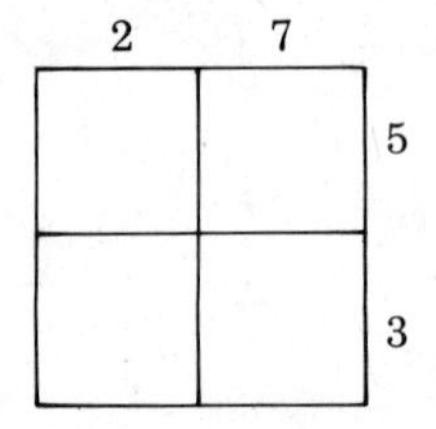

2 Draw a diagonal through each square (Figure 4.14).

FIGURE 4.14

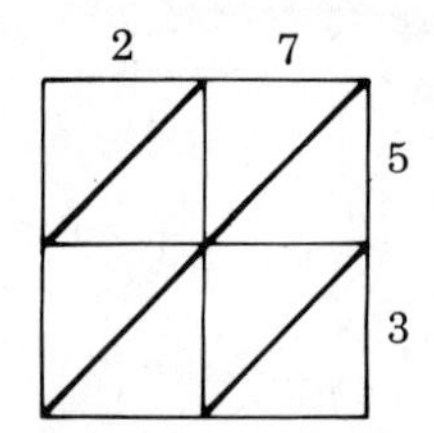

3 Beginning with the first digit on the right hand side, multiply it by each digit along the top. If the product is a single digit, write it *below* the diagonal. If the product is two digits, write the tens digit above and the units digit below the diagonal (Figure 4.15).

FIGURE 4.15

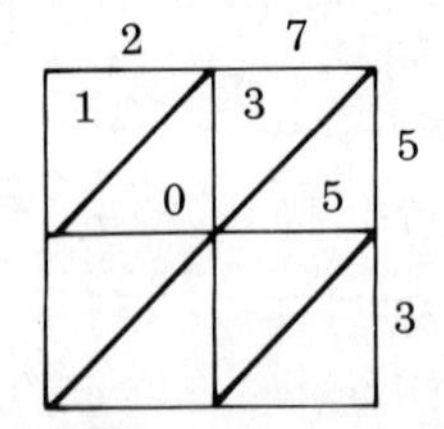

4 Then, multiply the second digit on the right by each digit along the top (Figure 4.16).

FIGURE 4.16

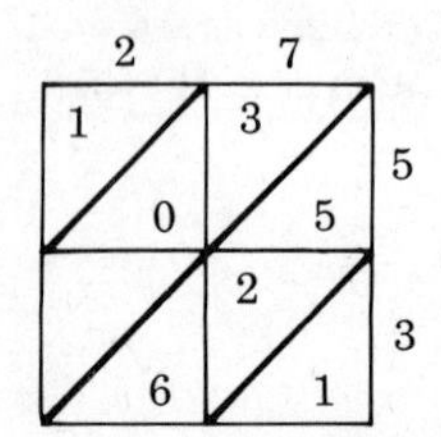

5 Then, beginning at the lower right corner and moving to the left, add the numbers appearing in each diagonal *corridor* and write down the sums as shown (Figure 4.17).

FIGURE 4.17

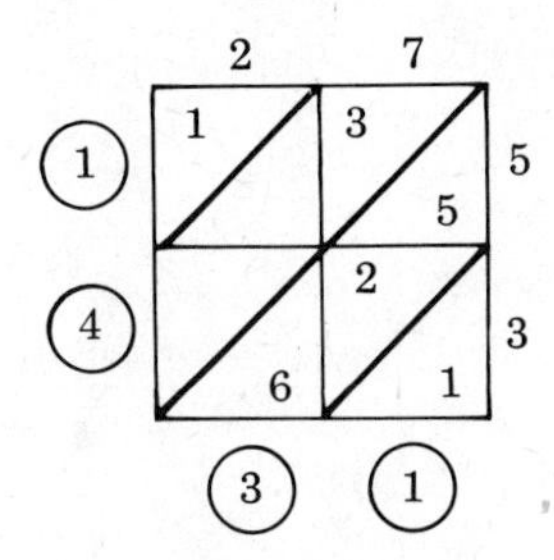

6 The product, reading from the left, is 1431.

The lattice method, which was introduced in Italy nearly 500 years ago as the *Gelosia method*, avoids the problem of carrying during multiplication. To see why the method works, compare the result in Figure 4.17 with the result obtained by using the traditional algorithm:

$$\begin{array}{rl} 27 & \\ \times 53 & \\ \hline 21 & \rightarrow (3 \times 7) \\ 60 & \rightarrow (3 \times 20) \\ 350 & \rightarrow (50 \times 7) \\ 1000 & \rightarrow (50 \times 20) \\ \hline 1431 & \end{array}$$

Note that the entries in each column in the algorithm correspond to the entries along the diagonal corridors in the lattice method.

NAPIER'S BONES

This method of multiplication derives its name from the English mathematician, John Napier (1550-1617), who developed a set of bones, or rods, to serve as a mechanical means for multiplication. Napier's Bones are considered by some to be the forerunner of the modern computer.

The idea of the bones is similar to that of the lattice method, except that with Napier's Bones we use rectangular strips prepared in advance. For each of the ten digits, a strip is prepared containing the various multiples of that digit. A special *index* strip numbered from 1 to 9 is also prepared. Figure 4.18(a) shows the *4* strip, and Figure 4.18(b) shows the special *index* strip.

To multiply 27 by 53 by this method, we select a *2*, a *7*, and an *index* strip and place them side by side (Figure 4.19). The partial products are read from the 2- and 7-strips opposite the two digits on the index strip representing 53. The partial products are then aligned according to place value and added. Note that we annex a 0 to the partial product 135 to give us 1350 since we multiplied the 27 by 50, not by 5.

FIGURE 4.18

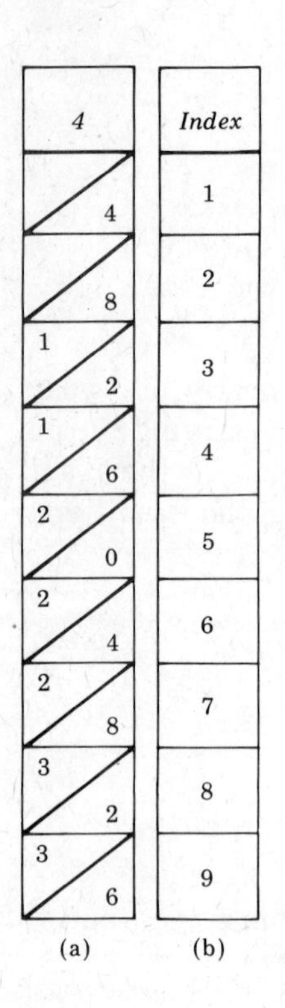

FIGURE 4.19

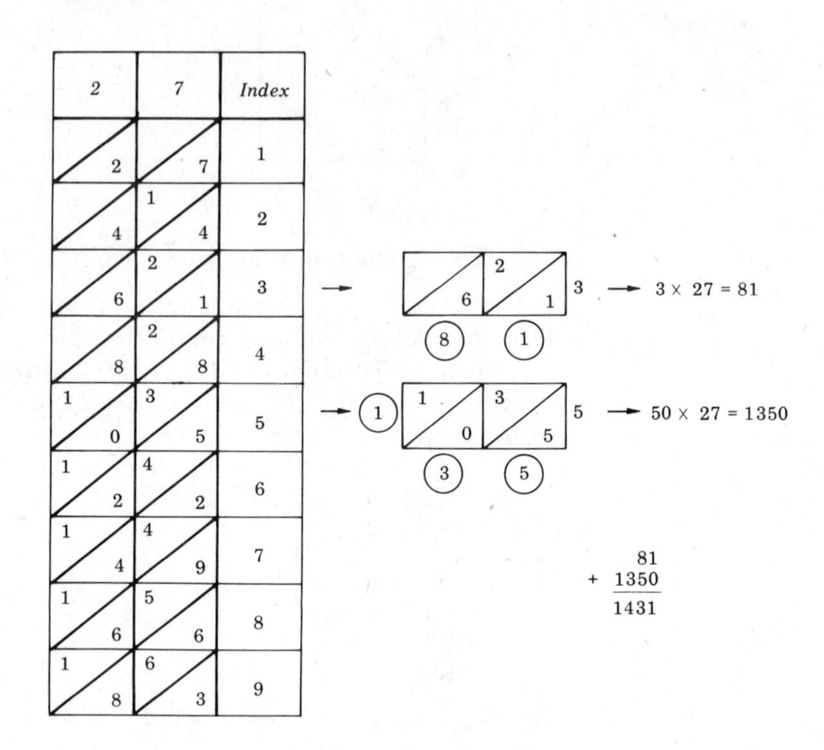

4.5 Nomographs

A *nomograph* is a simple kind of computer that can be used for finding sums, differences, products, and quotients. It is convenient to use graph paper for the construction of nomographs.

The nomograph in Figure 4.20 can be used for adding whole numbers from 0 to 10. Nomographs can be constructed for adding any range of numbers desired. The device consists of three scales, *A*, *B*, and *C*. Notice that the second scale *C* is midway between the two identical outer scales *A* and *B*. Also note that the distance between successive points in the middle scale *C* is exactly half the distance between successive points on each of the outer scales.

If you wish to find 3 + 5 on the nomograph, place any straight edge such as a ruler so that it touches 3 on the *A* scale and 5 on the *B* scale (Figure 4.21). The sum is at the point where the ruler crosses the C scale, 8. More generally, to find the sum $a + b$ of two numbers, place a ruler so as to connect a on the A scale and b on the B scale. The sum $a + b$ will then be the point where the ruler crosses the C scale.

A nomograph can also be used for subtraction. For example, to compute $9 - 4$, place the ruler so that it touches 9 on the C scale and 4 on the B scale (Figure 4.22). Then read the answer, 5, on the A scale.

To make a multiplication nomograph, a parabola (defined by the equation $y = x^2$) is used. For example, 3×4 is shown in Figure 4.23. The same nomograph can be used for division.

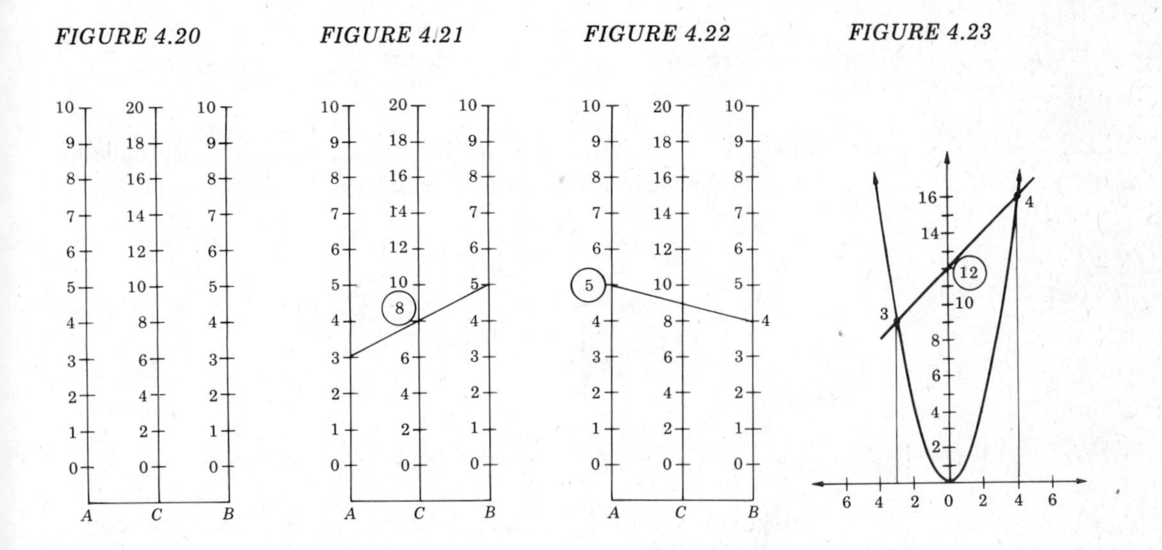

FIGURE 4.20 *FIGURE 4.21* *FIGURE 4.22* *FIGURE 4.23*

4.6 *Figurate Numbers*

Pythagoras was a Greek mathematician and philosopher who is believed to have lived between 569 and 500 B.C. After much travel, he settled down in Crotona, a Greek city of southern Italy, where he organized the Pythagorean Society. Membership in the society was restricted, and the members, who were called Pythagoreans, were pledged to secrecy. Although the greatest interest of the Pythagoreans was in moral reform, they were also interested in science and in mathematics and are credited with many important ideas in mathematics, astronomy, and the science of music.

Pythagoras and his followers were intensely interested in numbers and were the first to think of arithmetic as a science rather than as a tool in computation. They were the first to study the properties of whole numbers and to take the first steps in the development of number theory.

The Pythagoreans linked numbers with geometric patterns by originating the *figurate numbers*, also called *polygonal numbers*. These numbers can be thought of as the number of dots in certain geometrical configurations. For example, the numbers 1, 3, 6, 10, 15, ..., or, in general, numbers given by the formula $\frac{n(n+1)}{2}$, are called *triangular numbers* because they can form the triangular configurations shown in Figure 4.24.

FIGURE 4.24

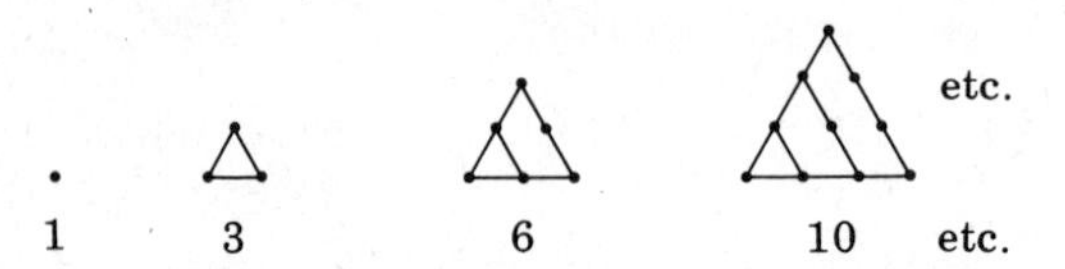

The numbers 1, 4, 9, 16, ..., given by the formula n^2, are called *square numbers* because they can form square configurations (Figure 4.25).

FIGURE 4.25

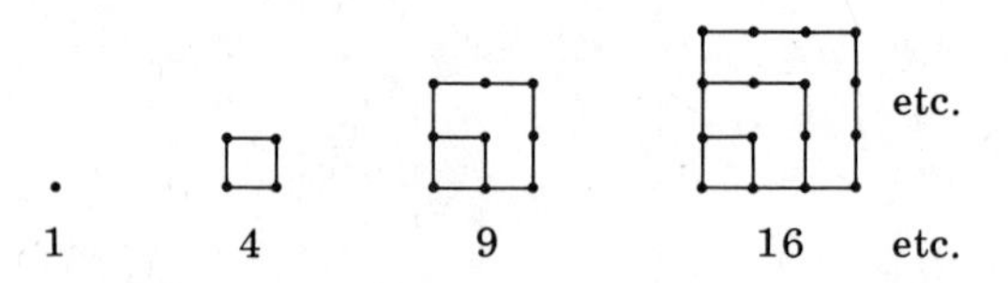

The numbers 1, 5, 12, 22, ..., given by the formula $\frac{n(3n-1)}{2}$, are called *pentagonal numbers* because they can form pentagonal configurations (Figure 4.26).

FIGURE 4.26

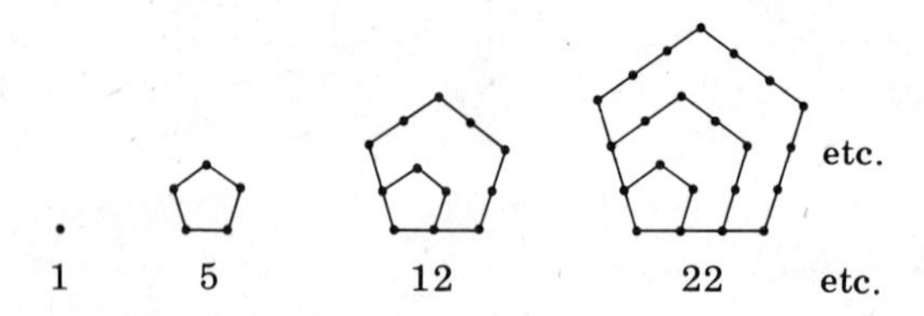

The *hexagonal numbers* 1, 6, 15, 28, ..., derived from the formula $2n^2 - n$, can form hexagonal configurations (Figure 4.27).

FIGURE 4.27

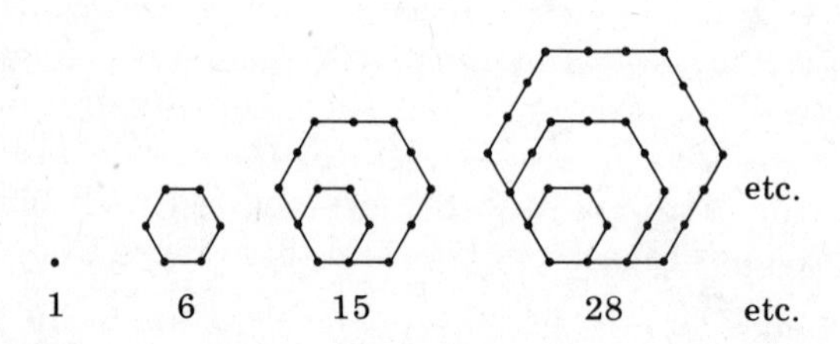

By means of these geometric configurations we can show some interesting properties of figurate numbers.

1 Any triangular number added to the next higher triangular number is a square number (Figure 4.28). Since any geometric configuration representing a square number can be divided as shown in Figure 4.28, we can conclude that any square number is the sum of two successive triangular numbers.

FIGURE 4.28

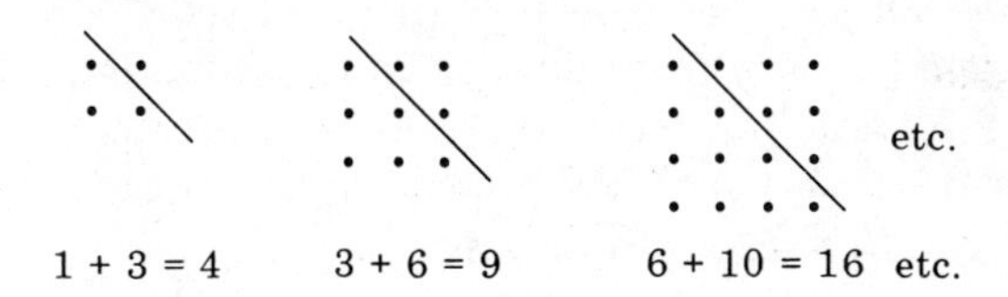

2 The sum of n consecutive odd integers, starting with 1, is always a square number, namely, n^2 (Figure 4.29).*

FIGURE 4.29

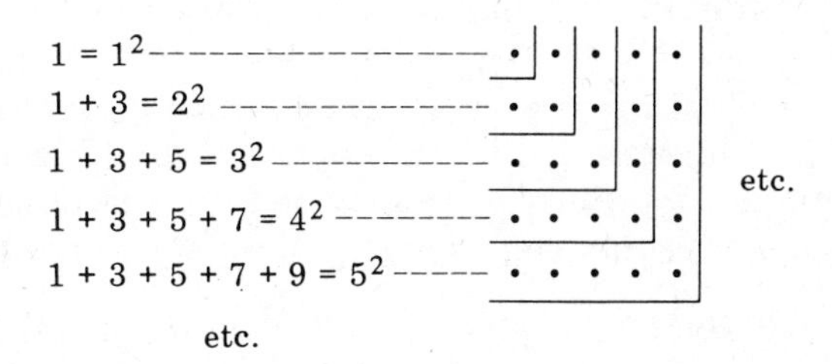

4.7 *Transfinite Numbers*

We have seen that the *cardinal number* of a set is the number of elements in the set. When we speak of finite sets, this idea of cardinality is easy enough to understand: the cardinal number can be zero, one, two, or any whole number as large as we wish to make it; if a finite set contains more elements than another finite set, then the two sets have different cardinal numbers and cannot be equivalent. The only way two finite sets can be equivalent is for them to have the same number of elements. Only then is it possible to have a one-to-one correspondence between the two sets of elements.

What, then, is the cardinal number of an *infinite* set? Are all infinite sets equivalent? Or, is there a hierarchy of infinite sets wherein some infinite sets are "larger" than others? The man who has contributed greatly to answering these questions and who is regarded as the founder of the *arithmetics of infinity* is Georg Cantor (1845-1918).

*For an interesting discussion of polygonal numbers see "Polygonal Numbers: A Study of Patterns," by Margaret A. Hervey and Bonnie H. Litwiller, *The Arithmetic Teacher,* January, 1970, pp. 33-38.

The cardinal number of an infinite set is called a *transfinite cardinal number.* The transfinite cardinal number of the set of counting numbers is denoted by the symbol $\aleph_0$ (read "aleph-null"). We can speak of $\aleph_0$ counting numbers the way we speak of 12 months in the year or 29 students in a class. The cardinal number of any set whose elements can be put into a one-to-one correspondence with the set of counting numbers is likewise $\aleph_0$. Any set whose cardinal number is $\aleph_0$ is said to be *denumerable* or *denumerably infinite.*

The property of equivalence is very different among infinite sets than it is among finite sets. With finite sets, if A is a *proper* subset of B, then the cardinal number of A is smaller than the cardinal number of B, and the two sets are not equivalent. But with infinite sets, it is possible for A to be a *proper* subset of B yet, at the same time, A and B can be equivalent.

Are *all* infinite sets, then, equivalent? The answer is no because it is possible to *prove* that $\aleph_0$ is only one of a hierarchy of alephs: $\aleph_0$, $\aleph_1$, $\aleph_2$. $\aleph_0$, the cardinal number of the set of counting numbers, can be shown to be of a lower order of infinity or "smaller," than $\aleph_1$, which is the cardinal number of *the set of points on a line;* and $\aleph_1$ can be shown to be "smaller" than $\aleph_2$, which is the cardinal number of *the set of all possible geometrical curves.* No one has yet defined the number $\aleph_3$. To prove that the three alephs are of different orders of infinity, we show that it is impossible to set up a one-to-one correspondence between the elements of any of these sets.

The number $\aleph_0$ has some unusual properties. For instance,

$$\aleph_0 + 1 = \aleph_0 \tag{1}$$

because if $\aleph_0$ is the cardinal number of set

$$C = \{1, 2, 3, ..., n, ...\},$$

then $\aleph_0 + 1$ is the cardinal number of the set C with one element, say 0, added. If we call the new set

$$W = \{0, 1, 2, ..., n, ...\},$$

then we can set up the following one-to-one correspondence between C and W:

$$\begin{array}{llllllllll} C = \{1, & 2, & 3, & 4, & 5, & ..., & n, & n+1, & ...\} \\ \ \ \ \ \ \ \updownarrow & \updownarrow & \updownarrow & \updownarrow & \updownarrow & & \updownarrow & \updownarrow & \\ W = \{0, & 1, & 2, & 3, & 4, & ..., & n-1, & n, & ...\}. \end{array}$$

C and W are, therefore, equivalent, making their cardinal numbers equal; that is, $\aleph_0 + 1 = \aleph_0$.

We can likewise show that

$$\aleph_0 + \aleph_0 = \aleph_0. \tag{2}$$

For any whole number n, $\quad n \times \aleph_0 = \aleph_0. \quad (3)$

$$(\aleph_0)^2 = \aleph_0 \times \aleph_0 = \aleph_0. \tag{4}$$

However, $(\aleph_0)^{\aleph_0}$ results in a transfinite number other than $\aleph_0$, and suspected to be $\aleph_1$.*

4.8 Peano's Postulates

In our discussion of the natural numbers, we said that these numbers are the fundamental entities on which all later number systems rest, and that even though their properties are easy to accept intuitively, they are actually based on the properties of sets.

Another and more elegant way of looking at natural numbers is through *axiomatics*. In this approach, we start with several undefined terms and axioms (or postulates) and then draw logical conclusions about the behavior of these undefined terms from the axioms. So when we use the axiomatic approach, we derive the characteristics of natural numbers not by relating them to the properties of sets but by deducing them logically from a few basic axioms. We can thereby prove the properties of the natural numbers rather than rely upon their intuitive acceptance.

To thė Italian mathematician Guiseppe Peano (1858-1932), we owe one of the most elegant and famous set of postulates from which we can derive the natural numbers and their properties. His system not only describes but actually generates the set of natural numbers. Peano based his system on five postulates and was able to prove that the entire theory of natural numbers follows from these postulates provided that suitable definitions of addition and multiplication are given.

THE PEANO POSTULATES

1 1 is a natural number.

2 For every natural number x, there is a unique natural number x' which we call the successor of x.

3 The natural number 1 is not the successor of any natural number.

4 If x and y are natural numbers such that $x' = y'$, then $x = y$.

5 If S is a set of natural numbers such that:
a) 1 is in S, and
b) if x is in S, then x' is in S;
then *all* the natural numbers are in the set S. That is, S must be the set of natural numbers.

*For an interesting discussion of transfinite numbers, read *Mathematics and the Imagination* by Edward Kasner and James Newman, Simon and Schuster, 1940, pp. 27-64.

Note that Peano's undefined terms are *natural number*, *set*, and *successor*. In Peano's system, the number 1 is his starting point for the set of natural numbers; i.e., 1 is the first natural number. His concept of *successor* guarantees that every natural number is always followed by another; i.e., it generates an infinite progression of natural numbers. Each natural number is smaller than its successor; no natural number lies between any other natural number and its successor; and two natural numbers are equal if they have the same successor.

Peano's fifth postulate is known as *The Principle of Mathematical Induction*, a principle that was stated 250 years earlier by the French mathematician Blaise Pascal (1623-1662). What the postulate says, in effect, is that a statement will be true for *every* natural number if: (a) the statement is true for the natural number 1; and (b) we can prove that *if* the statement is true for *any* natural number k, then it is also true for the succeeding natural number $k + 1$.

Imagine an *endless* row of dominoes, standing upright as shown in Figure 4.30, beginning with the first domino at the left and continuing indefinitely to the right. Suppose, further, that we wish to devise a system by which *all* the dominoes can be made to fall. How can we guarantee that *all* the dominoes will fall? We can't very well knock them down one at a time since the number of dominoes is endless. We can, however, accomplish our objective under these two conditions: (a) we knock down the first domino, and (b) the dominoes are so spaced that if any one domino falls it automatically knocks over the next domino (its successor). Conditions (a) and (b) just stated correspond roughly to conditions (a) and (b) in Peano's fifth postulate.

FIGURE 4.30

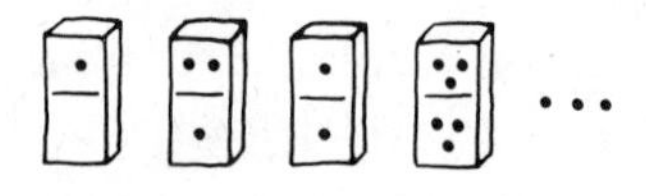

We shall now illustrate the use of the induction principle by using it to prove a well-known statement about natural numbers: the sum of the first n natural numbers can be found by using the formula

$$\frac{n(n+1)}{2}.$$

According to this formula, the sum of the first 5 natural numbers (i.e., $n = 5$) is

$$1 + 2 + 3 + 4 + 5 = \frac{5\,(5+1)}{2} = \frac{5 \times 6}{2} = 15;$$

and the sum of the first 9 natural numbers (i.e., $n = 9$) is:

$$1 + 2 + 3 + 4 + 5 + 6 + 7 + 8 + 9 = \frac{9\,(9 + 1)}{2}$$
$$= \frac{9 \times 10}{2} = 45.$$

We shall now use the principle of mathematical induction to prove that the above formula holds for *all* natural numbers. That is, we wish to prove that for *any* value of n,

$$1 + 2 + 3 + \cdots + n = \frac{n\,(n + 1)}{2} \tag{1}$$

Proof

1 Condition (a) holds since Equation 1 is true for $n = 1$:

$$1 = \frac{1(1 + 1)}{2} = \frac{2}{2} = 1.$$

2 We now show that condition (b) holds. That is, we show that if Equation 1 is true for $n = k$, then it will also be true for $n = k + 1$:

Given $$1 + 2 + 3 + \cdots + k = \frac{k\,(k + 1)}{2} \tag{2}$$

Prove $$1 + 2 + 3 + \cdots + k + (k + 1) = \frac{(k + 1)\,(k + 2)}{2} \tag{3}$$

If we add $(k + 1)$ to both sides of the given Equation 2, we get

$$1 + 2 + 3 + \cdots + k + (k + 1) = \frac{k\,(k + 1)}{2} + (k + 1)$$
$$= \frac{k^2 + k + 2k + 2}{2}$$
$$= \frac{k^2 + 3k + 2}{2}$$
$$= \frac{(k + 1)\,(k + 2)}{2},$$

which is the same as Equation 3 which we wanted to prove.

Chapter 5

NUMBER THEORY

Objectives

At the completion of this chapter, you should be able to:

1. Define a *prime* number.
2. Define a *composite* number.
3. Describe the sieve of Eratosthenes and be able to use it to find the primes in a given interval.

4. Write the *prime factorization* of a given counting number.
5. State and explain by example the Fundamental Theorem of Arithmetic.
6. State and apply the divisibility tests for 2, 3, 4, 5, 9, and 10.
7. Define the Greatest Common Divisor (GCD) and be able to find the GCD of two numbers.
8. Define the Least Common Multiple (LCM) and be able to find the LCM of two numbers.
9. Define an odd number and an even number in terms of divisibility by 2.
10. State the basic properties of odd and even numbers given in Theorems 5.2-5.7.

Prerequisites

1. You should know that in the expression ab,

   ab is called the *product* of a and b,

   a and b are called *factors* of ab,

   ab is said to be a *multiple of* a and a *multiple of* b,

   a is a *divisor* of ab, and b is a *divisor* of ab.

2. You should know the meaning of an *exponent*, e.g., that $2 \times 2 \times 2 = 2^3$, $3 \times 3 \times 5 \times 5 \times 5 = 3^2 \times 5^3$; and that an exponent is also called a *power*.

INTRODUCTION

If you have an unlimited supply of 5¢ and 8¢ stamps, what amounts of postage can you make with these stamps? Obviously, you can make postages in such amounts as 5¢, 8¢, 13¢, and 18¢. But can you make 27¢ postage or 96¢ postage or postage worth $3.41?

To answer this question fully, we have to investigate the relationship between the numbers 5 and 8. In *number theory* we study such relationships, and in this case we come out with the conclusion that it is possible to make only the following postages with 5¢ and 8¢ stamps: 0, 5, 8, 10, 13, 15, 16, 18, 20, 21, 23, 24, 25, 26, and all postages *above* 27¢. But it is not possible to make postage worth exactly 27¢.

Did you ever notice that when we add consecutive odd numbers we obtain this striking result:

$$1 + 3 = 4 = 2^2$$
$$1 + 3 + 5 = 9 = 3^2$$
$$1 + 3 + 5 + 7 = 16 = 4^2$$
$$1 + 3 + 5 + 7 + 9 = 25 = 5^2 \quad \text{etc.}$$

That is, the sum of the first n odd numbers is always equal to n^2. So the sum of the first 25 odd numbers is $(25)^2 = 625$, and the sum of the first 92 odd numbers is $(92)^2 = 8464$. Will this relationship *always* be true? In *number theory* we study such relationships and try to find out whether and why they are always true.

Numbers have fascinated people for thousands of years and have often, especially in the past, been associated with superstition and mysticism. Even today, many public buildings do not have a floor numbered 13! Pythagoras (ca. 540 *B.C.*), the Greek mathematician and philosopher, and his followers were intensely interested in numbers. According to Pythagorean number lore, odd numbers were considered male, and even numbers female. *One* was not considered a number itself but was taken to be the (divine) generator of all numbers. *Two*, the first even number, was associated with opinion. *Three*, the first true male number, was regarded as the number of harmony. *Four*, the first perfect square, was associated with justice. *Five*, representing the union of the first male and the first female numbers, was associated with marriage.

Despite the lore and mysticism surrounding the Pythagorean study of numbers, the Pythagoreans were the first to think of arithmetic as a science rather than as a tool in computation; they were the first to study the properties of whole numbers in a way that laid the foundation for the development of number theory. Within the last 400 years, three men who made very important contributions to number theory were Pierre Fermat (ca. 1600-1665), Leonhard Euler (1707-1783), and Carl Friedrich Gauss (1777-1855). Modern number theory has little to do with mysticism or religion but, instead, concerns itself primarily with the properties of the natural numbers and, more generally, with those of the integers and the relationships among them.

The fascination with numbers over the centuries was due to various factors. The subject matter is familiar, demands little previous knowledge, and appeals uniquely to human curiosity. There are many curious and odd facts about numbers that have been the basis of a great variety of number puzzles which have delighted people over the ages. It has been possible for those without any special mathematics background to ask, appreciate, and even to contribute answers to questions that are significant. Also, an aura of "mystery of the unknown" as well as challenge surrounds a great variety of number problems that are simple enough for the "average" person to understand but whose solutions continue to elude the efforts of the greatest mathematicians. In this chapter we shall pose some of these problems.

5.1 Primes and Composites

DEFINITION 1 A natural number other than 1 whose only factors are itself and 1 is called a *prime*. The first five primes are 2, 3, 5, 7, 11.

DEFINITION 2 A natural number that is neither 1 nor a prime is called *composite*. A composite number has itself and 1 as factors, but also has other factors. The first five composites are 4, 6, 8, 9, 10.

Comments

1. *There are infinitely many primes and infinitely many composites.* *(See page 127.)*

2. *Absent from the sets of prime and composite numbers are* 0 *and* 1. *Both* 0 *and* 1 *are neither prime nor composite.*

3. *Another way of defining a* prime *is to say that it is a whole number with exactly two, distinct factors. Since*

   $$0 = 0 \times 0 = 0 \times 1 = 0 \times 2, \quad \text{etc.},$$

   zero cannot be a prime because it has infinitely many factors. One *cannot be prime since it has only itself as a distinct factor; i.e.,* $1 = 1 \times 1$.

4. *Another reason for not considering* 1 *as a prime is that many theorems, including the Fundamental Theorem of Arithmetic, can be proved to hold for the primes if we exclude* 1 *as a prime.*

5. *The smallest prime is* 2, *and* 2 *is the only prime that is an even number. No other even number can be prime since every even number has* 2 *as a factor.*

6. *It is possible to define prime and composite numbers to include the negative integers by saying that a negative integer is prime if its additive inverse is prime, and that a negative integer is composite if it is not prime.*

7. *Is there a formula that could be used to generate all the primes? No such formula has yet been produced. The expression*

   $$n^2 - n + 41$$

   yields a prime for every natural number n between 1 *and* 41. *But if we let* $n = 41$, *we get*

   $$n^2 - n + 41 = (41)^2 - 41 + 41 = (41)^2$$

   which is not prime. The expression

   $$n^2 - 79n + 1601$$

   works for every natural number n between 1 *and* 79 *inclusive.*

5.2 *The Sieve of Eratosthenes*

There is an easy way of finding all the prime numbers less than some particular natural number, a method devised about 2200 years ago by the Greek mathematician Eratosthenes (ca. 230 B.C.). The method is called the *Sieve of Eratosthenes* because it sieves out from a list of numbers all the numbers that are *not* primes, leaving only the primes.

Suppose, for example, we wish to find all the primes less than 100. Write down the numbers from 1 to 100 and strike out the number 1 which is not a prime. Then draw a circle around 2, the first prime, and strike out all multiples of 2 that follow since each such number is divisible by 2 and so is not a prime. That is, strike out the numbers 4, 6, 8, 10, . . . , 100 (Figure 5.1).

Next draw a circle around 3, the next prime in the list, and strike out all multiples of 3 that follow; that is, strike out the numbers 6, 9, 12, 15, 18, . . . , 99. Some of these numbers, such as 6, 12, and 18, will already have been crossed out because they are also multiples of 2.

Next draw a circle around 5 and cross out all multiples of 5, and then continue the process with 7. When you reach 11, note that it is not necessary to cross out multiples of 11 since these numbers have already been crossed out in previous steps. This is also true of multiples of 13, 17, 19; in fact, this is true of all primes greater than 7. Therefore, all the remaining numbers in the list are prime and we draw circles around each of these numbers.

FIGURE 5.1

~~1~~	2	3	~~4~~	5	~~6~~	7	~~8~~	~~9~~	~~10~~
11	~~12~~	13	~~14~~	~~15~~	~~16~~	17	~~18~~	19	~~20~~
~~21~~	~~22~~	23	~~24~~	~~25~~	~~26~~	27	~~28~~	29	~~30~~
31	~~32~~	33	~~34~~	~~35~~	~~36~~	37	~~38~~	~~39~~	~~40~~
41	~~42~~	43	~~44~~	~~45~~	~~46~~	47	~~48~~	~~49~~	~~50~~
~~51~~	~~52~~	53	~~54~~	~~55~~	~~56~~	57	~~58~~	59	~~60~~
61	~~62~~	63	~~64~~	~~65~~	~~66~~	67	~~68~~	~~69~~	~~70~~
71	~~72~~	73	~~74~~	~~75~~	~~76~~	77	~~78~~	79	~~80~~
~~81~~	~~82~~	83	~~84~~	~~85~~	~~86~~	87	~~88~~	89	~~90~~
~~91~~	~~92~~	93	~~94~~	~~95~~	~~96~~	97	~~98~~	~~99~~	~~100~~

Comments

1 *In searching for the primes less than* 100, *we noted that the largest prime we need to use in our sieving process is* 7. *The next higher prime,* 11, *was not needed since all multiples of* 11 *were already eliminated.*

Why had all the multiples of 11 *already been eliminated? Because for a multiple of* 11 *to be less than* 100, *the number by which* 11 *is multiplied must be* less than 11, *and all the primes less than* 11 *had already been used in the sieving process.*

Therefore, to determine the largest prime needed to complete our sieving process, we need only find the prime number closest to, but not greater than, the square root of 100, *which is* 7.

More generally, if we wish to find all the prime numbers less than any natural number n, we need only eliminate all multiples of the prime p, where p is the greatest prime such that $p^2 \leq n$, and all multiples of the primes less than p.

2 *The ideas in 1 above provide us with a useful test to determine whether a given number is a prime:*

a) Divide the number by all the primes less than or equal to its square root.

b) If any one of these primes divides the given number, then the number is composite; otherwise it is prime.

For example, is 103 *a prime number? Since* $\sqrt{103}$ *is between* 10 *and* 11, *we need to test only primes not greater than 10. These primes are* 2, 3, 5, 7. *Since none of these primes divides* 103, 103 *is a prime number.*

Is 391 *a prime number?* $\sqrt{391}$ *is between* 19 *and* 20. *We test those primes not greater than* 19 *to see whether or not they are divisors of* 391. *The primes to be tested are:* 2, 3, 5, 7, 11, 13, 17, 19. *Since* 17 *divides* 391, 391 *is not a prime.*

5.3 *The Fundamental Theorem of Arithmetic*

When a composite number is written as a product of all its prime factors, such a product is called the *prime factorization* of that number. In the example $12 = 2 \times 2 \times 3$, the product $2 \times 2 \times 3$ is the prime factorization of 12 because the factors 2 and 3 are primes. If we wrote $12 = 4 \times 3$, the product 4×3 would be one factorization of 12 but not the *prime* factorization because the factor 4 is not a prime.

We can express *every* composite number as the product of prime numbers by selecting any factorization of the number (not including 1) and proceeding to factor those factors that are not prime. For example,

$$24 = 3 \times 8 = 3 \times 2 \times 2 \times 2 = 3 \times 2^3.$$

Since it is customary to arrange the prime factors in order from the smallest to the largest, we write $24 = 2^3 \times 3$.

Another way to find all the prime factors of a composite number is by repeated division, starting with the smallest prime that divides the given number and continuing until all prime factors have been obtained. For example, if we wish to find the prime factorization of 72, we note that 2 is the smallest prime that divides 72. So we proceed as follows:

$$72 = 2 \times 36 = 2 \times 2 \times 18 = 2 \times 2 \times 2 \times 9$$
$$= 2 \times 2 \times 2 \times 3 \times 3 = 2^3 \times 3^2.$$

Not only can every composite number be expressed as the product of prime factors, but there is only one such product possible if we write the factors in order from smallest to largest. In other words, the prime factorization of any composite number is *unique*. So many results in number theory are based on this property that it is known as the Fundamental Theorem of Arithmetic. This theorem, which we shall now state but not prove, is also known as the Unique Prime Factorization Theorem.

THEOREM 1 *(The Fundamental Theorem of Arithmetic)* Every composite number can be expressed uniquely as the product of prime factors, if the order of the factors is disregarded.

Example The prime factorization of 360 will be the same no matter what factorization we start with:

a) $360 = 2 \times 180 = 2 \times 2 \times 90 = 2 \times 2 \times 2 \times 45$
$= 2 \times 2 \times 2 \times 3 \times 15 = 2 \times 2 \times 2 \times 3 \times 3 \times 5$
$= 2^3 \times 3^2 \times 5$

b) $360 = 6 \times 60 = 6 \times 4 \times 15 = 2 \times 3 \times 2 \times 2 \times 3 \times 5$
$= 2 \times 2 \times 2 \times 3 \times 3 \times 5 = 2^3 \times 3^2 \times 5.$

Comment *The theorem tells us that every natural number can be obtained by the multiplication of specific primes and can never be the product of a different set of primes.*

EXERCISE SET 5.1–5.3

1. Using the Sieve of Eratosthenes, find all the primes (a) less than 150, (b) between 100 and 200.
2. Write each of the following as a product of primes.
a) 12 b) 45 c) 252 d) 180 e) 1925
3. Express each of the following as the sum of two (not necessarily distinct) odd primes.
a) 8 b) 14 c) 24 d) 36
4. Write 10 as the sum of primes in as many ways as possible.
5. Find the prime factorization of the following numbers.
a) 30 b) 54 c) 153 d) 620
6. Express each of the following numbers as the sum of three (not necessarily distinct) odd primes.
a) 11 b) 25 c) 49 d) 87

7 Test whether or not the following numbers are primes.
a) 89 b) 167 c) 837 d) 1171

8 Find five consecutive whole numbers none of which is prime.

9 Some primes can be expressed in the form $1 + n^2$ for some natural number n. For example, $5 = 1 + 2^2$, $17 = 1 + 4^2$. Find three more primes in the form $1 + n^2$.

10 Some primes can be expressed in the form $n^2 - 1$ for some natural number n. For example, $3 = 2^2 - 1$.
a) Can you find other primes of the form $n^2 - 1$?
b) Can you prove your answer?

11 Name a pair of prime numbers that differ by 1, and show that there is only one such pair possible.

12 What is the union of the set of prime numbers and the set of composite numbers? What is their intersection?

5.4 Odd and Even Numbers

DEFINITIONS A whole number is *even* if it is divisible by 2. A whole number is *odd* if it has a remainder of 1 when divided by 2.

Comments 1 *From these definitions it follows that:*

a) The even *whole numbers may be identified as the set*

$E = \{x | x = 2n, n \in W\}$; *that is* $E = \{0, 2, 4, 6, \ldots\}$.

b) The odd *whole numbers may be identified as the set*

$F = \{x | x = 2n + 1, n \in W\}$; that is, $F = \{1, 3, 5, 7, \ldots\}$.

Note that zero is an even number.

2 *Other ways of defining even and odd numbers are:*

a) A whole number is even *if and only if it has a factor of 2.*

b) A whole number is odd *if it is not even.*

PROPERTIES OF ODD AND EVEN NUMBERS

THEOREM 2 The sum of two even numbers is an even number.

Proof

1 Let $2n$ and $2p$ represent two even numbers ($n, p \in W$).

2 Then the sum of the two even numbers will be $2n + 2p$.

3 $2n + 2p = 2(n + p)$ (by the distributive property).

4 $n + p$ is a whole number (by the closure property).

5 $2(n + p)$ is an even number (since it has a factor of 2).

6 Therefore, $2n + 2p$ is an even number.

THEOREM 3 The product of two even numbers is an even number.

Proof

1. Let $2n$ and $2p$ represent two even numbers $(n, p \in W)$.
2. Then the product of the two even numbers will be $2n \times 2p$.
3. $2n \times 2p = 2(n \times 2p)$. (Why?)
   $= 2(2np)$
4. $2np$ is a whole number. (Why?)
5. Therefore, $2(2np)$ or $2n \times 2p$ is an even number. (Why?)

In a similar fashion we can prove the following other properties of odd and even numbers. (See Exercise Set 5.4, Exercises 1-4.)

THEOREM 4 The product of two odd numbers is an odd number.

THEOREM 5 The sum of two odd numbers is an even number.

THEOREM 6 The sum of an odd number and an even number is an odd number.

THEOREM 7 The product of an odd number and an even number is an even number.

All the properties expressed in Theorems 2-7 are summarized in Table 5.1 and 5.2. (*E* represents an *even* number and *O* represents an *odd* number.)

TABLE 5.1

+	E	O
E	E	O
O	O	E

TABLE 5.2

×	E	O
E	E	E
O	E	O

EXERCISE SET 5.4

1. Prove that the product of two odd numbers is an odd number.
2. Prove that the sum of two odd numbers is an even number.
3. Prove that the sum of an odd number and an even number is an odd number.
4. Prove that the product of an odd number and an even number is an even number.
5. a) The sum of any number of even numbers is ____.
   b) The sum of an even number of odd numbers is ____.
   c) The sum of an odd number of odd numbers is ____.
6. a) Is the set of even numbers closed under addition? Under multiplication?
   b) Is the set of odd numbers closed under addition? Under multiplication?

7 If c is a counting number, which of the following always represent even numbers and which represent odd numbers?
a) $2c$ b) $3c$ c) $2c + 1$
d) c e) $2c - 1$ f) $2c + 5$

8 a) Prove that if the square of a number is even, then the number is even.
b) Prove that if the square of a number is odd, then the number is odd.

9 What is the product of three even numbers and an odd number? Prove your answer.

10 Use the addition and multiplication tables (Tables 5.1 and 5.2) to prove that an even number added to the product of an even and odd number is an even number.

11 Prove that if the product of two natural numbers is even, then at least one of them is even.

5.5 *Tests for Divisibility*

We say that 15 is *divisible* by 5 because when 15 is divided by 5 we obtain a remainder of 0. When we divide 23 by 5, we do not get a remainder of 0; therefore, 23 is *not* divisible by 5. Another way to say that 15 is divisible by 5 is to say that 5 *divides* 15, denoted by $5 \mid 15$.

It is often desirable to have an easy way of telling whether a whole number is divisible by another whole number. So we shall now present ways for determining whether a whole number is divisible by 2, 3, 4, 5, 8, 9, and 10.

Before proceeding it will be useful to recall that the decimal expansion of the whole number 857 is

$$857 = (8 \times 10^2) + (5 \times 10) + 7$$

and that the expansion of 2569 is

$$2569 = (2 \times 10^3) + (5 \times 10^2) + (6 \times 10) + 9.$$

More generally, the decimal expansion of a whole number n is an expression of the form

$$n = \cdots + (e \times 10^4) + (d \times 10^3) + (c \times 10^2) + (b \times 10) + a \qquad (1)$$

where $\cdots$, e, d, c, b, and a are the digits in n.

DIVISIBILITY BY 2, 5, AND 10

Since every power of ten in Equation 1 is divisible by 2, the divisibility of the whole number n by 2 depends only on whether the last digit a is divisible by 2. Since a is divisible by 2 if and only if a is even, we have the following test for divisibility by 2:

THEOREM 8 A whole number is divisible by 2 if and only if its last digit is even.

Example 1 The numbers 156, 298, 5046638, and 499790 are divisible by 2 since the last digit of each of these numbers is even.

Comment *This and the other divisibility tests rest on the following theorem: For all whole numbers a, b, and c, $c \neq 0$, if a and b are divisible by c, then their sum $a + b$ is divisible by c. For example, if* 15 *and* 24 *are each divisible by* 3, *then their sum,* $15 + 24 = 39$, *is also divisible by* 3.

This theorem allows us to say that if each of the terms in Equation 1,

$(e \times 10^4), (d \times 10^3), (c \times 10^2), (b \times 10)$,

is divisible by 2, *then their sum,* $(e \times 10^4) + (d \times 10^3) + (c \times 10^2) + (b \times 10)$, *is divisible by* 2.

We obtain a divisibility test for 5 in a similar way. Since every power of ten is divisible by 5, the divisibility of the whole number n by 5 depends only on whether the last digit a is divisible by 5. Since a is divisible by 5 if and only if a is 0 or 5, we have the following test for divisibility by 5:

THEOREM 9 A whole number is divisible by 5 if and only if its last digit is 0 or 5.

Example 2 The numbers 60, 325, and 50895 are divisible by 5 since the last digit of each of these numbers is 0 or 5.

Similarly, 10 divides every power of ten and, therefore, n is divisible by 10 if and only if a is divisible by 10. But since a is only a single digit, a cannot be divisible by 10 unless a is 0. Therefore,

THEOREM 10 A whole number is divisible by 10 if and only if its last digit is 0.

Example 3 The numbers 150, 70, 133790, and 3200 are divisible by 10 since the last digit of each of these numbers is 0.

DIVISIBILITY BY 4 AND 8

Since 4 divides every power of ten from 10^2 on as shown in Equation 1, the divisibility of n by 4 depends only on whether $(b \times 10) +$

a; that is, the number formed by the last two digits, is divisible by 4. Therefore:

THEOREM 11 A whole number is divisible by 4 if and only if the number formed by its last two digits is divisible by 4.

Example 4 831612 is divisible by 4 since the number formed by its last two digits, 12, is divisible by 4. The numbers 23780, 59036, and 225709332 are also divisible by 4.

To test for divisibility by 8, we note that every power of ten from 10^3 on is divisible by 8. Therefore, the divisibility of n by 8 depends on whether $(c \times 10^2) + (b \times 10) + a$; that is, the number formed by the last three digits, is divisible by 8. Therefore:

THEOREM 12 A whole number is divisible by 8 if and only if the number formed by its last three digits is divisible by 8.

Example 5 19870816 is divisible by 8 since the number formed by its last three digits, 816, is divisible by 8. 51168, 2353240, and 732085728 are also divisible by 8.

DIVISIBILITY BY 3 AND 9

The decimal expansion of the whole number n in Equation 1 on page 115 can be expressed as

$$n = \cdots + 10000e + 1000d + 100c + 10b + a \tag{2}$$

where $\ldots, e, d, c, b$, and a are the digits in n.

Note that $10 = 9 + 1, 100 = 99 + 1, 1000 = 999 + 1, 10000 = 9999 + 1$, etc. Therefore,

$$\begin{aligned} n &= \cdots + (9999 + 1)e + (999 + 1)d + (99 + 1)c + (9 + 1)b + a \\ &= \cdots + 9999e + e + 999d + d + 99c + c + 9b + b + a \\ &= (\cdots + 9999e + 999d + 99c + 9b) + \\ &\quad (\cdots + e + d + c + b + a). \end{aligned} \tag{3}$$

Since each term in the first parentheses in Equation 3 is divisible by 3 or 9, the divisibility of n by 3 or 9 depends only on whether the second parentheses, the sum of the digits of n, is divisible by 3 or 9. Therefore:

THEOREM 13 A whole number is divisible by 3 or 9 if and only if the sum of its digits is divisible by 3 or 9 respectively.

Example 6 2391678 is divisible by 3 or 9 since

$$2 + 3 + 9 + 1 + 6 + 7 + 8 = 36,$$

which is divisible by 3 or 9. The numbers 238725 and 9870867 are also divisible by 3 or 9. The number 68496, however, is divisible only by 3 and not by 9 because the sum of its digits, 33, is divisible by 3 but not by 9.

Comments

1 *It should be noted that the divisibility tests discussed depend on the number being written in the decimal, or base-ten system. In other numeration systems the tests would be different.*

2 *The rule for divisibility by* 9 *states that a whole number is divisible by* 9 *if and only if the sum of its digits is divisible by* 9. *That is, dividing a whole number by* 9 *results in a remainder of zero if and only if dividing the sum of its digits by* 9 *likewise results in a remainder of zero. This rule can be extended to* any *remainder, not just zero. For example, when we divide* 256 *by* 9 *we get a quotient of* 28 *and a remainder of* 4; *when we divide the sum of the digits of* 256 *by* 9 *we also get a remainder of* 4:

$$256 = (28 \times 9) + 4$$
$$2 + 5 + 6 = 13 = (1 \times 9) + 4.$$

In other words, after all the 9s *are divided out of a number, we are left with the same remainder as that obtained when all the* 9s *are divided out of the sum of its digits.*

3 *The principle of finding the remainder for division by* 9 *by adding the digits is known as* casting out 9s. *There are two theorems related to casting out* 9s *that are of interest.*

THEOREM A The remainder for a sum of numbers *is the same as the* sum of the remainders of the addends.

For example, consider the sum

$$34 + 62 = 96.$$

By casting out 9s *we obtain the following remainders:*

$34 \rightarrow 7, \quad 62 \rightarrow 8, \quad 96 \rightarrow 6.$

Sum of the remainders of the addends $(7 + 8) \rightarrow 6$. Remainder for the sum $\rightarrow 6$.

THEOREM B The remainder for a product of numbers *is the same as the* product of the remainders for each of the factors.

For example, consider the product

$$23 \times 47 = 1081.$$

By casting out 9s *we obtain the following remainders:*

$23 \to 5, \quad 47 \to 2, \quad 1081 \to 1.$

Remainder for product $\to 1$.

Product of the remainders for each of the factors $(5 \times 2) \to 1$.

These two theorems provide the basis for quick and simple checks of the answers we obtain for addition, subtraction, multiplication, and division. If the remainders (as illustrated in Examples 7 and 8 agree, then the operation was probably *performed correctly. We say* probably *because casting out 9s is not an absolute test. If we had interchanged two digits in the answer, the check would still have worked, but this time for a wrong answer. Likewise, if in our computation, we make a mistake by a multiple of* 9, *our check would not indicate the error.*

4 *Casting out 9s may be used to check calculations with numbers other than whole numbers, such as the addition of, say,* 25.3 + 156.97. *In this case we simply ignore the decimal points and merely check on the correctness of the digits without considering the position of the decimal point.*

EXERCISE SET 5.5

1 By which of the numbers 2, 3, 4, 5, 8, 9, and 10 is each of the following divisible?
a) 576 b) 763 c) 3039 d) 15620 e) 3725056

2 Find digits for the blanks that will make:
a) 19_ divisible by 3.
b) 2_7 divisible by 9.
c) 53_ _ divisible by 4.
d) 451_6 divisible by 8.
e) 7_ _4 divisible by 8.

3 a) If a number is divisible by 2 and by 3, is it divisible by 6? Why?
b) If a number is divisible by 4 and by 3, is it divisible by 12? Why?

4 If a number is divisible by 8 will it be divisible by 2, by 4, by 16? Give a reason for each conclusion.

5 a) Write a five-digit number that will be divisible by 3 and by 5.
b) Write a five-digit number that will be divisible by 4 and by 9.
c) Write a six-digit number that will be divisible by 3, by 5, and by 8.

6 Find all the values of r and s such that the whole number 473rs is divisible by:
a) 3. b) 5. c) 8. d) 9. e) 10.

7 Devise a test for the divisibility of a whole number by 6.

8 Devise a test for the divisibility of a whole number by 11.

9 Devise a test for the divisibility of a whole number by 15.

10 Give three values of x for which 8^4 will divide 8^2x.

5.6 Greatest Common Divisor

Since 2 is a divisor of both 6 and 14, we say that 2 is a *common divisor* of 6 and 14. We say that 3 is a *common divisor* of 15 and 30 because 3 is a divisor of each of these numbers. However, 3 is not the only common divisor of 15 and 30; 5 and 15 are also common divisors of 15 and 30. Since 15 is the *largest* of these three common divisors, we call 15 the *greatest common divisor* or the *greatest common factor* of 15 and 30. More generally:

DEFINITION 1 *The greatest common divisor* (GCD) of two natural numbers a and b is the largest natural number that is a divisor of both a and b.

If n is the GCD of a and b, we often denote this fact by writing

$$\text{GCD}(a, b) = n.$$

Thus, GCD (36, 48) = 12 and GCD (90, 54) = 18.

HOW TO FIND THE GCD

We shall now show by several examples how to find the greatest common divisor of two natural numbers.

Example 1 Find the GCD of 18 and 42.

a) *Divisor Method*

1 We begin by writing down *all* the divisors of each number:
$A = \{1,2,3,6,9,18\}$ are the divisors of 18,
$B = \{1,2,3,6,7,14,21,42,\}$ are the divisors of 42.
2 The common divisors of 18 and 42 are:

$A \cap B = \{1,2,3,6\}$ are the common divisors.

3 The *greatest* of these common divisors is 6. Therefore,

GCD (18, 42) = 6.

b) *Prime Factorization Method*

1 Another way to find the greatest common divisor of 18 and 42 is to express each of these numbers as a product of its prime factors:

$18 = 2 \times 3 \times 3, \quad 42 = 2 \times 3 \times 7.$

2 When we examine these products for common factors, we find that 2 and 3 occur in each case.

3 Therefore, the greatest common factor of 18 and 42 is $2 \times 3 = 6$. That is, GCD (18, 42) = 6.

Example 2 Find the GCD of 48 and 60.

a) *Divisor Method*

1 $A = \{1,2,3,4,6,8,12,16,24,48\}$ are the divisors of 48, $B = \{1,2,3,4,5,6,10,12,15,20,30,60\}$ are the divisors of 60.

2 $A \cap B = \{1,2,3,4,6,12\}$ are the common divisors.

3 GCD(48, 60) = 12.

b) *Prime Factorization Method*

1 $48 = 2 \times 2 \times 2 \times 2 \times 3$, $60 = 2 \times 2 \times 3 \times 5$

2 The common factor of both numbers is $2 \times 2 \times 3 = 12$.

3 GCD(48, 60) = 12.

Comments

1 *Note that the reason we know that our lists of prime factors of* 48 *and* 60 *in Example 2 are complete is that by the Fundamental Theorem of Arithmetic there is only one prime factorization of* 48 *and only one prime factorization of* 60.

2 *The concept of the greatest common divisor may be extended to include more than two numbers. For example, to find the* GCD *of* 252, 90, *and* 54, *we write the prime factorization of each number:*

$252 = 2 \times 2 \times 3 \times 3 \times 7$, $90 = 2 \times 3 \times 3 \times 5$,
$54 = 2 \times 3 \times 3 \times 3$.

Since the common factor of all three numbers is $2 \times 3 \times 3 = 18$, *we conclude that*

GCD(252, 90, 54) = 18.

3 *To find the* GCD *of large numbers, a much more practical method than the two methods we have illustrated was devised by Euclid more than 2200 years ago. For Euclid's method* *see page 131.*

RELATIVELY PRIME NUMBERS

It is possible that two natural numbers have no common divisors other than 1. For example, 3 and 4 have no common divisor other than 1. In this case the numbers are called *relatively prime.*

DEFINITION 2 Two natural numbers a and b are *relatively prime* if and only if their only common divisor is 1.

Comments

1 *Note that when two numbers are relatively prime, neither number needs to be a prime. "Relatively" prime only means that both numbers are prime* with respect to each other.

2 *If* a *and* b *are relatively prime, then it follows from the definition that* GCD(a, b) = 1.

3 *If* GCD(a, b) = n, *then* $\frac{a}{n}$ *and* $\frac{b}{n}$ *are relatively prime because in dividing by* n *we have removed all common factors of* a *and* b *except* 1.

4 *When we reduce a fraction to its lowest terms, the numerator and the denominator become relatively prime. For example, when we reduce* $\frac{4}{6}$ *to* $\frac{2}{3}$, *the* 2 *and the* 3 *are relatively prime.*

5.7 Least Common Multiple

Consider the two natural numbers 2 and 3. What are their *multiples*? The multiples of 2 are the elements of the set

$$A = \{2, 4, 6, 8, 10, 12, \ldots\}.$$

The multiples of 3 are the elements of the set

$$B = \{3, 6, 9, 12, 15, \ldots\}.$$

The *common multiples* of 2 and 3 are the elements of the set

$$A \cap B = \{6, 12, 18, \ldots\}.$$

Apparently, there are infinitely many common multiples of 2 and 3. Since 6 is the *smallest* of these common multiples, we call 6 the *least common multiple* of 2 and 3. The numbers 2 and 3 are each a factor of the least common multiple 6 as well as of all the other common multiples.

DEFINITION The *least common multiple* (LCM) of two natural numbers a and b is the smallest natural number c such that a and b are factors of c.

We often denote the least common multiple by writing

$$\text{LCM}(a, b) = c.$$

For example, the least common multiple of 6 and 8 is 24; so we write

$$\text{LCM}(6, 8) = 24.$$

HOW TO FIND THE LCM

Prime Factorization Method The method we used to find the LCM of 2 and 3 can become cumbersome when working with large numbers. A better method is to use prime factorization.

Example 1 Find the LCM of 12 and 18.

1 First write the prime factorization of each number:

$12 = 2 \times 2 \times 3 = 2^2 \times 3, \qquad 18 = 2 \times 3 \times 3 = 2 \times 3^2.$

2 Because the LCM will be a multiple of 12, the LCM must contain the factors of 12: 2^2, 3. Because the LCM will also be a multiple of 18, the LCM must also contain the factors of 18: 2, 3^2.

3 Since we are looking for the *least* common multiple, we will include in our LCM only the *highest power* of the primes that occur in the two numbers and no other factors.

4 Thus, $\text{LCM}(12, 18) = 2^2 \times 3^2 = 36$.

Dividing Out Primes Method Another way of finding the LCM of two numbers is by dividing both numbers by primes, as many times as possible, until the row of answers consists only of ones and relatively prime numbers.

Example 2 Find the LCM of 24 and 30.

1 Divide both 24 and 30 by the prime number 3, getting 8 and 10 as shown.

$$\begin{array}{r|rr} 3 & 24 & 30 \\ \hline 2 & 8 & 10 \\ \hline & 4 & 5 \end{array}$$

2 Divide both 8 and 10 by the prime number 2, getting 4 and 5.

3 Since 4 and 5 are relatively prime, we have

$\text{LCM}(24, 30) = 3 \times 2 \times 4 \times 5 = 120.$

Why does this method work?

Trial and Error Method In this method we pick multiples of the larger number, in ascending order, and see whether the smaller number divides the multiple. If it does, then the multiple is the LCM.

Example 3 Find the LCM of 24 and 30.

1 The multiples of the larger number 30 are the elements of the set

$A = \{30, 60, 90, 120, 150, \ldots\}$.

2 We test the first element of A, 30, and find that it does not divide 24. This is also true of 60 and 90. But the next multiple, 120, is divisible by 24.

3 Therefore, LCM(24, 30) = 120.

Comments *1* *If* LCM(a, b) = c, *then* c *divides all the other common multiples of* a *and* b.

2 *The concept of the* LCM *may be extended to include more than two numbers. For example, the* LCM *of* 6, 9, *and* 20 *is*

$$6 = 2 \times 3, \quad 9 = 3^2, \quad 20 = 2^2 \times 5$$

$$\text{LCM}(6, 9, 20) = 2^2 \times 3^2 \times 5 = 180.$$

3 *We use the concept of* LCM *when we add or subtract fractions with different denominators. For example, to add* $\frac{1}{3}$ *and* $\frac{1}{4}$ *we change both denominators to* 12, *which is the* LCM *of* 3 *and* 4.

4 *An easy and interesting way to find the* LCM, *especially when the numbers are large, is to (a) find the* GCD *of the two numbers, and then (b) use the relationship,*

$$\text{GCD} \times \text{LCM} = ab \qquad \text{or} \qquad \text{LCM} = \frac{ab}{\text{GCD}},$$

where a *and* b *are the two numbers whose* LCM *we wish to find, and* GCD *is their greatest common divisor.*

For example, find the LCM *of* 12 *and* 18.

$$\text{LCM} = \frac{ab}{\text{GCD}} = \frac{12 \times 18}{\text{GCD}(12, 18)} = \frac{12 \times 18}{6} = 36.$$

Find the LCM *of* 144 *and* 180.

$$\text{LCM} = \frac{ab}{\text{GCD}} = \frac{144 \times 180}{\text{GCD}(144, 180)} = \frac{144 \times 180}{36} = 720.$$

5 *The number line can be used to illustrate common multiples as well as the* LCM. *For example, the common multiples of* 2 *and* 3 *can be shown as illustrated in Figure* 5.2. *Note that the upper and lower arcs meet at the points in the set*

$B = \{0, 6, 12, 18, \ldots\}$.

The first *point (other than* 0) *where they meet is* 6, *which is the* LCM. *All the points in B are common multiples of* 2 *and* 3.

FIGURE 5.2

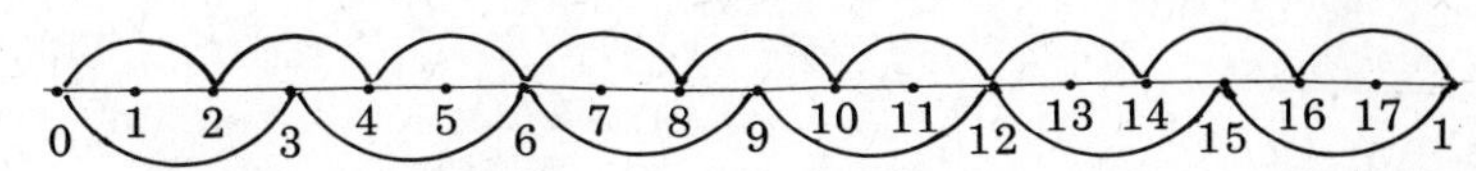

EXERCISE SET 5.6, 5.7

1 List the common divisors of (a) 16 and 40, (b) 36 and 78.

2 a) Since $18 = 3^2 \times 2$ and $15 = 3 \times 5$, the GCD of 18 and 15 is ___.
b) Since $72 = 2^3 \times 3^2$ and $60 = 2^2 \times 3 \times 5$, the GCD of 72 and 60 is ___.

3 Find the greatest common divisor of each of the following.
a) (36, 54) b) (20, 17) c) (0, 8) d) (12, 45)
e) (70, 252) f) (99, 363) g) $(n, 0)$
h) $5^2 \times 7^3$ and $3^2 \times 5 \times 7$ i) $11^2 \times 13^4$ and $2 \times 3 \times 11^3 \times 13^2$
j) (24, 90, 66) k) (195, 198, 126)

4 What is the GCD (a) of any two prime numbers? (b) of any two relatively prime numbers?

5 Why is GCD(a, b) = GCD(b, a)?

6 Since the GCD of 96 and 120 is 24, do we obtain prime numbers, composite numbers, or relatively prime numbers when we divide 96 by 24 and 120 by 24?

7 Why does every common divisor of a and b divide the *greatest* common divisor of a and b?

8 Which of the following pairs of numbers are relatively prime?
a) 7 and 8 b) 9 and 15 c) 57 and 38 d) 20 and 21
e) 25 and 24

9 What is the smallest natural number n, $n > 5$, that is relatively prime to each of the following?
a) 7 b) 30 c) 24 d) 31

10 a) Must two distinct prime numbers be relatively prime?
b) Must two relatively prime numbers be distinct prime numbers?
c) Give three examples of two composite numbers that are relatively prime?
d) Name three numbers that are relatively prime to 15.

11 Find the set of multiples of (a) 7, (b) 12, (c) 15.

12 Find the set of common multiples of:
a) 2 and 3. b) 5 and 8. c) 12 and 16.

13 Since $12 = 2^2 \times 3$ and $18 = 2 \times 3^2$, the lowest common multiple of 12 and 18 is ___.

14 Find the lowest common multiple of:
a) (6, 15). b) (8, 26). c) (11, 13). d) (18, 27).
e) (99, 363). f) $2^2 \times 3$ and $3^2 \times 5$. g) x^2y and xy^2.
h) m^2n and mr.

15 Given two natural numbers, n and $n + 1$, find their LCM.

16 a) If a and b are prime, what is their LCM?
b) If r and t are relatively prime, what is their LCM?

17 Prove that the LCM of a and b is ab if and only if the GCD$(a, b) = 1$.

18 Two school bells ring together at 8:30 A.M. Thereafter, one bell rings every half hour while the other bell rings every 45 minutes.
a) List the times between 8:30 A.M. and 2:30 P.M. when both bells ring together.
b) How is this problem related to the concept of common multiples?

PUZZLES

1 The number of germs in a jar doubles every minute. If the jar is filled with germs in one hour, how long will it take for the jar to be half full?

2 Your rich uncle offers you this choice:
a) He will give you \$1 now, \$2 next month, and each month there-he will double the amount of the previous month. He will continue paying you every month for 5 years.
b) He will give you \$1,000,000 in a lump sum.
Which would you take? Why?

3 If you have only a 3-quart measure and a 5-quart measure, how can you measure exactly 4 quarts?

BASICS REVISITED

1 Multiply: $4 \times 5 \times \frac{19}{31} \times .5 \times 0 \times 59$.

2 Round off \$299.59 to the nearest ten dollars.

3 From 69 take $8\frac{3}{8}$.

4 Subtract: $16.87 - 12.037$.

5 Find the product of 38.1 and .006.

6 Find $37\frac{1}{2}\%$ of \$89.50 to the nearest cent.

7 Write 3.7 as a percent.

8 $23 - 8\frac{7}{12}$

9 $\frac{3}{4} \div 2\frac{7}{9}$

10 Find the sum of 6.8, 1.005, and .39.

11 What percent of 75 is 5.1?

12 Divide 18.23 by 3.005. (Round to 3 places.)

13 Write in words: 5200908.

14 3% of what number is 250?

15 Find the least common denominator of $\frac{3}{8}$ and $\frac{5}{36}$.

16 Express $\frac{42}{300}$ as a percent.

17 Express without exponents:
a) 3^5. b) 12^0. c) 15^1.

18 Divide by multiplying by 1: $\frac{\frac{6}{11}}{\frac{5}{7}}$.

19 At a sports event, the number of men spectators is six times the number of women spectators. If there are 43,561 spectators present at the event, how many of them are men and how many of them are women?

20 Solve for n and check your answer:
a) $.03n = 14.5$. b) $\frac{23}{100}n = 92$.

21 A man invested $5,200 in stock and sold it for $3,140. What percent of his investment did he lose?

22 Mr. Baker bought a home for $19,500 and then spent $1575 to install a new heating system and $395 on miscellaneous repairs. He later sold the house for $27,450.
a) How much profit did he make?
b) What percent profit did he make on his cost?

23 A coat is purchased for $163. This is 65% of its original price. What was the original price of the coat?

24 Find the average thickness of five pieces of wood measuring $\frac{1}{8}$ inch, $\frac{3}{16}$ inch, $\frac{1}{4}$ inch, $\frac{3}{8}$ inch, and $\frac{5}{16}$ inch.

25 At what price should a store sell a pair of shoes costing $27.98 in order to make an 18% profit on cost?

EXTENDED STUDY

5.8 More on Primes

HOW MANY PRIMES ARE THERE?

One way to answer this question is to say that if we can be sure that *there is no greatest prime* then we could conclude that there must be infinitely many primes. How can we prove that no greatest prime exists? Let us reason as follows:

1 Suppose there does exist a greatest prime. Then we can list all the primes in their natural order: 2, 3, 5, 7, 11, . . . , p_n, where p_n is the greatest prime.

2 Let us now add 1 to the *product* of all our primes. We get a number

$$M = 2 \times 3 \times 5 \times 7 \times 11 \times \cdots \times p_n + 1.$$

Obviously, $M > p_n$.

3 Since M is a whole number greater than 1, M must be either prime or composite.

a) Suppose M is prime. Then we have found a prime M larger than p_n. This contradicts our assumption in *1* that p_n is the greatest prime. Therefore, M cannot be a prime.

b) If M is composite, then it must be divisible by some prime. But M is not divisible by any of the primes 2, 3, 5, 7, 11, ..., p_n since in each case there will be a remainder of 1. Therefore, M must be divisible by a prime greater than p_n. But this contradicts our assumption in *1* that p_n is the greatest prime.

4 So the assumption that p_n is the greatest prime leads to contradictions and must, therefore, be false.

5 Therefore we must conclude that there is no greatest prime. That is, that *there are infinitely many primes.*

HOW FAR APART CAN ADJACENT PRIMES BE?

Is it possible to find, say, 5 consecutive natural numbers such that not one of them is a prime?

Consider the 5 consecutive numbers

$$(6 \times 5 \times 4 \times 3 \times 2 \times 1) + 2,\quad (6 \times 5 \times 4 \times 3 \times 2 \times 1) + 3,$$
$$(6 \times 5 \times 4 \times 3 \times 2 \times 1) + 4,\quad (6 \times 5 \times 4 \times 3 \times 2 \times 1) + 5,$$
$$(6 \times 5 \times 4 \times 3 \times 2 \times 1) + 6.$$

We note that the first number is divisible by 2, the second is divisible by 3, the third by 4, the fourth by 5, and the fifth by 6. Therefore, these are 5 consecutive natural numbers none of which is a prime.

A shorter way to write these five numbers is to use the notation "6!" (read "6 factorial") to stand for

$$6 \times 5 \times 4 \times 3 \times 2 \times 1.$$

We can then write our five consecutive natural numbers as

$$6! + 2,\quad 6! + 3,\quad 6! + 4,\quad 6! + 5,\quad 6! + 6.$$

More generally, to find n consecutive natural numbers none of which is prime, we write

$$(n + 1)! + 2, (n + 1)! + 3, (n + 1)! + 4, \ldots, (n + 1)! + n,$$
$$(n + 1)! + (n + 1).$$

Here we see that the first number is divisible by 2, the second by 3, the third by 4, ..., and the last by $(n + 1)$. Since n can be any natural number, we have shown a way to produce as many consecutive composite numbers as we wish. That is, there are gaps as big as we please between adjacent primes.

Interestingly enough, the Russian mathematician Pafnuti Chebychev (1821-1894) proved that between every natural number n greater than 1 and its double $2n$, there exists at least one prime. That is, there exists a prime p such that $n < p \leqslant 2n$. For example, between 3 and 6 there is the prime 5; between 7 and 14 there are the primes 11 and 13.

5.9 A Way to Determine How Many Factors a Given Number Has

If we wish to determine how many factors the number 6 has, we can list all its factors and then count them: 1, 2, 3, 6. Thus, the number 6 has four factors. We can use the Greek letter ν (nu) to symbolize this fact by writing $\nu(6) = 4$, read "nu of 6 equals 4," which means that "the number of factors contained in 6 is 4."

Since the factors of 4 are: 1, 2, 4, we write $\nu(4) = 3$; and since the factors of 12 are: 1, 2, 3, 4, 6, 12, we write $\nu(12) = 6$.

With small numbers like 6, 4, or 12, this way of finding the number of factors is easy enough. But when we are dealing with large numbers, this method becomes more difficult. It is possible, however, to compute $\nu(n)$, where n is any counting number, by another method which relies on the Fundamental Theorem of Arithmetic and uses the factorization of the number into its prime factors. We shall now use this method to find $\nu(12)$.

We start by expressing 12 as a product of its prime factors. By the Fundamental Theorem of Arithmetic, this product can be expressed in only one way: $12 = 2^2 \times 3$. This product tells us that every factor of 12 is either 1 or a product of at most two 2s and at most one 3. More specifically, in the case of the 2s there are *three* possibilities:

1 The factor does not have 2 as a divisor.

2 The factor has 2 as a divisor.

3 The factor has 2^2 as a divisor.

In the case of 3, there are *two* possibilities:

1 The factor does not have 3 as a divisor.

2 The factor has 3 as a divisor.

Since there are *three* ways for a factor of 12 to have 2 as a divisor, and *two* ways for a factor of 12 to have 3 as a divisor, there are altogether 3×2, or 6 ways of getting a factor of 12 to contain a

2 or a 3. This is another way of saying that the number 12 has *six* possible factors; that is, $v(12) = 6$.

What we did, in effect, to find $\nu(12)$ was to increase the *exponent* of the factor 2 in the product $2^2 \times 3$ by 1 (to allow for the possibility that 2 will not occur at all), to increase the *exponent* of the 3 factor by 1 (to allow for the possibility that 3 will not occur at all), and then multiply the two sums.

$$(2 + 1)(1 + 1) = 3 \times 2 = 6.$$

To find $\nu(40)$, we first express 40 as a product of primes: $40 = 2^3 \times 5$. From this product we see that every factor of 40 is either 1 or a product of at most three 2s and at most one 5. In the case of the 2s, there are *four* possibilities:

1 The factor does not have 2 as a divisor.

2 The factor has 2 as a divisor.

3 The factor has 2^2 as a divisor.

4 The factor has 2^3 as a divisor.

In the case of the 5, there are *two* possibilities:

1 The factor does not have 5 as a divisor.

2 The factor has 5 as a divisor.

Since there are *four* ways for a factor of 40 to have 2 as a divisor, and *two* ways for a factor of 40 to have 5 as a divisor, the number 40 has, altogether, 4×2, or 8 factors. That is, $\nu(40) = 8$. Again, what we did, in effect, was to increase the *exponents* of the 2 and the 5 in the product $2^3 \times 5$ by 1, and then multiply the two sums:

$$(3 + 1)(1 + 1) = 4 \times 2 = 8.$$

By a similar procedure we find the values of $\nu(n)$ as shown in Table 5.3.

TABLE 5.3

n	Prime Factorization	$\nu(n)$
20	$2^2 \times 5$	$3 \times 2 = 6$
24	$2^3 \times 3$	$4 \times 2 = 8$
25	5^2	3
784	$2^4 \times 7^2$	$5 \times 3 = 15$
16,875	$3^3 \times 5^4$	$4 \times 5 = 20$

Again note that the ν value of a number does not depend on its prime factors but only *on the highest powers* of its prime factors. To find $\nu(n)$, we increase the exponents of the primes appearing in the prime factorization of n by 1 and then multiply the sums.

In general, if the prime factorization of a number n is

$$n = p_1^{a_1} \times p_2^{a_2} \times p_3^{a_3} \times \cdots \times p_k^{a_k}$$

for different primes $p_1, p_2, p_3, \ldots, p_k$, then

$$\nu(n) = (a_1 + 1)(a_2 + 1)(a_3 + 1) \cdots (a_k + 1).$$

Thus, since $1960 = 2^3 \times 5 \times 7^2$,

$$\nu(1960) = (3 + 1)(1 + 1)(2 + 1) = 4 \times 2 \times 3 = 24.$$

That is, the number 1960 has 24 factors.

5.10 Euclid's Algorithm

We have seen that to find the GCD of two numbers we can list the divisors of each number and then select from these divisors the greatest common divisor; or we can find the prime factorization of each number and then find the product of the greatest common factors. An interesting method called *Euclid's Algorithm* gives us another way to find the GCD and is particularly useful with large numbers. Before explaining Euclid's method we shall first state a theorem on which the method is based.

THEOREM 14 If y is divisible by x and if $y + z$ is divisible by x, then z is divisible by x.

Example Since 6 is divisible by 3 and since 6 + 12 is divisible by 3, 12 must also be divisible by 3.

Let us now find the GCD of 104 and 46. By the previous methods we know that GCD (104, 46) = 2. We shall now use Euclid's Algorithm to find this answer. We start by dividing 104 by 46. Since $104 = 46 \times 2 + 12$, we conclude, by Theorem 14, that any divisor of 104 and 46 must also be a divisor of 12. Therefore, in looking for GCD(104, 46), we can obtain the answer by finding GCD(46, 12), which is easier to do since it involves smaller numbers. That is,

$$\text{GCD}(104, 46) = \text{GCD}(46, 12).$$

Similarly, since $46 = 12 \times 3 + 10$, we conclude that

$$\text{GCD}(46, 12) = \text{GCD}(12, 10).$$

Also, since $12 = 10 \times 1 + 2$, we conclude that

$$\text{GCD}(12, 10) = \text{GCD}(10, 2) \quad \text{and} \quad \text{GCD}(10, 2) = \text{GCD}(2, 0).$$

Since GCD(2, 0) = 2, we conclude, by transitivity, that GCD(104, 46) = 2. What we have said, then, is that

GCD(104, 46) = GCD(46, 12) = GCD(12, 10) = GCD(10, 2) = GCD (2, 0) = 2.

Let us now use Euclid's Algorithm to find the greatest common divisor of 534 and 162:

Since $534 = 162 \times 3 + 48$, GCD(534, 162) = GCD(162, 48).
Since $162 = 48 \times 3 + 18$, GCD(162, 48) = GCD(48, 18).
Since $48 = 18 \times 2 + 12$, GCD(48, 18) = GCD(18, 12).
Since $18 = 12 \times 1 + 6$, GCD(18, 12) = GCD(12, 6).
Since $12 = 6 \times 2 + 0$, GCD(12, 6) = GCD(6, 0) = 6.

Therefore, GCD(534, 162) = 6.

5.11 Some Interesting Numbers

Within the set of whole numbers we can find subsets of whole numbers whose remarkable characteristics have fascinated people for a long time. Among these subsets are the following:

PERFECT NUMBERS

Before discussing *perfect numbers*, we shall first define the term *proper divisors*. The *proper divisors* of a number are all its divisors including 1 but excluding the number itself. For instance, all the divisors of 8 are 1, 2, 4, 8. The *proper divisors* of 8 are 1, 2, and 4.

A *perfect number* is a number which is equal to the sum of its proper divisors. For example, the divisors of 6 are 1, 2, 3, 6. The *proper* divisors are 1, 2, 3; their sum is

$$1 + 2 + 3 = 6.$$

So we call 6 a *perfect number*. In fact, 6 is the smallest perfect number. The next perfect number is 28 since

$$28 = 1 + 2 + 4 + 7 + 14,$$

the sum of all its proper divisors. The next perfect numbers are 496; 8128; 33,550,336.

It is not known how many perfect numbers exist. Only 23 such numbers have been discovered so far, the largest one having been found in 1964,

$$2^{11212}(2^{11213} - 1).$$

This largest perfect number found so far requires about 6800 digits to write without using exponents.

Note that all the perfect numbers listed above are even numbers. As a matter of fact, all the perfect numbers known at this time are even numbers. Although no odd perfect number has yet been found, mathematicians have not been able to prove that *no* odd perfect numbers exist. *If* it should turn out that odd perfect numbers do exist, then they would have to be of the form

$$n = p^a(2x + 1)^2,$$

where n is an odd perfect number, p is a prime, and both $p - 1$ and $a - 1$ are divisible by 4. Also, n must be greater than 10^{25}. These statements that describe what an odd perfect number must look like *if one exists* were proved by the Swiss mathematician Léonard Euler about 200 years ago.

Euler also proved that if a number is to be an *even* perfect number it must be expressible in the form

$$2^{p-1}(2^p - 1),$$

where the value of p makes $(2^p - 1)$ *a prime.* For example, if $p = 2$, then

$$2^p - 1 = 2^2 - 1 = 3,$$

which is a prime. So Euler's formula is applicable, and we get

$$2^{p-1}(2^p - 1) = 2(2^2 - 1) = 2(3) = 6,$$

which is a perfect number.

When $p = 3$, $2^p - 1 = 2^3 - 1 = 7$, which is a prime. Therefore Euler's formula is applicable and we get:

$$2^{p-1}(2^p - 1) = 2^2(7) = 4 \times 7 = 28,$$

which is a perfect number.

Another well-known property of even perfect numbers is that they always end in 28 or in 6 preceded by an odd digit. Note that if we express the first four perfect numbers in binary notation an interesting pattern emerges as shown in Table 5.4.

TABLE 5.4

$2^{p-1}(2^p - 1)$	*Binary*	*Decimal*
$p = 2$: $2(2^2 - 1)$	110	6
$p = 3$: $2^2(2^3 - 1)$	11100	28
$p = 5$: $2^4(2^5 - 1)$	111110000	496
$p = 7$: $2^6(2^7 - 1)$	1111111000000	8128

ABUNDANT NUMBERS AND DEFICIENT NUMBERS

If the sum of the proper divisors of a number is greater than the number itself, then the number is called *abundant*. For example, 12 is an abundant number since the sum of its proper divisors is $1 + 2 + 3 + 4 + 6 = 16$, which is greater than 12. If the sum of the proper divisors of a number is less than the number, then the number is called *deficient*. For example, 8 is a deficient number since the sum of its proper divisors is $1 + 2 + 4 = 7$, which is less than 8.

All primes are deficient since every prime has only one proper divisor, 1, which is less than any prime. Any power of a prime is deficient. For example,

$$2^3 = 8, \quad 3^2 = 9, \quad 5^3 = 125, \quad \text{and} \quad 7^2 = 49$$

are all powers of a prime and all are deficient.

Since there are infinitely many prime numbers and each prime number is a deficient number, it follows that there are infinitely many deficient numbers.

Every divisor of a deficient number is deficient, and every multiple of an abundant number is abundant.

Every natural number is either perfect, abundant, or deficient, depending on whether the sum of its proper divisors equals, exceeds, or is less than the number.

AMICABLE NUMBERS

Two numbers are *amicable* if each is the sum of the proper divisors of the other. The smallest pair of amicable numbers is 220 and 284. Other amicable pairs are 17,296 and 18,416; 1,184 and 1,210. The last pair was discovered by a 16-year old Italian boy in 1866 but overlooked by some of the greatest mathematicians such as Pierre de Fermat (ca. 1600-1665). At present there are 391 known pairs of amicable numbers. Among the properties of amicable numbers are the following:

> For each of the known amicable pairs, either *both* numbers are even or *both* are odd.

Each of the known amicable pairs has a common factor, and among these common factors is 2 *or* 3.

FIGURATE NUMBERS

For a discussion of these numbers see page 99.

5.12 Some Unsolved Problems in Number Theory

THE GOLDBACH CONJECTURE

Number theory contains some very interesting problems that remain unsolved. One of the best known of these problems is the *Goldbach Conjecture.* A conjecture in mathematics is a statement which one has reason to believe is true but for which no proof is known.

Christian Goldbach, an 18th century mathematician, conjectured that:

1 Every even number greater than 4 may be expressed as the sum of two odd primes; e.g., $6 = 3 + 3$, $8 = 3 + 5$.

2 Every odd number greater than 7 may be expressed as the sum of three odd primes; e.g., $9 = 3 + 3 + 3$, $21 = 3 + 5 + 13$.

Although a computer has shown that Goldbach's first conjecture is true for every even number up to 2,000,000, no mathematician has yet been able to either *prove* or *disprove* this conjecture. The difficulty seems to be that we are trying to break up a natural number into a sum of primes while the properties of primes are derived not from addition but from multiplication.

We can see that if conjecture *1* could be proved then conjecture 2 would also be true because

a) Assume N is an odd number and p is an odd prime smaller than N.

b) Then $N - p$ is even.

c) Then $N - p$ could be written, by conjecture *1*, as the sum of two odd primes, q and r; that is, $N - p = q + r$.

d) Therefore, $N = p + q + r$. This equation expresses the odd number N as the sum of three odd primes p, q, and r.

However conjecture *1* has yet to be proved.

FERMAT'S LAST THEOREM

Some squares can be written as the sum of two other squares. For example, $5^2 = 3^2 + 4^2$ and $13^2 = 5^2 + 12^2$. Any three natural numbers a, b, c that have the relationship

$$c^2 = a^2 + b^2$$

are called *Pythagorean triples.* Such triples are used in problems involving the Pythagorean Theorem.

The question arises whether it is possible to express a *cube* as the sum of two other cubes. That is, are there three natural numbers

a, b, c such that $c^3 = a^3 + b^3$? More generally, is there a natural number solution to the equation $c^n = a^n + b^n$ if n is any natural number greater than 2?

This problem occurred to the great number theorist Pierre de Fermat who wrote this note in the margin of the book he was using: "It is impossible to write a cube as the sum of two cubes, a fourth power as the sum of two fourth powers, and, in general, any power beyond the second as the sum of two similar powers. For this I have discovered a truly wonderful proof but the margin is too small to contain it."

This is Fermat's famous last theorem which most mathematicians believe to be true but which no one has been able either to prove or disprove. The theorem has been shown to be true for values of $n < 4003$. But what needs to be proved is that the theorem holds (or does not hold) for *every* n greater than 2. This is probably the most famous conjecture in all number theory.

TWIN PRIMES

Twin primes are primes that differ by 2. For example, (3, 5), (5, 7), (17, 19), (29, 31), (41, 43). The question is: How many twin primes are there? Are there infinitely many twin primes? This is one of the oldest problems in mathematics.

Twin primes are found everywhere in the table of primes, but they become scarcer the farther we move out in the table. The conjecture is that there are infinitely many twin primes, and this conjecture agrees with actual counts made by electronic computers. But since no one has yet *proved* this conjecture, the problem is still unsolved.

5.13 Note on the Postage Problem

In the introduction to this chapter, you were asked to state the amounts of postage you can make with 5¢ and 8¢ stamps if you had an unlimited supply of each of these stamps. Note that the numbers 5 and 8 are relatively prime.

This problem is one example of a more general problem that can be stated this way: If there is an unlimited supply of x¢ and y¢ stamps available, and if x and y are relatively prime, then what amounts of postage can be obtained? We shall state but not prove several theorems that apply to this problem:

1 You can make any postage *larger* than $xy - x - y$. In the case of 5¢ (x) and 8¢ (y) stamps,

$xy - x - y = 5 \times 8 - 5 - 8 = 40 - 13 = 27.$

That is, any postage larger than 27¢ can be made.

2 $xy - x - y$ itself cannot be made. That is, 27¢ postage cannot be made.

3 If a postage *less than* $xy - x - y$ can be made, it can be made in only one way. For example, 18¢ postage can be made in only one way: two 5¢ stamps and one 8¢ stamp.

4 Exactly half the postages from 1 through $xy - x - y - 1$ can be made. Thus, in the case of 5¢ and 8¢ stamps, we can make half the postages from 1¢ through 26¢, or 13 postages as follows: 5, 8, 10, 13, 15, 16, 18, 20, 21, 23, 24, 25, 26.

Chapter 6

ALGORITHMS FOR COMPUTATION WITH WHOLE NUMBERS

Objectives

At the completion of this chapter, you should be able to:

1. Use the *addition algorithm* to find the sum of two two-digit and three-digit numbers, with and without regrouping, and justify the procedure.
2. Demonstrate the steps in the addition algorithm with concrete materials.
3. Use the *subtraction algorithm* to find the difference of two-digit and three-digit whole numbers, with and without regrouping, and justify the procedure.
4. Demonstrate the steps in the subtraction algorithm with concrete materials.
5. Use the *multiplication algorithm* to find the product of two two-digit whole numbers and justify the procedure.

6 Use the *division algorithm* to find the quotient of two whole numbers and justify the procedure.

Prerequisites

1 You should have knowledge of the commutative, associative, and distributive properties of whole numbers.

2 You should understand the concept of place value in our decimal numeration system.

INTRODUCTION

As children in the elementary grades, we were shown methods for computing with numbers containing more than a single digit. We needed no such methods for computing with single digit numbers since to add, say, 2 and 3, we could form a set containing 2 objects and another set containing 3 objects; then, after joining the two sets, we could count the number of objects in the new set and come up with 5. Or, we could add 2 and 3 by counting on the number line, and again come up with 5. We can follow these procedures easily enough with any single digit numbers. But when we have to add large numbers, these methods are not only tedious but may also be impossible. What we need, therefore, is a convenient and efficient method for computing with large numbers when the result is not obvious. Such a method is called an *algorithm.*

The word *algorithm* is derived from the name of the Persian mathematician al-Khowârizmî (ca. 825) which, when translated, became *Algoritmi* and which, in turn, became algorithm.

The algorithms we use for addition, subtraction, multiplication, and division in everyday arithmetic are reasonably easy to learn, but the reason why these methods work are less obvious and certainly not understood by many people. In this chapter we shall analyze each of these algorithms and show why they produce the correct answers.

Generally, algorithms are based on the properties of our place-value system of numeration and on the properties of whole numbers, and assume that we know the basic addition and multiplication facts involving any pair of numbers in the set $\{0, 1, 2, 3, 4, 5, 6, 7, 8, 9\}$. The algorithms used in our decimal system can also be used in other place-value systems.

6.1 Addition Algorithm

The algorithm we usually use to add, say, 35 and 52 is

$$\begin{array}{r} 35 \\ +52 \\ \hline 87. \end{array}$$

In explaining why this algorithm produces the correct answer, we shall assume that we already know:

1. The sum of any pair of numbers in the set {0, 1, 2, 3, 4, 5, 6, 7, 8, 9}. We shall call such a sum a *basic addition fact*.
2. The properties of the set of whole numbers; i.e. the commutative, associative, and distributive properties.
3. The properties of our numeration system; for instance, we shall assume we know that

   $392 = (3 \times 100) + (9 \times 10) + (2 \times 1);$

   that

   $(7 \times 100) + (2 \times 10) + 5 = 725;$

   and that

   $12 = 10 + 2.$

To justify the algorithm we just used to add 35 and 52, we shall first analyze in detail what actually happens when we add these two numbers, and then show how we can obtain the same result in a shorter and simpler way. This shorter and simpler way turns out to be the algorithm. We shall demonstrate all this in three progressive stages—from expanded notation to partial sums and, finally, to the algorithm—and show how each stage is a simplification of the previous stage:

Expanded Notation

$$\begin{array}{llll}
1 & 35 = & (3 \times 10) + (5 \times 1) & \text{(Expanded notation)} \\
 & +52 = & \underline{(5 \times 10) + (2 \times 1)} & \\
2 & = & [(3+5) \times 10] + [(5+2) \times 1] & \text{(Distributive property)} \\
3 & = & (8 \times 10) + (7 \times 1) & \text{(Basic addition facts)} \\
4 & = & 87 & \text{(Place-value notation).}
\end{array}$$

Partial Sums

$$\begin{array}{rl} 35 & \\ +52 & \\ \hline 7 & (2 \text{ ones} + 5 \text{ ones}) \\ 80 & (5 \text{ tens} + 3 \text{ tens}) \\ \hline 87 & \end{array}$$

Algorithm

$$\begin{array}{r} 35 \\ +52 \\ \hline 87 \end{array}$$

Note that the *partial sums* example is a simplification of the *expanded notation*, and that the *algorithm* is a simplification of *partial sums*.

Now let us add 67 and 25, where there is a need to "carry," and see why the standard algorithm works. We shall again show the three stages of progression and how one stage is a simplification of the previous stage.

Expanded Notation

1	$67 = (6 \times 10) + (7 \times 1)$	(Expanded notation)
	$\underline{+25} = \underline{(2 \times 10) + (5 \times 1)}$	
2	$= [(6+2) \times 10] + [(7+5) \times 1]$	(Distributive property)
3	$= (8 \times 10) + (12 \times 1)$	(Basic addition facts)
4	$= (8 \times 10) + (10 + 2) \times 1$	(Decimal notation)
5	$= (8 \times 10) + (10 \times 1) + (2 \times 1)$	(Distributive property)
6	$= (8 \times 10) + (1 \times 10) + (2 \times 1)$	(Commutative property)
7	$= [(8+1) \times 10] + [2 \times 1]$	(Distributive property)
8	$= (9 \times 10) + (2 \times 1)$	(Basic addition facts)
9	$= 92$	(Place-value notation).

Partial Sums		*Algorithm*
67		67
+25		+25
12	(5 ones + 7 ones)	92
80	(2 tens + 6 tens)	
92		

Note that the partial sums method is a contraction of steps *2* through *9* in the expanded notation, while the algorithm is a contraction of partial sums. The algorithm is, of course, the shortest and easiest way to add the two numbers.

This justification of the addition algorithm for finding the sum of two two-digit addends can be extended to the sum of any number of addends with any number of digits.

Comments

1 *When we write* $35 = (3 \times 10) + (5 \times 1)$ we are not multiplying 10 *by* 3 *or* 1 *by* 5, *but merely saying that in our place-value system* 35 means 3 *tens* + 5 *ones; i.e.* $(3 \times 10) + (5 \times 1)$.

2 *The term* carry *in addition* is *sometimes replaced with* rename *or* regroup *which may better describe what is actually happening. When we write* 12×1 as $(1 \times 10) + (2 \times 1)$, *we are* renaming *the* 12 *ones; or, we are* regrouping 10 *of the* 12 *ones into a group of* 1 *ten and then adding the regrouped ten with the other tens in the problem.*

3 *We should note that in our justification of the addition algorithm in the expanded notation stage, we indicated only the principal steps in the proof. We have omitted some of the other steps for the sake of brevity and so that the reader does not lose sight of the forest for the trees. For instance, a more complete proof of the algorithm for* 35 + 52 *would have included the following statement between steps 1 and 2 in expanded notation:*

$= [(3 \times 10) + (5 \times 10)] + [(5 \times 1) + (2 \times 1)]$

(Commutative and associative properties for addition of whole numbers).

6.2 Subtraction Algorithm

We have defined subtraction in terms of addition (page 81). For example, $5 - 3 = 2$ because we already know from the addition table that $5 = 3 + 2$. Therefore, in explaining the subtraction algorithm, we shall assume that we already know the basic subtraction facts that correspond to the basic addition facts; for instance, that $13 - 5 = 8$ since $5 + 8 = 13$.

The algorithm for subtracting 13 from 48 is

$$\begin{array}{r} 48 \\ -13 \\ \hline 35. \end{array}$$

We shall now justify this subtraction algorithm by showing that it is a convenient simplification, first of expanded notation and, then, of partial differences:

Expanded Notation

1	$48 = (4 \times 10) + (8 \times 1)$	(Expanded notation)
	$-13 = (1 \times 10) + (3 \times 1)$	
2	$= [(4 - 1) \times 10] + [(8 - 3) \times 1]$	(Distributive property)
3	$= (3 \times 10) + (5 \times 1)$	(Basic subtraction facts)
4	$= 35$	(Place-value notation).

Partial Differences

$$\begin{array}{rl} 48 & \\ -13 & \\ \hline 5 & \text{(8 ones - 3 ones)} \\ 30 & \text{(4 tens - 1 ten)} \\ \hline 35 & \end{array}$$

Algorithm

$$\begin{array}{r} 48 \\ -13 \\ \hline 35 \end{array}$$

Let us now consider the subtraction $62 - 25$, where there is a need to "borrow," or to "regroup." Here, the usual algorithm is

$$\begin{array}{r} 62 \\ -25 \\ \hline 37. \end{array}$$

The justification for this subtraction algorithm can be seen from the following progression:

Expanded Notation

1	$62 = (6 \times 10) + (2 \times 1)$	$= (5 \times 10) + (12 \times 1)$
	-25	$= (2 \times 10) + (5 \times 1)$
2		$= [(5 - 2) \times 10] + [(12 - 5) \times 1]$
3		$= (3 \times 10) + (7 \times 1)$
4		$= 37$

Below are the reasons for each of the above steps.

1 Expanded notation; regrouping; also other steps involving the commutative, associative, and distributive properties.

2 Distributive property for multiplication over subtraction.

3 Basic subtraction facts.

4 Place-value notation.

Partial Differences		*Algorithm*
62 = 5 [1]2		62
−25 = 2 5		−25
7	(12 ones − 5 ones)	37
3 0	(5 tens − 2 tens)	
3 7		

6.3 Multiplication Algorithm

In justifying the standard multiplication algorithm, we shall assume that we already know:

1 The product of any pair of numbers is the set {0, 1, 2, 3, 4, 5, 6, 7, 8, 9}, which we shall call a *basic multiplication fact.*

2 The properties of the set of whole numbers; in particular, the distributive property for multiplication over addition, i.e.,

$a(b + c) = ab + ac.$

Also, two applications of the distributive property which enable us to write

$(a + b)(c + d) = a(c + d) + b(c + d) = ac + ad + bc + bd.$

3 The properties of our numeration system; in particular, that if $2 \times 3 = 6$, then

$20 \times 3 = 60, \quad 200 \times 3 = 600, \quad 20 \times 30 = 600,$ etc.

(see Comment on page 144). We shall refer to these products as *basic multiplication facts.*

4 The addition algorithm.

We shall begin by showing why the usual multiplication algorithm works when we multiply a two-digit number by a one-digit number.

Expanded Notation

1	$9 \times 27 = 9 \times (20 + 7)$	(Renaming)
2	$= (9 \times 20) + (9 \times 7)$	(Distributive property)
3	$= 180 + 63$	(Basic multiplication facts)
4	$= 243$	(Addition algorithm).

Partial Products		*Algorithm*
27		27
× 9		× 9
63	(9 × 7)	243
180	(9 × 20)	
243		

Note that the partial products method is a simplification of steps *2*, *3*, and *4* in the expanded notation and that the algorithm combines the two partial products, 63 and 180, into one step by "carrying" the 6 mentally after multiplying the 7 by the 9.

We can justify the usual multiplication algorithm used to multiply two two-digit numbers, say, 68 × 37, in the following way:

Expanded Notation

1	37 × 68	= (30 + 7) (60 + 8)
2		= 30(60 + 8) + 7(60 + 8)
3		= (30 × 60) + (30 × 8) + (7 × 60) + (7 × 8)
4		= 1800 + 240 + 420 + 56
5		= 2516

The reasons for each of the above steps are as follows:

1 Renaming.

2 Distributive property.

3 Distributive property.

4 Basic multiplication facts.

5 Addition algorithm.

Partial Products		*Algorithm*
68		68
×37		×37
56	(7 × 8)	476
420	(7 × 60)	204
240	(30 × 8)	2516
1800	(30 × 60)	
2516		

Note that in the algorithm, the 476 is the sum 56 + 420 in the partial products, and that the 204, which really means 2040, is the sum 240 + 1800 in the partial products.

Comment

In assumption 3 on page 143, we say that if we know that 2 × 3 = 6 *then we also know that* 20 × 3 = 60, *etc. The justification for this statement is as follows:*

1	20 × 3 =	3 × 20	*(Commutative property)*
2	=	3(2 × 10)	*(Place-value notation)*
3	=	(3 × 2) × 10	*(Associative property)*
4	=	6 × 10	*(Basic multiplication fact)*
5	=	60	*(Place-value notation).*

We can use the same reasoning to prove that $200 \times 3 = 600$, $20 \times 30 = 600$, *etc.*

6.4 *Division Algorithm*

We can think of division not only as the inverse operation of multiplication but also as the repeated subtraction of the divisor from the dividend. For example, to divide 6 by 2 we repeatedly subtract 2 from 6 until the remainder is less than 2:

$$
\begin{array}{r}
3 \\
6 \div 2 = 2\,\overline{)\,6} \\
\underline{-2^{*}} \\
4 \\
\underline{-2^{*}} \\
2 \\
\underline{-2^{*}} \\
0
\end{array}
$$

We have thus subtracted 2 three times until we obtained the remainder 0. Therefore, $6 \div 2 = 3$ or $6 = (2 \times 3) + 0$.

The procedure of removing *one* 2 at a time from 6 is easy enough with small numbers like 6 and 2. If we wish to divide, say, 96 by 2, this method would not be very efficient since we would have to go through the same procedure 48 times before we arrived at a remainder less than 2. The efficiency can be increased, however, by removing *more than one* 2 at a time. For example, we could remove random *multiples* of 2 at a time: first, 7×2, then 13×2, 9×2, 15×2, 4×2—or, altogether, 48×2. We thus have $96 = (2 \times 48) + 0$. So instead of subtracting *single* 2's *forty-eight times*, we subtracted *multiples* of 2 only *five* times.

$$
\begin{array}{rl}
48 & \\
2\,\overline{)\,96} & \\
\underline{-14} & (7 \times 2) \\
82 & \\
\underline{-26} & (13 \times 2) \\
56 & \\
\underline{-18} & (9 \times 2) \\
38 & \\
\underline{-30} & (15 \times 2) \\
8 & \\
\underline{-\ 8} & \underline{(4 \times 2)} \\
0 & (48 \times 2)
\end{array}
$$

We can improve the efficiency of our method still further if, instead of subtracting *random* multiples of the divisor, we subtract *particular* multiples of the divisor—units, tens, hundreds, thousands, etc.—to correspond to the place values in our numeration system. Let us use this method of repeated subtraction to divide 6379 by 4, and then compare what we do in this method with what we do in the usual division algorithm:

Repeated Subtraction			*Algorithm*
			1594
4) 6379			4) 6379
−4000	(1000 × 4)	*Step 1*	4
2379			23
−2000	(500 × 4)	*Step 2*	20
379			37
− 360	(90 × 4)	*Step 3*	36
19			19
−16	(4 × 4)	*Step 4*	16
3	(1594 × 4)		3

That is, 6379 = (1594 × 4) + 3.

Note that in the Algorithm, the "1" in the quotient "1594" corresponds to Step *1* in the repeated subtraction; the "5" corresponds to Step *2*; the "9" to Step *3*; and the "4" to Step *4*. The algorithm is a simpler and more compact form of the repeated subtraction method.

Comment

The partial products obtained in the algorithm can be used to check our answer. When we add the partial products to the remainder, we should get the dividend. *Why*?

In the last example, the partial products are

4000 + 2000 + 360 + 16 + 3 (*remainder*) = 6379.

EXERCISE SET 6.1–6.4

1 Use the addition algorithm to find the following sums and justify the procedure.
a) 36 + 42 b) 27 + 98 c) 145 + 339 d) 267 + 558
e) 145 + 67 + 237

2 Use the subtraction algorithm to find the following differences and justify the procedure.
a) 96 − 31 b) 75 − 57 c) 546 − 259 d) 403 − 46

3 Use the multiplication algorithm to find the following products and justify the procedure.
a) 83 × 9 b) 62 × 75 c) 384 × 56

4 In finding the product of 573 and 65, why is the partial product 3438 moved over one place to the left?

5 Use the division algorithm to find the following quotients and justify the procedure.
a) 3) 59 b) 12) 654 c) 9) 32957

PUZZLES

1 What numbers do the letters represent?

a)
```
  C R O S S
+ R O A D S
-----------
D A N G E R
```

b)
```
      T A R
    × T A R
    -------
      B A R
    A R E
  T A R
  ---------
  T R B A R
```

BASICS REVISITED

1 In a poker game there are 54,912 ways of getting three of a kind, 123,552 ways of getting two pairs, and 1,098,240 ways of getting one pair. In how many ways can a player get three of a kind, two pairs, or one pair?

2 How many times can 14 be subtracted from 250?

3 Using the short method, divide 295 by (a) 10, (b) 100, (c) 1000.

4 What is a *number*?

5 What number does each of the following represent if b stands for 3?

a) $3b$ b) b^3 c) $\frac{1}{b^2}$ d) b^1 e) b^0

6 Round off 560,853,401 to the nearest million.

7 Using the short method, multiply .05 by 3000.

8 Write $\frac{3}{7}$ as a decimal, rounded off to the nearest hundredth.

9 Express in words: 3,562.002.

10 Find 8.3% of $25 to the nearest cent.

11 Use a short method to multiply 96 by $37\frac{1}{2}$.

12 Multiply $2\frac{3}{4}$ by $4\frac{5}{6}$.

13 Subtract 18.94 from 50.

14 Find the product of 1.006 and 10.4.

15 Estimate the product of 35 and 502.

16 Which is larger and by how much: 3.12 or .9785?

17 What part of 650 is 25?

18 If 15% of a number is 4.5, find the number?

19 What part of $8 is $2.75?

20 $2\frac{1}{2}$% of what number is 34?

21 Divide, if possible. If it is not possible to divide, state a reason why it is not possible.

a) $\frac{10}{2}$ b) $\frac{3}{4}$ c) $\frac{4-4}{7}$ d) $\frac{3}{0}$ e) $\frac{0}{0}$ f) $\frac{a}{b-b}$

22 **A woman earns $13,600 a year and receives an 8% raise in salary. If the cost of living during the year rose 6.9%, how much of an increase in purchasing power did she receive?**

23 **The price of a piece of luggage is $40. If a man receives a 10% discount and another 25% discount on the balance, how much does he pay for the luggage?**

24 **If 3% of a city's population is 82,500, find the total population of the city.**

25 **How much is 950 decreased by $\frac{1}{2}$% (to the nearest whole number)?**

EXTENDED STUDY

6.5 Multiplication Short Cuts

Short cuts in arithmetic are nonconventional algorithms that some people consider shorter than the standard algorithms used in the same operations. Such short cuts, when learned with understanding, can serve as a source of enrichment as well as fun. We shall consider a few short cuts for multiplication.

MULTIPLYING BY 11

1 Let us see what happens when we multiply 34 by 11:

$$\begin{array}{r} 3\ 4 \\ \times\ \underline{1\ 1} \\ 3\ 4 \\ \underline{3\ 4\ \ } \\ 3\ 7\ 4 \end{array}$$

Note that the answer is obtained by adding the two digits in the multiplicand (3 + 4 = 7) and then placing their sum, 7, between the two digits, 3 and 4. So, *to multiply a two-digit number by* 11, *first add the two digits and then place their sum in the middle.* Thus,

34 × 11 = 374, 32 × 11 = 352, 61 × 11 = 671,
45 × 11 = 495.

2 Let us now see what happens if the sum of the digits is 10 or more, as in 58 × 11:

$$\begin{array}{r} 5\ 8 \\ \times\ \underline{1\ 1} \\ 5\ 8 \\ \underline{{}^{1}5\ 8\ \ } \\ 6\ 3\ 8 \end{array}$$

Note that here we again follow the procedure in *1*. Since 5 + 8 = 13, we write the 3 and add the 1 we carried with the 5, getting 6. *So, if the sum of the digits is* 10 *or more, we do what we did previously, except that we add the* 1 *we are carrying to the first digit*. Thus,

$58 \times 11 = 638$, $67 \times 11 = 737$, $89 \times 11 = 979$, $78 \times 11 = 858$.

3 Let us now multiply a three-digit number by 11, say, 398×11:

```
    3 9 8
    X 1 1
  ¹3 9 8
¹3 9 8
  4 3 7 8
```

a) Note that the ones digit, 8, in the 398 becomes the ones digit in the answer.

b) We add the ones digit with the tens digit, (8 + 9 = 17), writing the 7 and carrying the 1 to the hundreds place.

c) We add the hundreds digit with the tens digit and the 1 we carried (3 + 9 + 1 = 13), writing the 3 and carrying the 1 to the thousands place.

d) We add the thousands digit with the 1 we carried (3 + 1 = 4) and write the 4.

So, to multiply a three-digit number by 11, *add the digits as before but carry from the tens place to the hundreds, and from the hundreds place to the thousands.*

Examples $398 \times 11 = 4378$, $572 \times 11 = 6292$, $785 \times 11 = 8635$.

SQUARING NUMBERS THAT END IN 5

To square 75,

1 Disregard the 5. Then take the 7 and multiply it by the next higher number, 8: $7 \times 8 = 56$.

2 Now place 25 *after* the 56: 5625. Thus, $(75)^2 = 5625$.

So, to square a number that ends in 5, *first disregard the* 5. *Then take the number that is left and multiply it by the next higher number. Then place* 25 *after the product.*

Examples $(65)^2 = 4225$, $(45)^2 = 2025$, $(95)^2 = 9025$, $(105)^2 = 11025$.

The explanation for this short cut is that a two-digit number ending in 5 may be represented by $10a + 5$, where a is the tens digit in th given number. In the number 65, $a = 6$. Now, squaring the number $10a + 5$, we get:

$$\begin{aligned}(10a + 5)^2 &= 100a^2 + 100a + 25 \\ &= 100a(a + 1) + 25.\end{aligned}$$

That is, the square of any two-digit number is equal to 100 times the product of the tens digit, a, and the next higher number, $a + 1$, plus 25. By placing the 25 *after* the product, $a(a + 1)$, we are, in effect, multiplying the product by 100. To square three-digit numbers ending in 5, we follow the same procedure.

MULTIPLYING ANY TWO NUMBERS THAT END IN 5

Squaring numbers that end in 5 is a special case of the more general problem of multiplying *any* two numbers that end in 5. To multiply two such numbers, apply the following rules:

1 If the sum of the numbers to the left of the 5s is *even*, then the answer ends in 25.

2 If the sum of the numbers to the left of the 5s is *odd*, then the answer ends in 75.

3 To find the digits in the answer in front of the 25 or 75,

a) multiply the numbers to the left of the 5s; then

b) add to this product one-half the sum of the numbers to the left of the 5s, disregarding any fraction left over.

Example 1 35×75

1 Since 3 + 7, or 10, is even, the answer ends in 25.

2 The digits in front of 25 in the answer are

$$(3 \times 7) + \frac{1}{2}(3 + 7) = 21 + 5 = 26.$$

3 Therefore, $35 \times 75 = 2625$.

Example 2 45×95

1 Since 4 + 9, or 13, is odd, the answer ends in 75.

2 The digits in front of 75 in the answer are

$$(4 \times 9) + \frac{1}{2}(4 + 9) = 36 + 6\frac{1}{2} = 42\frac{1}{2} \rightarrow 42.$$

3 Therefore, $45 \times 95 = 4275$.

Example 3 155×35

1 Since 15 + 3, or 18, is even, the answer ends in 25.

2 The digits in front of 25 in the answer are

$$(15 \times 3) + \frac{1}{2}(15 + 3) = 45 + 9 = 54.$$

3 Therefore, $155 \times 35 = 5425$.

To understand why this shortcut works, let us see what happens when we multiply 2 two-digit numbers ending in 5, which we shall represent by $10a + 5$ and $10b + 5$:

$$\begin{aligned}(10a + 5)(10b + 5) &= 100ab + 50(a + b) + 25 \\ &= 100ab + \frac{100(a + b)}{2} + 25 \\ &= 100\left(ab + \frac{a + b}{2}\right) + 25.\end{aligned}$$

How does the last equation explain why the shortcut works?

MULTIPLYING NUMBERS THAT DIFFER BY 2

There is a short way of multiplying numbers that differ by 2, if the number that lies between them is easy to multiply by itself. For instance, to multiply 39 by 41,

1 Note that the number that lies between them is 40.

2 Squaring 40, we get $(40)^2 = 1600$.

3 Subtracting 1 from the 1600, we get 1599.

4 So, $39 \times 41 = 1599$.

The reason that this short cut works is that $41 = 40 + 1$ and $39 = 40 - 1$. To multiply 39 by 41, we have

$$41 \times 39 = (40 + 1)(40 - 1) = (40)^2 - 1.$$

We can generalize this short cut by noting that

$$(x + a)(x - a) = x^2 - a^2.$$

If x is any number that we can square easily, then we can use this short cut to multiply $(x + a)$ by $(x - a)$. For example, to multiply 47 by 53, square 50 and subtract 9: $50^2 - 9 = 2491$, since

$$53 \times 47 = (50 + 3)(50 - 3) = (50)^2 - 3^2.$$

MULTIPLYING BY 9 OR BY 99

To multiply any number by 9, first multiply it by 10, then subtract the given number. For instance,

$$9 \times 325 = 10 \times 325 = \begin{array}{r} 3250 \\ -\ 325 \\ \hline 2935. \end{array}$$

Similarly, to multiply any number by 99, first multiply it by 100, then subtract the given number.

Example 4

$$99 \times 576 = 100 \times 576 = \begin{array}{r} 57600 \\ -\ 576 \\ \hline 57024. \end{array}$$

Chapter 7

THE INTEGERS

Objectives

At the completion of this chapter, you should be able to:

1. Define an integer.
2. Define the set of integers.
3. Find the *additive inverse*, or opposite, of a given integer.
4. Find the *absolute value* of an integer.
5. Define the sum, difference, product, and quotient of any two integers.
6. Find the sum, difference, product, and quotient (where possible) of any two integers.
7. Find the sum and difference of two integers by using the number line.

8 Define and justify the various properties of addition, subtraction, multiplication, and division of integers.

9 Recognize and apply the various properties of addition, subtraction, multiplication, and division of integers.

10 Solve equations of the form $a + x = b$ where a, b, x are integers.

11 Evaluate expressions such as:

$$[3 - 7(^{-}4 + 9)] - [(^{-}7 + ^{-}2) \div ^{-}3].$$

12 Define $a < b$ and $a > b$.

13 Prove basic theorems concerning integers, such as:
a) If $a + c = b + c$, then $a = b$.
b) If $a < b$ and $b < c$, then $a < c$.

Prerequisites You should have familiarity with the properties of the whole numbers.

INTRODUCTION

On page 73 we mentioned some of the limitations of whole numbers. We noted that whole numbers do not make a distinction between a number that represents, say, a *$10 profit* and a number that represents a *$10 loss.* Both are written as $10. Whole numbers do not indicate the difference between an *above 0* reading in temperature and a *below 0* reading, or between walking 3 blocks *east* and walking 3 blocks *west.* These were practical shortcomings that had to be eliminated.

The set of whole numbers was further limited in a more mathematical sense. We could not subtract a whole number from a smaller whole number and get a whole number as an answer; that is, whole numbers are not closed with respect to subtraction. Another way of stating this limitation is to say that equations like $x + a = b$ are not always solvable in the set of whole numbers. They are not solvable when $a > b$.

Also, when we deal with numbers that can be represented by a line extending in either of two opposite directions, we need numbers that indicate not only magnitude but also *direction.* Whole numbers indicate only magnitude.

All these limitations led inevitably to the extension of the system of whole numbers to a larger system that included new numbers called *integers.* In this chapter we shall describe how these new numbers were created, how the new system relates to the old, and how it differs from the old. Then we will develop the operations and their properties in the system of integers.

7.1 Creating the Integers

We create the system of integers in the following way:

Corresponding to each whole number, such as 7, we create two new symbols: $^+7$ and $^-7$. The symbols preceded by a plus sign represent *positive* numbers; the symbols preceded by a minus sign represent *negative* numbers.

The number *zero* is a member of this system and is the integer which follows $^-1$ and precedes $^+1$. Zero is neither positive nor negative.

For every positive number ^+n, there is a negative number ^-n such that their sum is zero. That is,

$$(^+n) + (^-n) = 0.$$

Similarly, for every negative number ^-a, there is a corresponding positive number ^+a with the property that

$$(^-a) + (^+a) = 0.$$

For example, to the positive number $^+2$, there corresponds the negative number $^-2$ with the property that

$$(^+2) + (^-2) = 0;$$

to the negative number $^-5$, there corresponds the positive number $^+5$ with the property that

$$(^-5) + (^+5) = 0.$$

The integers ^+n and ^-n are called *opposites*: the opposite of a positive integer is a negative integer; the opposite of a negative integer is a positive integer; the opposite of 0 is 0 since $0 + 0 = 0$.

The integers ^+n and ^-n are also called *additive inverses.* Each of these integers is the additive inverse of the other.

DEFINITIONS

1 The set $\{^+1, {}^+2, {}^+3, \ldots\}$ is called the set of *positive integers.*

2 The set $\{^-1, {}^-2, {}^-3, \ldots\}$ is called the set of *negative integers.*

3 The union of the positive integers, the negative integers, and zero is called the set of *integers:* $\{\ldots, {}^-2, {}^-1, 0, {}^+1, {}^+2, \ldots\}$.

4 The set containing zero and the positive integers is called the set of *nonnegative integers:* $\{0, {}^+1, {}^+2, {}^+3, \ldots\}$.

The whole numbers are represented on the number line by the point 0 and points to the right of 0. The integers are represented on the number line by 0 and by points to the right *and left* of 0 (Figure 7.1).

FIGURE 7.1

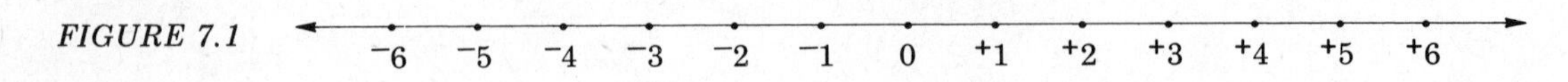

Comments

1 *In the set of integers, the positive numbers correspond to what we used to call the set of natural numbers. We shall, therefore, use* $^+2$ *and* 2, *or* $^+9$ *and* 9 *interchangeably. The nonnegative integers correspond to what we used to call the set of whole numbers.*

2 *The integers* ^+n *and* ^-n *are called each other's* negative. *Note, however, that* the negative of a number *is not necessarily a negative number*; the negative of $^-5$ *is* $^+5$.

3 *The negative of a negative is defined by the relation* $^-(^-a) = a$.

4 *The integers are sometimes called* signed numbers *and also* directed numbers.

7.2 Absolute Value

Let us look at $^+1$ and $^-1$ on the number line (Figure 7.2). We note that both integers are the same *distance* from 0 but lie in opposite *directions* from 0. This is also true of $^+2$ and $^-2$, $^+3$ and $^-3$, and, in fact, of every pair of additive inverses.

FIGURE 7.2

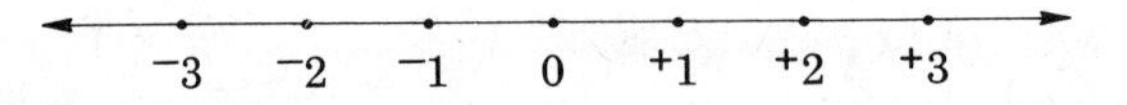

The distance of an integer x from 0, without regard to its direction, is called its *absolute value* and is denoted by the symbol $|x|$. For example, since the distance of $^+2$ from 0 is 2 units, we write $|^+2| = 2$, read "the absolute value of $^+2$ is 2." Since the distance of $^-2$ from the origin is also 2 units, we write $|^-2| = 2$. Similarly,

$$|19| = 19, \quad |^-7| = 7, \quad \text{and} \quad |0| = 0.$$

Thus we see that the absolute value of an integer is either zero or a positive integer. The absolute value of a positive integer is the number itself; the absolute value of a negative integer is its opposite; the absolute value of zero is zero. We put together all these ideas into:

DEFINITION

If x is any integer, then the *absolute value* of x, denoted by $|x|$, is defined as follows:

$$|x| = x \quad \text{if } x \geq 0,$$
$$|x| = {}^-x \quad \text{if } x < 0.$$

Comments

1 *We can define the* distance *between any two points a and b on the number line as the absolute value of the difference between the numbers associated with the points; that is,* $|a - b|$. (*Figure 7.3.*)

FIGURE 7.3

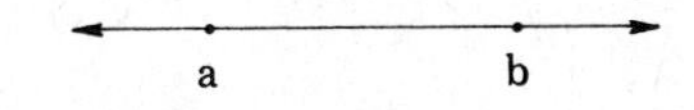

2 *The following are useful properties of absolute value:*

a) $|a - b| = |b - a|$

b) $|a \times b| = |a| \times |b|$

c) $|a + b| \leq |a| + |b|$.

Property c is true for the following reason. If a and b have the same sign, then

$$|a + b| = |a| + |b|. \quad (1)$$

If a and b have different signs, then

$$|a + b| < |a| + |b|. \quad (2)$$

Putting (1) *and* (2) *together, we get*

$$|a + b| \leq |a| + |b|.$$

7.3 Order Relation for the Integers

An order relation for integers can be defined in terms of the left-right order of points on the number line.

DEFINITION 1 If a and b are distinct integers, and the point that represents a is to the right of the point that represents b, then $b < a$, and $a > b$ (Figure 7.4).

FIGURE 7.4

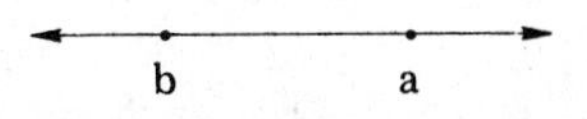

Example 1 $0 > {}^{-}3$ because the point representing 0 is to the right of the point representing ${}^{-}3$.

Comment *From this definition it follows that:*

1 *Every negative integer is less than* 0.

2 *Every positive integer is greater than* 0.

3 *Every positive integer is greater than every negative integer.*

Another definition of an order relation for integers can be given without reference to the number line.

DEFINITION 2 If a and b are any integers, then $a < b$ if and only if there exists a positive integer c such that $a + c = b$.

Example 2 $^{-}5 < {}^{+}3$ because there exists a positive integer $^{+}8$ such that $^{-}5 + {}^{+}8 = {}^{+}3$.

DEFINITION 3 If a and b are any integers, then a is greater than b if and only if b is less than a. That is, $a > b$ if and only if $b < a$.

Example 3 $^{-}2 > {}^{-}5$ since $^{-}5 < {}^{-}2$; and $0 > {}^{-}1$ since $^{-}1 < 0$.

PROPERTIES OF THE ORDER RELATION FOR INTEGERS

Trichotomy Property For any two integers a and b, precisely one of three possibilities holds: $a < b$, $a = b$, or $a > b$.

Transitivity Property If $a < b$ and $b < c$, then $a < c$.

Addition Property If $a < b$, then $a + c < b + c$.

Multiplication Property If $a < b$, then $ac < bc$ if c is positive, and $ac > bc$ if c is negative.

Comments

1 *The trichotomy, transitivity, and addition properties are the same as those in the system of whole numbers. The multiplication property, however, had to be modified to take into account the case where c is negative.*

2 *By the multiplication property for integers, if $a < b$, then $^{-}a > {}^{-}b$.*

EXERCISE SET 7.1–7.3

1 What are the additive inverses of the following integers?
a) $^{-}5$ b) 12 c) b d) 0 e) $^{-}(a + b)$ f) $x + y$

2 Show that there is a one-to-one correspondence between the elements of the set of positive integers and the elements of the set of negative integers.

3 What is the intersection of the set of positive integers and the set of negative integers?

4 Using the $<$ relation, order the following integers: 0, 4, $^{-}3$, 3, 15, $^{-}18$.

5 If x is an integer, what are the values of x given the following conditions?
a) $x < {}^{-}2$ b) $x \geq 0$ c) $x \leq {}^{+}3$ d) $x > {}^{+}10$

6 What must be true of a and b in the equation $x + a = b$:
a) for the solution to be a positive integer?
b) for the solution to be a negative integer?
c) for the solution to be zero?

7 What is the absolute value of each of the following?
a) $^{+}9$ b) $^{+}28$ c) $^{-}5$ d) $^{-}47$ e) 0 f) $^{+}129$

8 What is the value of each of the following?
a) $|^{+}25|$ b) $|^{-}4|$ c) $|^{-}121|$ d) $|0|$ e) $|^{+}36|$ f) $|^{-}2| + |^{-}5|$
g) $^{-}(|^{+}4| + |^{-}7|)$ h) $|^{-}3 + {}^{-}8|$ i) $|(^{-}2)(^{-}5)|$

9 Make each of the following a true statement by replacing the blank space with $<$, $=$, or $>$.
a) $^{+}3$ ___ $^{+}5$ b) $^{-}3$ ___ $^{+}3$ c) $^{-}15$ ___ $^{+}4$ d) $|^{+}3|$ ___ $|^{+}5|$
e) $|^{-}3|$ ___ $|^{+}3|$ f) $|^{-}15|$ ___ $|^{+}4|$ g) $|0|$ ___ $|^{-}1|$

10 If x is an integer, what are the values of x for each of the following?
a) $|x| = {}^{+}3$ b) $|x| = 0$ c) $|x| = {}^{+}21$ d) $x < {}^{-}2$
e) $x \geq {}^{-}5$ f) $x < 0$ g) $|x + 3| = 9$ h) $|x + 9| = 3$
i) $|x - 2| = 5$ j) $|x - 5| = 2$

11 Find the distance between each of the following pairs of points on the number line.
a) 3 and 5 b) $^{-}2$ and 1 c) 6 and 6 d) c and f e) $^{-}2$ and x

12 Is $|x - y| = |y - x|$? Why?

13 State whether each statement is true or false. Give a reason for your answer.
a) Every integer is a whole number.
b) Every counting number is an integer.
c) The opposite of every integer is a negative integer.
d) The absolute value of every integer is a whole number.
e) The set of whole numbers is a proper subset of the set of integers.
f) If $^{-}(^{-}a) + b = 0$, then b is the additive inverse of a.
g) The set of natural numbers is a subset of the set of integers.

14 Show that there is a one-to-one correspondence between the set of integers and the set of counting numbers.

7.4 Operations on Integers

Having created the integers, we must now decide what we mean by addition and multiplication. We will want to define these operations in such a way that all the properties of the whole numbers are preserved in the system of integers since the whole numbers are included in the integers.

Specifically, we will define addition and multiplication of integers in such a way as to insure the validity of the commutative, associative, and distributive properties. The only new basic property of the set of integers that does not hold for the set of whole numbers is the existence of an *additive inverse* for each integer. This property turns out to be of extreme importance.

ADDITION OF INTEGERS

We can think of the addition of integers as progressive moves along the number line. A positive integer describes a move to the right; a negative integer describes a move to the left. The starting point is always 0. So, $^{+}2 + ^{+}3$ means: starting at 0, move 2 units to the right followed by 3 more units to the right, landing on the point $^{+}5$ (Figure 7.5). Therefore, we say that $^{+}2 + ^{+}3 = ^{+}5$.

FIGURE 7.5

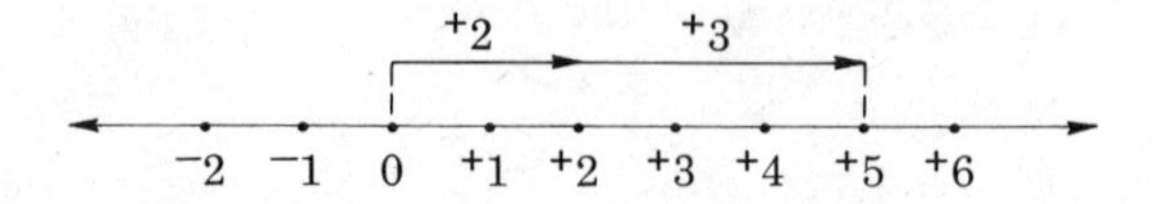

Similarly,

$$^{+}1 + ^{+}7 = ^{+}8 \quad \text{and} \quad ^{+}5 + ^{+}6 = ^{+}11.$$

These examples suggest that the sum of two positive integers can be found by adding their absolute values and that the result will always be a positive integer.

To add $^{-}3$ and $^{-}1$ on the number line, we start at 0, move 3 units to the left, and then move 1 more unit to the left, landing on the point $^{-}4$ (Figure 7.6). Thus, $^{-}3 + ^{-}1 = ^{-}4$.

FIGURE 7.6

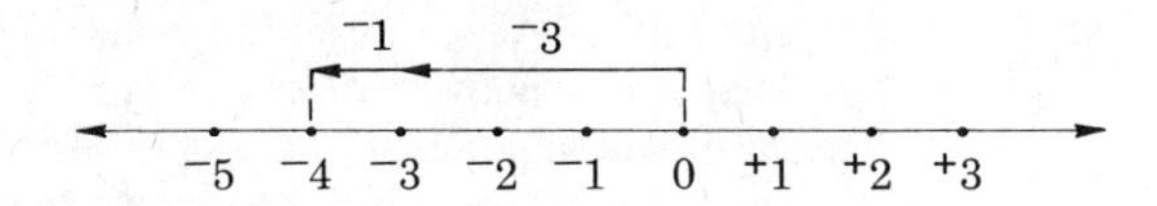

Similarly,

$$^{-}5 + ^{-}2 = ^{-}7, \quad \text{and} \quad ^{-}2 + ^{-}1 = ^{-}3.$$

These examples suggest that the sum of two negative integers can be found by adding their absolute values and that the result will always be a negative integer.

To add $^{+}5$ and $^{-}2$ on the number line, we start at 0, move 5 units to the right, and then 2 units to the left, landing on point $^{+}3$ (Figure 7.7). Thus, $^{+}5 + ^{-}2 = ^{+}3$.

FIGURE 7.7

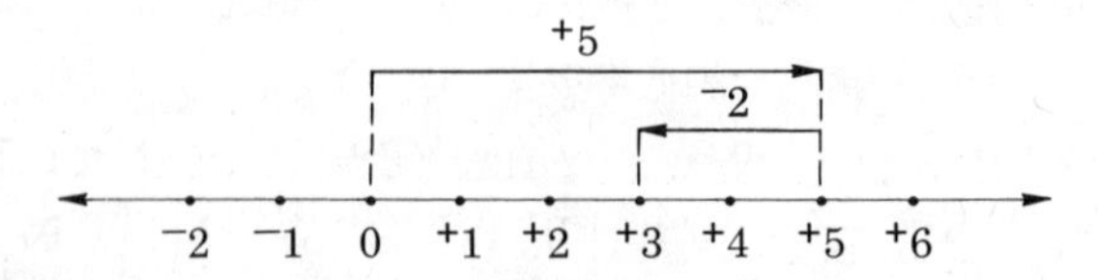

Similarly,

$$^{+}3 + ^{-}8 = ^{-}5, \quad ^{-}4 + ^{+}1 = ^{-}3, \quad \text{and} \quad ^{+}2 + ^{-}2 = 0.$$

These example suggest that the sum of a positive integer and a negative integer may be positive, negative, or zero depending on which of the addends has the greater absolute value or whether both addends have the same absolute value. To add a positive and a negative integer, subtract the smaller absolute value from the larger, and use the sign of the integer that has the larger absolute value.

Summarizing the above, we have

DEFINITION 1

a) $a + b = {}^{+}(|a| + |b|)$ if $a > 0$ and $b > 0$. For example, $^{+}3 + {}^{+}4 = {}^{+}(3 + 4) = {}^{+}7$.

b) $a + b = {}^{-}(|a| + |b|)$ if $a < 0$ and $b < 0$. For example, $^{-}2 + {}^{-}6 = {}^{-}(2 + 6) = {}^{-}8$.

c) $a + b = {}^{+}(|a| - |b|)$ if $a > 0$, $b < 0$, and $|a| > |b|$. For example, $^{+}5 + {}^{-}1 = {}^{+}(5 - 1) = {}^{+}4$.

d) $a + b = {}^{-}(|b| - |a|)$ if $a > 0$, $b < 0$, and $|b| > |a|$. For example, $^{+}2 + {}^{-}9 = {}^{-}(9 - 2) = {}^{-}7$.

e) $a + b = 0$ if $b = {}^{-}a$. For example, $^{+}3 + {}^{-}3 = 0$.

Comments

1 *Note that the + and − signs in the symbols* $^{+}5$ *and* $^{-}5$ *do not indicate the operation of addition or subtraction. They are part of the numeral used to distinguish one direction from another. To emphasize this distinction, the + and − have been written as a superscript when they are part of the numeral:* $^{-}7$, $^{+}12$.

2 *It is useful to first think of signed numbers through practical situations such as profit and loss, temperatures above and below zero, and gains and losses on a football field before examining their basic properties as numbers.*

3 *Note that, in contrast with other number systems, every integer is preceded by a particular integer and followed by a particular integer. Also, there is no smallest and no largest integer.*

4 *If a number line is drawn across a sheet of paper and the sheet is then folded through the origin, each point will fall on its opposite, or additive inverse; e.g.* $^{+}1$ *will fall on* $^{-}1$, $^{+}2$ *on* $^{-}2$, *and so on. Zero will fall on itself; i.e., zero is its own additive inverse.*

5 *Although negative numbers were first created by the Hindus many centuries ago, the concept of negative numbers was resisted by mathematicians until only a few hundred years ago. This was because numbers were associated with physical quantities rather than with abstract ideas. As a physical quantity, a negative number, "less than nothing," was too strange an idea to accept. But when people began to see negative numbers as abstract ideas created to make possible the solution of certain problems, such as solving* $2 - 5 = ?$ *negative numbers gained acceptance.*

PROPERTIES OF ADDITION OF INTEGERS For all integers a, b, and c.

1 *Closure* $a + b$ is a unique integer.

2 *Commutative* $a + b = b + a$

3 *Associative* $(a + b) + c = a + (b + c)$

4 *Additive Identity* There is a unique integer 0, called the *additive identity*, such that for all integers a,

$a + 0 = 0 + a = a.$

5 *Additive Inverse* Every integer a has a unique opposite ^{-}a, called the *additive inverse* of a, such that

$a + {}^{-}a = {}^{-}a + a = 0.$

The following is another important property of addition, which we shall prove.

THEOREM (*Cancellation Property*) If $a + c = b + c$, then $a = b$.

Proof

$$\begin{aligned} a + c &= b + c && \text{(Given)} \\ (a + c) + {}^{-}c &= (b + c) + {}^{-}c && \text{(Addition)} \\ a + (c + {}^{-}c) &= b + (c + {}^{-}c) && \text{(Associative)} \\ a + 0 &= b + 0 && \text{(Additive inverse)} \\ a &= b && \text{(Additive identity)} \end{aligned}$$

SUBTRACTION OF INTEGERS

In the system of whole numbers, we defined subtraction in terms of addition. We said that "7 - 3 = ?" asks the question "what whole number must be added to 3 to get 7?" In the system of integers, we define subtraction the same way: "$^{+}5 - {}^{+}2 = ?$" asks the question "what integer must be added to $^{+}2$ to get $^{+}5$?" We conclude that $^{+}5 - {}^{+}2 = {}^{+}3$ because $^{+}2 + {}^{+}3 = {}^{+}5$. Similarly, $^{+}3 - {}^{+}5 = {}^{-}2$ because $^{+}5 + {}^{-}2 = {}^{+}3$. More generally:

DEFINITION 2 If a and b are any two integers, then the *difference* $a - b$ is defined to be the integer c such that

$a = b + c.$

The operation of finding the difference is called *subtraction.*

On the number line, "$^{+}5 - {}^{+}2 = ?$" can be interpreted as asking the following question: "If we start at the point $^{+}2$, how far, and in which direction must we move in order to land on the point $^{+}5$?" (Figure 7.8). Obviously, we must move 3 units to the right; that is, $^{+}5 - {}^{+}2 = {}^{+}3$.

FIGURE 7.8

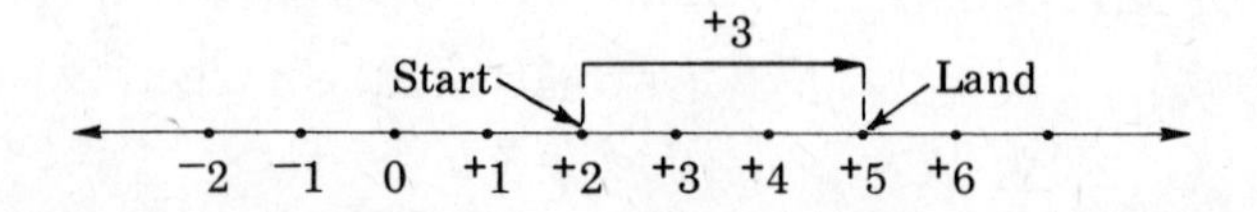

Similarly, $^{-}3 - {}^{+}5$ represents the distance and direction we must move on the number line if we start at $^{+}5$ and wish to land on $^{-}3$ (Figure 7.9). That is, $^{-}3 - {}^{+}5 = {}^{-}8$.

FIGURE 7.9

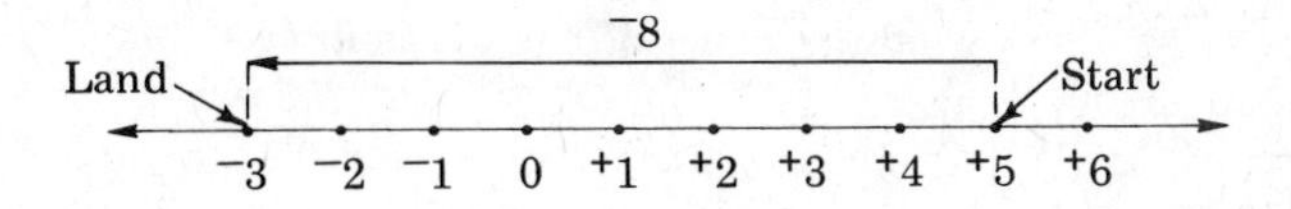

Similarly,

$$^{+}2 - {}^{-}3 = {}^{+}5,\quad {}^{-}3 - {}^{+}4 = {}^{-}7,\quad {}^{-}5 - {}^{-}2 = {}^{-}3,\quad \text{and}\quad {}^{-}2 - {}^{-}3 = {}^{+}1.$$

Since subtraction is the opposite of addition, *subtracting an integer is the same thing as adding its inverse.* For example, if we wish to subtract $^{+}2$ from $^{+}5$ we can add the inverse of $^{+}2$, which is $^{-}2$, to $^{+}5$. In other words,

$$^{+}5 - {}^{+}2 = {}^{+}5 + {}^{-}2 = {}^{+}3 \quad \text{and} \quad {}^{-}4 - {}^{-}1 = {}^{-}4 + {}^{+}1 = {}^{-}3.$$

We can now restate the definition of subtraction of integers.

DEFINITION 3 If a and b are integers, then $a - b = a + {}^{-}b$.

Comments

1 *Because every integer has an additive inverse, the set of integers is closed under subtraction. That is, subtraction is defined for every pair of integers. However, subtraction is neither commutative nor associative.*

2 *We can now say that if a and b are integers, then every equation of the form $a + x = b$ has a unique integer as its solution. To prove that the solution is* unique, *we first assume that the solution is* not *unique; i.e. we first assume that there are two solutions, say, x_1 and x_2. We then show that x_1 and x_2 are really the same solution:*

If x_1 and x_2 were two solutions, then

$$a + x_1 = b \quad \text{and} \quad a + x_2 = b,$$

making $a + x_1 = a + x_2$. By the cancellation property, $x_1 = x_2$.

EXERCISE SET 7.4a

1 Add each of the following.

a) $^{+}9 + {}^{+}2$ b) $^{+}5 + {}^{-}3$ c) $^{+}3 + {}^{-}19$
d) $^{-}5 + {}^{-}8$ e) $^{-}25 + {}^{+}14$ f) $^{+}2 + {}^{-}5 + {}^{+}6$
g) $^{-}14 + {}^{-}7 + {}^{+}12$ h) $^{-}54 + {}^{+}37 + {}^{-}15$

2 Subtract each of the following

a) $^{+}6 - {}^{+}2$ b) $^{+}4 - {}^{-}5$ c) $^{-}12 - {}^{-}9$

d) $^{-}37 - {}^{+}22$ e) $^{-}52 - {}^{+}67$

3 Perform the indicated operations.

a) $9 + 7 - {}^{-}2$ b) $^{-}5 + {}^{-}9 - {}^{-}2$ c) $35 - {}^{-}42 + {}^{-}67$

4 Find the values of x if x is an integer.

a) $x + 2 = 7$ b) $x + {}^{-}3 = 5$ c) $^{-}2 + x = {}^{-}7$

d) $7 - x = {}^{-}5$ e) $3 + x = 2$ f) $x + 7 = 0$

5 a) If m is a positive integer less than a, is $a + {}^{-}m$ positive or negative?

b) If $x < y$ and y is a negative number, is $x + {}^{-}y$ positive or negative?

6 What property of addition of integers is illustrated by each of the following?

a) $^{-}5 + ({}^{+}2 + {}^{-}3) = {}^{-}5 + ({}^{-}3 + {}^{+}2)$ b) $a + (0 + {}^{-}a) = a + {}^{-}a$

c) $a + ({}^{-}b + {}^{-}c) = (a + {}^{-}b) + {}^{-}c$

MULTIPLICATION OF INTEGERS

Let us recall that since the whole numbers are included in the integers we must define multiplication of integers in such a way as to preserve all the properties of whole numbers such as the commutative, associative, and distributive properties, and the multiplication property of zero. We shall now examine each of the four situations that we can encounter when we multiply two integers:

1 Since the set of positive integers is the same as the set of natural numbers, *the product of two positive integers will be the same as the product of two natural numbers.* So,

$({}^{+}2)({}^{+}3) = {}^{+}6$ and $({}^{+}a)({}^{+}b) = {}^{+}(ab)$.

2 The product of a positive integer and a negative integer, say, $({}^{+}2)({}^{-}3)$, may be seen as repeated addition of $^{-}3$; that is,

$({}^{+}2)({}^{-}3) = {}^{-}3 + {}^{-}3 = {}^{-}6$.

Since the repeated addition of a negative integer always results in a negative integer, *the product of a positive integer and a negative integer is a negative integer.*

3 The product of a negative integer and a positive integer, say, $({}^{-}2)({}^{+}5)$, will yield the same result as $({}^{+}5)({}^{-}2)$ since we wish to preserve the commutative property for multiplication. That is,

$({}^{-}2)({}^{+}5) = ({}^{+}5)({}^{-}2) = {}^{-}10$.

Therefore, *the product of a negative integer and a positive integer is a negative integer.*

4 Let us now examine the case of the product of two negative integers, say, $({}^{-}2)({}^{-}5)$. We shall show that $({}^{-}2)({}^{-}5)$ is the *additive inverse* of $({}^{-}2)({}^{+}5)$. Since we already know that

$(^-2)\ (^+5) = {}^-10$, its additive inverse, by definition, must be $^+10$. To show that $(^-2)\ (^-5)$ and $(^-2)\ (^+5)$ are additive inverses, we need only show that their sum is zero:

$$\begin{aligned} (^-2)\ (^+5) + (^-2)\ (^-5) &= {}^-2(^+5 + {}^-5) && \text{(Distributive prop.)} \\ &= {}^-2(0) && \text{(Additive inverse)} \\ &= 0 && \text{(Mult. prop. of zero)} \end{aligned}$$

More generally, we can prove that $(^-a)\ (^-b)$ is the additive inverse of $(^-a)\ (^+b)$. Since

$$(^-a)\ (^+b) = {}^-(ab),\ (^-a)\ (^-b) = {}^+(ab).$$

Therefore, *the product of two negative integers is a positive integer.* (See Comment 2 below.)

The above considerations lead to the following definition of multiplication of integers.

DEFINITION 4

a) $ab = {}^+(|a| \times |b|)$ if $a > 0$ and $b > 0$. For example, $^+5 \times {}^+3 = {}^+(5 \times 3) = {}^+15$.

b) $ab = {}^-(|a| \times |b|)$ if $a > 0$ and $b < 0$. For example, $^+2 \times {}^-7 = {}^-(2 \times 7) = {}^-14$.

c) $ab = {}^-(|a| \times |b|)$ if $a < 0$ and $b > 0$. For example, $^-4 \times {}^+6 = {}^-(4 \times 6) = {}^-24$.

d) $ab = {}^+(|a| \times |b|)$ if $a < 0$ and $b < 0$. For example, $^-2 \times {}^-4 = {}^+(2 \times 4) = {}^+8$.

Comments

1 *From the definition of multiplication of integers we conclude that (a) the product of two integers having the same sign is a positive integer, and (b) the product of two integers having opposite signs is a negative integer.*

2 *To prove that $(^-a)\ (^-b)$ and $(^-a)\ (^+b)$ are additive inverses, we show that their sum is zero:*

$$\begin{aligned} (^-a)\ (^+b) + (^-a)\ (^-b) &= {}^-a\ (^+b + {}^-b) && \textit{(Distrib. prop.)} \\ &= {}^-a\ (0) && \textit{(Additive inverse)} \\ &= 0 && \textit{(Mult. prop. of zero)} \end{aligned}$$

3 *To show that the product of a positive integer and a negative integer is a negative integer, we regard multiplication as repeated addition. We can give another justification for this product by use of the distributive property:*

$$\begin{aligned} {}^+b + {}^-b &= 0 && \textit{(Additive inverse)} \\ {}^+a(^+b + {}^-b)\ {}^+a(0) &= 0 && \textit{(Mult. prop. of zero)} \\ (^+a)\ (^+b) + (^+a)\ (^-b) &= 0 && \textit{(Distributive property)} \end{aligned}$$

Since the sum of $(^+a)\ (^+b)$ and $(^+a)\ (^-b)$ is zero, the two products are additive inverses; since $(^+a)\ (^+b) = {}^+(ab)$, then $(^+a)\ (^-b)$ must be equal to $^-(ab)$.

PROPERTIES OF MULTIPLICATION OF INTEGERS For all integers a, b, and c,

1. *Closure* ab is a unique integer.
2. *Commutative* $ab = ba$
3. *Associative* $(ab)c = a(bc)$
4. *Multiplicative Identity* There is a unique integer $^+1$, called the *multiplicative identity*, such that for all integers a,

   $a(^+1) = {}^+1(a) = a.$
5. *Distributive* $a(b + c) = ab + ac$
6. *Cancellation* If $ac = bc$, $c \neq 0$, then $a = b$.
7. *Multiplication Property of Zero* For every integer n,

   $n \times 0 = 0 \times n = 0.$

Comment *We can prove the cancellation property and the multiplication property of zero from the other properties. For instance, we can prove that* $n \times 0 = 0$ *as follows:*

$$\begin{aligned} 0 + 0 &= 0 && \textit{(Additive prop. of 0)} \\ n(0 + 0) &= n(0) && \textit{(Substitution)} \\ n(0) + n(0) &= n(0) && \textit{(Distributive prop.)} \\ n(0) + n(0) &= n(0) + 0 && \textit{(Additive prop. of 0)} \\ n(0) &= 0 && \textit{(Cancellation prop. of addition).} \end{aligned}$$

DIVISION OF INTEGERS

We define *division* of integers in terms of multiplication in the same way as in the system of whole numbers.

DEFINITION 5 Let a, b, and c be integers, $b \neq 0$. Then $a \div b = c$, if and only if $a = bc$.

Example $^+6 \div {}^-2 = {}^-3$ because $(^-2)(^-3) = {}^+6$.

Comments

1. *In the system of integers, as in the system of whole numbers, division is not always possible. For instance,* $^+3 \div {}^+2$ *does not exist since there is no integer* c *such that* $^+2 \times c = {}^+3$.
2. *From the definitions of division and multiplication of integers, we get the following rules of sign for the quotient. If (+) represents a positive integer and (−) represents a negative integer, and if the quotient exists, then*

   a) $(+) \div (+) = (+)$ *because* $(+) \times (+) = (+)$. *That is, the quotient of two positive integers is a positive integer. For example,* $^+6 \div {}^+2 = {}^+3$.

b) $(-) \div (-) = (+)$ *because* $(-) \times (+) = (-)$. *That is, the quotient of two negative integers is a positive integer. For example,* $^{-}10 \div {}^{-}5 = {}^{+}2$.

c) $(+) \div (-) = (-)$ *because* $(-) \times (-) = (+)$. *That is, the quotient of a positive integer and a negative integer is a negative integer. For example,* $^{+}8 \div {}^{-}4 = {}^{-}2$.

d) $(-) \div (+) = (-)$ *because* $(+) \times (-) = (-)$. *That is, the quotient of a negative integer and a positive integer is a negative integer. For example,* $^{-}12 \div {}^{+}3 = {}^{-}4$.

3 *In the set of integers, division by zero is undefined for the same reason that it is undefined in the set of whole numbers.*

4 *Division of integers is neither commutative nor associative.*

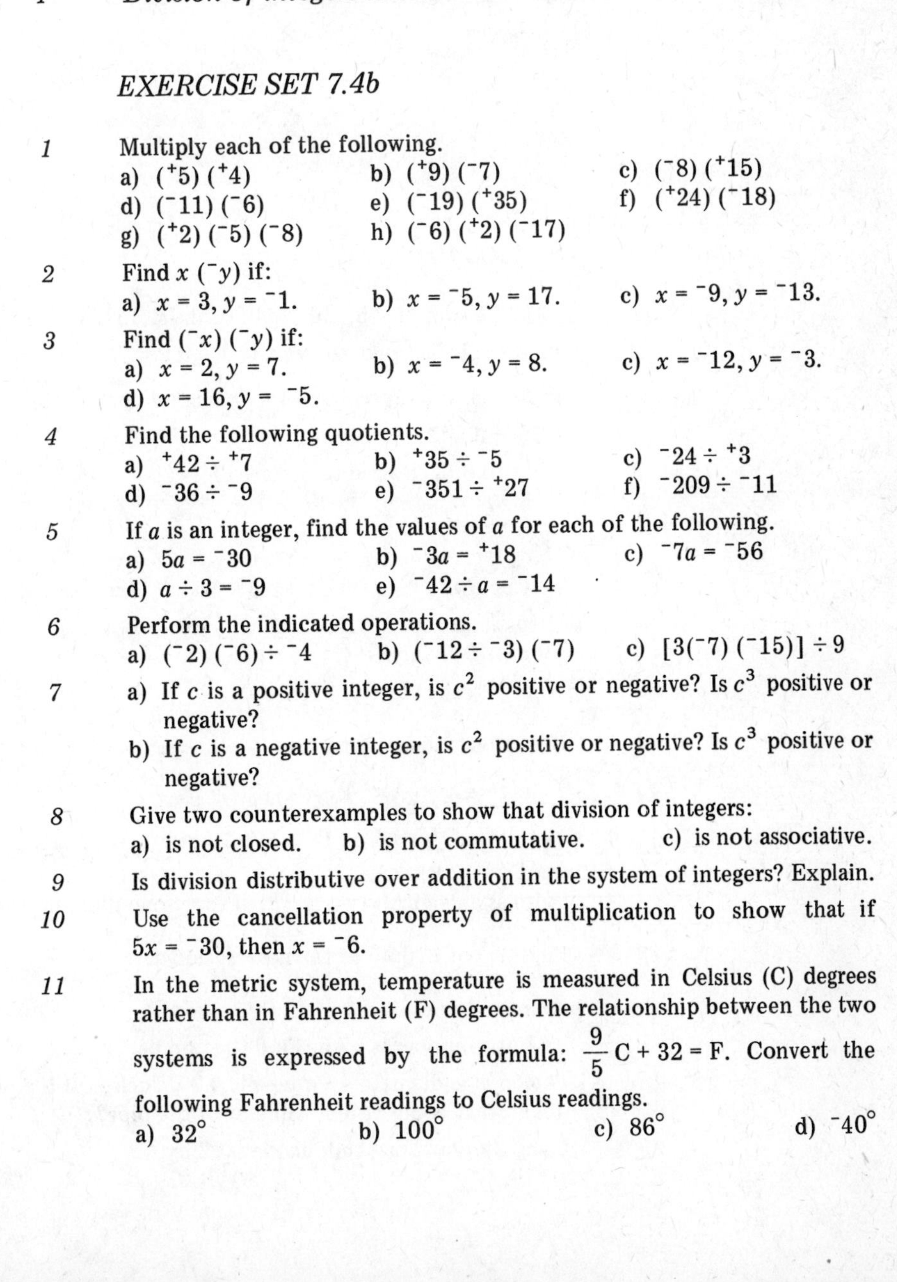

EXERCISE SET 7.4b

1 Multiply each of the following.
a) $({}^{+}5)({}^{+}4)$ b) $({}^{+}9)({}^{-}7)$ c) $({}^{-}8)({}^{+}15)$
d) $({}^{-}11)({}^{-}6)$ e) $({}^{-}19)({}^{+}35)$ f) $({}^{+}24)({}^{-}18)$
g) $({}^{+}2)({}^{-}5)({}^{-}8)$ h) $({}^{-}6)({}^{+}2)({}^{-}17)$

2 Find $x\,({}^{-}y)$ if:
a) $x = 3, y = {}^{-}1$. b) $x = {}^{-}5, y = 17$. c) $x = {}^{-}9, y = {}^{-}13$.

3 Find $({}^{-}x)({}^{-}y)$ if:
a) $x = 2, y = 7$. b) $x = {}^{-}4, y = 8$. c) $x = {}^{-}12, y = {}^{-}3$.
d) $x = 16, y = {}^{-}5$.

4 Find the following quotients.
a) ${}^{+}42 \div {}^{+}7$ b) ${}^{+}35 \div {}^{-}5$ c) ${}^{-}24 \div {}^{+}3$
d) ${}^{-}36 \div {}^{-}9$ e) ${}^{-}351 \div {}^{+}27$ f) ${}^{-}209 \div {}^{-}11$

5 If a is an integer, find the values of a for each of the following.
a) $5a = {}^{-}30$ b) ${}^{-}3a = {}^{+}18$ c) ${}^{-}7a = {}^{-}56$
d) $a \div 3 = {}^{-}9$ e) ${}^{-}42 \div a = {}^{-}14$

6 Perform the indicated operations.
a) $({}^{-}2)({}^{-}6) \div {}^{-}4$ b) $({}^{-}12 \div {}^{-}3)({}^{-}7)$ c) $[3({}^{-}7)({}^{-}15)] \div 9$

7 a) If c is a positive integer, is c^2 positive or negative? Is c^3 positive or negative?
b) If c is a negative integer, is c^2 positive or negative? Is c^3 positive or negative?

8 Give two counterexamples to show that division of integers:
a) is not closed. b) is not commutative. c) is not associative.

9 Is division distributive over addition in the system of integers? Explain.

10 Use the cancellation property of multiplication to show that if $5x = {}^{-}30$, then $x = {}^{-}6$.

11 In the metric system, temperature is measured in Celsius (C) degrees rather than in Fahrenheit (F) degrees. The relationship between the two systems is expressed by the formula: $\frac{9}{5}C + 32 = F$. Convert the following Fahrenheit readings to Celsius readings.
a) 32° b) 100° c) 86° d) ${}^{-}40^\circ$

12 Using the formula in Exercise 11, convert the following Celsius readings to Fahrenheit readings.

a) 3° b) 28° c) 70° d) $^{-}5°$

PUZZLES

1 The device below is called a *nomograph* and can be used to add and subtract integers. Can you figure out:

a) how the device can be used to add and subtract integers, and

b) why it works?

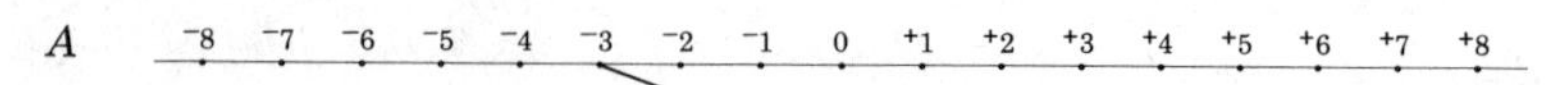

S $^{-}16$ $^{-}14$ $^{-}12$ $^{-}10$ $^{-}8$ $^{-}6$ $^{-}4$ $^{-}2$ 0 $^{+}2$ $^{+}4$ $^{+}6$ $^{+}8$ $^{+}10$ $^{+}12$ $^{+}14$ $^{+}16$

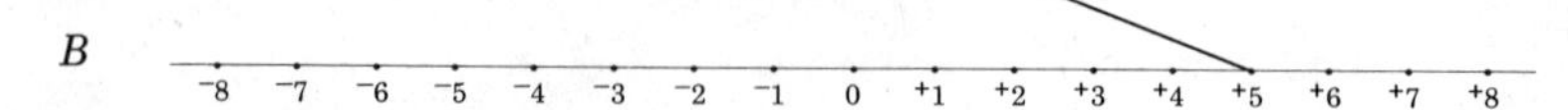

BASICS REVISITED

1 Find the sum of .45, .303, .0089, and 3.987.

2 Find the difference between 180 and 25.17.

In Exercises 3–6, perform the indicated operations.

3 $6.19 \times .0042$

4 $2.6\overline{)\,10.003}$ (Round to 3 places.)

5 $\frac{2}{3} \times \frac{6}{25} \times 12\frac{1}{2}$

6 $\left(2\frac{2}{3} \div 1\frac{1}{2}\right) \div \frac{3}{4}$

7 Five pounds of sugar cost $1.39. What is the cost per pound?

8 Carpeting costs $19.75 a square yard. How much will it cost to carpet a room 270 square feet?

9 What percent of 45 is 31.5?

10 Which is smaller and by how much: $\frac{7}{17}$ or $\frac{15}{34}$?

11 Find 72% of $20.55 to the nearest cent.

12 $14\frac{3}{16} - 1\frac{7}{8} = ?$

13 What part of 247 is 39?

14 Write as a decimal: two hundred forty-one thousandths.

15 Find $2\frac{5}{6}$ of $10.95 to the nearest dollar.

16 Estimate the quotient of $42.09 \div 5.806$.

17 25% of a number is 17.5. Find the number.

18 The outside diameter of a pipe is 3.295 inches. If the pipe is .078 inches thick, what is the inside diameter of the pipe?

19 Using the short method, divide 0.02 by

a) 10. b) 100. c) 1000.

20 A salesman earns $125 a week plus 14% commission on the amount of his sales. If his sales in one week are $795, what are his total earnings that week?

21 Verify on the number line that $15 \div 3 = 5$.

22 Show on the number line that $2 \times \frac{3}{8} = \frac{3}{4}$.

23 State whether *true or false*: The highest common factor of two numbers can be no greater than the smaller of the two numbers.

24 If 23 people make up 5% of an audience, how many people are in the audience?

25 A student needs an average of 92 or better on four tests and a final examination to receive an A in a course. The final examination counts $\frac{1}{3}$, and the four tests count $\frac{2}{3}$ of the grade. If a student's marks on the four tests are 83, 92, 90, and 99, what is the lowest mark he can receive on the final examination to qualify for an A in the course?

EXTENDED STUDY

7.5 Integers as Ordered Pairs

We used the concept of an additive inverse to construct the integers. Another approach to the construction of the integers is by the use of ordered pairs of whole numbers.

Let us start with the set of whole numbers

$$W = \{0, 1, 2, 3, \ldots\}$$

and then consider the Cartesian product $W \times W$. This Cartesian product is the set of all ordered pairs of whole numbers such as (2, 0), (1, 15), (3, 3), etc.

We now define an integer a to be an ordered pair of whole numbers (r, s) such that $a = r - s$. For example, we define the integers:

$$3 \text{ as } (4,1) \text{ or } (5,2) \text{ or } (6,3), \text{ etc.;}$$
$$^{-}1 \text{ as } (0,1) \text{ or } (3,4) \text{ or } (1,2), \text{ etc.;}$$
$$0 \text{ as } (1,1) \text{ or } (2,2) \text{ or } (3,3), \text{ etc.}$$

Two ordered pairs of whole numbers (r_1, s_1) and (r_2, s_2) represent the same integer if and only if

$$r_1 - s_1 = r_2 - s_2 \qquad [r_1 \geq s_1, r_2 \geq s_2];$$

that is, if and only if $r_1 + s_2 = r_2 + s_1$.

DEFINITION 1 Two ordered pairs of whole numbers (r_1, s_1) and (r_2, s_2) are *equivalent* (denoted by $\sim$) if and only if

$$r_1 + s_2 = r_2 + s_1.$$

Example 1 Thus, $(3, 1) \sim (4, 2)$ since $3 + 2 = 4 + 1$, and
$(2, 5) \sim (7, 10)$ since $2 + 10 = 5 + 7$.

We should note that the ordered pair (3, 1) is equivalent not only to the ordered pair (4, 2) but to the infinitely many ordered pairs in the set

$$A = \{(2, 0), (3, 1), (4, 2), (5, 3), (6, 4), ...\},$$

and the ordered pair (2, 5) is likewise equivalent to the infinitely many ordered pairs in the set

$$B = \{(0, 3), (1, 4), (2, 5), (3, 6), (4, 7), ...\}.$$

Set A is a subset of the Cartesian product $W \times W$. Because it consists of all ordered pairs that are *equivalent* to each other, set A is called an *equivalence class* of $W \times W$. There are infinitely many such equivalence classes, all of which are disjoint subsets of $W \times W$. Every ordered pair appears in one and only one of these equivalence classes, and each equivalence class represents a different *integer*. We can now define an *integer* as follows:

DEFINITION 2 An *integer* is an equivalence class of ordered pairs of whole numbers where

$$(r_1, s_1) \sim (r_2, s_2)$$

if and only if $r_1 + s_2 = r_2 + s_1$.

We name the integers as follows:

$$\begin{aligned} 0 &= \{(0, 0), (1, 1), (2, 2), ...\}, \\ 1 &= \{(1, 0), (2, 1), (3, 2), ...\}, \\ {}^-1 &= \{(0, 1), (1, 2), (2, 3), ...\}, \\ 2 &= \{(2, 0), (3, 1), (4, 2), ...\}, \\ {}^-2 &= \{(0, 2), (1, 3), (2, 4), ...\}, \quad \text{etc.} \end{aligned}$$

An integer can be named by any ordered pair which belongs to the equivalence class with which the integer is identified.

From this definition of an integer, we move to the definitions of addition and multiplication of integers.

DEFINITION 3 The *sum* of two integers (r_1, s_1) and (r_2, s_2) is defined as follows:

$$(r_1, s_1) + (r_2, s_2) = (r_1 + r_2, s_1 + s_2).$$

Example 2

a) $(3, 1) + (2, 5) = (3 + 2, 1 + 5) = (5, 6)$

b) $(0, 3) + (1, 5) = (0 + 1, 3 + 5) = (1, 8)$

This definition of addition of integers is consistent with Definition 2 on page 161. Under the earlier approach, Example 2 would be expressed this way:

a) $^{+}2 + {}^{-}3 = {}^{-}1$. The answer $^{-}1$ corresponds to the answer (5, 6).

b) $^{-}3 + {}^{-}4 = {}^{-}7$. The answer $^{-}7$ corresponds to the answer (1, 8).

DEFINITION 4 The *product* of two integers (r_1, s_1) and (r_2, s_2) is defined as follows:

$$(r_1, s_1) \times (r_2, s_2) = (r_1 r_2 + s_1 s_2, r_1 s_2 + s_1 r_2).$$

Example 3

a) $$\begin{aligned}(3, 1) \times (1, 6) &= (3 \times 1 + 1 \times 6, 3 \times 6 + 1 \times 1)\\ &= (3 + 6, 18 + 1)\\ &= (9, 19).\end{aligned}$$

b) $$\begin{aligned}(1, 4) \times (3, 5) &= (1 \times 3 + 4 \times 5, 1 \times 5 + 4 \times 3)\\ &= (3 + 20, 5 + 12)\\ &= (23, 17).\end{aligned}$$

Under the earlier approach, Example 3 would be expressed as follows:

a) $(^{+}2) \times (^{-}5) = {}^{-}10$. The answer $^{-}10$ corresponds to (9, 19).

b) $(^{-}3) \times (^{-}2) = {}^{+}6$. The answer $^{+}6$ corresponds to (23, 17).

Note how easily and naturally the product of two negative integers comes out to be a positive integer. It derives from the definition of multiplication.

We can easily prove that with these definitions of addition and multiplication, integers have the same properties that they have under our earlier approach to integers: addition and multiplication are closed, are commutative, and are associative; multiplication is distributive over addition; the set of integers contains an additive identity (0, 0), and a multiplicative identity (1, 0); and every integer (a, b) has an additive inverse (b, a). The integers (a, b) and (b, a) are additive inverses since their sum is the additive identity; i.e.

$$(a, b) + (b, a) = (a + b, b + a) = (0, 0).$$

7.6 Integers as Vectors

Another way to look at integers is by using a geometric figure called a *vector.* For our purpose, we can think of a vector as an arrow along the number line. Once we know the length of the arrow and the direction in which it points, the vector is fully determined.

To represent an integer by a vector, draw an arrow from the point 0 to the point representing the integer. Arrows pointing to the right represent positive integers; arrows pointing to the left represent negative integers. For example, $^{+}2$ may be represented by an arrow pointing to the right; this arrow is the same length as the line segment from 0 to $^{+}2$ (Figure 7.10). $^{+}2$ can also be represented by an arrow that begins at $^{+}3$ and ends at $^{+}5$ since this arrow, like the first one, points to the right and is 2 units long. In the same way, $^{-}3$ can be a vector extending from 0 to $^{-}3$ or from $^{-}2$ to $^{-}5$ (Figure 7.10). We draw the vectors above the number line for easy identification.

FIGURE 7.10

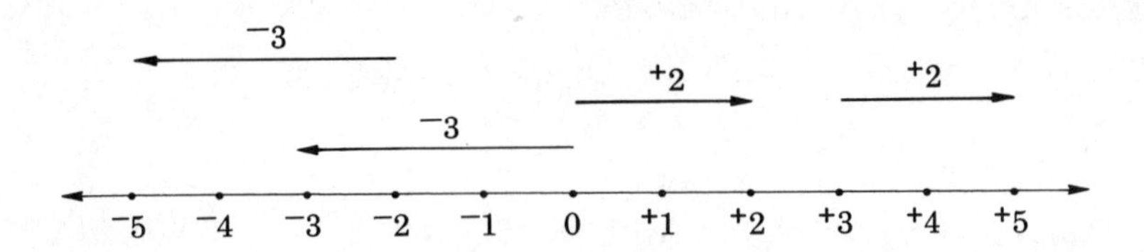

We shall denote the vector that extends from $^{+}3$ to $^{+}5$ by the symbol $[^{+}3, {}^{+}5]$. The points determining the vector are in brackets, not in parentheses, because $(^{+}3, {}^{+}5)$ represents an ordered pair of numbers, whereas $[^{+}3, {}^{+}5]$ represents a *directed segment*, or an arrow, that starts at the point $^{+}3$ and ends at the point $^{+}5$.

Two vectors are *equal* if their lengths and directions are the same. For example, $[^{-}2, {}^{-}5]$ and $[^{+}1, {}^{-}2]$ are equal vectors because each vector points to the left and is 3 units long as shown in Figure 7.11. They differ only in their position on the number line.

FIGURE 7.11

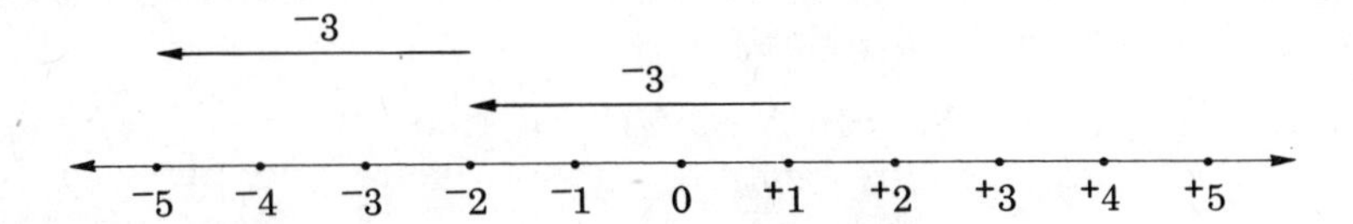

Vectors such as $[^{+}2, {}^{+}2]$, $[^{-}3, {}^{-}3]$, and $[0, 0]$, having neither magnitude nor direction, are called *null vectors.* Null vectors, however, have positions on the number line since they start and end at a specified point.

ADDITION OF INTEGERS THROUGH VECTORS

Vector addition may be performed on two vectors *if the terminal point of the first is the initial point of the second.* An example of two such vectors is $[^{-}1, {}^{+}3]$ and $[^{+}3, {}^{+}6]$ (Figure 7.12). In this instance, the terminal point of the first vector is $^{+}3$, and the initial point of the second vector also is $^+3$. We call vectors that are related to each other in this way *abutting vectors.*

Example 1 To add $[^-1, ^+3]$ and $[^+3, ^+6]$ we draw the vectors as shown in Figure 7.12. Their sum, indicated by the dashed arrow, is defined to be:

$$[^-1, ^+3] + [^+3, ^-6] = [^-1, ^+6].$$

Translated into integers, this statement becomes: $^+4 + ^+3 = ^+7$.

FIGURE 7.12

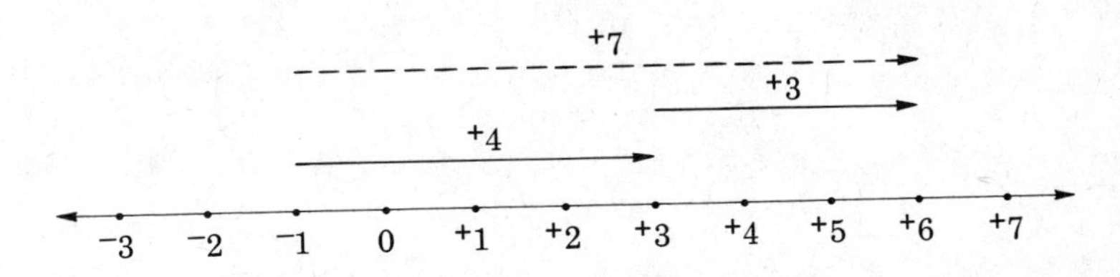

Example 2 The sum of $[^-2, ^+4]$ and $[^+4, ^-5]$ is shown by the dashed arrow in Figure 7.13,

$$[^-2, ^+4] + [^+4, ^-5] = [^-2, ^-5].$$

In integers, this statement becomes: $^+6 + ^-9 = ^-3$.

FIGURE 7.13

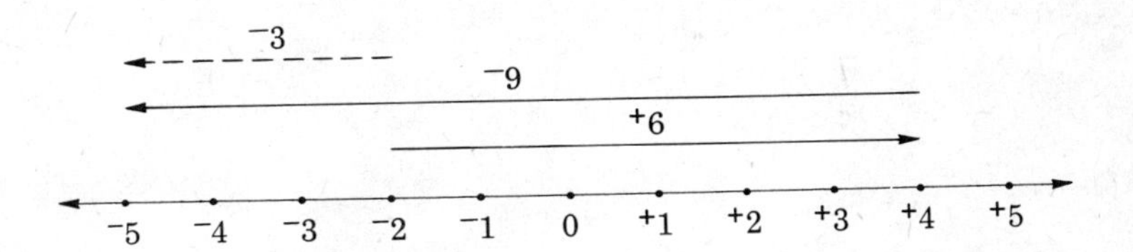

To find the sum of two integers by means of vectors, we first represent the two integers as *two abutting vectors*, and then proceed as above.

Example 3 To find the sum of $^-2$ and $^+5$, express $^-2$ as, say, $[0, ^-2]$, and express $^+5$ as the abutting vector $[^-2, ^+3]$ and then proceed as before (Figure 7.14). By adding the vectors, we see that $^-2 + ^+5 = ^+3$.

FIGURE 7.14

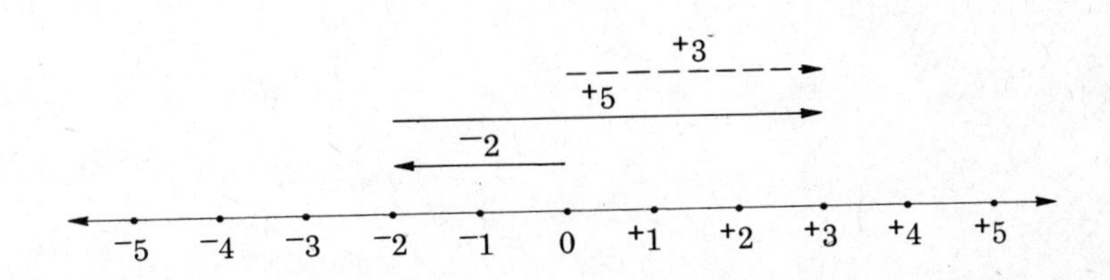

SUBTRACTION OF INTEGERS THROUGH VECTORS

If a, b, and c are any three integers, then

$$[a, b] + [b, c] = [a, c] \tag{1}$$

Since we define subtraction in terms of addition, addition statement (1) yields two subtraction facts:

$$[a, c] - [a, b] = [b, c], \tag{2}$$

and $$[a, c] - [b, c] = [a, b]. \tag{3}$$

Note that in (2) the two vectors that are subtracted have the same initial points, while in (3) the two vectors have the same terminal points. This suggests that for subtraction of any two vectors to be possible, the two vectors must have either the same initial points or the same terminal points.

Let us now use vectors to perform the subtraction $^{+}2 - {}^{-}3$. We begin by translating the $^{+}2$ and the $^{-}3$ into two vectors *with the same initial points*:

$$^{+}2 \rightarrow [0, {}^{+}2], \qquad {}^{-}3 \rightarrow [0, {}^{-}3].$$

We now draw these vectors as shown in Figure 7.15.

FIGURE 7.15

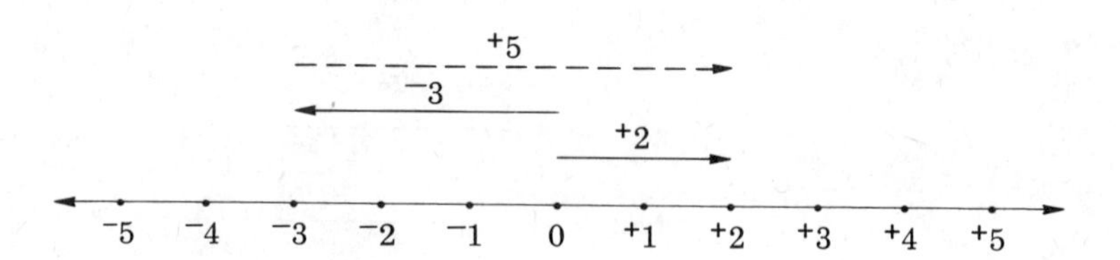

We recall that in the case with whole number, "5 - 2 = ?" means "what whole number must we add to 2 to get 5?" With vectors, "$[0, {}^{+}2] - [0, {}^{-}3] = ?$" means "what vector must we add to $[0, {}^{-}3]$ to get $[0, {}^{+}2]$?" From Figure 7.15 we see that the dashed arrow is the required vector. That is,

$$[0, {}^{+}2] - [0, {}^{-}3] = [{}^{-}3, {}^{+}2]$$

since $$[0, {}^{-}3] + [{}^{-}3, {}^{+}2] = [0, {}^{+}2].$$

Translating the vectors back to integers, we get $^{+}2 - {}^{-}3 = {}^{+}5$.

Chapter 8

THE RATIONALS

Objectives

Upon the completion of this chapter, you should be able to:

1. Define a *rational number*.
2. Define *equality* for rational numbers.
3. Define *equivalence* of fractions.
4. Define a rational number as an equivalence class of fractions.
5. Define and perform the operations of addition, subtraction, multiplication, and division on rational numbers.
6. State and identify the following properties of the rational numbers with respect to addition and multiplication: closure,

commutativity, associativity, identity, inverse; also, the distributive property for multiplication over addition.

7 Determine the order relation of any set of rational numbers.

8 Define and illustrate the *property of denseness* of the rational numbers.

9 Solve equations and inequalities over the rational numbers.

10 Define a rational number in terms of its decimal representation.

11 Express rational numbers as *repeating decimals.*

12 Express repeating decimals as rational numbers.

13 State two shortcomings of the integers that were eliminated by the rationals.

Prerequisites

You should know and understand:

1 The meaning of an equivalence class.

2 The properties of the integers.

3 The distinction between a *number* and a *numeral.*

INTRODUCTION

We created the integers in order to eliminate certain shortcomings in the system of whole numbers. We shall now create *rational numbers* in order to eliminate certain shortcomings in the system of integers.

There are two major defects in the system of integers. First, integers cannot be used to express *parts* of things such as half an inch or two-thirds of a pie. Second, division of integers is not always possible. For instance, there is no answer in the system of integers to a division problem like $5 \div 2$; that is, there does not exist a solution in the system of integers for some equations of the form $ax = b$, such as $5x = 2$. More formally, we say that the system of integers is not closed with respect to division.

Using the integers, we shall now create a new number system, called the *rational number system*, in which there will be a solution to every equation $ax = b$ $(a \neq 0)$; that is, division will always be possible, except division by 0. In the process of creating this new system, we shall make sure that all the properties of the integers are preserved.

8.1 Creating the Rationals

We shall use the integers as building blocks with which to create the rational numbers.

DEFINITION 1 A *rational number* is a number which can be expressed in the form $\frac{a}{b}$, where a and b are integers, $b \neq 0$.

Example 1 Examples of rational numbers are $\frac{3}{4}, \frac{^-2}{5}, \frac{^-4}{^-7}, \frac{0}{5}$.

DEFINITION 2 Two rational numbers $\frac{a}{b}$ and $\frac{c}{d}$ are *equal* if and only if $ad = bc$.

Example 2 $\frac{2}{3} = \frac{6}{9}$ because $2(9) = 3(6)$, and $\frac{^-2}{5} = \frac{6}{^-15}$ because $(^-2)(^-15) = 5(6)$.

The set of integers and the set of whole numbers are proper subsets of the set of rational numbers since any integer and any whole number can be expressed in the form $\frac{a}{b}$. For example, $^-2 = \frac{^-2}{1}$, $8 = \frac{8}{1}$, $0 = \frac{0}{1}$.

The set of rational numbers may be partitioned into three disjoint subsets: the set of *positive* rational numbers, the set of *negative* rational numbers, and the set containing 0. A rational number $\frac{a}{b}$ is *positive* if and only if ab is a positive integer, e.g., $\frac{1}{3}, \frac{^-2}{^-5}$; $\frac{a}{b}$ is *negative* if and only if ab is a negative integer, e.g., $\frac{^-1}{3}, \frac{2}{^-5}$; $\frac{a}{b}$ is *zero* if and only if $a = 0$, e.g., $\frac{0}{2}, \frac{0}{^-5}$.

Comments

1 *The expression* "$\frac{a}{b}$" *is a* numeral *that represents a rational number. A numeral that represents a rational number is called a* fraction *with* numerator *a and* denominator *b. The distinction between a* fraction *and a* rational number *is the same as the distinction between a* numeral *and a* number. *The word* fraction *is derived from the Latin word* fractio *(to break). A fraction was first thought of as a broken number.*

2 *A rational number may have several different meanings. For example,* $\frac{2}{3}$ *may be thought of as:*

a) *two-thirds of a complete unit. For instance, in Figure 8.1 the rectangular region has been divided into thirds. Two of these regions have been shaded to denote* $\frac{2}{3}$ *of the original rectangle.*

FIGURE 8.1

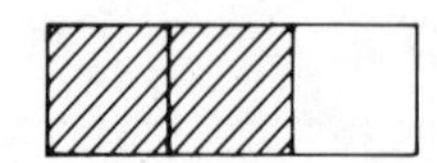

b) *a* ratio *of* 2 *to* 3. *This means that: given two sets of objects, for every* 2 *elements in one set, there are* 3 *elements in the other set. The sets may contain* 2 *and* 3 *elements respectively, or* 4 *and* 6 *elements, or* 10 *and* 15. *The essential idea is that if the first set contains* $2n$ *elements then the second set contains* $3n$ *elements.*

c) *the operation of* division; *that is,* $\frac{2}{3}$ *means* $2 \div 3$.

3 *The early Greeks were the first to use two integers to write fractions. Some Greek writers wrote the two numerals side by side, e.g.,* 2, 3; *some wrote the numerator above the denominator, e.g.,* $\frac{2}{3}$; *others wrote the denominator above the numerator, e.g.,* $\frac{3}{2}$.

4 *The Egyptians had an interesting way of representing rational numbers between* 0 *and* 1. *They represented all such rational numbers as a* sum of rational numbers with a numerator of 1. *For example,*

$$\frac{1}{2} = \frac{1}{4} + \frac{1}{4};\ \frac{1}{3} = \frac{1}{4} + \frac{1}{12};\ \frac{2}{5} = \frac{1}{3} + \frac{1}{15}.$$

About a hundred years ago the English mathematician J. J. Sylvester proved that it is possible to express any *rational number between* 0 *and* 1 *as a sum of rational numbers with a numerator of* 1.

8.2 Rational Numbers and the Number Line

The rational numbers can be represented as points on the number line (Figure 8.2). We associate just one point of the line with each rational number and call such a point a *rational point.* We shall see later that not all points on the number line represent rational numbers; in fact, many more points on the number line do not represent rational points.

FIGURE 8.2

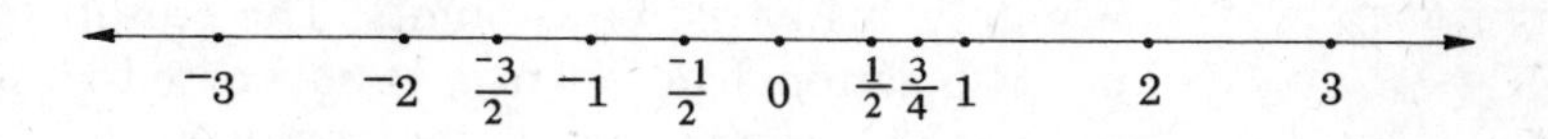

8.3 Rational Numbers as Equivalence Classes

In Figure 8.3 we note that every rational point on the number line has infinitely many fraction names. For example, the point $\frac{^-1}{2}$ is the same as the points $\frac{^-2}{4}, \frac{^-3}{6}, \ldots$. Fractions that name the same point on the number line are called *equivalent*. They are merely different names for the same rational number. Each set of equivalent fractions, such as

$$\left\{\frac{^-1}{2}, \frac{^-2}{4}, \frac{^-3}{6}, \ldots\right\} \quad \text{or} \quad \left\{\frac{0}{1}, \frac{0}{2}, \frac{0}{3}, \ldots\right\},$$

forms an *equivalence class*. There are infinitely many such equivalence classes of fractions, and each equivalence class represents a different rational number. We may, therefore, regard a rational number as a number represented by any one of the fractions in an equivalence class of fractions. Two rational numbers are *equal* if they are represented by fractions from the same equivalence class. We talk of *equal rational numbers*, but *equivalent fractions*.

$^-1$ $\frac{^-1}{2}$ $\frac{0}{1}$ $\frac{3}{4}$ 1

$\frac{^-2}{4}$ $\frac{0}{2}$ $\frac{6}{8}$

$\frac{^-3}{6}$ $\frac{0}{3}$ $\frac{9}{12}$

⋮   ⋮   ⋮

FIGURE 8.3

Comment

The fraction in an equivalence class whose numerator and denominator are relatively prime is said to be reduced to lowest terms. *So when we reduce a fraction to lowest terms, we are finding the* reduced representative *of the equivalence class of which the given fraction is a member.*

8.4 The Property of Denseness

Between any pair of distinct rational points a and b there are infinitely many rational points. The reason for this is that between any two rational numbers a and b, there is at least one other rational number, the average of the two numbers,

$$\frac{a+b}{2} \qquad \text{(Figure 8.4).}$$

There is at least one other rational number between a and $\frac{a+b}{2}$, the average of the two numbers,

$$\frac{a+\frac{a+b}{2}}{2} \qquad \text{or} \qquad \frac{3a+b}{4}.$$

Since we can repeat this process of taking averages indefinitely, the result is infinitely many rational numbers between any two rational numbers and infinitely many rational points on every segment of the number line no matter how small the segment. We describe this property of the rational number system by saying that it is *dense*. The property of *denseness* is not possessed by the whole numbers or by the integers.

FIGURE 8.4

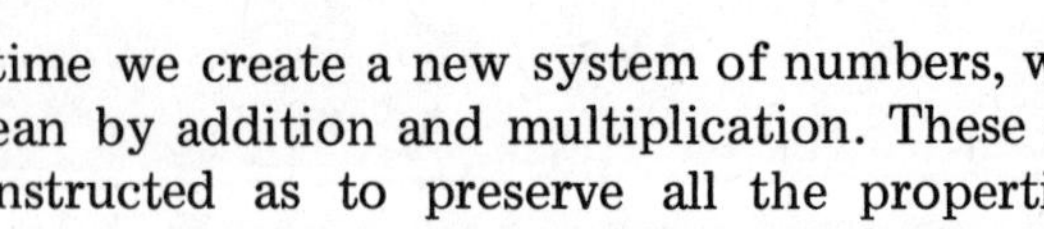

$\frac{3a+b}{4}$

$a \qquad \frac{a+b}{2} \qquad b$

8.5 Operations on the Rationals

Each time we create a new system of numbers, we must define what we mean by addition and multiplication. These definitions must be so constructed as to preserve all the properties of the previous systems since these are subsets of the new system. Specifically, we shall define addition and multiplication of rational numbers to preserve all the properties of the integers. Therefore, given integers a, b, c, d ($b \neq 0$, $d \neq 0$), we have the following:

DEFINITIONS

Addition of rational numbers $\frac{a}{b} + \frac{c}{d} = \frac{ad+bc}{bd}$. For example,

$$\frac{2}{3} + \frac{4}{7} = \frac{(2\cdot 7)+(3\cdot 4)}{3\cdot 7} = \frac{14+12}{21} = \frac{26}{21}.$$

Subtraction of rational numbers $\frac{a}{b} - \frac{c}{d} = \frac{ad-bc}{bd}$. For example,

$$\frac{2}{3} - \frac{4}{7} = \frac{(2\cdot 7)-(3\cdot 4)}{3\cdot 7} = \frac{14-12}{21} = \frac{2}{21}.$$

Multiplication of rational numbers $\frac{a}{b} \times \frac{c}{d} = \frac{ac}{bd}$. For example,

$$\frac{2}{3} \times \frac{4}{7} = \frac{2 \cdot 4}{3 \cdot 7} = \frac{8}{21}.$$

Division of rational numbers $\frac{a}{b} \div \frac{c}{d} = \frac{ad}{bc}$. For example,

$$\frac{2}{3} \div \frac{4}{7} = \frac{2 \cdot 7}{3 \cdot 4} = \frac{14}{12} = \frac{7}{6}.$$

Comments

1 *Using these definitions, it is possible to prove that the rational number system obeys all the properties of the integers. For example, to prove the commutative property for addition, we have, from the definition of addition,*

$$\begin{aligned} \frac{a}{b} + \frac{c}{d} &= \frac{ad + bc}{bd} \\ &= \frac{da + cb}{db} && \textit{(Commutativity of mult. of integers)} \\ &= \frac{cb + da}{db} && \textit{(Commutativity of add. of integers)} \\ &= \frac{c}{d} + \frac{a}{b}. && \textit{(Definition of add. for rational numbers)} \end{aligned}$$

The other properties can be proved in a similar way.

2 *The definition of division states that a division problem may be written as an equivalent multiplication problem; for example,*

$$\frac{2}{5} \div \frac{3}{4} = \frac{2}{5} \times \frac{4}{3}.$$

The reason we can do this is as follows: $\frac{2}{5} \div \frac{3}{4}$ *can be rewritten as* $\frac{\frac{2}{5}}{\frac{3}{4}}$. *We shall now change* $\frac{\frac{2}{5}}{\frac{3}{4}}$ *to an equivalent fraction whose denominator is* 1. *The denominator* $\frac{3}{4}$ *becomes* 1 *if we multiply it by its reciprocal* $\frac{4}{3}$. *To obtain an equivalent fraction, we must likewise multiply the numerator by* $\frac{4}{3}$. *Thus, we get*

$$\frac{2}{5} \div \frac{3}{4} = \frac{\frac{2}{5}}{\frac{3}{4}} = \frac{\frac{2}{5} \times \frac{4}{3}}{\frac{3}{4} \times \frac{4}{3}} = \frac{\frac{2}{5} \times \frac{4}{3}}{1} = \frac{2}{5} \times \frac{4}{3}.$$

More generally,

$$\frac{a}{b} \div \frac{c}{d} = \frac{\frac{a}{b}}{\frac{c}{d}} = \frac{\frac{a}{b} \times \frac{d}{c}}{\frac{c}{d} \times \frac{d}{c}} = \frac{\frac{a}{b} \times \frac{d}{c}}{1} = \frac{a}{b} \times \frac{d}{c} \text{ or } \frac{ad}{bc}.$$

This explains the rule we were given when we first learned division of fractions to "change ÷ to × and invert."

EXERCISE SET 8.1–8.5

1 Locate the following points on the number line.

a) $\frac{3}{4}$ b) $\frac{^{-}3}{5}$ c) $\frac{0}{10}$ d) $\frac{^{-}8}{3}$ e) $\frac{13}{4}$

2 Which of the following pairs of rational numbers are equal?

a) $\frac{3}{^{-}4}, \frac{^{-}4}{7}$ b) $\frac{^{-}3}{^{-}6}, \frac{4}{8}$ c) $\frac{5}{9}, \frac{^{-}15}{27}$ d) $\frac{0}{17}, \frac{0}{^{-}5}$

3 Add each of the following.

a) $\frac{^{-}3}{5} + \frac{^{-}4}{7}$ b) $\frac{^{-}1}{^{-}3} + \frac{5}{8}$ c) $\frac{^{-}5}{6} + \frac{^{-}2}{^{-}3}$

d) $\frac{^{-}9}{4} + \frac{^{-}2}{5}$ e) $\frac{^{-}c}{d} + \frac{m}{^{-}n}$ f) $\frac{3}{4} + \frac{^{-}5}{3} + \frac{^{-}3}{5}$

g) $\frac{2}{a} + \frac{3}{^{-}5a}$

4 Subtract each of the following.

a) $\frac{^{-}2}{5} - \frac{^{-}2}{3}$ b) $\frac{^{-}3}{8} - \frac{^{-}2}{7}$ c) $\frac{^{-}1}{4} - \frac{^{-}1}{4}$

d) $\frac{1}{6} - \frac{^{-}2}{^{-}9}$ e) $\frac{p}{^{-}q} - \frac{^{-}c}{t}$ f) $2\frac{3}{5} - 3\frac{1}{4}$

5 Multiply each of the following.

a) $\frac{^{-}4}{7} \times \frac{^{-}5}{^{-}3}$ b) $\frac{4}{^{-}5} \times \frac{3}{^{-}16}$ c) $\frac{7}{2} \times \frac{^{-}5}{14}$

d) $\frac{^{-}a}{t} \times \frac{^{-}b}{s}$ e) $\frac{a^2}{b} \times \frac{b^3}{a}$ f) $\frac{^{-}9}{^{-}2} \times \frac{5}{^{-}9} \times \frac{^{-}2}{3}$

6 Divide each of the following.

a) $\frac{^{-}3}{4} \div \frac{2}{3}$ b) $\frac{3}{2} \div \frac{^{-}5}{8}$ c) $\frac{8}{^{-}3} \div \frac{7}{^{-}2}$

d) $2\frac{1}{3} \div \frac{^{-}3}{5}$ e) $\frac{^{-}m}{^{-}a} \div \frac{^{-}r}{s}$

7 Simplify by completing all the indicated operations.

a) $\left(\frac{2}{3} - \frac{1}{3}\right) + \left(\frac{5}{3} + \frac{2}{3}\right)$ b) $\left(3\frac{1}{3} + 2\frac{1}{6}\right) - 2\frac{1}{12}$

c) $\dfrac{\frac{2}{3}}{\frac{5}{9}}$ d) $\dfrac{\frac{1}{2} + \frac{1}{3}}{\frac{3}{8}}$ e) $\dfrac{\frac{2}{3} + \frac{1}{6}}{\frac{1}{2} + \frac{3}{5}}$

f) $\dfrac{\frac{x}{y}}{\frac{a}{b} + \frac{c}{d}}$

8 Find the value of c which will make each of the following pairs of fractions equivalent.

a) $\frac{2}{3}, \frac{c}{9}$ b) $\frac{c}{^{-}16}, \frac{3}{8}$ c) $\frac{^{-}2}{c}, \frac{^{-}5}{15}$

d) $\frac{^{-}3}{16}, \frac{6}{^{-}c}$ e) $\frac{6}{^{-}3c}, \frac{^{-}8}{16}$

9 Find a rational number midway between:

a) $\frac{^{-}1}{2}$ and $\frac{3}{8}$. b) $\frac{2}{3}$ and $\frac{^{-}4}{5}$. c) $\frac{^{-}c}{d}$ and $\frac{^{-}m}{n}$.

10 Solve for x in the set of rational numbers.

a) $x + \frac{2}{3} = \frac{5}{6}$ b) $\frac{^{-}2}{5} - x = \frac{^{-}3}{4}$ c) $\left(\frac{^{-}2}{3}\right)x = \frac{5}{9}$

d) $\frac{^{-}x}{5} = \frac{^{-}1}{^{-}3}$ e) $\left(\frac{3}{5}\right)x - \frac{2}{3} = \frac{^{-}7}{8}$

11 If a is a negative rational number, then is a^2 positive or negative?

12 If W is the set of whole numbers, I the set of integers, and R the set of rational numbers, indicate which of the following are true.

a) $R \subseteq I$ b) $I \subset R$ c) $I \cap W \subseteq R$ d) $I \cap R = W \cup I$

13 Is it ever true that for $a > 0$, $b > 0$,

$$\frac{1}{a} + \frac{1}{b} = \frac{1}{a+b}?$$

Prove your answer.

14 If $ab = \frac{0}{c}$, what is true of a or b?

15 From the definition of subtraction (the inverse of addition), prove that

$$\frac{a}{b} - \frac{c}{d} = \frac{ad - bc}{bd}.$$

16 Show that the equivalence of fractions is an equivalence relation.

17 a) The negative of a positive rational number is a ____ number.
b) The negative of a negative rational number is a ____ number.
c) The negative of 0 is ____.
d) If the negative of a certain number is equal to the number, then the number is ____.
e) If the opposite of a certain number is a negative number, then the number is a ____ number.
f) On the number line, the rational numbers assigned to two points equally distant from the origin are called ____ of each other.

18 A man spends $\frac{1}{5}$ of his salary for rent, $\frac{1}{3}$ for food, $\frac{1}{6}$ for clothing, and $\frac{1}{4}$ for miscellaneous items, and saves the rest. How much does he save out of every dollar he earns?

19 a) If n represents a negative rational number and p represents a nonnegative rational number, then np represents a ____ rational number.
b) The reciprocal of the reciprocal of $\frac{^{-}1}{3}$ is ____.
c) The reciprocal of zero is ____.

20 Is the number represented by a fraction changed if you subtract the same number from both numerator and denominator? Why?

21 If the numerator of a fraction is greater than twice its denominator, show that the number represented by the fraction is greater than 2.

22 Show some process for dividing $\frac{2}{3}$ by $\frac{3}{5}$ other than by the invert-and-multiply rule.

23 Answer true or false for each statement.

a) Whole numbers are not rational numbers.

b) In adding rational numbers, if the denominators of the fractions are equal, we add the numerators.

c) The fractions $\frac{0}{a}$ and $\frac{0}{b}$ represent the same rational number if neither a nor b is zero.

d) The sum of two rational numbers whose fractions have equal numerators may be found by adding their denominators.

e) In the division problem $\frac{1}{2} \div \frac{1}{3}$, we are looking for a number which when multiplied by $\frac{1}{3}$ gives $\frac{1}{2}$.

f) $\frac{a}{c} + \frac{b}{c} = \frac{a+b}{2c}$.

g) There is no greatest number on the number line.

h) All negative rational numbers are associated with points on the number line to the right of zero.

i) The point $^{+}3$ on the number line and the point $^{-}3$ are equidistant from the point 0.

j) The rational number $^{-}4$ is greater than the rational number $^{-}3$.

k) The number zero is not a rational number.

l) A rational number may be expressed as an integer divided by a counting number.

m) Zero is a nonnegative number.

n) Zero is a nonpositive number.

8.6 Properties of the Rationals

The properties of the rationals correspond to the properties of the integers with one important addition. The new property is the *multiplicative inverse* which states that for every rational number x $(x \neq 0)$, there exists a unique rational number x^{-1} such that $x \cdot x^{-1} = x^{-1} \cdot x = 1$. The numbers x and x^{-1} are called *multiplicative inverses*, or *reciprocals*, of each other. For example, $\frac{^{-}2}{3}$ and $\frac{^{-}3}{2}$ are multiplicative inverses since

$$\left(\frac{^{-}2}{3}\right)\left(\frac{^{-}3}{2}\right) = 1;$$

the rational numbers b and $\frac{1}{b}$ $(b \neq 0)$ are multiplicative inverses of each other since

$$(b)\left(\frac{1}{b}\right) = 1.$$

Similarly, $\frac{a}{b}$ and $\frac{b}{a}$ $(a \neq 0,\ b \neq 0)$ are multiplicative inverses since

$$\left(\frac{a}{b}\right)\left(\frac{b}{a}\right) = 1.$$

The presence of multiplicative inverses makes the rational number system closed with respect to division (except, of course, by zero) and makes possible the solution of all equations of the form $ax = b$ ($a \neq 0$).

We shall now list eleven basic properties of the rational numbers with respect to multiplication and addition, letting x, y, z be any rational numbers.

Closure If $x + y = r$, then r is a rational number; if $xy = s$, then s is a rational number.

Commutativity $x + y = y + x$; $xy = yx$.

Associativity $x + (y + z) = (x + y) + z$; $x(yz) = (xy)z$.

Distributivity (multiplication over addition) $x(y + z) = xy + xz$.

Additive Identity There exists a unique rational number 0, called the *additive identity*, such that

$$x + 0 = 0 + x = x$$

for every x.

Multiplicative Identity There exists a unique rational number 1 such that

$$1 \cdot x = x \cdot 1 = x$$

for every rational number x.

Additive Inverse For every rational number x, there exists a unique rational number ^-x such that

$$x + (^-x) = {}^-x + x = 0.$$

The numbers x and ^-x are called *additive inverses*, or *negatives*, of each other. For example: $^+5 + {}^-5 = {}^-5 + {}^+5 = 0$.

Multiplicative Inverse For every rational number x ($x \neq 0$), there exists a unique rational number x^{-1} such that

$$xx^{-1} = x^{-1}x = 1.$$

For example: The multiplicative inverse of 2 is $\frac{1}{2}$ since $2 \times \frac{1}{2} = 1$; the multiplicative inverse of $\frac{3}{4}$ is $\frac{4}{3}$ since $\frac{3}{4} \times \frac{4}{3} = 1$; and the multiplicative inverse of $\frac{^-5}{6}$ is $\frac{6}{^-5}$ since $\frac{^-5}{6} \times \frac{6}{^-5} = 1$.

Any system which has all these eleven properties is called a *field*, and the properties are called *field properties*. The rational numbers form a field, but the whole numbers and the integers do not.

ADDITIONAL PROPERTIES OF THE RATIONALS

There are additional properties of the rationals that are important because they are the basis for much of the computational work with fractions. From the definition of equality of rational numbers, we obtain the following theorem.

THEOREM 1 If a, b, and c are integers ($b \neq 0$, $c \neq 0$), then

$$\frac{a}{b} = \frac{ac}{bc} \quad \text{since} \quad a(bc) = b(ac).$$

That is, if the numerator and the denominator of a fraction are multiplied by the same nonzero integer, the resulting fraction is equivalent to the given fraction. For example,

$$\frac{3}{4} = \frac{3 \times 2}{4 \times 2} = \frac{6}{8}.$$

We use this property when adding and subtracting fractions.

THEOREM 2 If a, b, and c are integers ($b \neq 0$, $c \neq 0$), then

$$\frac{a}{b} = \frac{a \div c}{b \div c} \quad \text{since} \quad a \cdot \frac{b}{c} = b \cdot \frac{a}{c}.$$

That is, if the numerator and the denominator of a fraction are divided by the same nonzero integer, the resulting fraction is equivalent to the given fraction. For example,

$$\frac{10}{15} = \frac{10 \div 5}{15 \div 5} = \frac{2}{3}.$$

This property is used for reducing fractions to their lowest terms. Theorems 1 and 2 might be considered the fundamental laws of fractions.

Other properties of the rationals that correspond to properties of the integers are the following:

Addition Let x, y, z be rational numbers. If $x = y$, then $x + z = y + z$.

Multiplication If $x = y$, then $xz = yz$.

Cancellation For addition: if $x + z = y + z$, then $x = y$. For multiplication: if $xz = yz$ ($z \neq 0$), then $x = y$.

Multiplication by Zero For every rational number x, $x \cdot 0 = 0 \cdot x = 0$.

Zero Product If $xy = 0$, then $x = 0$ or $y = 0$.

Distributive (multiplication over subtraction) $x(y - z) = xy - xz$.

Comment

We can prove that the rationals are distributive for multiplication over subtraction in the following way. You should be able to supply the reasons for each step.

1 $x(y - z) = x\,(y + {}^{-}z)$

2 $= xy + x({}^{-}z)$

3 $= xy + {}^{-}xz$

4 $= xy - xz$

EXERCISE SET 8.6

1 What is meant by (a) an additive inverse of a number x, (b) a multiplicative inverse of a number x?

2 Write the additive inverse of each of the following in two ways.

a) $\frac{{}^{-}3}{4}$ b) $\frac{2}{5}$ c) $\frac{{}^{-}4}{{}^{-}9}$ d) $\frac{5}{{}^{-}7}$ e) $\frac{{}^{-}r}{s}$ f) $\frac{0}{m}$

3 Write the multiplicative inverse (where possible) of each of the following.

a) 6 b) $\frac{0}{3}$ c) $\frac{{}^{-}2}{3}$ d) $\frac{-r}{{}^{-}t}$ e) $\frac{c}{{}^{-}d}$ f) $\frac{0}{{}^{-}a}$

4 Each of the following is true by one of the properties of rational numbers. Name the property in each case.

a) $\frac{2}{3}\left(\frac{{}^{-}5}{6}\right) = \frac{{}^{-}5}{6}\left(\frac{2}{3}\right)$ b) $\frac{5}{2}\left(\frac{3}{2} + \frac{{}^{-}2}{5}\right) = \frac{5}{2} \times \frac{3}{2} + \frac{5}{2} \times \frac{{}^{-}2}{5}$.

c) $1 \times \frac{4}{3} = \frac{4}{3}$ d) $\frac{1}{4} \times \frac{6}{6} = \frac{1}{4}$ e) $0 + \frac{3}{5} = \frac{3}{5}$

f) $\frac{{}^{-}7}{8} \times \frac{8}{{}^{-}7} = 1$ g) $\frac{5}{3} + \frac{{}^{-}5}{3} = 0$.

5 Show by counterexample that:

a) subtraction of rational numbers is neither commutative nor associative;

b) division of rational numbers is neither commutative nor associative.

6 Does the distributive property of division over addition hold for rational numbers? Prove your answer.

7 What are the differences between the structure of the rational numbers and that of the integers?

8 Name the largest and smallest rational number (where they exist) of each of the following sets.

a) The set of positive rational numbers.

b) The set of negative rational numbers.

c) The set of nonnegative rational numbers.

d) The set of nonpositive rational numbers.

e) The set of rational numbers greater than $\frac{1}{3}$ and less than $\frac{4}{5}$.

f) The set of rational numbers greater than $\frac{1}{3}$ and less than or equal to $\frac{4}{5}$.

9 **Check all the properties that apply to each number system.**

Property	*Counting Numbers*	*Whole Numbers*	*Integers*	*Rationals*
Closure +				
Closure ×				
Commutative +				
Commutative ×				
Associative +				
Associative ×				
Identity +				
Identity ×				
Inverse +				
Inverse ×				
Distributive				

10 **State the reason for each step in the proof of $^-(x + y) = {}^-x + {}^-y$.**

a) $^-(x + y) + (x + y) = \frac{0}{1}$

b) $^-x + {}^-y + (x + y) = {}^-x + [{}^-y + (x + y)]$

c) $= {}^-x + [{}^-y + x] + y$

d) $= {}^-x + [x + {}^-y] + y$

e) $= [{}^-x + x] + [{}^-y + y]$

f) $= \frac{0}{1}$

g) $^-(x + y) = {}^-x + {}^-y.$

11 **Prove that (a) the additive identity and (b) the multiplicative identity in the set of rational numbers are unique.**

12 **Prove that (a) the additive inverse and (b) the multiplicative inverse in the set of rational numbers are unique.**

8.7 Order Relation of the Rationals

The order relation of the rationals must conform with the order relation of the integers since the integers are part of the rationals. So the ideas that follow agree with those on page 157.

We can define an order relation for rationals in several ways. We can define it in terms of the left-right order of points on the number line.

DEFINITION 1 If $\frac{a}{b}$ and $\frac{c}{d}$ are distinct rationals, and the point that represents $\frac{a}{b}$ is to the *left* of the point that represents $\frac{c}{d}$, then

$$\frac{a}{b} < \frac{c}{d} \quad \text{and} \quad \frac{c}{d} > \frac{a}{b} \qquad \text{(Figure 8.5)}$$

FIGURE 8.5

$\frac{a}{b}$ $\frac{c}{d}$

Example 1 Since $\frac{^-1}{2}$ is to the *left* of $\frac{^+1}{3}$ on the number line,

$$\frac{^-1}{2} < \frac{^+1}{3} \text{ and } \frac{^+1}{3} > \frac{^-1}{2} \text{ (Figure 8.6).}$$

FIGURE 8.6

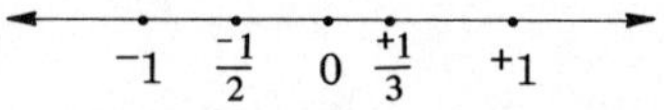

$^-1 \quad \frac{^-1}{2} \quad 0 \quad \frac{^+1}{3} \quad ^+1$

We can also define an order relation for rationals without reference to the number line just as we defined the order relation for integers.

DEFINITION 2 If $\frac{a}{b}$ and $\frac{c}{d}$ are any rationals, then $\frac{a}{b} < \frac{c}{d}$ if and only if there exists a positive rational number $\frac{e}{f}$ such that $\frac{a}{b} + \frac{e}{f} = \frac{c}{d}$.

Example 2 $\frac{^-1}{2} < \frac{^+1}{3}$ because $\frac{^-1}{2} + \frac{^+5}{6} = \frac{^+1}{3}$.

Perhaps the most useful definition of an order relation for the rationals is the following:

DEFINITION 3 If $\frac{a}{b}$ and $\frac{c}{d}$ are rational numbers expressed with positive denominators, then $\frac{a}{b} < \frac{c}{d}$ if and only if $ad < bc$. If $ad < bc$, then $bc > ad$.

Example 3 $\frac{7}{13} < \frac{9}{16}$ because $(7)(16) < (13)(9)$; i.e., $112 < 117$. Or, we can say that $\frac{9}{16} > \frac{7}{13}$ because $(9)(13) > (16)(7)$.

Example 4 $\frac{^-6}{23} < \frac{^-4}{17}$ because $(^-6)(17) < (23)(^-4)$. Or, $\frac{^-4}{17} > \frac{^-6}{23}$ because $(^-4)(23) > (17)(^-6)$.

Comment *We can show that Definition 2 and Definition 3 are equivalent. You should be able to supply the reason for each step in the proof below:*

1 $\frac{a}{b} + \frac{e}{f} = \frac{c}{d}$

2 $\frac{af + be}{bf} = \frac{c}{d}$

3 $(af + be)d = (bf)c$

4 $(af)d + (be)d = (bf)c$.

5 But $(be)d > 0$.

6 $(af)d < (bf)c$

7 $f(ad) < f(bc)$

8 $ad < bc$.

If we combine Definition 3 with the definition of equality of rational numbers, i.e.,

$$\frac{a}{b} = \frac{c}{d} \text{ if and only if } ad = bc,$$

we obtain a general rule for determining the order relation of any two positive rational numbers:

1 $\frac{a}{b} < \frac{c}{d}$ if and only if $ad < bc$.

2 $\frac{a}{b} > \frac{c}{d}$ if and only if $ad > bc$.

3 $\frac{a}{b} = \frac{c}{d}$ if and only if $ad = bc$.

PROPERTIES OF THE ORDER RELATION

The following properties of *less than* are derived from the corresponding properties of integers. We shall not prove them here, but merely state them. For simplicity, we shall denote rational numbers by x, y, and z rather than by $\frac{a}{b}$, $\frac{c}{d}$, and $\frac{e}{f}$.

Trichotomy If x and y are any two rational numbers, then one and only one of three possibilities holds:

$$(1)\ x = y, \qquad (2)\ x < y, \qquad \text{or} \qquad (3)\ x > y.$$

Transitivity If $x < y$ and $y < z$, then $x < z$. For example, since $\frac{^-1}{5} < \frac{1}{10}$ and $\frac{1}{10} < \frac{2}{7}$, $\frac{^-1}{5} < \frac{2}{7}$.

Addition If $x < y$, then $x + z < y + z$. For example, since $\frac{^-2}{5} < \frac{^-1}{3}$, $\frac{^-2}{5} + \frac{1}{2} < \frac{^-1}{3} + \frac{1}{2}$.

Multiplication If $x < y$, then

$xz < yz$ if z is positive,
$xz > yz$ if z is negative.

For example, since $\frac{^-1}{3} < \frac{1}{4}$, $\frac{^-1}{3} \times \frac{1}{2} < \frac{1}{4} \times \frac{1}{2}$; and

since $\frac{^-1}{3} < \frac{1}{4}$, $\frac{^-1}{3} \times \frac{^-1}{2} > \frac{1}{4} \times \frac{^-1}{2}$.

EXERCISE SET 8.7

1. What is meant by *the order relation* of the rationals?

2. If a and b are positive integers and $a < b$, then what is the order relation between $\frac{1}{a}$ and $\frac{1}{b}$? What is the order relation between $\frac{1}{a}$ and $\frac{1}{b}$ if $a > b$?

3. Determine the order relation between each of the following pairs of numbers.

   a) $\frac{3}{2}$ and $\frac{7}{5}$ b) $\frac{^-1}{3}$ and $\frac{2}{^-5}$ c) $\frac{26}{5}$ and $\frac{^-54}{^-12}$

   d) $\frac{^-3}{37}$ and $\frac{^-6}{75}$

4. If r, s, and t are rational numbers ($r > 0$, $t > 0$), then is $rt > st$ or is $rt < st$?

5. If the numerators of two fractions are the same but their denominators are different, how can you tell which fraction is greater?

6. Prove that if a and b are rational numbers and $a < b$, then $a < \frac{a+b}{2} < b$.

7. How can you tell at a glance that:

   a) $\frac{4238}{4237} > 1$. b) $\frac{2893}{2894} < 1$.

8. Insert three rational numbers between each of the following pairs of numbers.

   a) $\frac{5}{6}$ and $\frac{7}{8}$ b) $\frac{^-1}{3}$ and $\frac{2}{^-7}$ c) $\frac{^-2}{^-15}$ and $\frac{1}{16}$

9. Show a method by which you can name as many rational points as desired between the points $\frac{1}{1000}$ and $\frac{2}{1000}$. Name 6 such points.

10. Why is each of the following true for all counting numbers?

    a) $\frac{n+1}{n} > 1$ b) $\frac{n-1}{n} < 1$.

11. Which is less expensive: five oranges for 67¢ or four oranges for 53¢?

12. When is the reciprocal of a number (a) greater than the number, (b) less than the number, (c) equal to the number?

13. Arrange the following sets of rational numbers in ascending order.

    a) $\left\{\frac{2}{3}, {}^-1, 0, \frac{^-5}{6}\right\}$ b) $\left\{\frac{9}{19}, \frac{4}{7}, \frac{17}{35}, \frac{8}{17}\right\}$

    c) $\left\{\frac{^-3}{13}, \frac{8}{^-29}, \frac{^-17}{^-80}, \frac{7}{30}\right\}$

14. Find the solution set for each of the following if x is a rational number.

    a) $x - \frac{2}{3} < \frac{3}{4}$ b) $\frac{x}{5} - \frac{1}{2} > \frac{3}{^-5}$ c) $2x + \frac{3}{8} < \frac{^-5}{6}$

    d) $\frac{1}{6} - x < \frac{2}{3} \cdot \frac{^-4}{7}$

8.8 *Decimal Representation of the Rationals*

Any rational number can be expressed as a decimal. In particular, fractions that have denominators that are powers of 10 are easily converted to decimals by using the ideas of place value. For instance,

$$\frac{2}{10} = 0.2, \qquad \frac{2}{100} = 0.02, \qquad \text{and} \qquad \frac{2}{1000} = 0.002.$$

Conversely, a decimal stands for a sum of fractions in which each denominator is a power of 10 and each numerator is a one-digit number. For example,

$$.125 = \frac{1}{10} + \frac{2}{100} + \frac{5}{1000} \qquad \text{or}$$

$$.125 = \frac{1}{10} + \frac{2}{10^2} + \frac{5}{10^3}.$$

We can also express the answer with negative exponents:

$$.125 = (1 \times 10^{-1}) + (2 \times 10^{-2}) + (5 \times 10^{-3})$$

where $10^{-1} = \frac{1}{10}$, $10^{-2} = \frac{1}{10^2}$, $10^{-3} = \frac{1}{10^3}$, $10^{-k} = \frac{1}{10^k}$.

We can now express in expanded notation any number given in decimal notation. For example,

$$6187.302 = (6 \times 10^3) + (1 \times 10^2) + (8 \times 10^1) + (7 \times 10^0) + (3 \times 10^{-1}) + (0 \times 10^{-2}) + (2 \times 10^{-3}).$$

REPEATING DECIMALS

We can convert any rational number with *any* denominator into an equivalent decimal by dividing the numerator by the denominator. For example,

$$\frac{3}{4} = 4\,\overline{)\,3.00}\ = 0.75 \qquad \begin{array}{r} 0.75 \\ 4\,)\,3.00 \\ \underline{28}\ \\ 20 \\ \underline{20} \\ 0 \end{array}$$

$$\frac{2}{3} = 3\,\overline{)\,2.000} = 0.666 \qquad \begin{array}{r} 0.666 \\ 3\,)\,2.000 \\ \underline{18}\ \ \\ 20\ \\ \underline{18}\ \\ 20 \\ \underline{18} \\ 2 \end{array}$$

$$\frac{3}{11} = 11\,\overline{)\,3.0000} = 0.2727 \qquad \begin{array}{r} 0.2727 \\ 11\,)\,3.0000 \\ \underline{2\,2}\ \ \ \\ 80\ \ \\ \underline{77}\ \ \\ 30\ \\ \underline{22}\ \\ 80 \\ \underline{77} \\ 3 \end{array}$$

In the case of $\frac{3}{4}$, we obtain a remainder of 0 after carrying the division to two decimal places. So we say that the decimal *terminates.* In the case of $\frac{2}{3}$ and $\frac{3}{11}$, the remainder is never zero regardless of the number of places to which we carry the division. But although these decimals never terminate, certain blocks of digits keep repeating after a certain point. For example, for $\frac{2}{3}$, the digit 6 keeps repeating, and for $\frac{3}{11}$, the digits 27 keep repeating. We call such a decimal a *repeating decimal.* The block of digits that keeps repeating is indicated by a bar over the block; e.g.,

$$\frac{2}{3} = 0.\overline{6}, \qquad \frac{3}{11} = 0.\overline{27}, \qquad \frac{1}{7} = 0.\overline{142857}.$$

Comments

1 *Note that if division were continued in the case of* $\frac{3}{4}$, *we would get* 0.750000..., *indicating that even a terminating decimal may be considered* repeating *since* 0 *can be thought of as the repeating digit.*

2 *Repeating decimals are really* geometric series. *For example,* 0.3333... *means*

$$\frac{3}{10} + \frac{3}{10^2} + \frac{3}{10^3} + \cdots$$

This is a geometric series with the first term, $a = \frac{3}{10}$, *and the common ratio,* $r = \frac{1}{10}$.

How can we tell by looking at a fraction whether or not it can be represented by a terminating decimal? There is a very simple way to tell. Every fraction with a denominator that is the product of 2s and/or 5s will produce a terminating decimal. Conversly, if the denominator contains any factor other than 2s and/or 5s, then the fraction cannot be expressed as a terminating decimal. For example, $\frac{7}{30}$ cannot be expressed as a terminating decimal because the denominator $30 = 2 \times 3 \times 5$ contains a 3. However, $\frac{11}{40}$ can be so expressed because the denominator $40 = 2^3 \times 5$ contains only 2s and 5s as factors. We say this more formally with a theorem.

THEOREM 3 A rational number $\frac{a}{b}$ (expressed in lowest terms) can be written as a terminating decimal if and only if the prime factorization of b is of the form $b = 2^x 5^y$, where x and y are whole numbers. (Why is this theorem true?)

RATIONAL NUMBERS AS REPEATING DECIMALS

We shall now show that a rational number may be defined as a number that can be represented by a repeating decimal.

THEOREM 4 Every rational number can be represented by a repeating decimal.

To see why this theorem is true, let us convert $\frac{9}{7}$ to a decimal. There are seven possible remainders when we divide by 7: 0, 1, 2, 3, 4, 5, 6. After we carry the division to six decimal places, we get a remainder of 2. The block of digits 285714 will now repeat since the remainder 2 has now appeared for the second time. The next digit in the quotient will be 2, and the sequence of digits will start repeating.

```
     1.285714
  7 ) 9.000000
      7
      2 0
      1 4
        60
        56
         40
         35
          50
          49
           10
            7
           30
           28
            2
```

More generally, let us convert the rational number $\frac{a}{b}$ to a decimal.

Division by b can result in b possible remainders: 0, 1, 2, 3, ..., $b - 1$. If, in the process of dividing a by b, we obtain a 0 remainder, then the decimal will terminate at that point and will be a repeating decimal with 0 repeating. If the division does not produce a 0 remainder, then the remainder must be one of the numbers 1, 2, 3, ..., $b - 1$. This means that after not more than $b - 1$ steps, we must obtain a remainder that has already appeared earlier. When this happens, the repetition has started.

THEOREM 5 Every repeating decimal represents a rational number.

To see why this theorem is true we pick any repeating decimal, such as 0.737373... with the digits 73 repeating, and see how we can find the rational number represented by this decimal.

Example 1 Let $0.\overline{73}$ represent the rational number n; that is,

$$n = 0.737373... \tag{1}$$

Multiplying both sides of Equation 1 by 100, we get

$$100n = 73.737373... \tag{2}$$

Subtracting (1) from (2) and solving for n, we get

$$\begin{aligned} 100n &= 73.737373... \\ n &= .737373... \\ \hline 99n &= 73 \\ n &= \frac{73}{99}, \text{ which is a rational number.} \end{aligned}$$

This procedure can be followed with every repeating decimal.

Note that we multiplied n by 100 (or 10^2) because there are two digits in this repeating block. If there were three digits, we would multiply by 1000 (or 10^3); if there were five digits, we would multiply by 10^5.

Example 2 Let us find the rational number for $1.\overline{273}$.

$$\begin{aligned} \text{Let } n &= 1.273273273... \\ 1000n &= 1273.273273... \\ n &= 1.273273... \\ \hline 999n &= 1272 \\ n &= \frac{1272}{999} = \frac{424}{333}. \end{aligned}$$

Since we have shown that:

1 for every rational number, there is a repeating decimal, and

2 for every repeating decimal, there is a rational number,

there exists a one-to-one correspondence between the set of rational numbers and the set of repeating decimals. The set of rational numbers can, therefore, be defined as the set of numbers that can be represented by repeating decimals.

EXERCISE SET 8.8

1 Express the following rational numbers in decimal form.

a) $\frac{5}{8}$ b) $\frac{13}{40}$ c) $\frac{15}{16}$ d) $\frac{41}{3}$ e) $2\frac{3}{8}$

2 Express the following decimal fractions in expanded notation.

a) 2.53 b) 15.108 c) 196.0125 d) *abc.xy*

3 Determine the rational number that corresponds to each of the following.

a) 0.24 b) 0.017 c) 5.3070 d) $0.\overline{1}$ e) $0.\overline{05}$

f) $6.0\overline{8}$ g) $3.0\overline{12}$ h) 0.131131131131... i) $17.0\overline{3492}$

4 Which of the following fractions can be represented by a terminating decimal.

a) $\frac{7}{32}$ b) $\frac{37}{100}$ c) $\frac{7}{30}$ d) $\frac{13}{80}$ e) $\frac{9}{28}$ f) $\frac{13}{150}$

5 Define the integer r for which $\frac{1}{r}$ can be expressed as a terminating decimal.

6 Show by three examples that if p is a prime number, then the number of digits in the repeating part of the decimal value of $\frac{1}{p}$ is a divisor of $p - 1$.

7 If a mile is approximately equivalent to 1.6 kilometers in the metric system, convert the following numbers of miles to kilometers.

a) 10 b) 20 c) 75 d) 250 e) 1776

8 A kilogram in the metric system is equivalent to about 2.2 pounds. Convert the following numbers of pounds to kilograms.

a) 5 b) 12 c) 35 d) 220

9 A square centimeter in the metric system is equivalent to about .155 square inches. How many square centimeters are there in a sheet of glass containing 10 square inches?

10 If a liter in the metric system is approximately equivalent to 1.06 quarts, how many liters of milk are there in 250 gallons?

PUZZLES

1 A man spends $\frac{1}{3}$ of his money and loses $\frac{2}{3}$ of the remainder. He then has $12 left. How much money did he have at the beginning?

2 A bolt of cloth is colored as follows: $\frac{1}{3}$ and $\frac{1}{4}$ of it is blue and the remaining 5 yards are gray. How long is the bolt?

3 Three boys, A, B, and C, gathered some sea shells and agreed to divide them in accordance with the following procedure: (a) they would pick lots to determine the order in which each boy would take his share of shells, (b) each boy would take exactly $\frac{1}{3}$ of the pile.

A went first and found that if he threw away one of the shells he could then divide those remaining into three equal shares. He put his share in his pocket, and now it was B's turn. B also found that if he discarded one of the shells that remained in the collection after A removed his share he could divide the remainder into three equal shares. B put his share in his pocket and now it was C's turn. C did exactly what A and B did: he had to discard one shell in order to be able to divide the remaining shells into three equal shares. C did so and put his share in his pocket.

What is the *least* number of shells the boys must have gathered originally so that there would be a whole number of shells left after each boy removed his share?

BASICS REVISITED

1 Illustrate the associative property for addition of whole numbers on the number line.

2 $\frac{2}{3} \cdot \frac{1}{2} + \frac{3}{5} \cdot \frac{2}{5} = ?$

3 Answer *true or false*: 136 is a factor of 17.

4 Insert a decimal point to make the following statement true: $.72003 \div .03 = 24001$.

5 A roast loses $\frac{1}{12}$ of its weight in cooking. How many pounds are needed to serve 8 people if each person is to be served a half-pound of roast?

6 Divide 5.007 by 26. (Round to 3 places.)

7 Find $87\frac{1}{2}\%$ of 18.4.

8 Take $2\frac{7}{8}$ from 12.

9 Write $\frac{1}{4}\%$ as a decimal and as a fraction.

10 Which is larger and by how much: .715 or .8?

11 Estimate the product of 306×792.

12 Divide: $5\frac{3}{4} \div 2\frac{1}{5}$.

13 Add: .029, .70358, and 14.

14 Multiply: $.865 \times .013$.

15 If paint costs $6.95 a gallon, estimate the cost of 21 gallons of paint.

16 Round off $999.12 to the nearest ten dollars.

17 Express .088 as a percent.

18 125% of what number is 85?

19 Divide: $.62\frac{1}{2} \div \frac{7}{12}$.

20 Find the lowest common denominator of $\frac{5}{12}$, $\frac{7}{8}$, and $\frac{11}{20}$.

21 Which is greater and by how much: $\frac{21}{59}$ or $\frac{24}{67}$?

22 Subtract $\frac{1}{8}$ from .125.

23 Find the percent of profit on cost if the cost of an item is $57 and it is sold for $76.

24 If an ore contains 16% copper, how many tons of ore are needed to get 30 tons of copper?

25 Mr. Smith works a regular work week of 40 hours, earning $\$5.34\frac{1}{2}$ an hour. One week he works 6 hours overtime at time-and-a-half. If there is a federal income tax of 24% and a city wage tax of $1\frac{5}{8}\%$ withheld from Mr. Smith's gross earnings, compute his take-home pay for this week.

EXTENDED STUDY

8.9 The Rational Number System is a Field

We defined the binary operations of addition and multiplication on the set of rational numbers and have seen that:

1 Both operations are closed.

2 Both operations are commutative.

3 Both operations are associative.

4 There are identity elements for each operation.

5 Every element has an additive inverse and every nonzero element has a multiplicative inverse.

6 Multiplication is distributive over addition.

A set of at least two elements on which two binary operations have been defined and which possess the above properties (*1–6*) forms a mathematical structure called a *field*. If the system also has the order properties, then it is called an *ordered field*. By the order properties, we mean that if a, b, and c are elements of the set, then:

1 Exactly one of the following relations is true:

$$a < b, \quad a = b, \quad \text{or} \quad a > b.$$

2 If $a < b$, then $a + c < b + c$.

3 If $a < b$ and $c > 0$, then $ac < bc$.

4 If $a < b$ and $c < 0$, then $ac > bc$.

5 If $a < b$ and $b < c$, then $a < c$.

The set of rational numbers is the first number system we have encountered that is not only a *field* but is also an *ordered field*. The set of whole numbers is not a field because it does not contain additive inverses (except for 0) or multiplicative inverses (except for 1); thus the whole numbers do not satisfy condition *5*. The set of integers is not a field because it lacks multiplicative inverses (except for 1 and $^-1$).

Notice that the definition of a field is completely abstract. All that is required is a set of elements (regardless of what they are) with two binary operations so defined that they possess properties *1–6*. By making our definition completely abstract, we achieve great economy since any theorems we prove on the basis of properties *1–6* will be true of every field.

Another, and interesting, example of a field is a system based on a three-hour clock numbered 0, 1, 2, as in Figure 8.7. "Addition" on this clock may be performed by rotations of the hand of the clock in a clockwise direction. For example, 1 + 2 means that, starting at the 0 position, we rotate the hand 1 hour, then 2 more hours. The effect of these two consecutive rotations is to leave the hand of the clock on the 0 position. That is, $1 + 2 = 0$. Similarly, $1 + 1 = 2$ and $2 + 2 = 1$. We interpret 0 to mean *no* rotation as well as a position on the clock. Thus, $1 + 0 = 1$ and $0 + 2 = 2$.

FIGURE 8.7

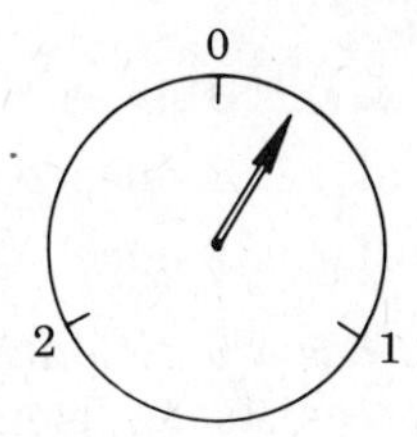

"Multiplication" on this clock is defined as repeated addition. Thus,

$$2 \times 1 = 1 + 1 = 2 \quad \text{and} \quad 2 \times 2 = 2 + 2 = 1.$$

"Addition" and "multiplication" tables based on these definitions are shown in Figure 8.8. Note that we are denoting "addition" by $\oplus$ and "multiplication" by $\otimes$ since in this system addition and multiplication have special meanings.

FIGURE 8.8

$\oplus$	0	1	2
0	0	1	2
1	1	2	0
2	2	0	1

$\otimes$	0	1	2
0	0	0	0
1	0	1	2
2	0	2	1

Let us now examine the properties of the mathematical system formed by the set $S = \{0, 1, 2\}$ with the binary operations $\oplus$ and $\otimes$ as defined in Figure 8.8.

Both operations are closed. For any pair of elements of S, there is a unique element which represents their *sum* and which is also a number of S. We can easily verify this statement by looking at the addition table and noting that every sum results in either 0, 1, or 2. This is also true of the *product* of any pair of elements of S.

Both operations are commutative. That is, for any pair of elements a and b of S,

1 $a \oplus b = b \oplus a$. For example, $1 \oplus 2 = 2 \oplus 1 = 0$.

2 $a \otimes b = b \otimes a$. For example, $1 \otimes 2 = 2 \otimes 1 = 2$.

We can verify commutativity for both operations by examining Figure 8.8 and noting the symmetry with respect to the main diagonals.

Both operations are associative. (Verification is left to the reader.)

There are identity elements for each operation.

1 0 is the identity element for addition since $0 \oplus 0 = 0$, $0 \oplus 1 = 1 \oplus 0 = 1$, and $0 \oplus 2 = 2 \oplus 0 = 2$.

2   1 is the identity element for multiplication since $1 \otimes 0 = 0 \otimes 1 = 0$, $1 \otimes 1 = 1$, and $1 \otimes 2 = 2 \otimes 1 = 2$.

Every element in S has an additive inverse. That is, for each element a of S, there exists an element ^{-}a of S such that $a \oplus {}^{-}a = 0$, the identity element for addition. Thus:

The additive inverse of 0 is 0 since $0 \oplus 0 = 0$.
The additive inverse of 1 is 2 since $1 \oplus 2 = 0$.
The additive inverse of 2 is 1 since $2 \oplus 1 = 0$.

Every nonzero element of S has a multiplicative inverse. That is, for each element b of S, there exists an element b' of S such that $b \otimes b' = 1$, the identity element for multiplication. Thus:

The multiplicative inverse of 1 is 1 since $1 \otimes 1 = 1$.
The multiplicative inverse of 2 is 2 since $2 \otimes 2 = 1$.

Multiplication is distributive over addition. For example,

$$1(0 \oplus 2) = (1 \otimes 0) \oplus (1 \otimes 2)$$

since $1(0 \oplus 2) = 1 \otimes 2 = 2$, and $(1 \otimes 0) \oplus (1 \otimes 2) = 0 \oplus 2 = 2$.

Thus we see that the mathematical system based on a three-hour clock is another example of a field. It can be shown that an m-hour clock is a field if m is a prime number; if m is a composite number, then the system is not a field.

EXERCISE SET 8.9

1. Write the addition and multiplication tables for a five-hour clock, and then show that such a system is a field.
2. Write the addition and multiplication tables for a four-hour clock, and then show why this system is not a field.

8.10 Zeno's Paradoxes

On page 180 we noted that an important property of the rational numbers is that they are *dense*—that between any two rational numbers there is always another rational number such as the average of the two numbers. On the number line this property means that between any two points A and B (Figure 8.9), there is another point C midway between them; between C and B, there is the midpoint D; between D and B, there is the midpoint E; and so on, ad infinitum. This property of denseness lies behind some of the famous paradoxes of the Greek philosopher Zeno (ca. 450 B.C.).

FIGURE 8.9

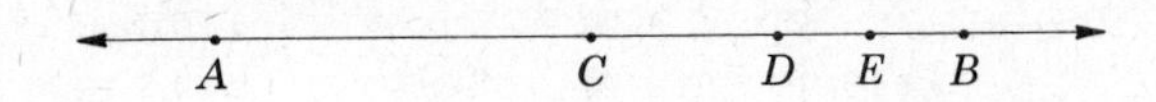

In one of these paradoxes, known as the *Dichotomy*, Zeno asserts that it is impossible to traverse *any* given distance. His reasoning is as follows. Before a man can get from A to B (Figure 8.9), he must first cover half the distance to C; then he must cover half the remaining distance to D; then, half the remaining distance to E; and so on, ad infinitum. There *always* remains some part of the distance yet to be covered. Therefore, the man can never get from A to B. Zeno argued, in effect, that you cannot traverse infinitely many points in a finite length of time. If follows that motion is impossible; that, in fact, it can never even begin.

This paradox can be explained in the following way. Zeno's argument rests on the assumption that if a segment contains infinitely many points, then the segment itself must be infinite; therefore, the segment cannot be traversed in a finite length of time. Let us see why this assumption is not correct.

The successive distances to be covered between A and B (Figure 8.10) form the infinite geometric series

$$\frac{1}{2} + \frac{1}{4} + \frac{1}{8} + \frac{1}{16} + \cdots + \frac{1}{2^n} + \cdots$$

where each term is half of the one preceding it. To find the sum of these distances, we use the standard formula for the sum of an infinite geometric series:

$$s = \frac{a}{1-r},$$

where s is the sum, a is the first term, and r is the ratio. In this case, $a = \frac{1}{2}$ and $r = \frac{1}{2}$, giving

$$s = \frac{a}{1-r} = \frac{\frac{1}{2}}{1-\frac{1}{2}} = \frac{\frac{1}{2}}{\frac{1}{2}} = 1.$$

So we see that although this series has infinitely many terms, its *sum* is *finite* and equal to 1. This result means that even though the segment $\overline{AB}$ has infinitely many points, its *length* is *finite* and can therefore be traversed in a finite length of time.

FIGURE 8.10

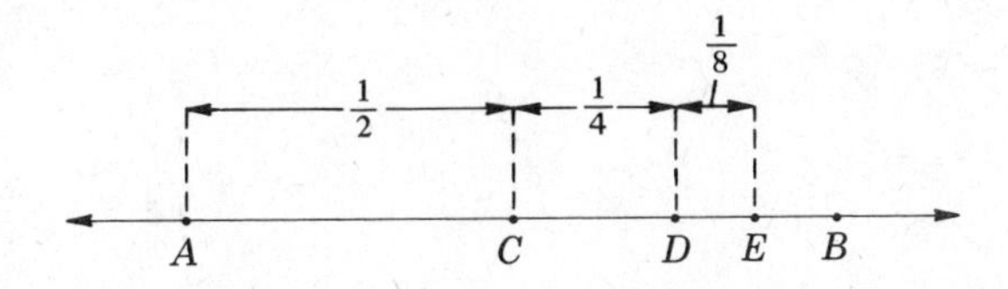

Another of Zeno's paradoxes is the famous *Achilles* paradox. Here, Achilles is racing against a tortoise that was given a headstart. Zeno asserts that Achilles, no matter how fast he runs, can never overtake the tortoise no matter how slowly the tortoise runs. His reasoning is as follows. By the time Achilles reaches the starting point of the tortoise, the tortoise will have advanced beyond that point to a new position; by the time Achilles reaches this new position, the tortoise will have advanced to a newer position. As Achilles arrives at each new point in the race, the tortoise will have advanced another short distance to still another position. This goes on ad infinitum and, therefore, the swift Achilles can never overtake the slow tortoise.

The explanation of the Achilles paradox lies in the incorrect inference in Zeno's argument that since Achilles must occupy the same number of positions as the tortoise he cannot cover a greater distance than the tortoise in the same amount of time. But there is no relation between the number of points on a segment and its length. We can prove, for instance, that a 2-inch segment and a 5-inch segment contain exactly the same number of points.

Let $\overline{AB}$ represent a 5-inch segment and $\overline{A'B'}$ a 2-inch segment (Figure 8.11). We can establish a one-to-one correspondence between the points of $\overline{A'B'}$ and the points of $\overline{AB}$ in the following way. If C is the intersection of $\overrightarrow{AA'}$ and $\overrightarrow{BB'}$, where $\overline{A'B'}$ is placed parallel to and nonincident with $\overline{AB}$, then to any point P of $\overline{AB}$ there will correspond a point P' of $\overline{A'B'}$ which is on CP. Similarly, to any point P' of $\overline{A'B'}$, there will correspond a point P of $\overline{AB}$ which is on $\overline{CP}$. So there exists a one-to-one correspondence between the set of points on the 5-inch segment and the set of points on the 2-inch segment. From the definition of the equivalence of sets, we conclude that both segments contain exactly the same number of points although they are of different lengths.

FIGURE 8.11

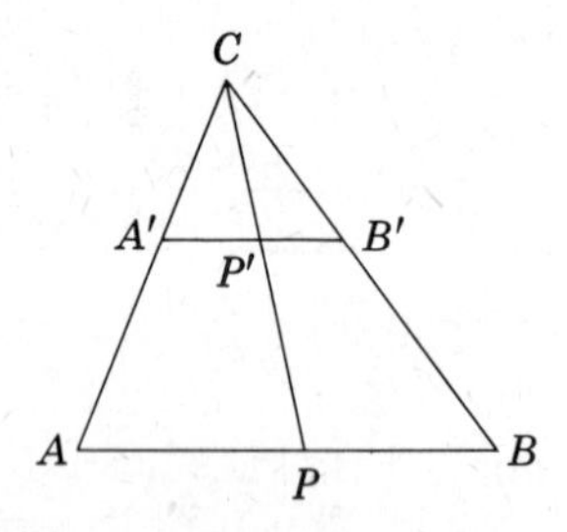

Therefore, even if Achilles occupies the same number of positions as the tortoise, he can still traverse a greater distance than the tortoise in the same amount of time.

Chapter 9

THE REALS

Objectives

At the completion of this chapter, you should be able to:

1. State why there was a need to expand the rational number system to the *real number system*.
2. Define an *irrational number*.
3. Define the set of real numbers in terms of its decimal representation.
4. State and demonstrate the field properties of the set of real numbers.
5. Distinguish between the reals and the rationals in terms of the *property of completeness*.
6. Compute a decimal approximation of an irrational number like $\sqrt{5}$.

7 Locate irrational numbers like $\sqrt{2}$ on the number line.

8 Solve equations of the form $x^2 = a$ where a is any non-negative real number.

Prerequisites

You should know:

1 The meaning of repeating decimals.

2 The properties of the rational numbers.

INTRODUCTION

Although the rational number system is an improvement over the integers, this system, too, has limitations. One such limitation is that there does not always exist a solution in the rational number system for an equation of the form $x^2 = a$, where a is a rational number. For instance, no rational number exists that satisfies the equation $x^2 = 5$. We call the positive number whose square is 5 the *square root* of 5, and we denote it by $\sqrt{5}$.

A second limitation is that the rational number system does not contain enough numbers to correspond to every point on the number line. We can demonstrate this deficiency by locating a point on the number line for which there is no rational number:

On the number line, construct a right triangle such that each leg, $\overline{AC}$ and $\overline{CB}$, is 1 unit long (Figure 9.1). The hypotenuse $\overline{AB}$ will therefore be $\sqrt{2}$ units long by the Pythagorean Theorem (see page 259). We now locate the point D on the number line such that $AD = \sqrt{2}$. D is located by drawing an arc of a circle with center A and radius $\overline{AB}$, intersecting the number line at point D. The point D corresponds to the number $\sqrt{2}$. But $\sqrt{2}$ is *not* a rational number (see page 212). This means that we have located a point on the number line which does not correspond to a rational number. In fact, we can locate infinitely many such points on the number line.

FIGURE 9.1

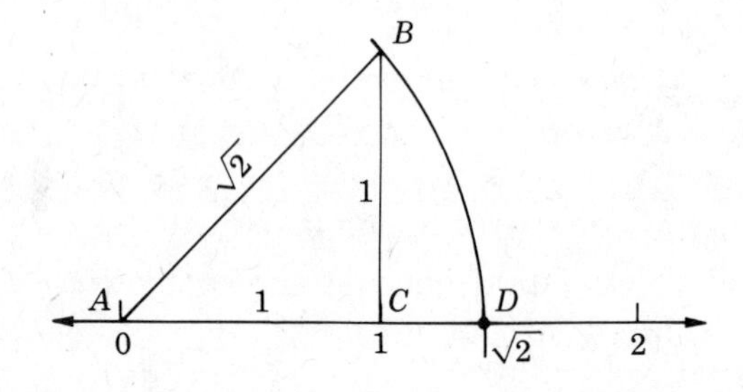

To remove these limitations of the rational number system, we expand the system in such a way that there is a number that satisfies *every* equation of the form $x^2 = a$, where a is a positive rational number, and there is a number that corresponds to *every* point on the number line. We call this expanded system the *real number system.* We shall now create this new system in such a way that the rationals are a subset of the reals, and all the properties of the rational number system also hold for the real number system.

9.1 Creating the Reals

To create the *real numbers*, we start with the definition of rational numbers as numbers that can be expressed as repeating decimals. We then ask whether there exist decimal expressions that are *non-*repeating. We can show that such decimals do exist by devising a way in which they can be constructed. Take any repeating decimal such as 0.3333... and convert it into another decimal by inserting any digit such as 1 after the first 3, 11 after the second 3, 111 after the third 3, and so on. This procedure converts 0.3333... to 0.313113111..., an obviously nonrepeating decimal. Since, as we have seen in the last chapter, repeating decimals and only repeating decimals represent rational numbers, *nonrepeating* decimals must represent numbers that are *not* rational. We call such numbers *irrational numbers.*

DEFINITION 1 An *irrational number* is any number represented by a nonrepeating decimal.

Example 0.23233233323333... is a nonrepeating decimal and is, therefore, an irrational number.

DEFINITION 2 The *real number system* consists of all rational and irrational numbers.

Comments

1 *It follows from the definition that any real number can be represented by a decimal. If the decimal is repeating, then the number is rational; if it is nonrepeating, then the number is irrational.*

2 *In addition to irrational numbers such as $\sqrt{2}$ that arise from finding roots of rational numbers, there are other irrational numbers, like π, that are called* transcendental *numbers and that are not the roots of rational numbers. In 1767, the Swiss mathematician J. H. Lambert (1728-1777) proved that π is*

irrational. The approximate value of π *is* 3.141592...
Another important example of an irrational number is e which is defined as the limit as n tends to infinity *of*

$$\left(1+\frac{1}{n}\right)^n.$$

The approximate value of e is 2.71828....

9.2 Decimal Representation of Irrational Numbers

TRIAL AND ERROR METHOD

Given an irrational number like $\sqrt{2}$, how do you find its decimal approximation? There are several methods for doing this besides, of course, using a square root table. One of these methods is essentially a *trial and error method* which we illustrate in Example 1.

Example 1 We know that $\sqrt{2}$ must fall somewhere between 1 and 2:

$$\sqrt{1}=1 \qquad \sqrt{2}=? \qquad \sqrt{4}=2.$$

That is,

$$1<\sqrt{2}<2.$$

Since $(1.4)^2 = 1.96$ and $(1.5)^2 = 2.25$, we know that

$$1.4<\sqrt{2}<1.5.$$

Since $(1.41)^2 = 1.9881$ and $(1.42)^2 = 2.0164$, we know that

$$1.41<\sqrt{2}<1.42.$$

Since $(1.414)^2 = 1.999396$ and $(1.415)^2 = 2.002225$, we know that

$$1.414<\sqrt{2}<1.415.$$

We can continue with this process to as many decimal places as we wish, each time getting a closer approximation to the irrational number $\sqrt{2}$. If we carry the approximation to five decimal places we get $\sqrt{2} = 1.41421$.

Comment *Note that* $\sqrt{a}+\sqrt{b} \neq \sqrt{a+b}$. *For example,* $\sqrt{2}+\sqrt{3} \neq \sqrt{5}$ *since*

$$\sqrt{2}+\sqrt{3} = 1.414 + 1.732 = 3.146,$$

whereas $\sqrt{5} = 2.236$.

DIVIDE AND AVERAGE METHOD

Another method for finding the decimal approximation of an irrational number is the *divide and average method.* This method is based on the idea that if we divide a number by its exact square root, we get the same square root as the quotient; for example,

$$5\,\overline{)\,25}^{\;5}, \qquad 7\,\overline{)\,49}^{\;7}.$$

Suppose we don't know what the square root of 25 is, and we estimate it to be, say, 4. A way to check whether our estimate is correct is to divide the 25 by 4:

$$4\,\overline{)\,25.00}^{\;6.25}.$$

Since we get a quotient of 6.25, our estimate of 4 is obviously too small. So we conclude that the answer lies somewhere between our first estimate, 4, and 6.25. By averaging the 4 and the 6.25, we obtain a second estimate which is a closer approximation of $\sqrt{25}$:

$$\frac{4 + 6.25}{2} = 5.125.$$

To obtain a still better approximation of $\sqrt{25}$, we divide the 25 by our second estimate:

$$5.125\,\overline{)\,25.0000}^{\;4.8780}$$

and then average the 5.125 and the 4.8780 to obtain a third estimate:

$$\frac{5.125 + 4.8780}{2} = 5.0015.$$

We continue this process as many times as necessary until we obtain the desired degree of accuracy. Each time we divide, we retain more digits in the quotient than in the previous approximation, thereby increasing the degree of accuracy with each succeeding approximation. Let us use this method to obtain a decimal approximation of $\sqrt{2}$ as illustrated in Example 2.

Example 2 We note that $1 < \sqrt{2} < 2$.

1st Estimate Since $\sqrt{2}$ is closer to 1 than to 2, let us estimate $\sqrt{2} \approx 1.4$, where $\approx$ means *approximately equal to.*

Divide $1.4\,\overline{)\,2.00}^{\;1.42}$

2nd Estimate $\dfrac{1.4 + 1.42}{2} = 1.41.$

$$\text{Divide}\quad 1.41\,\overline{)\,2.0000}\quad 1.4184$$

$$\text{3rd Estimate}\quad \frac{1.41 + 1.4184}{2} = 1.4142.$$

9.3 Properties of the Real Numbers

Let a, b, and c be any real numbers.

Closure If $a + b = r$, then r is a unique real number; if $ab = s$, then s is a unique real number.

Commutativity $a + b = b + a$; $ab = ba$.

Associativity $a + (b + c) = (a + b) + c$; $a(bc) = (ab)c$.

Distributivity $a(b + c) = ab + ac$.

Additive Identity There exists a unique real number 0, called the *additive identity*, such that

$$a + 0 = 0 + a = a$$

for every real number a.

Multiplicative Identity There exists a unique real number 1, called the *multiplicative identity*, such that

$$1 \times a = a \times 1 = a$$

for every real number a.

Additive Inverse For every real number a, there exists a real number (^-a) such that

$$a + (^-a) = {^-a} + a = 0.$$

Multiplicative Inverse For every real number a ($a \neq 0$), there exists a real number a^{-1} such that

$$a \times a^{-1} = a^{-1} \times a = 1.$$

In addition to these field properties, the real number system, like the rational number system, also has the property of *denseness;* that is, between any two distinct real numbers, there is always another real number. Also, the real number system has the *completeness* property which means, as we have seen, that there is a one-to-one correspondence between the set of real numbers and the set of points on the number line. The real number system is also *ordered;* that is, for any two real numbers a and b, $a < b$ if and only if there is a positive number c such that $a + c = b$. The trichotomy property also operates in the real number system; that is, only one of the following relations holds: $a < b$, $a = b$, or $a > b$.

Comments

1 *The difference between the* denseness *property and the* completeness *property can be seen by comparing the rationals with the reals. Although the rational points on the number line are a* dense set, *they are not a* continuous set. *The set contains "holes" which are filled by the irrational points. When a set is* complete, *there are no "holes" in it. The* real *points form such a set.*

2 *The* completeness *property is the only one of all the properties of real numbers listed which does not hold for the rational numbers. Since we call the rationals an* ordered field, *we call the reals, because of its completeness property,* a complete ordered field.

EXERCISE SET 9.1–9.3

1 Which of the following are irrational numbers?

a) $\sqrt{49}$ b) $\sqrt{\frac{4}{81}}$ c) $\sqrt{8}$ d) $\sqrt{\frac{9}{13}}$ e) π

f) $2\sqrt{5}$ g) $2\sqrt{7} + 5\sqrt{7}$ h) $2\sqrt{3} \times 7\sqrt{3}$ i) $\sqrt{6} \times \sqrt{5}$

2 Which of the following decimals represent rational numbers and which represent irrational numbers?
a) 0.7 b) 0.222... c) 0.3141441444... d) 0.57629762976297...

3 Locate each of the following points on the number line.
a) $\sqrt{3}$ b) $\sqrt{5}$ c) $\sqrt{8}$ d) $2\sqrt{3}$

4 Are the rational numbers adequate to handle any measurement problems? Explain your answer.

5 a) If a is a rational number, is a^2 always rational? Why?
b) If c is an irrational number, is c^2 always irrational? Why?

6 Prove that the product of an irrational number and a nonzero rational number is irrational.

7 Prove that the sum of an irrational number and a rational number is irrational.

8 Is the product of two irrational numbers necessarily irrational? Explain.

9 Find the decimal approximation, correct to three places, of each of the following.
a) $\sqrt{5}$ b) $\sqrt{21}$ c) $\sqrt{.08}$ d) $\sqrt{29.25}$

10 Find two rational numbers between $\sqrt{3}$ and $\sqrt{5}$.

11 If a and b are nonzero rational numbers, show that $a\sqrt{2} + b\sqrt{3}$ is irrational.

12 Which set of numbers does each of the following represent?
a) The union of the rationals and the irrationals.
b) The intersection of the rationals and the irrationals.
c) The union of the reals and the rationals.
d) The union of the reals and the irrationals.
e) The intersection of the reals and the rationals.

f) The intersection of the irrationals and the reals.
g) The complement with respect to the rationals.
h) The complement with respect to the irrationals.
i) The complement with respect to the reals.

13 Name all the number systems in which each of the following equations has a solution.

a) $x + 5 = 9$ b) $x^2 + 1 = 18$ c) $x^2 = 25$ d) $3x = 10$
e) $x + 7 = 2$ f) $x^2 + 1 = 0$

14 State: (a) two shortcomings of the integers that were overcome by the rationals and (b) two shortcomings of the rationals that were overcome by the reals.

15 In the table below check all the properties possessed by each set.

Property

Property	*Counting Numbers*	*Whole Numbers*	*Integers*	*Rationals*	*Reals*
Closure (+)					
Closure (X)					
Inverse (+)					
Inverse (X)					
Order					
Denseness					
Completeness					

PUZZLES

1 Write two 2s so that, together with standard mathematical symbols, the result is 5.

2 Three men rented a suite of rooms in a hotel for $60. After they are taken to their rooms, the manager discovers that he overcharged them since the suite only rents for $50. He then sends a bellhop upstairs with the $10 change. The dishonest bellhop decides to keep $4 and returns only $6 to the men. Now the rooms originally cost $60, but the men received $6 back. This means that they only paid $54 for the suite, and the bellhop kept $4. Since $54 + $4 = $58, what happened to the extra two dollars?

3 Ask a person to write down the following four numbers.

a) The year of his birth.
b) The year of some important event in his life.
c) His age (as of December 31 of that year).
d) The number of years since the important event took place.

What is the sum of these four numbers?

BASICS REVISITED

1 Find $4\frac{1}{4}\%$ of \$360.

2 9.21 is what percent of 9210?

3 If you receive $62 a year interest on a $1000 investment, how much must you invest to earn $350 interest that year?

4 $1\frac{3}{10} + 2\frac{3}{5} + \frac{3}{4} = ?$

5 Find the average, by inspection, of 16, 19, 22, 25, 28.

6 $.075\overline{)25.6855}$ (Round to 2 places.)

7 Find the product, by inspection, of

$\frac{2}{3} \cdot \frac{7}{8} \cdot 5 \cdot \frac{8}{7} \cdot \frac{3}{2}.$

8 If you have 2007 feet of wire, how many inches of wire do you have?

9 Take $13\frac{7}{8}$ from 100.

10 How many hundreds are there in a million?

11 Find the highest common factor of 36, 45, and 96.

12 Arrange in order of size; $\frac{3}{5}, \frac{1}{3}, \frac{2}{7}, \frac{5}{14}.$

13 By paying cash for a television set you receive a 5% discount. If your bill is $199.50, what is the regular price of the set?

14 $\left(\frac{3}{8} \times \frac{1}{10}\right) \div \frac{5}{6} = ?$

15 $6\frac{2}{5} - 2\frac{7}{8} = ?$

16 Round off to the nearest thousandth: 2.0035942.

17 Use a short way to divide 500 by $16\frac{2}{3}$.

18 Square 2.16 and round the answer to 2 places.

19 At a speed of 60 miles per hour, how long will it take to travel 23 miles?

20 A man buys a $100 bond for $95. If he receives $7 in dividends at the end of the year, what is his rate of income from his investment?

21 Insert a decimal point to make the following statement true: $1.003 \times .47 = 47141.$

22 Answer true or false: $(48 \div 12) \div 4 = 4 \div (48 \div 12).$

23 Using the short method, divide 3.049 by
a) 10, b) 100, c) 1000.

24 Tom travels 175 miles at 50 miles per hour, and 80 miles at 42 miles per hour. What is Tom's average speed (to the nearest mile) over the 255-mile trip?

25 Wendy went on a diet and reduced her weight from 70 kilograms to 61.6 kilograms. What was the percent of decrease in her weight?

EXTENDED STUDY

9.4 Proof: $\sqrt{2}$ is Irrational

We shall use an *indirect proof*, first used by Euclid, to show that $\sqrt{2}$ is irrational. We shall start by *assuming* that $\sqrt{2}$ is *rational*, and then show that this assumption leads to a contradiction.

Proof

1 Assume that $\sqrt{2} = \frac{a}{b}$ is a rational number, where a and b are relatively prime; that is, they have no common factors other than 1.

2 Then $2 = \frac{a^2}{b^2}$

3 and $2b^2 = a^2$.

4 Since $2b^2$ is an even number, a^2 must likewise be even; therefore, a is even. (If a were odd, a^2 would also be odd since the product of two odd numbers is odd.)

5 Since a is even, it can be written as $a = 2n$, where n is some integer.

6 From *5* we get $a^2 = 4n^2$.

7 Substituting in *3*, we get $2b^2 = 4n^2$, or $b^2 = 2n^2$.

8 Since b^2 is even, b must likewise be even.

9 From *4* and *8*, we see that both a and b are even, which contradicts our assumption that a and b are relatively prime.

10 Therefore, our assumption that $\sqrt{2}$ is rational must be false. So $\sqrt{2}$ must be irrational.

We can use the fact that $\sqrt{2}$ is irrational to prove the existence of many other irrational numbers. For example, we can now prove that the sum of *any* rational number and $\sqrt{2}$ is irrational. We again use an indirect proof.

Let n be any rational number. We wish to prove that $n + \sqrt{2}$ is irrational.

Proof

1 Assume that $n + \sqrt{2}$ is rational; that is that

$$n + \sqrt{2} = \frac{a}{b},$$

where a and b are integers ($b \neq 0$).

2 Then $\sqrt{2} = \frac{a}{b} - n$.

3 $\frac{a}{b} - n$ is a rational number by the closure property of subtraction for rational numbers.

4 Therefore, by *2*, $\sqrt{2}$ would have to be a rational number.

5 But since we have proven that $\sqrt{2}$ is an irrational number, our original assumption in *1* that $n + \sqrt{2}$ is rational leads to a contradiction and must, therefore, be false.

6 Therefore, $n + \sqrt{2}$ is irrational.

The proof we used to establish that $\sqrt{2}$ is irrational can be extended to establish the irrationality of $\sqrt{3}$; except that in the case of $\sqrt{3}$, the proof hinges on the question of divisibility by 3 instead of divisibility by 2 as in the case of $\sqrt{2}$.

Note The discovery that the square root of 2 is not a rational number is said to have been celebrated by the Greek philosophers with the sacrifice of 100 oxen.

9.5 *Another Way of Creating the Reals: Dedekind Cuts*

We used the concept of repeating and nonrepeating decimals to define rational and irrational numbers. Then we called the system that includes both the rationals and irrationals the *real number system*. There are other ways of creating the real number system. One such way is through *Dedekind cuts*, each *cut* being identified with a real number.

The *Dedekind cuts* approach was developed by the German mathematician, Richard Dedekind (1831-1916), who started with the system of rational numbers and then used the rational numbers to define a new kind of number, an irrational number, that did not belong in that system. The method he used was that of making a "cut" in the set of rational numbers, separating the set into two disjoint subsets. The characteristics of this separation determines whether the "cut" is to be identified with a rational number or with an irrational number. The "rational cuts" and the "irrational cuts" together make up the *real number system*.

Dedekind began with the following three properties of rational numbers:

1 Let a, b, and c be any three distinct rational numbers. If $a > b$ and $b > c$, then $a > c$.

2 Between any two rational numbers there are infinitely many other rational numbers.

3 Any rational number n divides the set of rational numbers into two disjoint subsets A and B such that A contains all those rational numbers that are *smaller* than n, and B contains all those rational numbers that are *greater* than n. The number n may be assigned either to A or to B. But no matter where we assign n, it will still be true that any number of A is smaller than any number of B.

If we assign n to A, then A has a *greatest* number, namely n, while set B has no *least* number since, by property *2*, there are infinitely many other rational numbers between n and any member of B. On the other hand, if we assign n to set B, then B has a *least* number, namely n. A has no *greatest* number since between any supposedly greatest number in A and the number n of B, there are infinitely many rational numbers.

We can, therefore, conclude that every rational number n produces a *cut*, or a separation, that possesses the property that either set A contains a *greatest* number *or* set B contains a *least* number.

Dedekind then exhibits a *cut* that differs from the other *cuts* in that it produces a separation of the rational numbers where set A has no greatest number *and* set B has no least number. Such a *cut* cannot be produced by a rational number since, as we have just seen, any rational number produces a *cut* in which either A has a greatest number *or* B has a least number. Dedekind calls this new *cut* an *irrational number.* By this definition, rational numbers and irrational numbers differ in just one respect: the presence or absence of a *least* or *greatest* number in sets A and B. We shall now illustrate this distinction.

Example 1 Let us choose the number 3 with which to produce a separation in the set of rational numbers. The *cut* produces two disjoint sets A and B that may be described in two different ways, depending on where we wish to assign the number 3:

$$A = \{x \mid x < 3\}, \qquad B = \{x \mid x \geqslant 3\} \tag{1}$$

or

$$A = \{x \mid x \leqslant 3\}, \qquad B = \{x \mid x > 3\}. \tag{2}$$

In (1), the number 3 was assigned to set B; in (2), the number 3 was assigned to set A.

In (1), A has no greatest number but B has a least number, namely 3. In (2), A has a greatest number, namely 3, while B has no least number. Since the number 3 separates the set of rational numbers in such a way that the *cut* has either a greatest number or a least number, the *cut* identified with the number 3 is called a *rational cut*.

Example 2 Suppose we choose the number $\sqrt{2}$ with which to separate the set of rational numbers. We can then describe the two sets produced by the *cut* as

$$A = \{x \mid x^2 < 2\}, \qquad B = \{x \mid x^2 \geqslant 2\} \tag{3}$$

or
$$A = \{x \mid x^2 \leqslant 2\}, \qquad B = \{x \mid x^2 > 2\}. \tag{4}$$

Since there is no rational number whose square is 2, no set in (3) or (4) contains a greatest or least number. Therefore, $\sqrt{2}$ is not identified with any rational number and is called an *irrational cut. Any cut* is called a *real number. Rational cuts* are called *rational real numbers; irrational cuts* are called *irrational real numbers.*

It can be shown that there is a one-to-one correspondence between the equivalence classes of all the *cuts* just described and the points on the number line.

9.6 From Real to Complex Numbers

As we expanded the counting numbers to the whole numbers, to the integers, to the rationals, and then to the reals, we noted that each extension removed shortcomings of the previous system. Although we might now consider the system of real numbers as adequate for our daily lives, we should note that even this system has shortcomings. We have seen that a number like $\sqrt{2}$, or the square root of any positive number, has meaning in the real number system. But what does $\sqrt{^{-}1}$, or the square root of any negative number mean? There is no real number which multiplied by itself will result in $^{-}1$ since the product of two positive numbers or of two negative numbers is always a positive number. So an equation like $x^2 + 1 = 0$ has no solution in the real number system.

To correct this inadequacy, we expand our number system once again. We introduce a new symbol, i, with this definition:

$$i = \sqrt{-1} \quad \text{and} \quad i^2 = \sqrt{^{-}1} \times \sqrt{^{-}1} = {^{-}1}.$$

We call i an *imaginary number*, and the system that contains both the real numbers and the imaginary numbers is called the *complex number system.*

When these new numbers were first introduced in the sixteenth century they were called "imaginary" because people were accustomed to think only in terms of positive and negative numbers. Today we realize that imaginary numbers are no less real than the real numbers since all numbers are abstractions invented by man to meet certain needs.

Any *complex number* may be expressed in the form

$$a + bi,$$

where a and b are real numbers and $i = \sqrt{-1}$ such that $i^2 = -1$. In the number $a + bi$, a represents the *real* part and bi represents the *imaginary* part. If $a = 0$ and $b \neq 0$, the number is a *pure imaginary number;* if $b = 0$ and $a \neq 0$, the number is a *real* number. Every real number may be regarded as a complex number of the form $a + bi$ in which $b = 0$. For example, $3 = 3 + 0i$. Two complex numbers $a + bi$ and $c + di$ are *equal* if and only if $a = c$ and $b = d$.

GEOMETRIC INTERPRETATION OF COMPLEX NUMBERS

Complex numbers can be pictured geometrically just like real numbers. We pictured real numbers by all the points on a line. We picture complex numbers by all the points in a plane. To do this, we use a pair of perpendicular axes, where the horizontal axis represents the scale of *real* numbers and the vertical axis represents the scale of *imaginary* numbers (Figure 9.2).

FIGURE 9.2

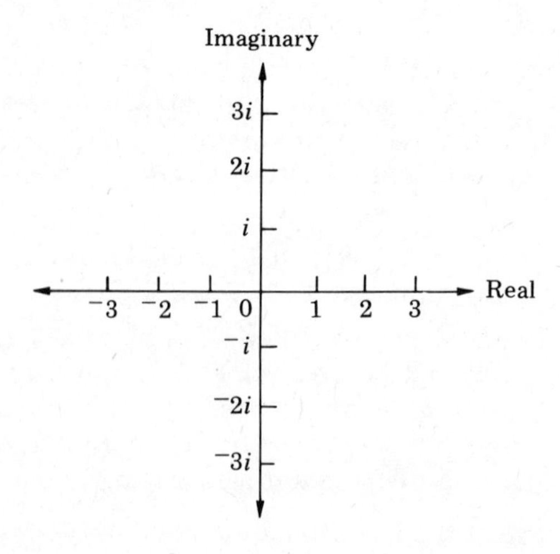

A complex number like $2 + 3i$ would, therefore, be represented by point P (Figure 9.3).

FIGURE 9.3

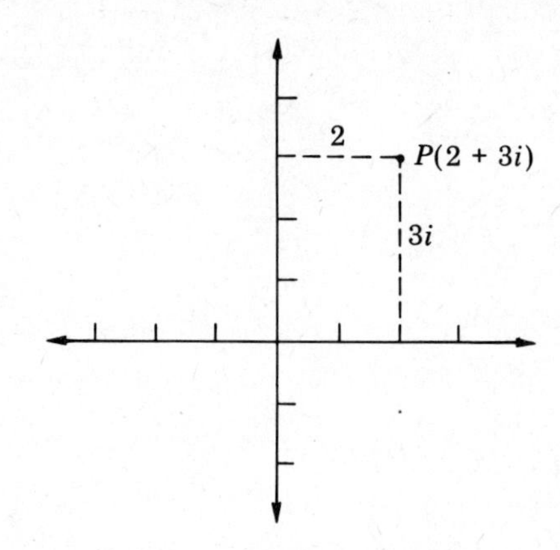

To assign a number to any point R in the plane (Figure 9.4), we draw a horizontal line and a vertical line through the point R. The sum of these two numbers is the complex number assigned to this point; in this case: $^-1 + 4i$.

FIGURE 9.4

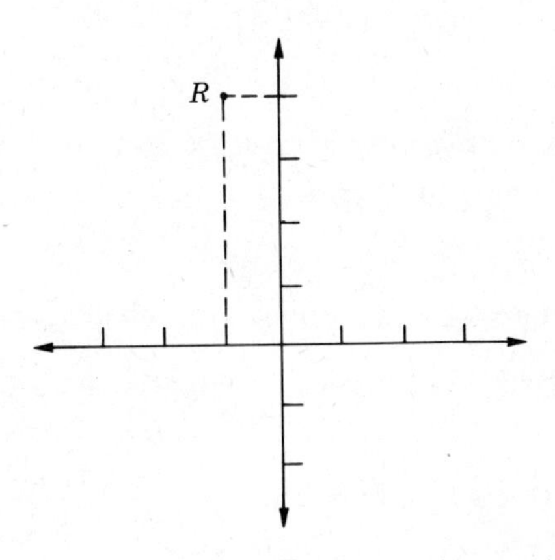

The complex number $a + bi$ can be represented geometrically not only by the point (a, b) but also by a directed line segment from the origin to this point. So that the point $3 + 2i$ is represented by the point P and also by the directed line segment $\overrightarrow{OP}$ (Figure 9.5). The length of $\overrightarrow{OP}$ is called the *modulus* of the complex number; the angle θ is called the *amplitude* of the complex number. The directed line segment $\overrightarrow{OP}$ is called a *vector*.

FIGURE 9.5

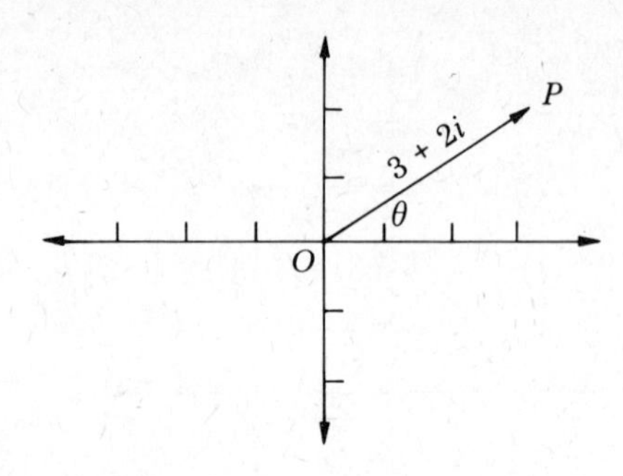

OPERATIONS WITH COMPLEX NUMBERS

Since we want the complex number system to contain all the previous number systems, we must define addition and multiplication of complex numbers in such a way as to preserve the properties of the earlier systems. It can be shown that the following definitions meet this requirement.

DEFINITION 1 The *sum* of two complex numbers $a + bi$ and $c + di$ is the complex number

$$(a + c) + (b + d)i.$$

Example 1 $(2 + 5i) + (7 - 3i) = (2 + 7) + (5 - 3)i = 9 + 2i$. If we arrange these additions vertically, we get:

$$\begin{array}{r} a \ + bi \\ + c \ + di \\ \hline (a + c) + (b + d)i \end{array} \qquad \begin{array}{r} 2 + 5i \\ + 7 - 3i \\ \hline 9 + 2i \end{array}$$

The second addition is shown geometrically in Figure 9.6.

FIGURE 9.6

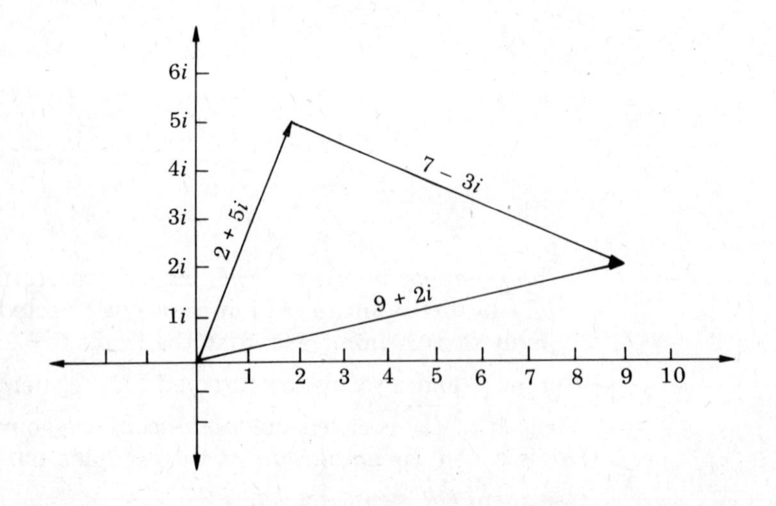

DEFINITION 2 The *product* of two complex numbers $a + bi$ and $c + di$ is the complex number $(ac - bd) + (ad + bc)i$.

That is,

$$\begin{array}{lll} & a + bc & \\ \times & c + di & \\ \hline & ac + bci & \\ & \quad + adi & + bd(i)^2 \\ \hline & ac + (bc + ad)i & + bd(i)^2 \end{array}$$

Since $i^2 = {}^{-}1$, we have

$$ac + (bc + ad)i + bd(i)^2 = ac + (bc + ad)i - bd$$

or $$(ac - bd) + (ad + bc)i.$$

Example 2 $(3 + 2i)\,(2 + 5i) = (3 \times 2 - 2 \times 5) + (3 \times 5 + 2 \times 2)i = {}^{-}4 + 19i.$

Arranged vertically, we have

$$\begin{array}{ll} & 3 + 2i \\ \times & 2 + 5i \\ \hline & 6 + 4i \\ & \quad + 15i + 10(i)^2 \\ \hline & 6 + 19i + 10(i)^2 = 6 - 10 + 19i = -4 + 19i. \end{array}$$

As in the case of real numbers, *subtraction* of complex numbers is defined in terms of addition, and *division* is defined in terms of multiplication. But in extending the real number system to the complex number system, we have to pay a price. We have to sacrifice the *order relation*, greater than and less than, for complex numbers. We cannot say, for example, that $1 + 3i < 3 + 2i$; but by the definition of equality, we can say that

$$1 + 3i \neq 3 + 2i.$$

The set of complex numbers forms a field (but not an ordered field) since it can be shown to satisfy all the field properties on page 185.

Although the complex numbers are sufficient for most practical applications, still other numbers have been created. For example, *quaternions* are ordered quadruples (a, b, c, d) where a, b, c, and d are real numbers; *matrices* are rectangular arrays of real numbers; *vectors* are ordered triples of real numbers; and *Gaussian integers* are numbers of the form $a + bi$ where a and b are restricted to being integers.

The creation of all new numbers beyond the reals required the sacrifice of properties that held for earlier number systems. In the case of the complex numbers, we saw that we sacrificed the order relation. Quaternions, vectors, and matrices all sacrifice the commutative property for multiplication; that is, in all these systems, it is not necessarily true that $a \times b = b \times a$. Gaussian integers fail to satisfy the Fundamental Theorem of Arithmetic, which states that every integer can be uniquely factored into prime factors.

Chapter 10

GEOMETRY

Chapter 10A

GEOMETRIC CONCEPTS

Objectives

At the completion of this chapter, you should be able to:

1. Recognize the abstract character of *points*, *lines*, and *planes* and be able to illustrate each concept with physical examples.
2. State the *incidence properties* of points and lines in the plane.
3. State the *incidence properties* of points, lines, and planes in space.
4. State the *separation properties* of a line by a point; of a plane by a line.
5. Recognize and define *closed curves*, *simple curves*, and *simple closed curves.*
6. Define and identify: *half-line*; *ray*; *segment*; *polygon*, regular polygon, polygonal region; *angle*, right angle, acute angle, obtuse angle; dihedral angle; complementary angles, supplementary angles, vertical angles, linear pair.
7. Define and identify the following types of triangles: *equilateral*, *isoscles*, *scalene*, *right*, *acute*, *obtuse*, *equiangular.*
8. Distinguish between *intersecting lines*, *parallel lines*, and *skew lines.*

9. Define and state some basic properties of a trapezoid, parallelogram, rhombus, rectangle, square, circle.
10. Give the general definition of *congruence.*
11. State the basic congruence theorems for triangles.
12. State the Pythagorean theorem and its converse, and identify the kinds of problems that can be solved by each theorem.
13. Define *similarity* and state its basic properties.
14. Define a *sphere.*
15. Define and identify a *polyhedron*, a *prism*, a *pyramid.*

Prerequisites

You should

1. Understand the concepts of set, subset, union and intersection of sets, one-to-one correspondence.
2. Know the definition of *real number* and of *absolute value.*
3. Know the meaning of *ratio.*
4. Be able to use a ruler and protractor to draw and measure segments and angles.
5. Be able to find the square root of a number.
6. Be able to solve simple fractional equations such as $\frac{2}{3} = \frac{5}{x}$.

INTRODUCTION

The earliest study of geometry goes back four thousand years when it began with measurement. Two thousand years later the Greeks took this physical concept of geometry and transformed it from a science of measurement to a system of reasoning. In this system we arrive at conclusions not by measuring but by reasoning.

Before we can understand what geometry is all about we must first understand the building blocks on which it rests—ideas such as *point*, *space*, *curve*, *line*, *segment*, *ray*, and *plane*. Although we use these ideas over and over again in geometry, we do not define all of them. Some of these ideas remain *undefined*, but acquire meaning through a description of some of their characteristics. We then use these undefined terms as a basis for *defining* other terms.

Why do we use *undefined terms*? Why don't we *define every* term? We can see the reason for this by a comparison with the way we learn a foreign language. Suppose you want to learn French but do not know *any* French words. Would it help you to use an all-French dictionary? If you looked up a French word in such a dictionary, you would find a definition that only uses other French words, and this would not help you very much. But if you first learn the meanings of a few French words *before* using such a dictionary, you can then use these words as a basis for learning other words.

This is also true in mathematics. For example, we could define a "triangle" in terms of "line segments"; but what are line segments? We could then define a line segment in terms of "points"; but what are points? We can see that sooner or later this process must come to an end because with each new definition, we can pose the same question: what are the definitions of the terms used in the definition? We must, therefore, use *some* terms without definitions and then use these *undefined terms* to define other terms.

How do we know the meaning of a term if we do not define it? We can obtain a good idea of its meaning from a description of some of its characteristics. For example, the term *point* in geometry is often taken as undefined. But we can obtain a good idea of what it means by saying that a point is suggested by a dot on a piece of paper and represents an exact location. With these assumed properties of a point in mind, we proceed to *define* such terms as segment and triangle.

Another ingredient in a mathematical system, besides undefined terms and definitions, is a set of *postulates.* A *postulate* is a statement that we *assume* to be true. Postulates are also called *assumptions* and *axioms.* By applying deductive reasoning to the undefined terms, definitions, and postulates, we prove other statements, called *theorems*, to be true. This, then, is the essence of a mathematical system such as geometry: it starts with a few *undefined terms*, *definitions*, and *postulates* and from these, by the use of logical reasoning, are derived *theorems.*

Where do postulates come from? For more than two thousand years, the postulates of geometry were thought to be "self-evident truths" about physical space. But with the development of the non-Euclidean geometries about two hundred years ago (see page 349) this view was shown to be untenable. Today we regard postulates as nothing more than *assumptions.* Although postulates are often suggested by the physical world, this need not be their origin. The only *necessary* condition for our choice of postulates is that the system be internally *consistent.* This means that no postulate within the system contradicts any other postulate in the system. This condition ensures against proving two contradictory theorems from the same set of postulates.

In this chapter most of the discussion will be informal and intuitive in nature. Some ideas, however, will be treated more formally in order to provide a broader and deeper insight into the nature of mathematics.

10A.1 Points, Lines, and Planes

We begin our geometry with three undefined terms: *point*, *line*, and *plane*. These undefined terms may be described informally in the following way.

POINT

The basic ingredient of all geometric figures is a *point*. A point may be thought of as an exact location in space. A point is suggested by the head of a pin, the tip of a pencil, or a speck of dust, and can be represented by a dot on a paper. A point does not move. When a dot, representing a point, is erased, the point, that is, the location, still remains. A point has no size; it has position only. It is named by a capital letter.

LINE

A *line* is a set of points which may be thought of as a string tightly stretched, infinitely far, in both directions. In all our discussions, a line will mean a *straight line*. When we draw a picture of a line, we can draw a picture of only *part* of the line. We indicate that a picture represents a line by attaching arrows to both ends of the line (Figure 10A.1).

FIGURE 10A.1

A line has neither width nor thickness, only length. We usually name a line either by a single lowercase letter, such as m, or by any two points on the line, such as A and B, denoted by $\overleftrightarrow{AB}$ (Figure 10A.2).

FIGURE 10A.2

A *piece* of a line, with definite *endpoints* A and B, as shown in Figure 10A.2, is called a *line segment* and is denoted by $\overline{AB}$.

A line that stretches infinitely far in only *one* direction is called a *ray*. A good physical representation of a ray is a beam of light emanating from a pin-point source. If the source is at A (Figure 10A.3) and the beam shines in the direction of B, then we denote ray AB by $\overrightarrow{AB}$.

FIGURE 10A.3

A B

PLANE

A *plane* is a set of points suggested by any flat surface such as a wall, floor, or desk top if you can imagine that the surface stretches infinitely far in every direction. A plane has no thickness. It is usually named by a capital letter or by three of its points all of which do not lie on the same line (Figure 10A.4).

FIGURE 10A.4

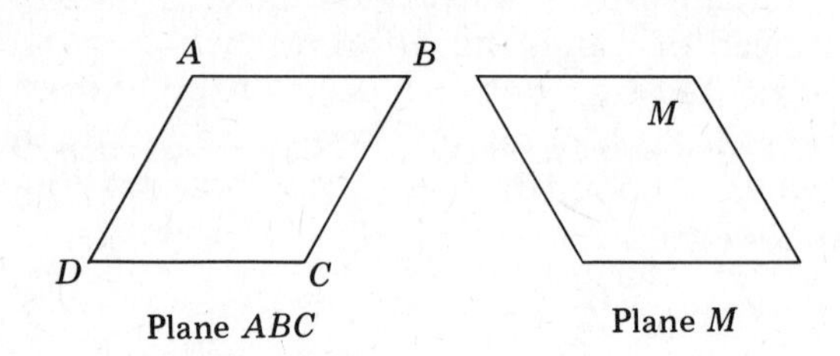

POSTULATES OF INCIDENCE

We shall now move to a more formal level in the development of our geometry by stating six *postulates* that describe the relationships that exist among points, lines, and planes.

POSTULATE 1 All lines and planes are sets of points.

DEFINITION The set of all points is called *space.*

POSTULATE 2 Given any two different points, there is exactly one line containing them.

If the points are A and B, then the line containing them is denoted by $\overleftrightarrow{AB}$.

DEFINITIONS

1. Points lying on one line are *collinear.*
2. Points lying in one plane are *coplanar.*

Let us agree that:

1. If a line is a subset of a plane, we will say that the line "lies in" the plane, or that the plane "contains" the line.

2 If a point belongs to a line, we will say that the point "lies on" the line, that the line "passes through" the point, or that the line "contains" the point.

3 If a point belongs to a plane, we will say that the point "lies in" the plane, that the plane "passes through" the point, or that the plane "contains" the point.

POSTULATE 3 Three different noncollinear points are contained in exactly one plane (Figure 10A.5).

FIGURE 10A.5

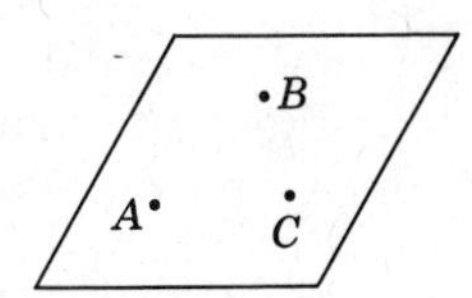

POSTULATE 4 If a plane contains two distinct points, then the line containing them lies in the plane (Figure 10A.6).

FIGURE 10A.6

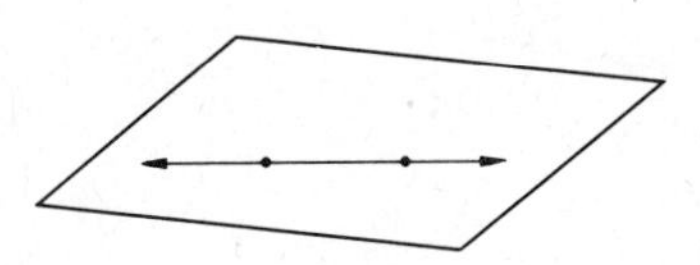

POSTULATE 5 If two distinct planes intersect, then their intersection is a line (Figure 10A.7).

FIGURE 10A.7

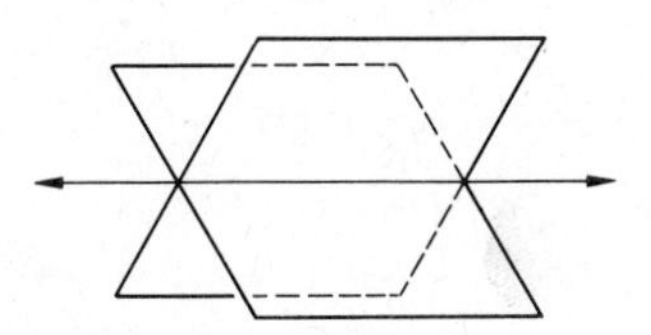

POSTULATE 6 Every line contains at least two distinct points. Every plane contains at least three noncollinear points. Space contains at least four noncoplanar points.

Comments

1 *Postulates 1–6 are called* postulates of incidence. *They express the basic positional relationships of points, lines, and planes. Euclid never spelled out these postulates in his list of axioms, although he assumed them in some of his proofs. In a tight deductive system, unstated assumptions are not permitted.*

2 *Another way in which Postulate 2 is often expressed is "Two distinct points determine a line."*

3 *Postulate 3 can be expressed this way: three noncollinear points determine a plane.*

4 *Postulate 4 suggests itself intuitively by the straightness of the line and the flatness of the plane.*

5 *Postulate 5 can be "seen" in the classroom through the intersection of the floor and the walls and the intersection of the ceiling and the walls.*

THEOREMS OF INCIDENCE

From the undefined terms and postulates which we have just introduced, we can derive four theorems that tell us more about the relationships among points, lines, and planes. We shall state but not prove these theorems.

THEOREM 1 If two distinct lines intersect, then their intersection contains exactly one point (Figure 10A.8).

FIGURE 10A.8

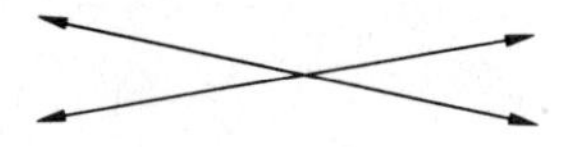

THEOREM 2 If a line intersects a plane not containing it, then the intersection is a single point (Figure 10A.9).

FIGURE 10A.9

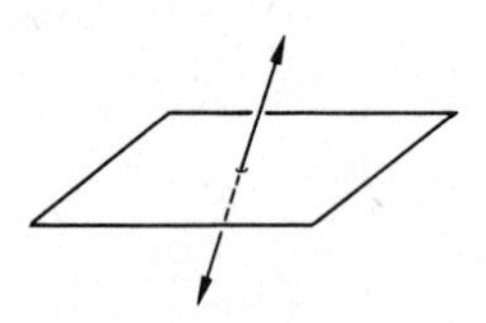

THEOREM 3 If l is a line and P is a point not on l, then there is one and only one plane containing l and P (Figure 10A.10).

FIGURE 10A.10

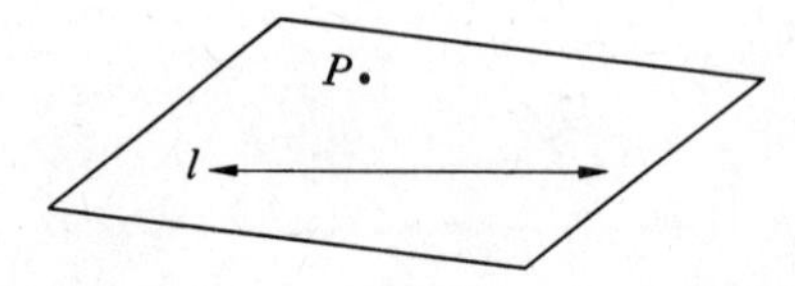

THEOREM 4 If two distinct lines intersect, then there is exactly one plane containing them (Figure 10A.11).

FIGURE 10A.11

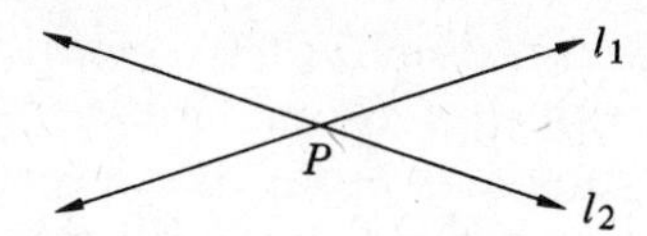

Comments

1 *Just as Postulates 1–6 are called* incidence postulates, *so are Theorems 1–6 called* incidence theorems. *They deal with such questions as whether and how points, lines, and planes intersect and whether one contains the other.*

2 *By the postulates and theroems given so far, we have established the following ways of determining a plane:*

a) By any three noncollinear points (Postulate 3)

b) By any line and any point not on the line (Theorem 3).

c) By any two distinct intersecting lines (Theorem 4).

PARALLEL LINES AND PLANES

Although the geometry of parallel lines will be discussed later, the definition of parallelism will be given here because the concept is an important one in the theory of incidence.

DEFINITIONS See Figure 10A.12.

1 Two *lines are parallel* if they are contained in the same plane and do not intersect. If line m is parallel to the line k, we write $m \parallel k$.

2 Two lines are said to be *skew lines* if and only if they are not in the same plane.

3 Two *planes are parallel* if they do not intersect.

4 *A line and a plane are parallel* if they do not intersect.

FIGURE 10A.12

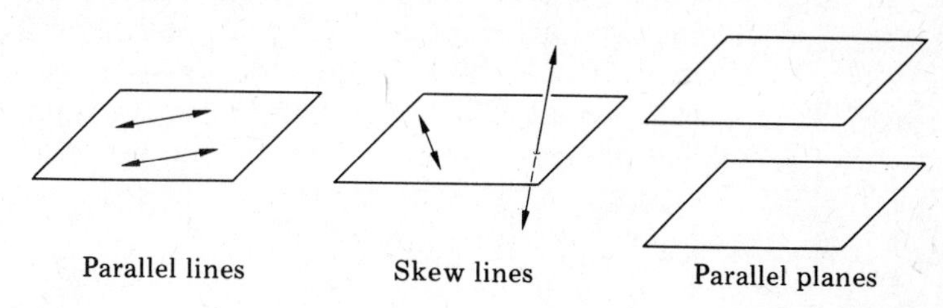

SEPARATION

We shall now see how basic geometric figures can be divided, or *separated*, into other figures. Before examining this idea, we note that the ideas of *separation* find frequent application in everyday situations. For example, a room divider (screen) "separates" the room into the set of points in front of the screen and the set of points behind the screen. An aisle in an auditorium separates the seats on the left side from the seats on the right side. Any mark on the ruler, say the 5-inch mark, separates the set of points before the 5-inch mark from the set of points after the 5-inch mark.

Although the idea of *separation* is suggested by these physical illustrations, it is important to remember that *geometric separation* involving abstractions such as points, lines, and planes is not a physical process but an interrelationship among sets, as we shall now see.

Line Separation Consider line l and a point P on l (Figure 10A.13). We say that *P separates l* into the following two sets of points:

1 The set L_1 on one side of P.

2 The set L_2 on the other side of P.

FIGURE 10A.13

L_1 L_2 l

P

The point P itself is not in either set.

L_1 and L_2 are called *half-lines.* A half-line is like a ray without the endpoint.

Note that a line segment has two endpoints, a ray has one endpoint, and a half-line has no endpoints.

If two points on line l belong to the same half-line, we say they are *on the same side of P.* If they are on different half-lines, then they are *on opposite sides of P.*

Plane Separation The separation of a plane by a line is analogous to the separation of a line by a point. Just as a point separates a line into two half-lines, so does a line separate a plane into two *half-planes.* The line serves as a boundary between the two half-planes but does not belong to either. In Figure 10A.14, l separates the plane X into two *opposite half-planes,* P_1 and P_2. l is called the *edge* of each half-plane.

FIGURE 10A.14

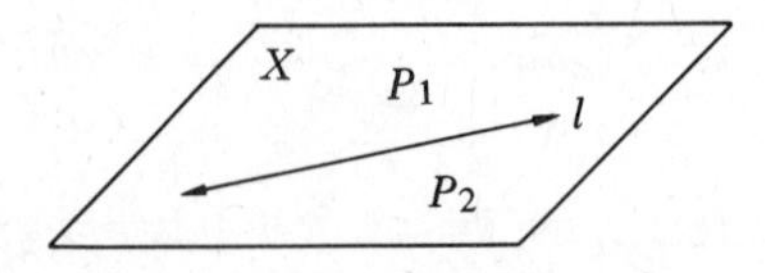

What we are saying, in effect, is that any line of a plane separates the plane into three disjoint sets of points:

1 The set of points on one side of the line.

2 The set of points on the other side of the line.

3 The set of points of the line itself.

Comments

1 *If we do not restrict ourselves to one plane, we can have infinitely many half-planes having l as an edge. For example, the various positions of a revolving door or the pages of a book represent different half-planes having the same edge.*

2 *A half-plane can be seen, intuitively, as a generalization of a* ray. *Think of a* ray *as the set of points of a line that is on a given side of a given point of the line. Now think of a* half-plane *as the set of points of a plane that is on a given side of a given line of the plane.*

Space Separation We extend the concept of separation to the separation of space. We say that *any plane separates space* into three disjoint sets of points:

1 The set of points on one side of the plane.

2 The set of points on the other side of the plane.

3 The set of points of the plane itself.

The set of points on each side of the plane is called a *half-space* and the plane is called the *face* of each half-space.

SIMPLE CLOSED CURVES

A *curve* is a set of points that can be represented by a pencil drawing without lifting the pencil off the paper. Curves are, therefore, *connected*, or *continuous* sets of points.

Examples of curves are given in Figure 10A.15. Examples of sets of points that are *not* curves are given in Figure 10A.16.

FIGURE 10A.15

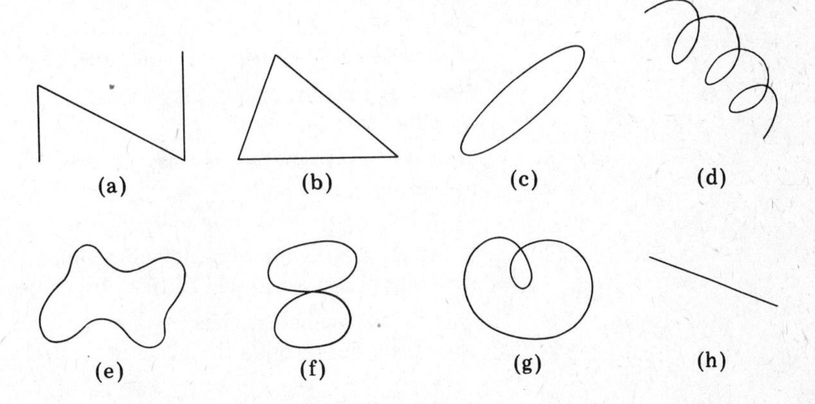

FIGURE 10A.16

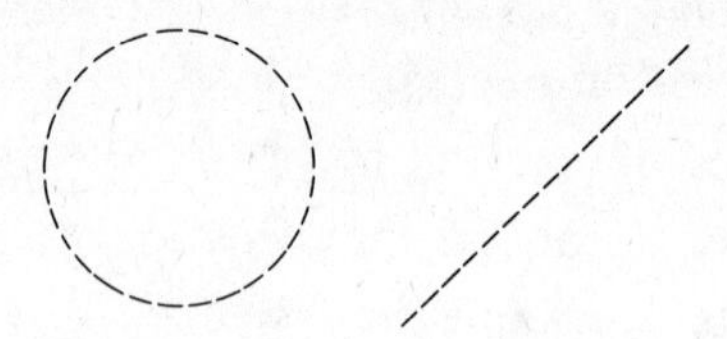

A *closed* curve is a curve that returns to its starting point. In the curves shown in Figure 10A.15, (b), (c), (e), (f), and (g) are closed curves, and (a), (d), and (h) are not closed curves.

A planar closed curve that does not cross itself is called a *simple closed curve.* In Figure 10A.15, (b), (c), and (e) are simple closed curves, and (a), (d), (f), (g), and (h) are not simple closed curves. Since (f) and (g) are closed but cross themselves, they are *nonsimple* closed curves.

A planar simple closed curve *and* its interior is called a *region.* The curve itself is called the *boundary* of the region. (See Figure 10A.17.)

FIGURE 10A.17

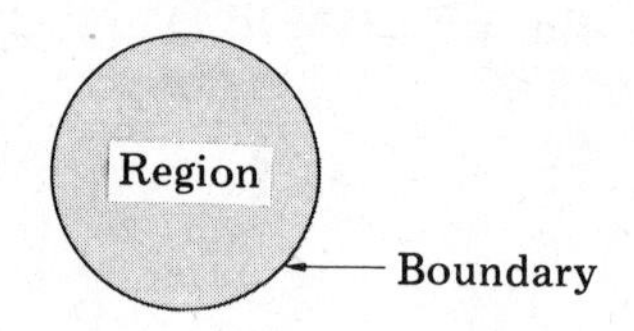

EXERCISE SET 10A.1

1. a) If three points are coplanar, must they also be collinear?
   b) If three points are collinear, must they also be coplanar?
2. Draw a simple closed curve. Locate point A inside the curve and point B outside the curve such that the intersection of $\overline{AB}$ and the curve is three points.
3. Given: four points of which no three are on the same line, and the points are not all in one plane. If planes are passed through every set of three points, how many planes are there altogether?
4. What is the most number of planes determined by the ends of four legs of a table if no two legs have the same length?
5. How many different lines may contain:
   a) One common point? b) A pair of common points?
6. How many different planes may contain:
   a) One common point? b) A pair of common points?
   c) A set of three common noncollinear points?
7. If A, B, and C are three distinct points and all are in each of two different planes, what will be true about these points?
8. Give a mathematical reason for the fact that a four-legged table sometimes rocks, but a three-legged table is always stable.

9 If two planes have a common point, does it follow that they have more than one common point? Why?

10 Restate the following postulates in the *if . . . then* form.
a) Three different noncollinear points are contained in exactly one plane.
b) Every line contains at least two points.

11 Is the statement "If a figure does not contain at least three noncollinear points, then it is not a plane" equivalent to the statement "Every plane contains at least three noncollinear points"? Why?

10A.2 Angles and Triangles

WHAT IS AN ANGLE?

Past experience with angles may lead us to associate an *angle* with the amount of rotation or difference of rotation between two rays, or with a region of the plane bounded by two intersecting lines. But the idea of *rotation* is really related to *measurement* of an angle, while the idea of a region between two intersecting lines really relates to the *interior* of an angle.

We shall make a careful distinction between an *angle*, its *measure*, and its *interior*. We shall think of an angle as a geometric figure, that is, as a set of points, rather than as an amount of rotation. Although such a view of an angle is inadequate in trigonometry and in coordinate geometry, it is adequate for Euclidean geometry.

DEFINITION An *angle* is the union of two distinct rays that have the same endpoint but do not lie on the same line (Figure 10A.18). The endpoint is called the *vertex* of the angle, and the two rays are called the *sides* of the angle.

FIGURE 10A.18

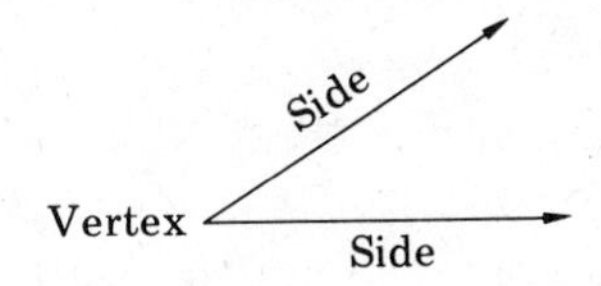

Example The angle formed by $\overrightarrow{BA}$ and $\overrightarrow{BC}$, as shown in Figure 10A.19, is called angle ABC, denoted by $\angle ABC$, or angle CBA, denoted by $\angle CBA$. Note that the middle letter always names the vertex. Other ways of naming the angle shown are $\angle B$ and $\angle 1$.

FIGURE 10A.19

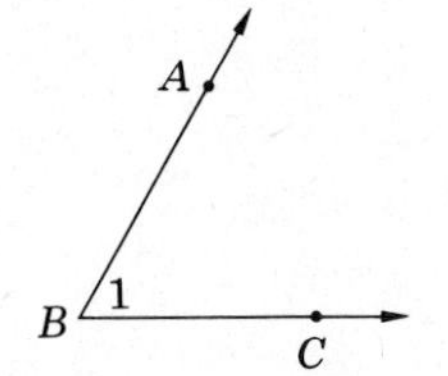

Comments

1 *The sides of an angle extend indefinitely in two directions because the sides are rays rather than segments.*

2 *We always have* $\angle ABC = \angle CBA$ *because each is the union of the rays* $\overrightarrow{BA}$ *and* $\overrightarrow{BC}$.

3 *Points* between *the rays are not part of the angle.*

We can easily extend our definition of an angle on a plane to define a *dihedral angle* in space. Two rays with the same endpoint, not on the same line, define a planar angle. Two noncoplanar half-planes with a common edge form a *dihedral angle.* More precisely:

DEFINITION A *dihedral angle* is the union of a line and two noncoplanar half-planes having the line as an *edge* (Figure 10A.20).

FIGURE 10A.20

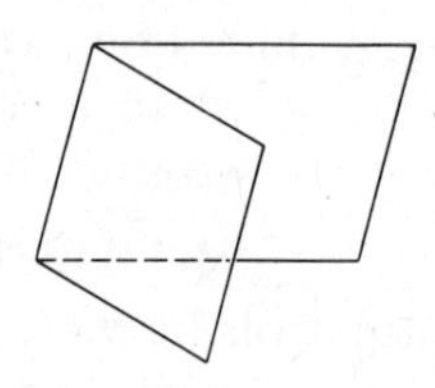

MEASUREMENT OF ANGLES

1 To measure an angle, we first select an arbitrary angle, say $\angle a$ (Figure 10A.21), to serve as a *unit* and agree that its measure is the number 1. We write this as $m\angle a = 1$, which we read as "the measure of $\angle a$ is 1."

FIGURE 10A.21

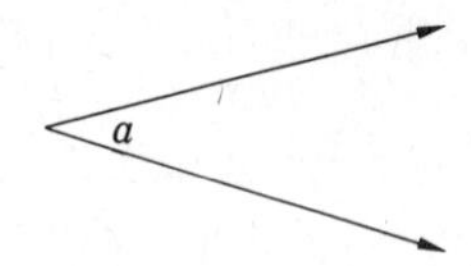

2 We can now form an $\angle ABC$ with measure 2 by laying off the unit angle $\angle a$ twice about a common vertex B (Figure 10A.22). In terms of the unit $\angle a$, the measure of $\angle ABC$ is 2. We write this as $m\angle ABC = 2$.

FIGURE 10A.22

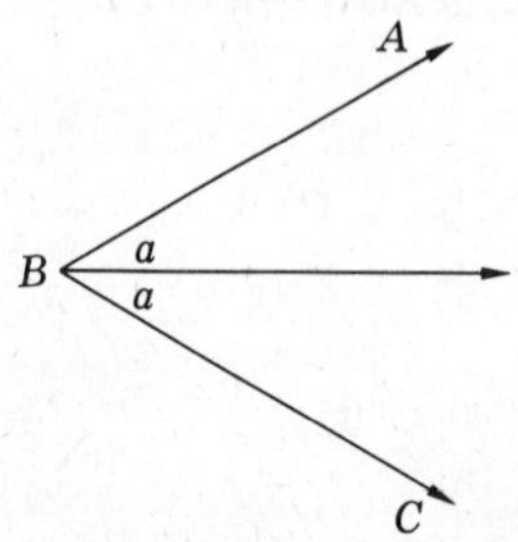

3 In a similar way we can form angles whose measures are 3, 4, etc.

4 A *standard* unit of measure is as useful in measuring angles as in measuring segments. The most common *standard* unit of angle measure is called a *degree.* One degree is written 1°.

5 If we lay off 360 unit angles about a common vertex B, then they, together with their interiors, cover the entire plane (Figure 10A.23).

FIGURE 10A.23

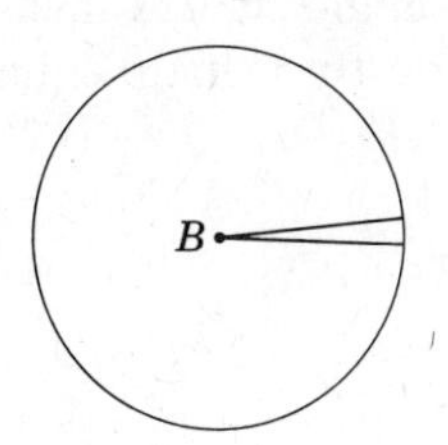

6 The number of degrees in an angle is called its *measure.*

7 An instrument used to determine the measure of an angle is called a *protractor.* The protractor is a scale of degree units marked, usually in a semicircle, from 0 to 180. In Figure 10A.24,

$m\angle ABC = 40$, $m\angle ABD = 90$,

$m\angle ABE = 120$, $m\angle ABF = 135$,

$m\angle CBD = 90 - 40 = 50$.

FIGURE 10A.24

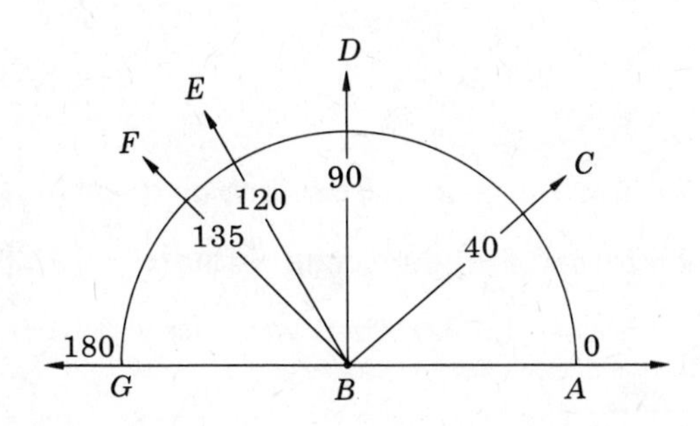

Comments

1 *Note that an* angle *is a* set of points, *while its* measure *is a* number.

2 *A protractor is used for finding the measures of given angles, but it is also used for drawing angles with a given measure.*

3 *We write* $m\angle ABC = 40$, *not* $m\angle ABC = 40^\circ$, *because the measure is a* number. *But in labeling figures, we use the*

degree sign (°) to indicate that the letters or numbers are meant to be the degree measures of the angle rather than the name of the angle (Figure 10A.25).

FIGURE 10A.25

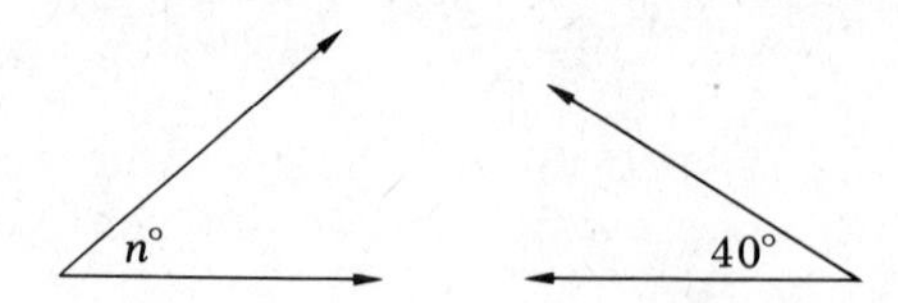

DEFINITIONS

1 A *right angle* is an angle whose measure is 90. We sometimes indicate a right angle by a little square at the vertex (Figure 10A.26).

FIGURE 10A.26

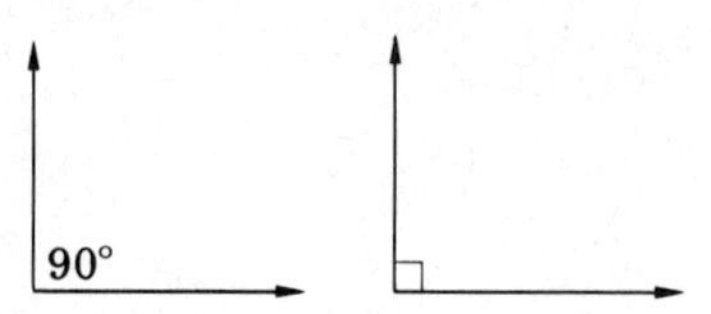

2 An *acute angle* is any angle whose measure is less than 90 (Figure 10A.27).

FIGURE 10A.27

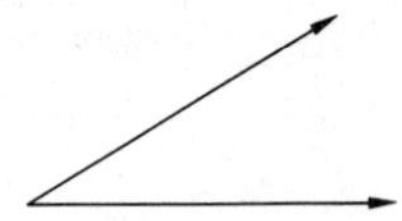

3 An *obtuse angle* is any angle whose measure is greater than 90 and less than 180 (Figure 10A.28).

FIGURE 10A.28

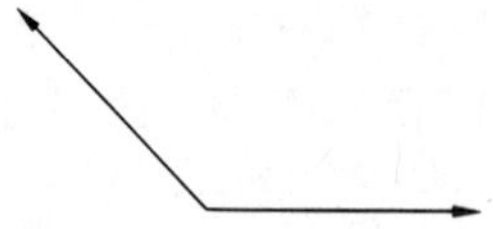

4 Two lines are *perpendicular* to each other if their intersection forms a right angle (Figure 10A.29). If $\overleftrightarrow{AB}$ and $\overleftrightarrow{CD}$ are perpendicular, then we write $\overleftrightarrow{AB} \perp \overleftrightarrow{CD}$.

FIGURE 10A.29

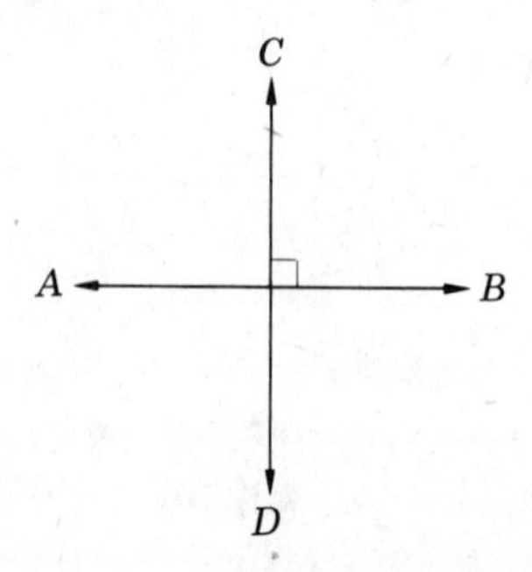

Note The definition of perpendicularity applies to any two sets, either or both of which is a line, a ray, or a segment.

ANGLE RELATIONSHIPS

DEFINITIONS

1 If the sum of the measures of any two angles is 90, then the angles are *complementary*, and each angle is said to be the complement of the other.

2 If $\overrightarrow{AB}$ and $\overrightarrow{AC}$ are opposite rays and $\overrightarrow{AD}$ is any third ray, then $\angle DAB$ and $\angle DAC$ form a *linear pair* (Figure 10A.30).

FIGURE 10A.30

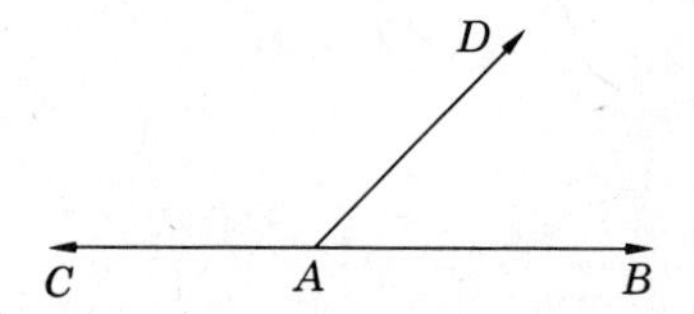

3 If the sum of the measures of two angles is 180, then the angles are called *supplementary*, and each angle is said to be the supplement of the other.

4 Two angles in a plane are said to be *adjacent* if they have a side in common and their interiors do not intersect (Figure 10A.31).

FIGURE 10A.31

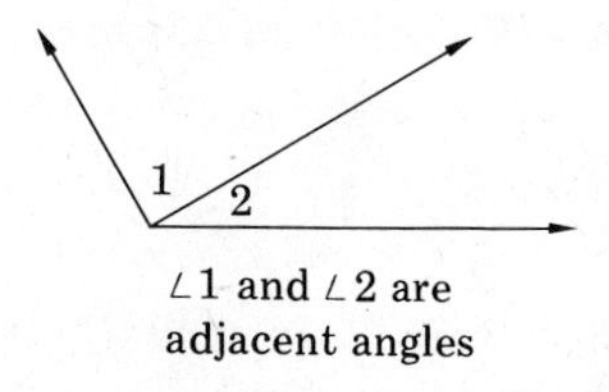

5 Two angles are *vertical* angles if their union is the union of two intersecting lines (Figure 10A.32).

FIGURE 10A.32

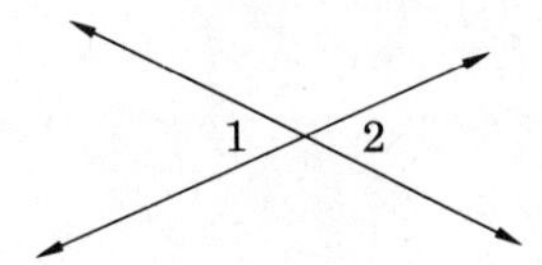

Comments

1 *If two angles form a linear pair, then they are adjacent; but if two angles are adjacent, they do not necessarily form a linear pair.*

2 *If two angles form a linear pair, then they are supplementary.*

WHAT IS A TRIANGLE?

DEFINITION If A, B, and C are three noncollinear points, then the set $\overline{AB} \cup \overline{BC} \cup \overline{AC}$ is called a *triangle* (Figure 10A.33).

FIGURE 10A.33

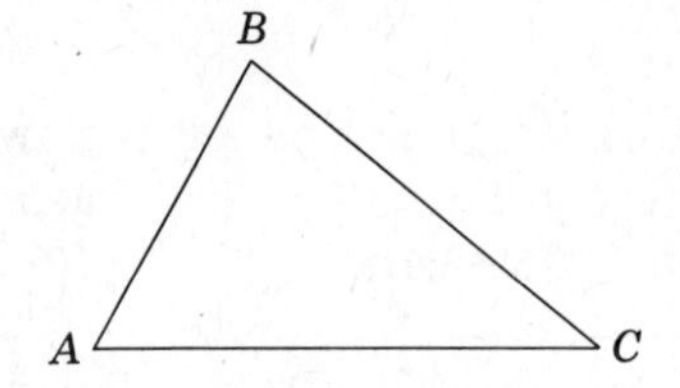

The triangle is usually denoted by the symbol $\triangle ABC$.

The points A, B, and C are called the *vertices* of the triangle. (Each point is called a *vertex.*)

The segments $\overline{AB}$, $\overline{BC}$, and $\overline{CA}$ are called the *sides* of the triangle.

$\angle A$, $\angle B$, and $\angle C$ are called the *angles* of the triangle.

Comments

1 *Since three noncollinear points determine a unique plane, a triangle also determines a unique plane.*

2 *Note that while* $\triangle ABC$ determines *three angles, it does not actually* contain *them because the sides of a triangle are segments while the sides of an angle are rays. Although a triangle does not contain its angles, it does contain its sides.*

CLASSIFICATION OF TRIANGLES

Triangles can be classified by comparing their *sides* or their *angles.*

1 Comparing the *sides* of a triangle (Figure 10A.34),

a) If all three sides are congruent, the triangle is *equilateral.*

b) If two sides are congruent, the triangle is *isosceles.*

c) If no two sides are congruent, the triangle is *scalene.*

FIGURE 10A.34

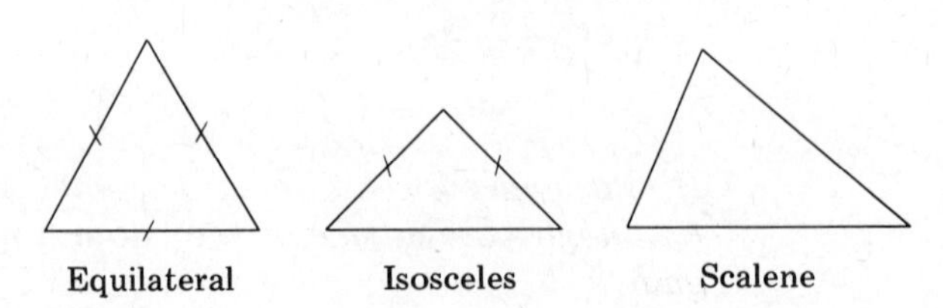

2 Comparing the *angles* of a triangle (Figure 10A.35),

a) One angle in the triangle may be a right angle. We call such a triangle a *right* triangle.

b) All the angles in the triangle may be acute. We call such a triangle an *acute* triangle.

c) One angle in the triangle may be obtuse. We call such a triangle an *obtuse* triangle.

d) All the angles in the triangle may be congruent. We call such a triangle an *equiangular* triangle.

FIGURE 10A.35

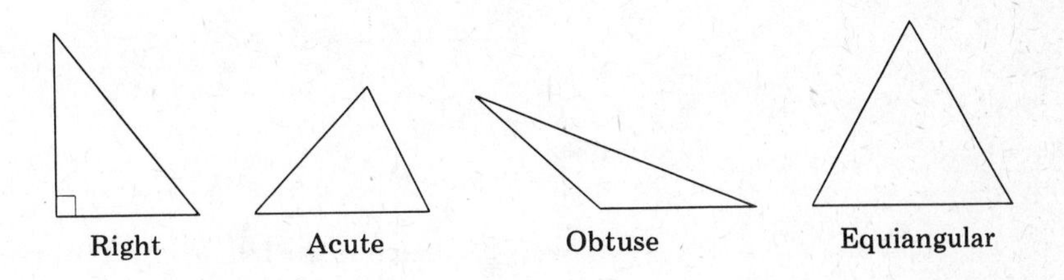

EXERCISE SET 10A.2

1 Name all the angles determined by the accompanying figure.

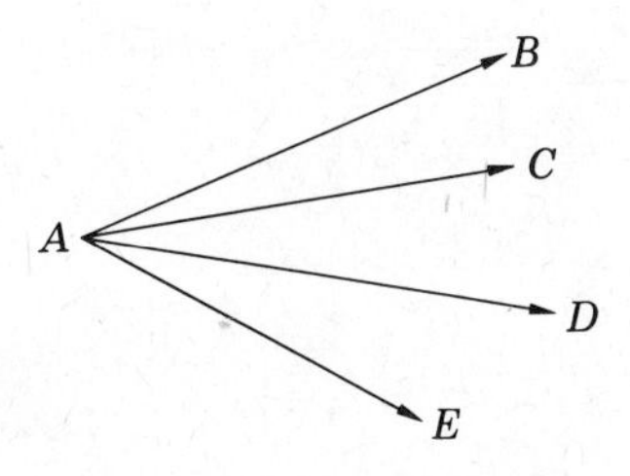

2 Make *three* true statements that begin with "If two lines intersect . . . ."

3 If $\angle A$ and $\angle B$ are complementary and $\angle A$ has a measure of 30° more than $\angle B$, what is the measure of $\angle A$ and $\angle B$?

4 If $\angle X$ and $\angle Y$ are supplementary and $\angle X$ is twice the measure of $\angle Y$, what is the measure of $\angle X$ and $\angle Y$?

5 State true or false, and give a reason for your answer.

a) If two angles are complementary, then each of them is acute.

b) Not every angle determines a unique plane.

c) If two angles are supplementary, then one of them must be obtuse.

d) An angle is the union of any two rays.

e) Vertical angles need not be in the same plane.

f) Vertical angles cannot be supplementary.

6 What is the measure of an angle whose measure is twice the measure of its complement?

7 The measure of an angle is five times the measure of its supplement. Find the measure of each angle.

8 If the measure of an angle is a, what is the measure of its supplement?

9 If the measure of one of a pair of vertical angles is m, what are the measures of the other three angles formed?

10 Describe the intersection of:
a) $\triangle XYZ$ and $\triangle TXZ$.
b) $\triangle XYZ$ and $\triangle RSZ$.
c) $\triangle TXZ$ and $\triangle RSZ$.

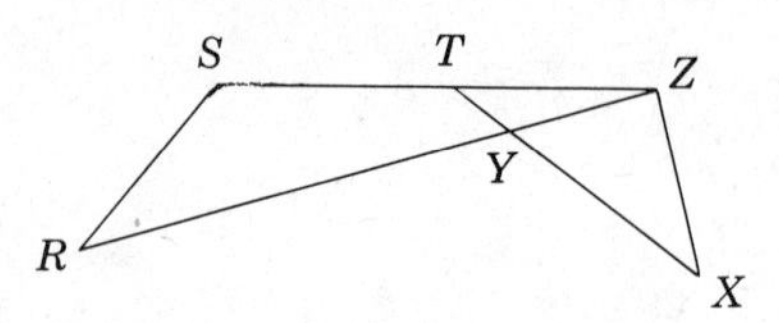

11 Given $\triangle XYZ$, describe the following sets of points:
a) $\triangle XYZ \cap \overrightarrow{ZY}$.
b) $\overline{XZ} \cap \angle Y$.
c) (Interior of $\triangle XYZ$) $\cap$ (exterior of $\triangle XYZ$).

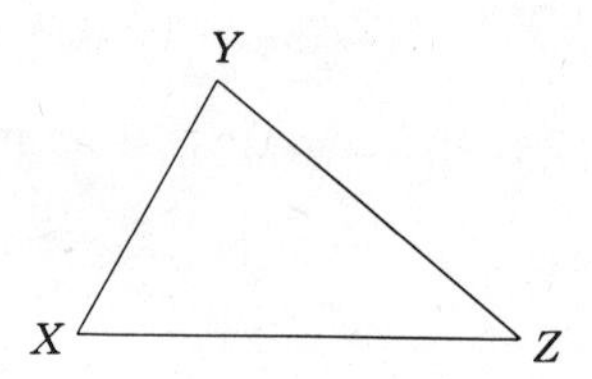

10A.3 Congruence

If we take a pile of nickels and stack one on top of the other, they would match perfectly because all the nickels have the same size and shape. A deck of cards, too, can be stacked perfectly without any corners sticking out because all the cards in the deck have the same size and the same shape.

Objects that have the same size and shape are of interest and importance not only in the world of mathematics but in the physical world around us. In mathematics we use a special word to describe this condition of "same size and shape." We say that such objects are *congruent* to each other. Assuming that all our nickels have the same thickness and all our cards have the same thickness, then all the nickels in our pile are congruent to each other, and all the cards in our deck are congruent to each other.

We see many examples of congruent figures all around us. When a manufacturer engages in "mass production," he is producing congruent objects. The same model cars are congruent to each other and so are the same make and model radios and basketballs.

We just suggested that a practical test of congruence is to place one object over the other (superimpose) and see whether the two objects match. If the fit is perfect, then they are congruent. We can easily do this with a deck of cards or a pile of nickels, but how can we match, by this method, two automobiles or a couple of basketballs?

Obviously, we must have better ways of telling whether figures are congruent than placing one on top of the other and seeing whether the two figures match.

We shall see that certain properties of congruence make it unnecessary to physically superimpose one figure on the other to test their congruence. We shall see that we do not even have to *measure* all the parts of two figures to find out whether they are congruent.

We shall first define congruence for segments, then for angles and triangles, and then for any two sets of points.

CONGRUENCE OF SEGMENTS

DEFINITIONS

1. The *length* of a segment is the distance between its endpoints.
2. Two segments are *congruent* if they have the same length. If $\overline{AB}$ is congruent to $\overline{CD}$, we write $\overline{AB} \cong \overline{CD}$, or $AB = CD$.

CONGRUENCE OF ANGLES

DEFINITION Two angles are *congruent* if they have the same measure. We use the symbol $\cong$ to mean *is congruent to*.

Comments

1. *Just as we defined congruence of segments in terms of distance, so do we define congruence of angles in terms of measure. That is, if* $m\angle ABC = m\angle DEF$, *then the angles are* congruent, *and we write* $\angle ABC \cong \angle DEF$. *Therefore, the two statements* $m\angle ABC = m\angle DEF$ *and* $\angle ABC \cong \angle DEF$ *are equivalent.*
2. *It is important to make the distinction between* equal *and* congruent *angles:*

   a) $\angle APB = \angle CPD$ *refers to the same angle with different names (Figure 10A.36).*

FIGURE 10A.36

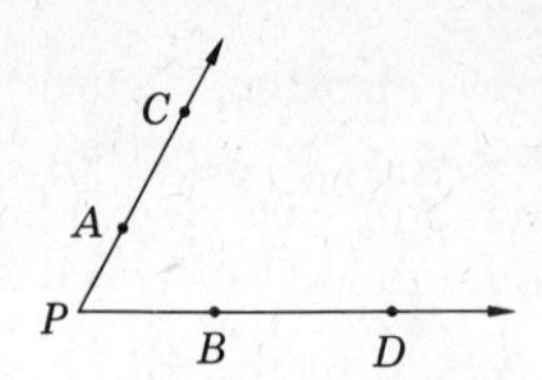

b) $\angle ABC \cong \angle DEF$ refers to two angles having the same measure, say 30° (Figure 10A.37).

FIGURE 10A.37

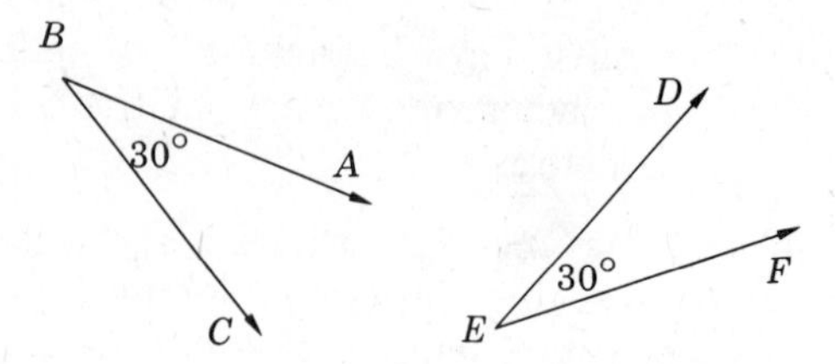

The definition of congruent angles leads to the following basic theorems.

THEOREM 5 Any two right angles are congruent.

THEOREM 6 If two angles are both congruent and supplementary, then each of them is a right angle.

THEOREM 7 Supplements of congruent angles are congruent.

THEOREM 8 Complements of congruent angles are congruent.

THEOREM 9 Vertical angles are congruent.

CONGRUENCE OF TRIANGLES: INFORMAL IDEAS

Comment

The intuitive idea of congruence is the same for triangles, or any other geometric figures, as for segments and angles. It means that two figures are congruent if the first figure can be moved so as to coincide with the second figure. Two triangles are congruent if one triangle can be moved onto the other triangle in such a way that the triangles fit exactly.

Although this is not a definition of congruence, it will soon lead us to such a formal definition.

Consider $\triangle ABC$ and $\triangle DEF$ in Figure 10A.38. Either triangle can be moved onto the other in such a way as to fit exactly.

FIGURE 10A.38

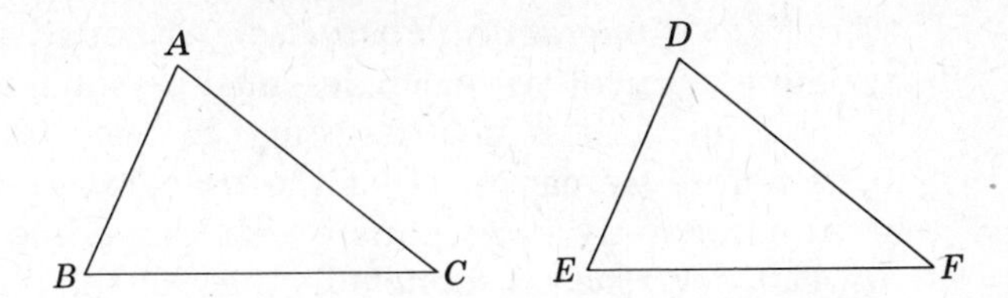

To move $\triangle ABC$ onto $\triangle DEF$ so as to fit exactly, we move $\triangle ABC$ in such a way that point A falls on point D, point B on point E, and point C on point F. That is, we must set up the following one-to-one correspondence between the vertices of the two triangles:

$$A \longleftrightarrow D, \quad B \longleftrightarrow E, \quad \text{and} \quad C \longleftrightarrow F.$$

A shorter way to show this correspondence is to write

$$ABC \longleftrightarrow DEF,$$

where the first letter on the left (A) corresponds to the first letter on the right (D), the second (B) to the second (E), and the third (C) to the third (F).

Given a correspondence between the vertices of two triangles, $ABC \longleftrightarrow DEF$ (Figure 10A.38), we speak of the *corresponding sides* as

$$\overline{AB} \longleftrightarrow \overline{DE}, \quad \overline{BC} \longleftrightarrow \overline{EF}, \quad \overline{AC} \longleftrightarrow \overline{DF},$$

and we speak of the *corresponding angles* as

$$\angle A \longleftrightarrow \angle D, \quad \angle B \longleftrightarrow \angle E, \quad \angle C \longleftrightarrow \angle F.$$

To test whether a correspondence between the vertices will produce a perfect fit, we need to check whether the corresponding sides and angles are congruent. Our formal definition of congruence is based on this idea.

CONGRUENCE OF TRIANGLES: FORMAL TREATMENT

DEFINITION Given $\triangle ABC$, $\triangle DEF$, and a one-to-one correspondence $ABC \longleftrightarrow DEF$ between their vertices (Figure 10A.39), if every pair of corresponding sides is congruent and every pair of corresponding angles is congruent, then the correspondence is a *congruence.* We then say that $\triangle ABC$ and $\triangle DEF$ are *congruent*, and we write $\triangle ABC \cong \triangle DEF$.

FIGURE 10A.39

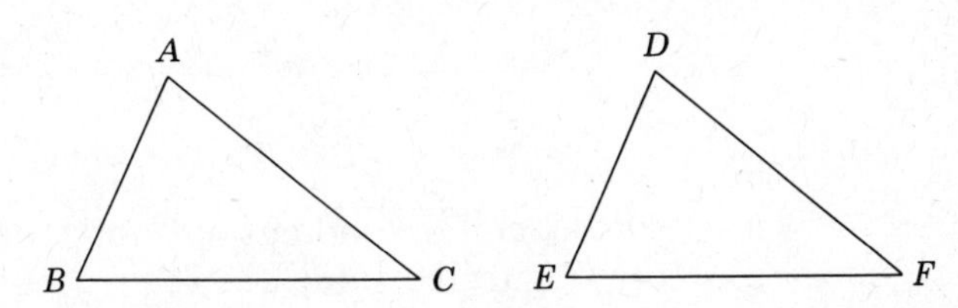

The Basic Congruence Postulate Although the six pairs of corresponding parts of two triangles must be congruent for the triangles to be congruent, it is not necessary to *test* the congruence of *all* six pairs before we can say that the triangles are congruent. Sometimes, if we know only *three* pairs of corresponding parts to be congruent, we can nevertheless, conclude that all the remaining corresponding parts are also congruent—if the three pairs are the right combination of parts.

One such "right" combination of three parts is any two sides and the included angle, that is, if two sides and the included angle of one triangle are congruent to the corresponding parts of another triangle (Figure 10A.40).

FIGURE 10A.40

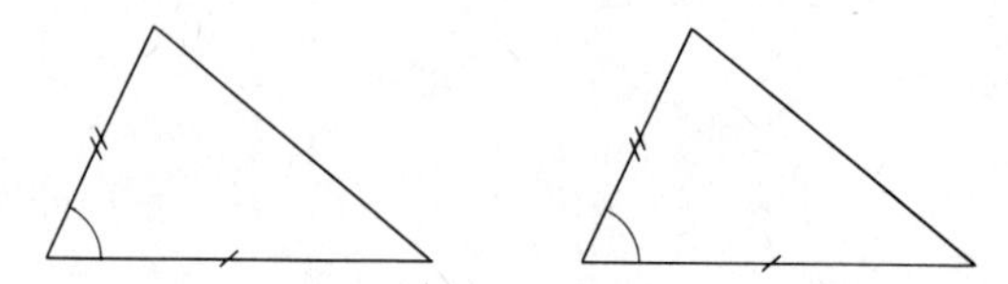

We can check informally whether this combination is "right" to produce a congruence by a simple experiment.

Example

1 Draw two 40° angles on a sheet of paper (Figure 10A.41).

FIGURE 10A.41

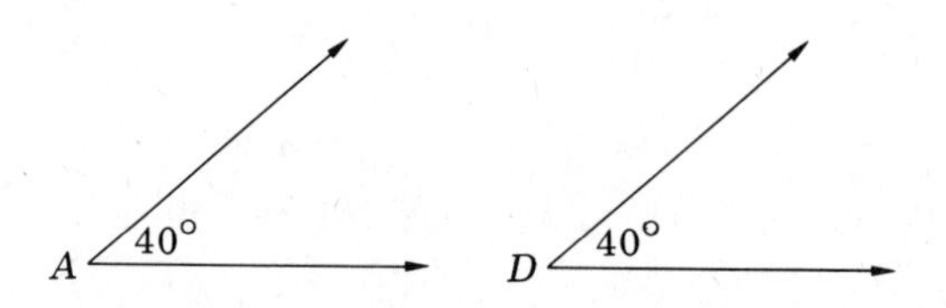

2 Mark off on $\angle A$ points B and C such that $AB = 3$ and $AC = 4$, and mark off on $\angle D$ points E and F such that $DE = 3$ and $DF = 4$ (Figure 10A.42).

FIGURE 10A.42

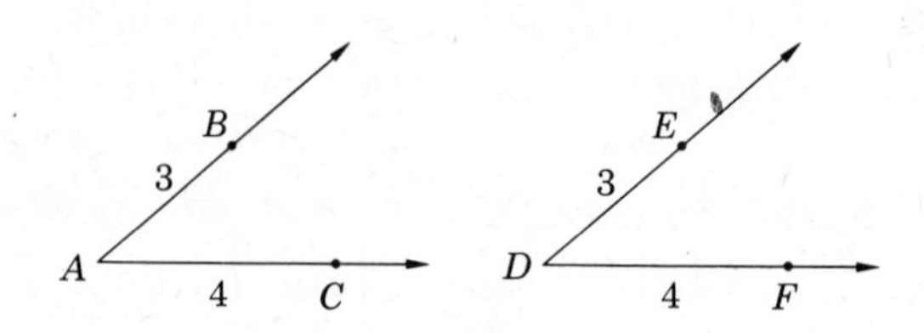

3 Draw $\overline{BC}$ and $\overline{EF}$ and cut out both triangles from the sheet of paper (Figure 10A.43).

FIGURE 10A.43

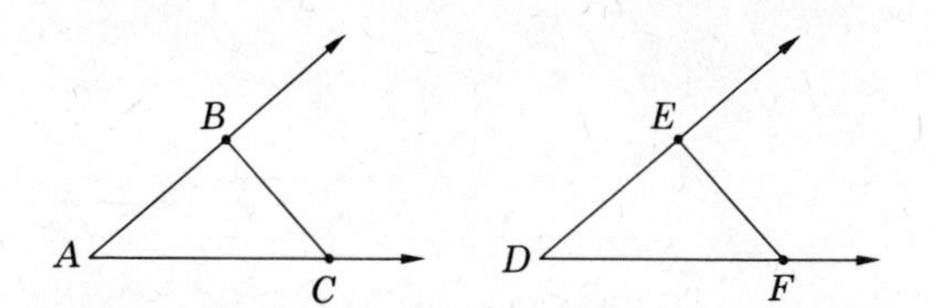

4 Superimpose $\triangle ABC$ onto $\triangle DEF$ so that $ABC \longleftrightarrow DEF$.

5 The two triangles will be seen to coincide.

We formalize this idea by stating it as our basic congruence postulate.

POSTULATE 7 (*Side Angle Side Postulate*) Given a correspondence between two triangles (or between a triangle and itself), if two sides and the included angle of the first triangle are congruent to the corresponding parts of the second triangle, then the correspondence is a congruence.

That is, if $\overline{AB} \cong \overline{DE}$, $\overline{BC} \cong \overline{EF}$, and $\angle B \cong \angle E$, then $\triangle ABC \cong \triangle DEF$ (Figure 10A.44). We sometimes abbreviate the statement of this postulate by using the letters SAS (side angle side).

FIGURE 10A.44

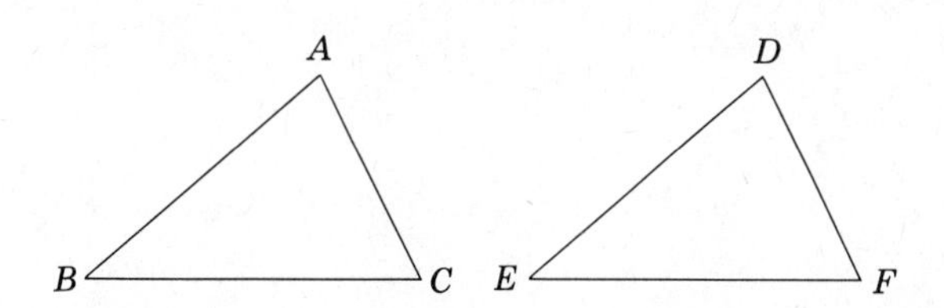

Postulate 7 can be used to develop other ways for determining whether or not two triangles are congruent. Two such methods are given in the following basic congruence theorems.

THEOREM 10 (*Basic Congruence Theorem*) If two angles and the included side of one triangle are congruent to the corresponding parts of a second triangle, then the triangles are congruent (ASA). (See Figure 10A.45.)

FIGURE 10A.45

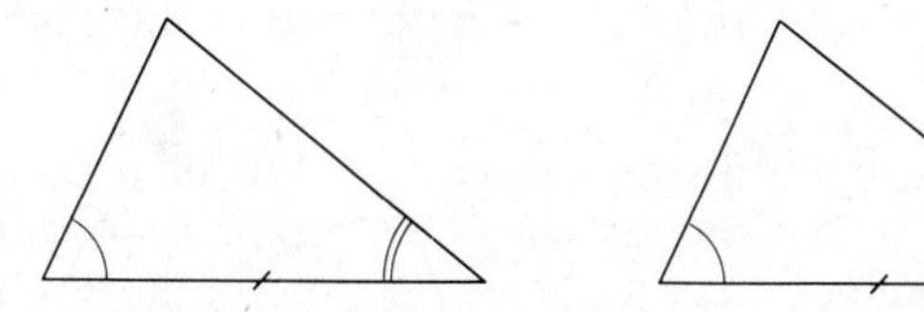

THEOREM 11 (*Basic Congruence Theorem*) If three sides of one triangle are congruent to three sides of another triangle, then the triangles are congruent (SSS). (See Figure 10A.46.)

FIGURE 10A.46

Comment

We can check informally whether the conditions in Theorems 10A.10 and 10A.11 are sufficient to produce congruent triangles by experiments like the one described for SAS.

A GENERAL CONCEPT OF CONGRUENCE

We said that two figures are congruent if they have the same size and shape. They will have the same size and shape if their corresponding sides and angles are congruent. There is no difficulty accepting this definition of congruence when we are talking about figures like segments, angles, or triangles.

But suppose we are talking about figure A and its replica B as shown in Figure 10A.47. What is the "size" of A? What is the "shape" of B? Here, we obviously need a more precise definition of size and shape. We need to define precise conditions that determine when two figures, or any two sets of points, have the same size and shape.

FIGURE 10A.47

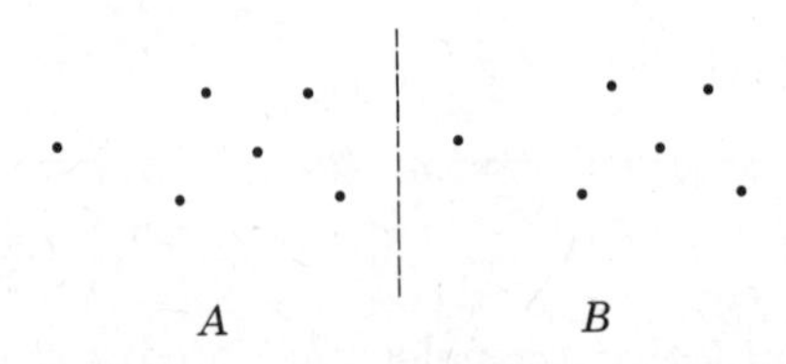

This need has been met by the following *general* definition of congruence that applies to any finite or to any infinite set of points:

DEFINITION

Two sets of points A and B are congruent if and only if there exists a one-to-one correspondence between their points such that the segment determined by any two points of one set is congruent to the segment determined by the corresponding points of the other set.

In other words, two figures, or sets of points, will have the same size and shape if the distance between any two points of the first figure is the same as the distance between their corresponding points of the second figure. That is, a correspondence between points that *preserves distance* assures us the same size and shape. These correspondences are called *isometries.*

EXERCISE SET 10A.3

1 Given $\angle 1 \cong \angle 2$, $\overline{BC} \cong \overline{DE}$, $\overline{AD} \cong \overline{CF}$, prove $\angle A \cong \angle F$.

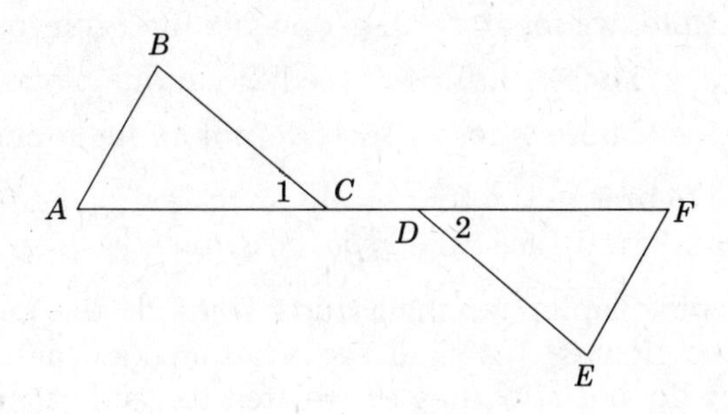

2 Given $\overline{BC} \perp \overline{AC}$, $\overline{BC} \perp \overline{CD}$, $\overline{AC} \cong \overline{CD}$, prove $\overline{AB} \cong \overline{BD}$.

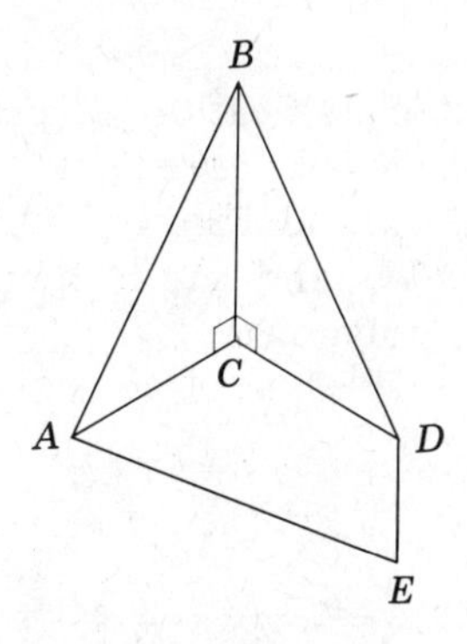

3 If $AB = CD$ and $AC = BD$, prove that $\triangle AED$ is isosceles.

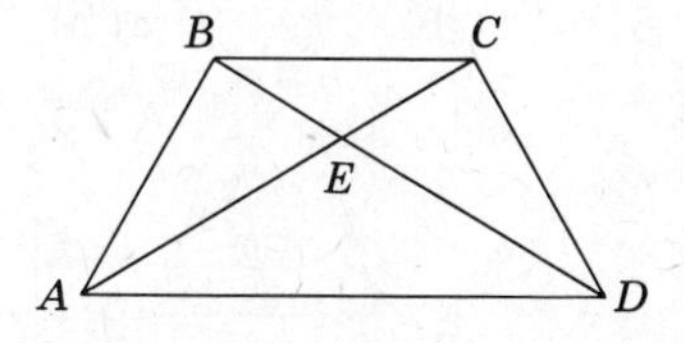

4 Given that $\angle A, \angle B, \angle C$, and $\angle D$ are right angles, and $AB = BC = CD = AD$, prove:

a) $AC = BD$.
b) $\overline{AC}$ and $\overline{BD}$ bisect each other.
c) $\overline{AC} \perp \overline{BD}$.

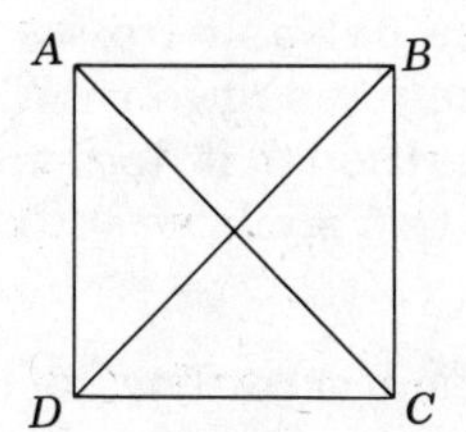

10A.4 Plane and Space Figures

To make it easier to classify the endless variety of geometric figures possible, we separate all geometric figures into two groups:

1 Those whose points all lie in one plane.

2 Those whose points do not all lie in one plane.

The figures in the first group are called *plane figures*, and those in the second group are called *space figures.*

In this chapter we shall study some of the more common plane and space figures. We shall see what makes these figures look the way they do and how they are related to each other.

PLANE FIGURES

Polygons An important classification of plane figures is the set of *polygons.* A *polygon* is a closed figure consisting entirely of segments and looks like those shown in Figure 10A.48. The segments of a polygon are called its *sides*, and the endpoints of the segments are called the *vertices* of the polygon.

FIGURE 10A.48

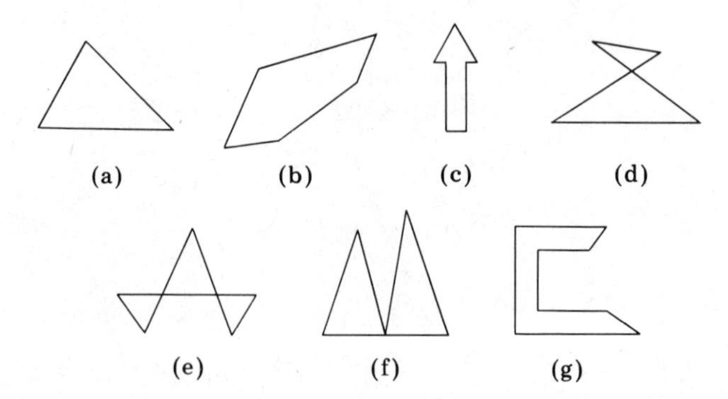

We wish to distinguish between two types of polygons: *simple* and *nonsimple.* A polygon is *simple* if it does not cross itself, such as in (a), (b), (c), and (g). If the polygon crosses itself, such as in (d), (e), and (f), we call it a *nonsimple* polygon. [We shall agree to think of (f) as "crossing" itself even though it does not "cross" itself the same way as (d) and (e) do.] In this book we shall confine our attention to simple polygons.

Polygons are often classified according to their number of sides as shown in Table 10A.1.

TABLE 10A.1

Name of Polygon	*Number of Sides*
Triangle	3
Quadrilateral	4
Pentagon	5
Hexagon	6
Heptagon	7
Octagon	8
Nonogon	9
Decagon	10
Dodecagon	12

DEFINITIONS

1 A polygon is called a *regular polygon* if all its sides are congruent and all its angles are congruent.

2 The sum of the lengths of the sides of a polygon is called the *perimeter* of the polygon.

3 The union of a polygon and its interior is called a *closed polygonal region.*

Quadrilaterals Certain quadrilaterals are of great importance because of special properties they possess. One such property relates to parallelism. We shall now define these special quadrilaterals.

DEFINITIONS

1 A quadrilateral that has *at least one* pair of parallel sides is called a *trapezoid* (Figure 10A.49).

2 A quadrilateral that has two pairs of parallel sides is called a *parallelogram* (Figure 10A.49).

FIGURE 10A.49

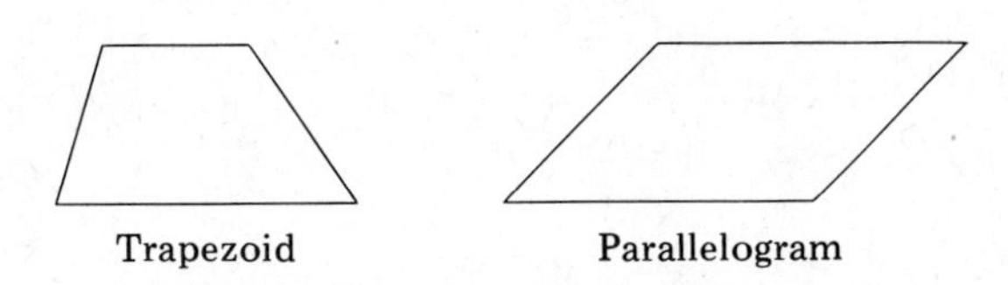

3 A *rhombus* is a parallelogram all of whose sides are congruent. (Or, if two adjacent sides of a parallelogram are congruent, then the quadrilateral is a *rhombus*.) See Figure 10A.50.

FIGURE 10A.50

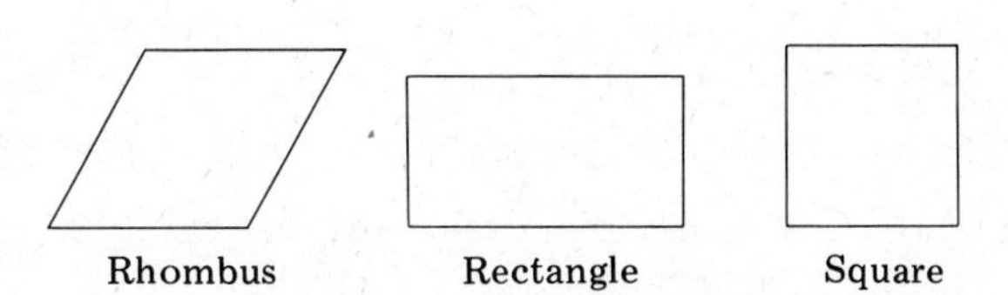

4 A parallelogram containing at least one right angle is called a *rectangle* (Figure 10A.50).

5 A rhombus with at least one right angle is called a *square*. (Or, a *square* is a rectangle all of whose sides are congruent.)

Theorems About Parallelograms

THEOREM 12 In a parallelogram, any two opposite sides are congruent.

THEOREM 13 Any two opposite angles of a parallelogram are congruent.

THEOREM 14 Any two consecutive angles of a parallelogram are supplementary.

THEOREM 15 The diagonals of a parallelogram bisect each other.

Conditions That Make a Quadrilateral a Parallelogram

THEOREM 16 If, in a given quadrilateral, both pairs of opposite sides are congruent, then the quadrilateral is a parallelogram.

THEOREM 17 If two sides of a quadrilateral are both parallel and congruent, then the quadrilateral is a parallelogram.

THEOREM 18 If the diagonals of a quadrilateral bisect each other, then the quadrilateral is a parallelogram. (This is the converse of Theorem 15.)

Circles

DEFINITIONS 1 A *circle* is the set of all points in a plane at a given distance from a given point in the plane. The given point P (Figure 10A.51) is called the *center*, and the given distance r is called the *radius* of the circle. If A is any point of the circle, then the segment $\overline{PA}$ is a radius of the circle and A is called its *outer end*.

FIGURE 10A.51

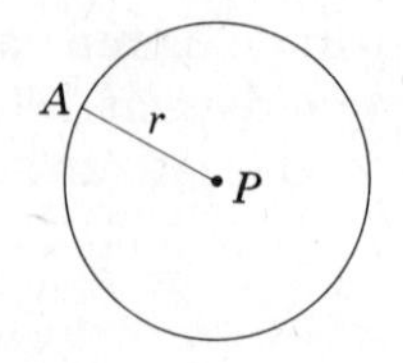

2 A *chord* is a segment having both endpoints on a circle, for example $\overline{AB}$, $\overline{CE}$ (Figure 10A.52).

FIGURE 10A.52

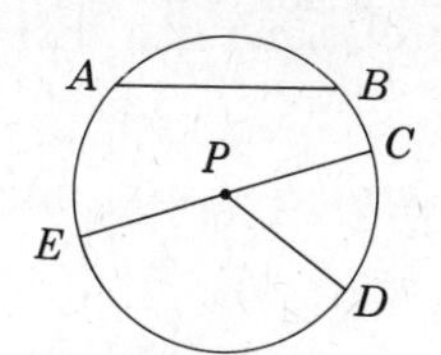

3 If the chord contains the center of a circle, it is called a *diameter* of the circle, for example $\overline{CE}$ (Figure 10A.52). The length of a diameter is the number $2r$.

4 A *tangent* to a circle is a line in the plane of the circle which intersects the circle in only one point, for example $\overleftrightarrow{AB}$ as shown in Figure 10A.53. This point is called the *point of tangency*, or the *point of contact*. If the line intersects the circle at more than one point, it is called a *secant* line, for example $\overleftrightarrow{CD}$.

FIGURE 10A.53

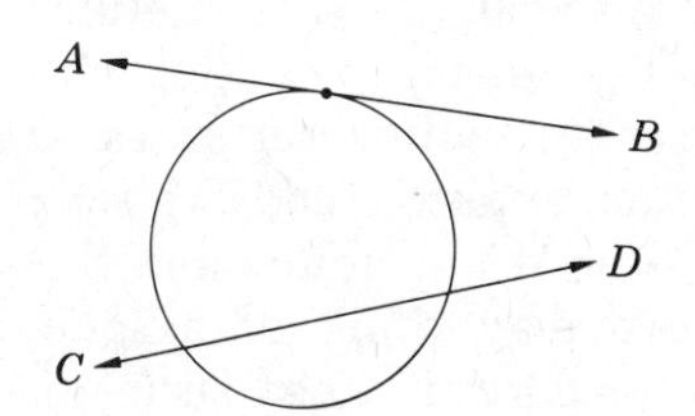

5 Circles that lie in the same plane and have the same center are called *concentric circles* (Figure 10A.54).

FIGURE 10A.54

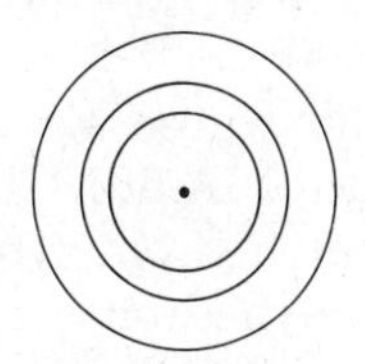

6 The *interior of a circle* is the set of all points of the plane whose distance from the center is less than the radius (Figure 10A.55). The *exterior of a circle* is the set of all points of the plane whose distance from the center is greater than the radius.

FIGURE 10A.55

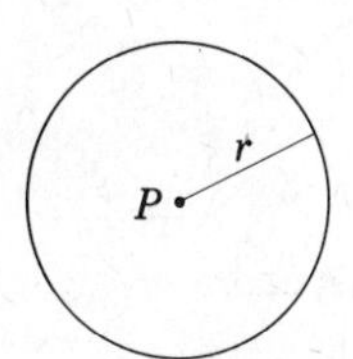

7 The union of a circle and its interior is a *closed circular region.*

8 Two circles are *congruent* if and only if they have congruent radii.

Comments

1 *Unlike previous figures we talked about, a circle is an example of a set of points* not *made up of lines, rays, and segments.*

2 *Note that the center is* not *a point of the circle, and a radius is* not *a part of the circle.*

SPACE FIGURES

The geometry of space figures is sometimes called *solid* geometry. The figures in solid geometry, like those in plane geometry, are sets of points. If a set of points separates space into three distinct parts—an interior, an exterior, and the set of points itself—then we call the union of this set of points and the points of the interior a *solid.* Therefore we may think of a *solid* as a portion of space bounded by plane or curved closed surfaces or by a combination of such surfaces. Physical objects such as a sugar cube, a block of wood, or a dowel are examples of solids. Geometric solids are abstractions of such physical objects.

Polyhedra

DEFINITIONS

1 A *polyhedron* is a solid whose surface consists entirely of polygonal regions (Figure 10A.56). The polygonal regions are called *faces* of the polyhedron. The intersection of two adjacent faces of the polyhedron is a line segment called an *edge* of the polyhedron. The points of intersection of the edges are the *vertices* of the polyhedron (Figure 10A.57).

FIGURE 10A.56

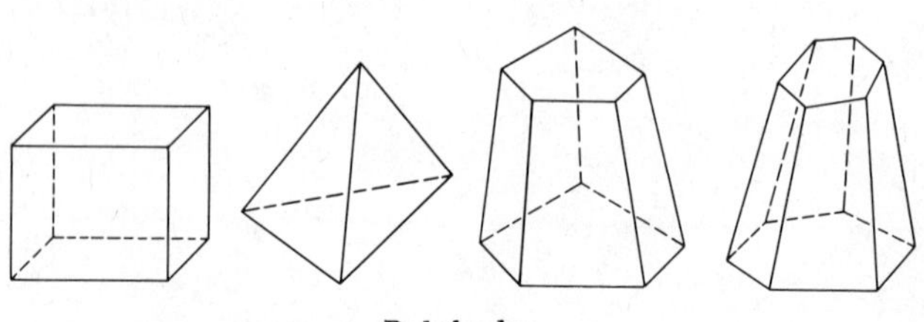

Polyhedra

FIGURE 10A.57

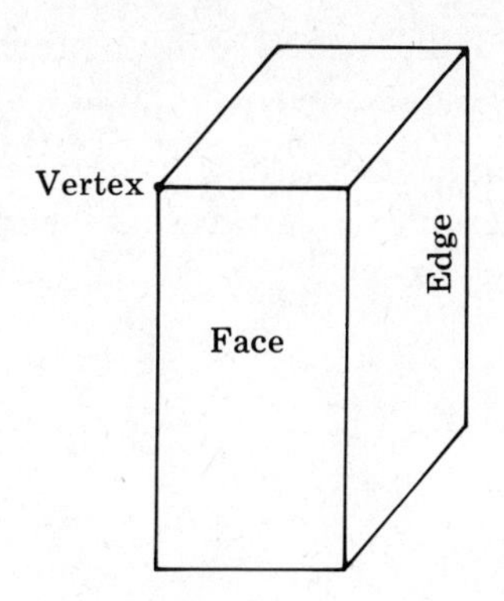

2 The figure formed by the intersection of a solid and a plane is called a *section* of the solid (Figure 10A.58).

FIGURE 10A.58

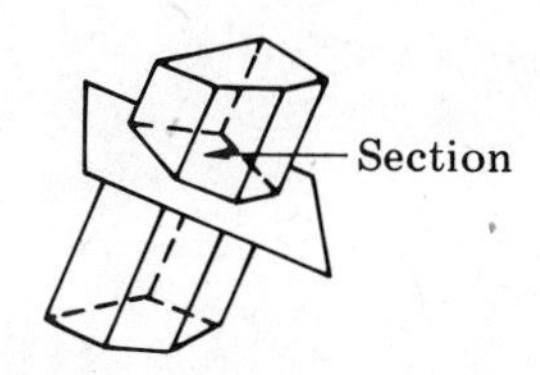

Polyhedra are classified according to the number of their faces as shown in Table 10A.2. The smallest number of faces possible is four.

TABLE 10A.2

Name	*Number of Faces*
Tetrahedron	4
Pentahedron	5
Hexahedron	6
Octahedron	8
Decahedron	10
Dodecahedron	12
Icosahedron	20

Prisms

DEFINITIONS 1 A *prism* is a polyhedron of which two faces, called *bases*, are congruent polygons in parallel planes (ABC and DEF), and the other faces, called *lateral faces*, are parallelograms (Figure 10A.59). Each edge determined by the intersection of two lateral faces is called a *lateral edge* ($\overline{AD}$, $\overline{BE}$, $\overline{CF}$).

FIGURE 10A.59

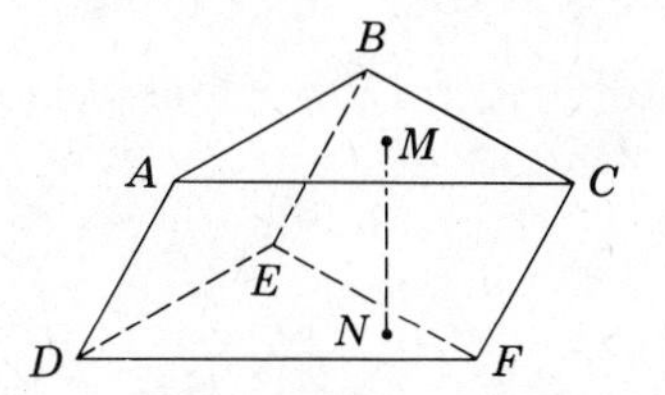

2 An *altitude* of the prism is any segment that has endpoints in the two bases and is perpendicular to them ($\overline{MN}$). We call the length of any altitude the *height* of the prism.

3 If the lateral edges of a prism are perpendicular to its bases, the prism is called a *right prism* (Figure 10A.60).

FIGURE 10A.60

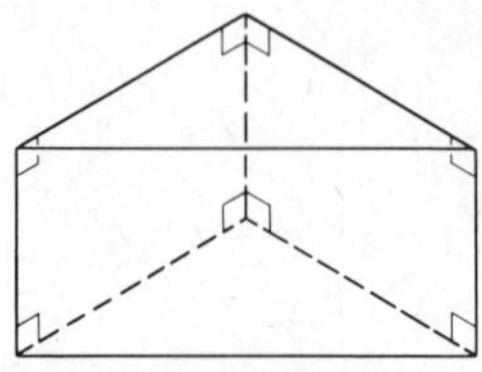

Right Prism

4 A *parallelepiped* is a prism whose faces are parallelograms. If each of the faces is a rectangle, then the prism is a *rectangular parallelepiped* or a *rectangular solid* or box (Figure 10A.61).

FIGURE 10A.61

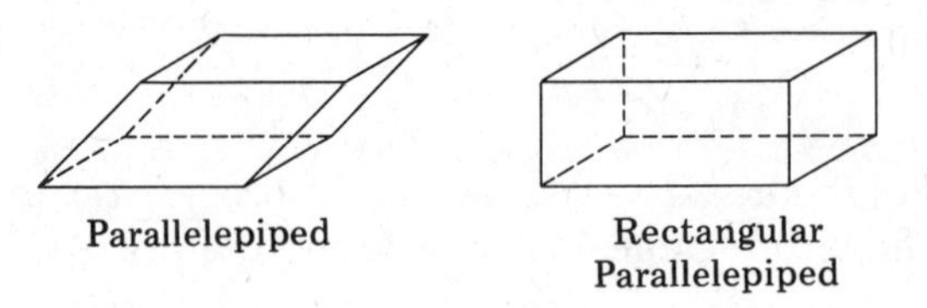

Parallelepiped

Rectangular Parallelepiped

Prisms are classified according to the shape of their bases, that is, triangular, quadrangular, pentagonal, etc. (see Figure 10A.62).

FIGURE 10A.62

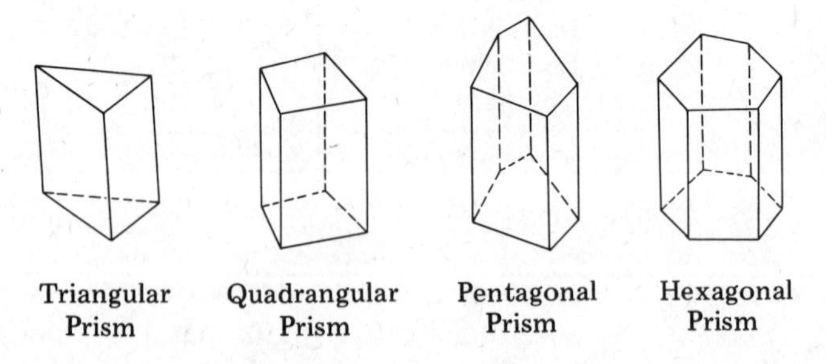

Triangular Prism

Quadrangular Prism

Pentagonal Prism

Hexagonal Prism

Pyramids

DEFINITIONS

A *pyramid* is a polyhedron of which one face is called the *base* (*ABCD*), and the other faces, called *lateral faces*, are triangles (*PAD*, *PCD*, *PBC*, *PAB*) having a common vertex called the *apex*, (*P*) (Figure 10A.63).

FIGURE 10A.63

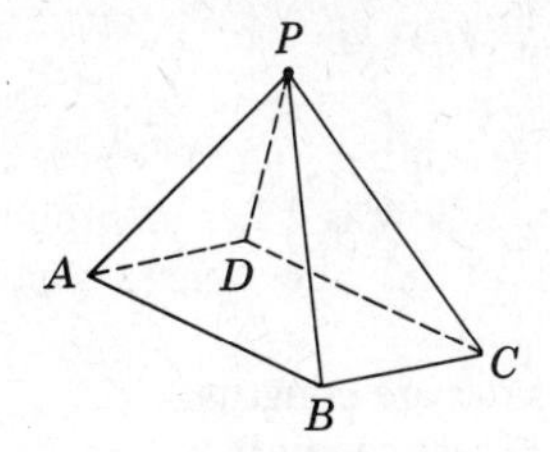

Pyramids are classified according to the shape of their base, that is, triangular, quadrangular, pentagonal, etc. (see Figure 10A.64).

FIGURE 10A.64

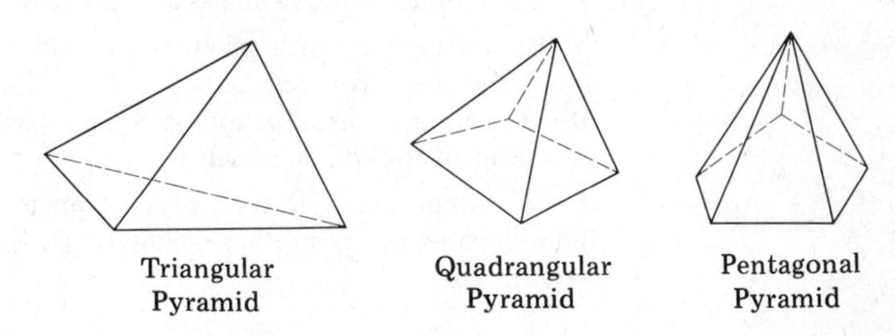

Spheres

DEFINITIONS

1 A *sphere* is the set of all points at a given distance from a given point in space. The given distance r is called the *radius* of the sphere; the given point P is called the *center* of the sphere (Figure 10A.65). If A is any point of the sphere, then the segment $\overline{PA}$ also is called a *radius* of the sphere. Any segment through the center and having endpoints in the sphere is a *diameter* of the sphere. The two endpoints of the diameter are called *antipodal points* of the sphere.

FIGURE 10A.65

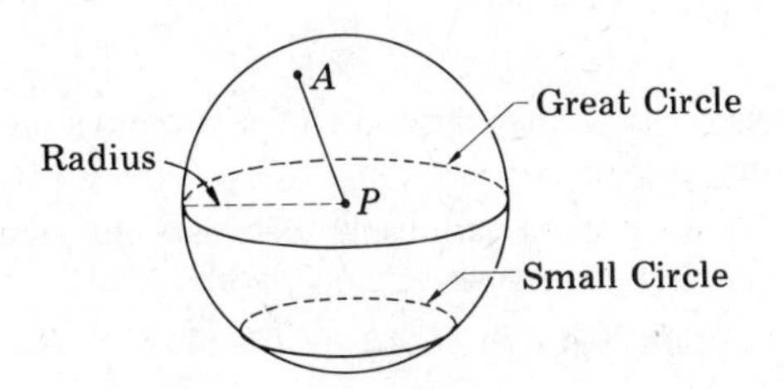

2 A *great circle* on a sphere is any intersection of the sphere with a plane through the center of the sphere. All circles on a sphere which are not great circles are called *small circles.*

EXERCISE SET 10A.4

1 Here is a list of geometric figures: rectangle, parallelogram, rhombus, square, trapezoid. Below is a list of properties. Next to each property write the names of *all* the figures (listed above) about which the property is true.

a) The opposite sides are congruent.
b) The diagonals bisect each other.
c) The diagonals are congruent.
d) The angles are all right angles.
e) The diagonals are perpendicular to each other.
f) At least one pair of opposite sides is parallel.
g) All the sides are congruent.
h) At least one angle is a right angle.
i) Any two consecutive angles are supplementary.

2 a) The intersection of a sphere with a plane passing through the center of the sphere is called ____.
b) The intersection of a sphere with a plane not passing through the center of the sphere is called ____.

3 If the midpoints of the sides of a rectangle are joined in order, will the figure formed be (a) another rectangle? (b) a square? (c) a rhombus? (d) a trapezoid?

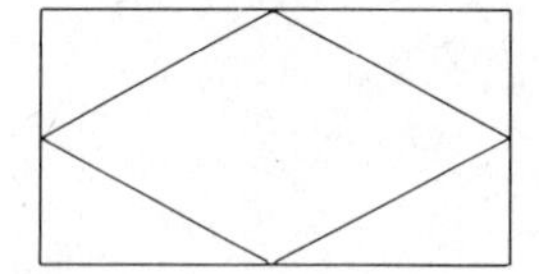

4 Why do the diagonals of a square form four isosceles triangles?

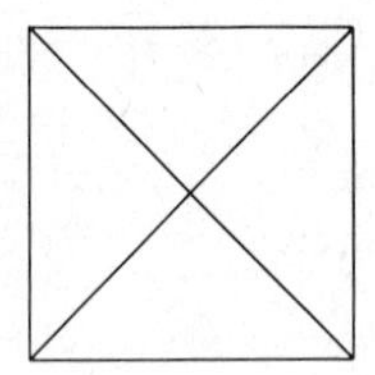

5 Prove that the diagonals of a rhombus are perpendicular bisectors of each other.

6 What is the relationship between the number of diagonals and the number of vertices of a polygon?

7 In trapezoid $ABCD$, $AB = CD$ and $\overline{BC} \parallel \overline{AD}$. Prove that $\angle A \cong \angle D$.

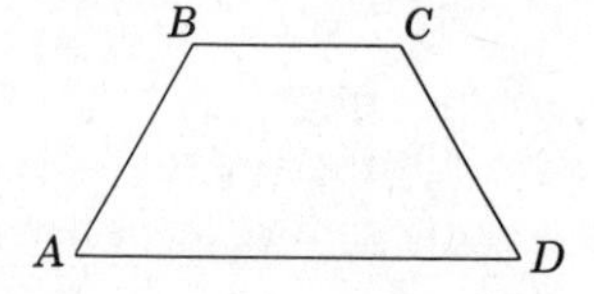

8 Imagine a 3-inch cube painted blue.

a) How many cuts will be required to divide the cube into 1-inch cubes?
b) How many 1-inch cubes would there be altogether?
c) How many of these cubes would have *four* blue faces?
d) How many would have *three* blue faces?
e) How many would have *two* blue faces?
f) How many would have *one* blue face?
g) How many would have *no* blue faces?

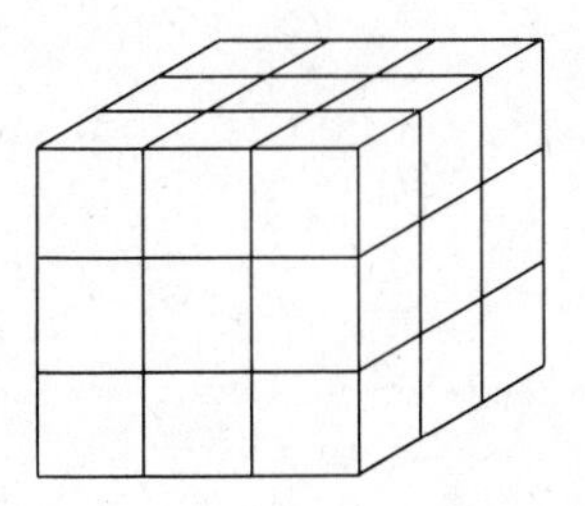

10A.5 The Pythagorean Theorem

In ancient Egypt, every spring, the waters of the Nile River overflowed their banks and washed out the boundaries that separated one man's property from his neighbor's. So every year, after the floods, the Egyptians were faced with the problem of reestablishing their boundaries. But to do this they had to construct right angles, and the construction of right angles was not well understood in those days. So whenever the washed-out boundaries had to be redrawn, special consultants known as "rope stretchers" had to be called in.

These specialists took a rope and tied 13 knots in it in such a way as to get 12 equal spaces as shown in Figure 10A.66. They then drove stakes into the ground in this manner: A stake was driven through the first and thirteenth knots. Another stake was then driven through the fourth knot. A third stake was driven through the eighth knot. In each case the rope was stretched as tightly as possible before the stake was driven into the ground. They found that the angle opposite the side containing the 5 spaces turned out to be a right angle. The rope stretchers did not understand why stretching a rope in this manner resulted in a right angle, but they knew it worked and so were perfectly happy.

FIGURE 10A.66

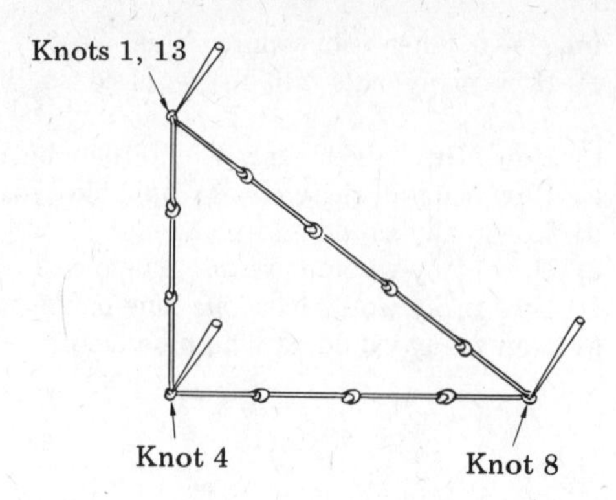

At about the same time in India, the Hindus also needed to construct right angles. They went a step further than the Egyptians. They discovered that in addition to the number combination (3, 4, 5) forming a right angle, there were other combinations that also resulted in a right angle: (5, 12, 13), (15, 20, 25), (12, 35, 37), (8, 15, 17), (12, 16, 20), (15, 36, 39). But like the Egyptians, the Hindus did not understand why these number combinations yielded a right angle.

According to tradition, the answer to this question was finally found about 2500 years ago by the Greek mathematician Pythagoras. His important discovery has been called ever since the Pythagorean Theorem.

THE PYTHAGOREAN THEOREM AND ITS CONVERSE

The Pythagorean theorem gives us a way of finding the length of any side of a right triangle if we know the lengths of the other two sides. We use the converse of the theorem to test whether or not a triangle is, in fact, a right triangle.

Before stating the theorem, we shall define the parts of a right triangle.

DEFINITIONS A *right triangle* is a triangle that has one right angle. The side opposite the right angle is called the *hypotenuse.* The sides that include the right angle are called the *legs.* (See Figure 10A.67.)

FIGURE 10A.67

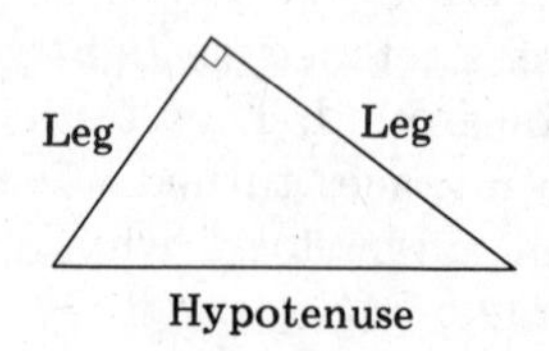

THEOREM 19 (*Pythagorean Theorem*) In any right triangle, the square of the length of the hypotenuse is equal to the sum of the squares of the lengths of the legs (Figure 10A.68).

FIGURE 10A.68

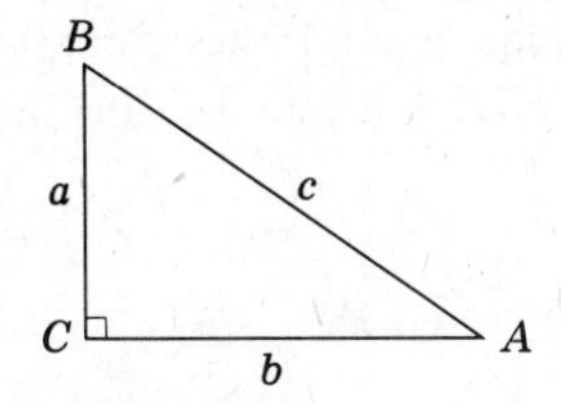

That is, if ΔABC is a right triangle with right angle at C, then $c^2 = a^2 + b^2$.

THEOREM 20 (*Converse of the Pythagorean Theorem*) If the square of the length of one side of a triangle is equal to the sum of the squares of the lengths of the other two sides, then the triangle is a right triangle whose right angle is opposite the longest side (Figure 10A.68).

That is, if in ΔABC, $c^2 = a^2 + b^2$, then ΔABC is a right triangle with its right angle at C.

Comment

As a consequence of Theorem 20, we can test whether a triangle is a right triangle if we know the lengths of its three sides. We substitute the values of the lengths in $c^2 = a^2 + b^2$, where c is the largest of the three sides. If the three lengths satisfy this relationship, then they form a right triangle.

EXERCISE SET 10A.5

1 Combine the Pythagorean Theorem and its converse into one statement:
   a) Using the phrase "if and only if."
   b) Using the phrase "necessary and sufficient condition." (See pages 458–459 in Appendix A.)

2 Why is each leg of a right triangle shorter than the hypotenuse?

3 State true or false. Give a reason for your answer.
   a) If you know the length of the diagonal of a square, you can find the length of the side of the square.
   b) If you know the length of the diagonal of a rectangle, you can find the lengths of the sides of the rectangle.
   c) The sides of a right triangle can have lengths of 3, 5, and 7.
   d) Tripling the lengths of the sides of a right triangle changes the triangle to an obtuse triangle.
   e) The Pythagorean Theorem was known to the Hindus by a different name.

4 Why can the sides of three squares of equal area never form a right triangle?

5 The distance around a square is 800. Find the diagonal.

6 The area of a square is 25 square inches. Find its perimeter.

7 The diagonal of a square is 8 inches. Find the side of the square.

8 Find the length of $\overline{AD}$ in the 3-inch cube shown.

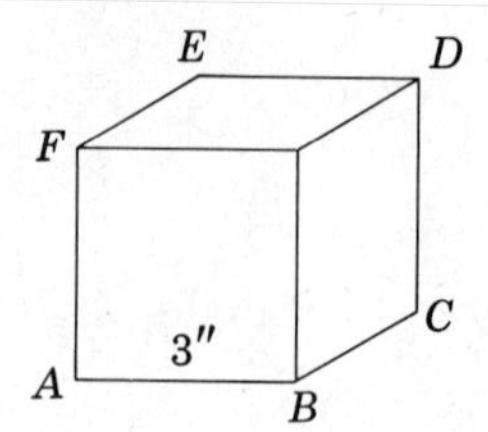

9 The distance around a rectangle is 68 and a diagonal is 26. Find the dimensions of the rectangle.

10 Find the length of $\overline{BD}$ in the triangle shown.

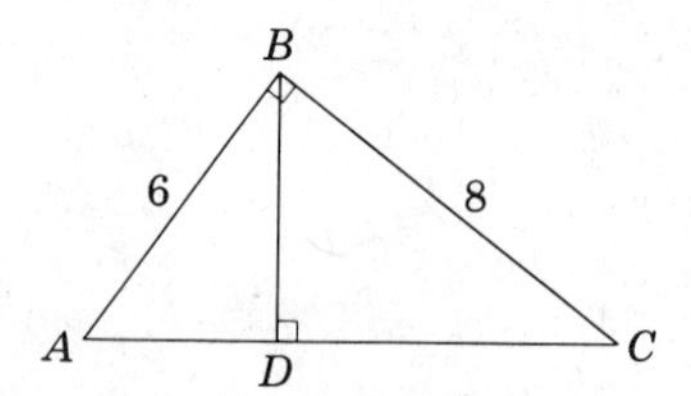

11 Two cars leave from your house at the same time. One travels north at a rate of 36 miles per hour, and the other travels east at a rate of 48 miles per hour. How far apart are they at the end of 30 minutes?

12 On a square baseball field measuring 90 feet on each side, how far is a ball thrown from the shortstop to home plate if the shortstop is on the baseline midway between second and third base?

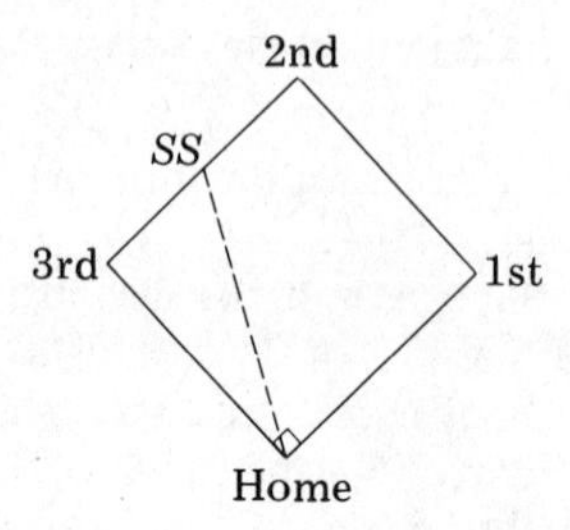

13 In the figure shown, find the length of (a) $\overline{PB}$, (b) $\overline{PC}$, (c) $\overline{PD}$, (d) $\overline{PE}$.

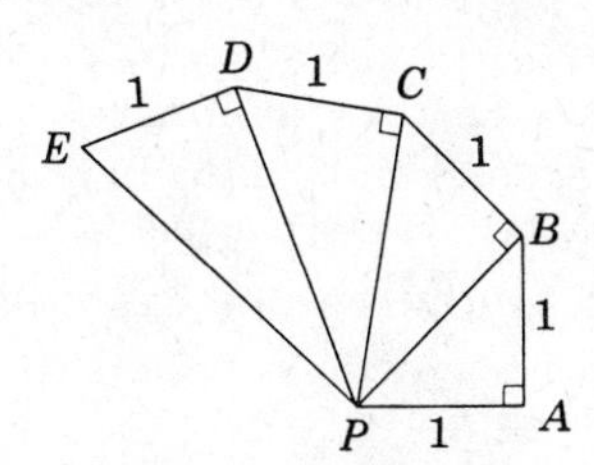

14 If we continue the pattern shown in Exercise 13, and if we think of $\overline{PA}$ as the first segment, $\overline{PB}$ as the second segment, $\overline{PC}$ as the third segment, etc.,

a) What would be the length of the sixth segment? Of the seventh segment? Of the thirteenth segment?

b) What would be the length of the kth segment?

15 Find the area of the region bounded by an equilateral triangle whose sides are of length (a) 6, (b) 10, (c) a.

16 Consider the right $\triangle ABC$. Why is each of the following statements true?

a) $AC = 25$.

b) Area of $\triangle ABC = 150$.

c) Area of $\triangle ABC = \frac{1}{2} BD \times AC$.

d) Area of $\triangle ABC = \frac{1}{2} BD \times 25$.

e) $BD = 12$.

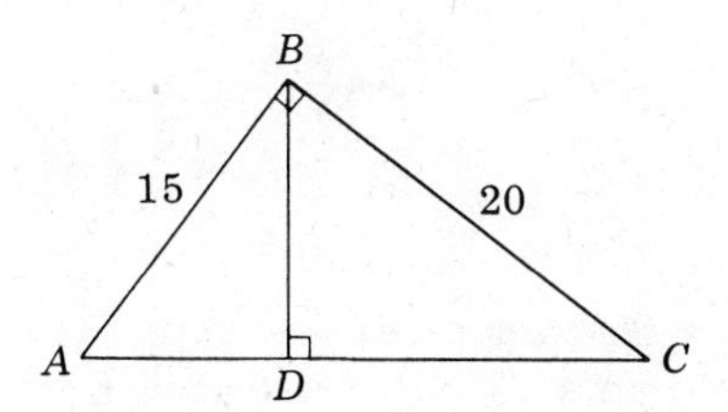

17 What method does Exercise 16 suggest for finding the altitude to the hypotenuse of a right triangle? Use this method to find the altitude to the hypotenuse in $\triangle LMN$.

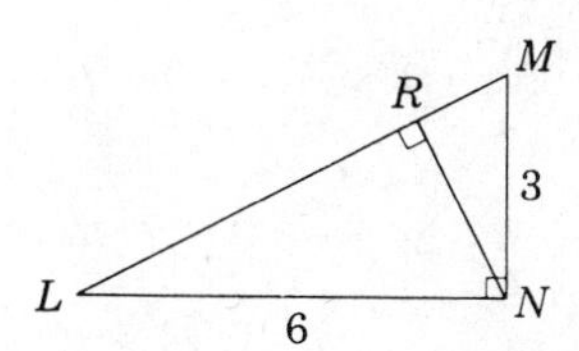

10A.6 Similarity

We already know that figures which have the same shape and size are called *congruent*. We shall now consider figures that have the same shape but not necessarily the same size. Such figures are called *similar*. Figure 10A.69 shows several examples of similar figures.

FIGURE 10A.69

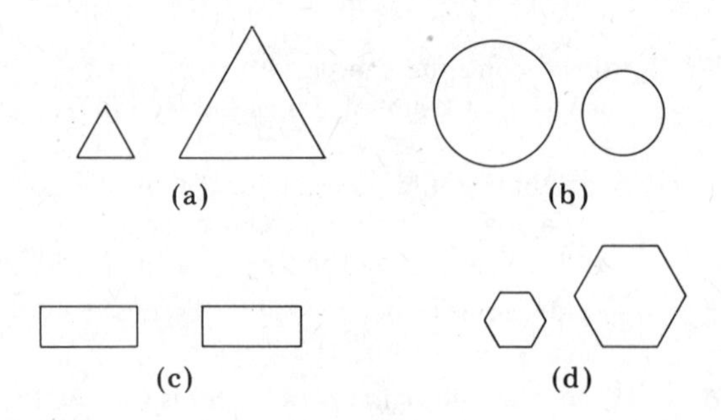

Another way to describe similarity is to say that two figures are *similar* if one of them is a photographic enlargement or reduction of the other (Figure 10A.70).

FIGURE 10A.70

Times Times

Similarity is the mathematical concept behind blueprints, maps, and picture enlargements. Scale models of houses, ships, planes, spacecraft, and other large structures are often built to assist in their planning and construction. The mathematics of similarity guarantees the similarity of the model and the actual structure.

PROPORTION

Consider the two triangles in Figure 10A.71. Note that the ratios of the corresponding sides are

$$\frac{a}{a'} = \frac{3}{6} = \frac{1}{2}, \quad \frac{b}{b'} = \frac{4}{8} = \frac{1}{2}, \quad \frac{c}{c'} = \frac{5}{10} = \frac{1}{2}$$

or $$\frac{a}{a'} = \frac{b}{b'} = \frac{c}{c'} = \frac{1}{2}$$

That is, the sides of the smaller triangle are half the corresponding sides of the larger triangle; or, the sides of the larger triangle are twice the sides of the smaller triangle.

FIGURE 10A.71

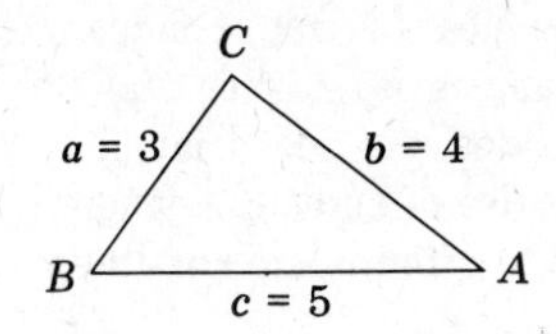

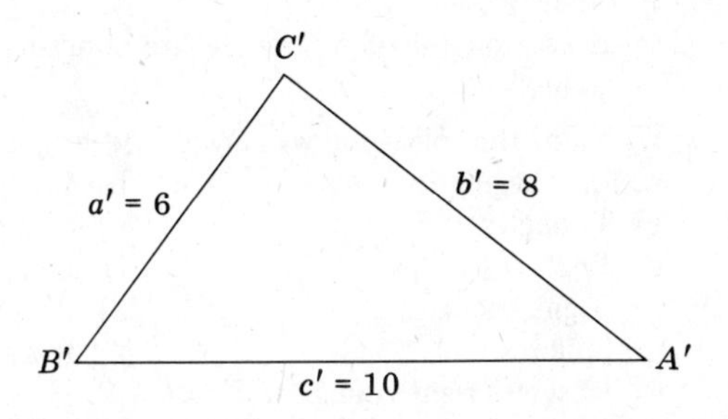

The statement that two ratios are equal is called a *proportion*. For example, $\frac{a}{a'} = \frac{b}{b'}$ is a *proportion*; or, *a and a' are proportional to b and b'*. We can extend the concept of proportionality to more than two ratios. Thus we can say of $\triangle ABC$ and $\triangle A'B'C'$ in Figure 10A.71 that since

$$\frac{a}{a'} = \frac{b}{b'} = \frac{c}{c'},$$

their corresponding sides are proportional.

CONDITIONS THAT ASSURE SIMILARITY

Two conditions must be met for geometric figures to be similar. If both conditions are met, then the figures are similar; if two figures are similar, then both of these conditions will hold.

DEFINITION

Two geometric figures are *similar* if

1 their corresponding angles are congruent, and

2 their corresponding sides are proportional.

Thus, the triangles in Figure 10A.71 are *similar* since

1 $\angle A \cong \angle A'$, $\angle B \cong \angle B'$, $\angle C \cong \angle C'$, and

2 $\frac{a}{a'} = \frac{b}{b'} = \frac{c}{c'}$

EXERCISE SET 10A.6

1. Tell whether the figures mentioned in each statement below are similar, congruent, or neither similar nor congruent.
   a) If the three angles of one triangle are congruent to the three corresponding angles of another triangle.
   b) If the three sides of one triangle are congruent to the three corresponding sides of another triangle.
   c) If all the angles of a figure are congruent to the corresponding angles of another figure.
   d) If one angle of a triangle is congruent to an angle of another triangle.
   e) If two angles of a triangle are congruent to two angles of another triangle.

2. Which of the following will always be similar? A pair of
   a) Line segments.
   b) Circles.
   c) Triangles.
   d) Squares.
   e) Rectangles.
   f) Equilateral triangles.
   g) Right triangles.
   h) Isosceles triangles.
   i) Spheres.
   j) Regular hexagons.
   k) Isosceles right triangles.

3. State true or false. Give a reason for your answer.
   a) If an acute angle of one right triangle is congruent to an acute angle of another right triangle, then the triangles are similar.
   b) If the corresponding angles of two quadrilaterals are congruent, then the ratios of their corresponding sides are equal.
   c) If the ratios of the corresponding sides of two quadrilaterals are equal, then the corresponding angles are congruent.
   d) If the ratios of the corresponding sides of two triangles are equal, then the corresponding angles are congruent.
   e) Two congruent figures are always similar.
   f) Congruence is a special kind of similarity.

4. In an architect's drawing, the scale is given as $\frac{1}{4}$ inch to 1 foot. What should be the measurements of the drawing of a room measuring 18 by 25 feet?

5. The scale of a map is $\frac{1}{4}$ inch to 10 miles. What is the distance from city X to city Y if the map shows a distance of $4\frac{3}{8}$ inches?

6. From the measurements shown in the figure, determine the distance AB across a stream.

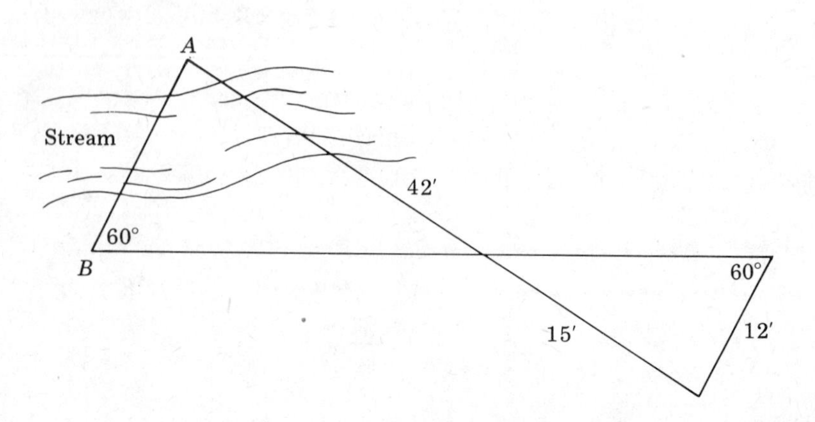

7 Two rectangular fences are situated as shown. If the width of the larger rectangle (w') is twice the width of the smaller rectangle (w), and the length of the smaller fence is 60 feet, what is the length of the larger fence?

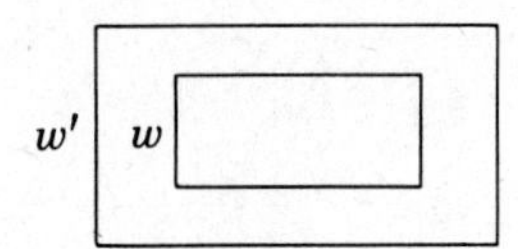

8 If the midpoints of the sides of a triangle are joined, a second triangle is formed. Prove that the second triangle is similar to the original triangle.

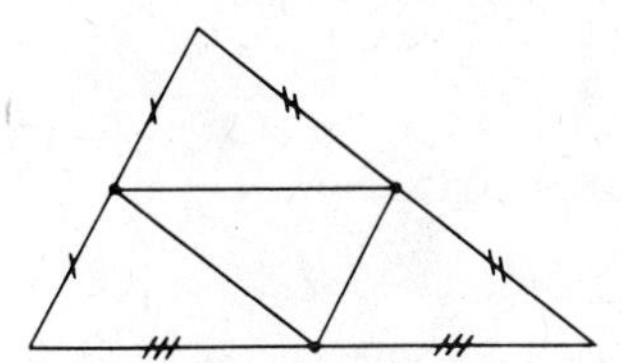

9 Prove that two regular polygons with the same number of sides are similar.

10 Prove that if two triangles are similar, then the ratio of their perimeters is the same as the ratio of two corresponding sides.

Chapter 10B

CONSTRUCTIONS

Objectives

At the completion of this chapter, you should be able to:

1 Define a *geometric construction.*

2 Perform the following constructions:

a) Copy a given line segment on a given line.
b) Bisect a line segment.
c) Construct the perpendicular bisector of a given segment.
d) Copy a given angle on a given side of a given ray.
e) Bisect a given angle.
f) Copy a given triangle on a given side of a given ray.

Prerequisites

You should:

1 Understand the concept of *congruence.*

2 Be familiar with methods of proving triangles congruent by SAS, ASA, and SSS.

3 Know that corresponding parts of congruent triangles are congruent.

INTRODUCTION

In making drawings, a modern draftsman or artist uses such instruments as a ruler, compass, protractor, T-square, draftman's triangle, French curve, parallel ruler, proportional divider, and linkages.

In this chapter, however, we shall see how to construct a variety of figures using only a compass and straightedge. A straightedge is an unmarked ruler. We shall use the straightedge only to draw lines; we shall use the compass only to draw a circle or part of a circle with a given center and radius. These are the instruments the Greeks used many years ago, and in their efforts to find solutions to certain problems, they discovered many important and unforeseen ideas in mathematics.

Geometric drawings that are made only by compass and straightedge are called *constructions.*

10B.1 Construction 1:

TO COPY A GIVEN LINE SEGMENT ON A GIVEN LINE

Example Given line segment $\overline{AB}$ and line l (Figure 10B.1), how can you construct on l a segment $\overline{CD}$ that is congruent to $\overline{AB}$?

FIGURE 10B.1

A • ——— • B

←——————→ l

Solution

1. Place the spike of the compass at point A and the pencil tip at point B (Figure 10B.2). The distance between the spike and the pencil tip is called the *radius of the compass.*

FIGURE 10B.2

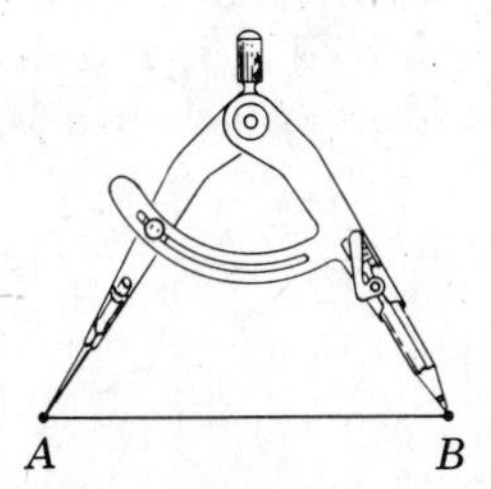

2. Keeping the radius fixed, place the spike at any point C on l and draw a short arc intersecting l at D (Figure 10B.3). Then $CD = AB$.

FIGURE 10B.3

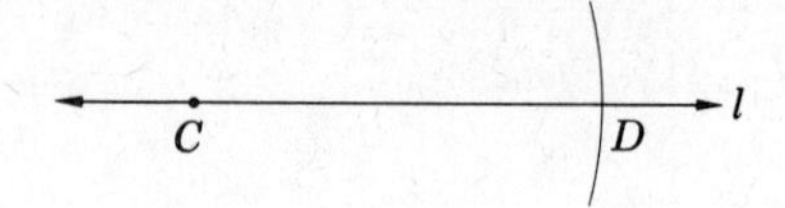

Proof $\overline{CD}$ and $\overline{AB}$ have the same measure since the measure of each is the radius of the compass. Therefore, by the definition of congruent line segments, $\overline{CD} \cong \overline{AB}$.

Note When we construct $\overline{CD}$ congruent to $\overline{AB}$, we say that we have "copied" segment $\overline{AB}$.

10B.2 Construction 2:

TO BISECT A LINE SEGMENT

Example Given line segment $\overline{AB}$ (Figure 10B.4), how can you locate a point C on $\overline{AB}$ that will *bisect* $\overline{AB}$, that is, divide $\overline{AB}$ into two congruent parts such that $\overline{AC} \cong \overline{CB}$?

FIGURE 10B.4

A ———————— B

Solution

1 Place the spike of the compass at either endpoint A or B with the radius set at more than half the length of AB. Draw arcs above and below $\overline{AB}$ (Figure 10B.5).

FIGURE 10B.5

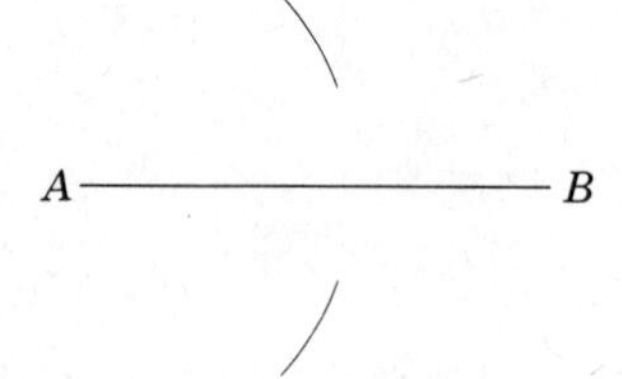

2 Without changing the radius of the compass, place the spike at the other endpoint of $\overline{AB}$, and draw arcs that intersect the arcs in 1 at points R and S (Figure 10B.6).

FIGURE 10B.6

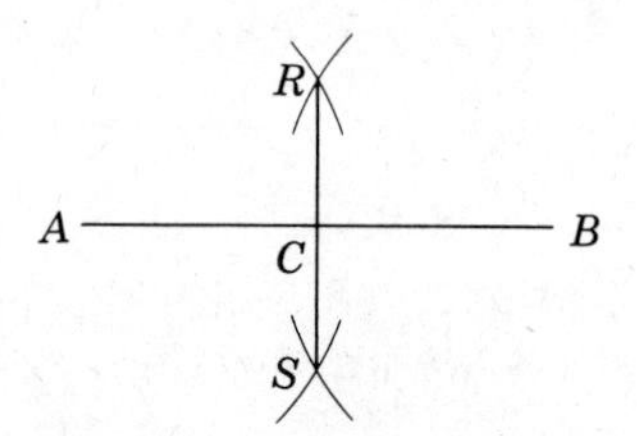

3 Draw a line segment through points R and S, intersecting $\overline{AB}$ at C. Point C bisects $\overline{AB}$ and is called the *midpoint* of $\overline{AB}$.

Proof

1 $AR = RB = BS = SA$ since the measure of each segment is the same radius (Figure 10B.7).

FIGURE 10B.7

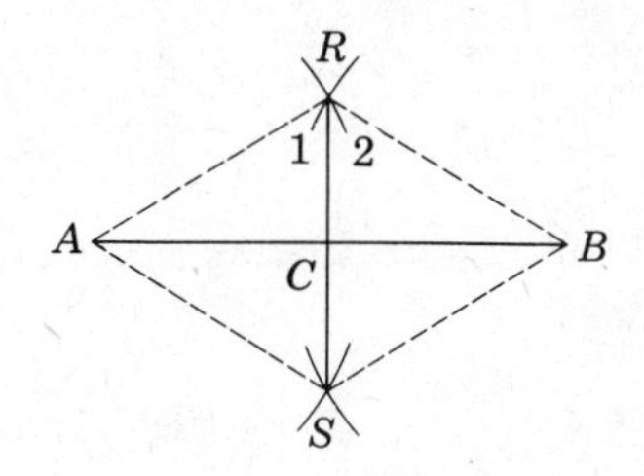

2 $\triangle ARS \cong \triangle BRS$ (SSS) and the corresponding angles $\angle 1$ and $\angle 2$ are congruent.

3 $\triangle ARC \cong \triangle BRC$ (SAS) and $\overline{AC} \cong \overline{CB}$ since they are corresponding sides of congruent triangles.

10B.3 Construction 3:

TO CONSTRUCT THE PERPENDICULAR BISECTOR OF A GIVEN SEGMENT

Example Given line segment $\overline{AB}$ (Figure 10B.8), how do you construct $\overleftrightarrow{RS}$ (1) to be perpendicular to $\overline{AB}$, and (2) to bisect $\overline{AB}$?

FIGURE 10B.8

A———————————B

Solution Bisect $\overline{AB}$ as in Construction 2. $\overleftrightarrow{RS}$ is perpendicular to $\overline{AB}$ at point C and also bisects $\overline{AB}$ at point C (Figure 10B.9).

FIGURE 10B.9

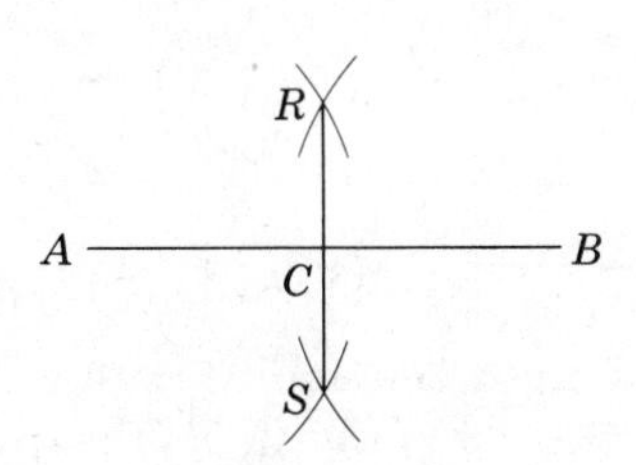

Proof

1 $\triangle ARC \cong \triangle BRC$ by the proof in Construction 2; $\overline{AC} \cong \overline{CB}$.

2 $\angle 3 \cong \angle 4$ since they are corresponding angles of congruent triangles (Figure 10B.10).

FIGURE 10B.10

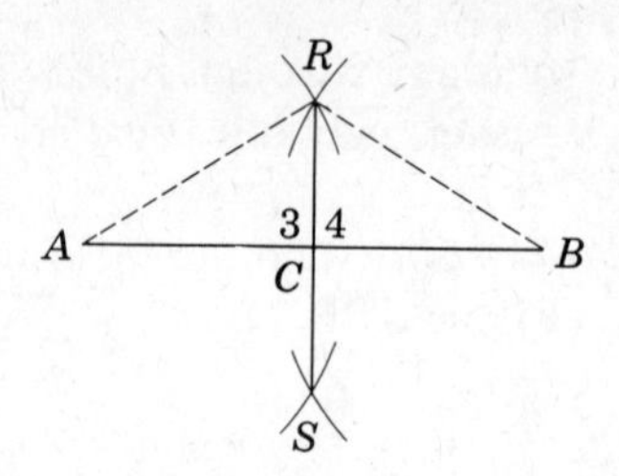

3 $\angle 3$ and $\angle 4$ are right angles since they are congruent and supplementary.

4 Therefore, $\overleftrightarrow{RS} \perp \overline{AB}$ at C, and also bisects $\overline{AB}$ at C.

10B.4 Construction 4:

TO COPY A GIVEN ANGLE ON A GIVEN SIDE OF A GIVEN RAY

Example Given $\angle A$ and a ray with endpoint at R (Figure 10B.11), how do we construct an angle with vertex R and having the given ray as a side, congruent to $\angle A$?

FIGURE 10B.11

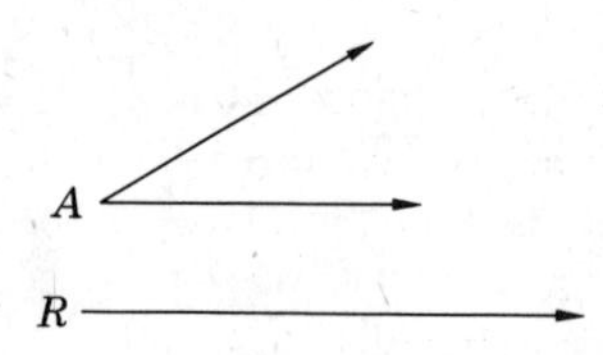

Solution 1 With A as center, draw an arc intersecting the sides of the given angle in points B and C (Figure 10B.12).

FIGURE 10B.12

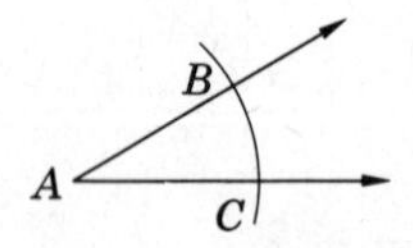

2 Without changing the radius, draw an arc with R as center, intersecting the given ray in T (Figure 10B.13).

FIGURE 10B.13

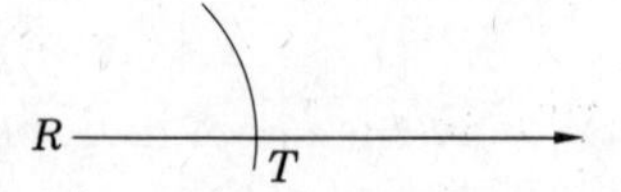

3 With T as center and BC as radius, draw an arc intersecting the arc in step 2 in S (Figure 10B.14).

FIGURE 10B.14

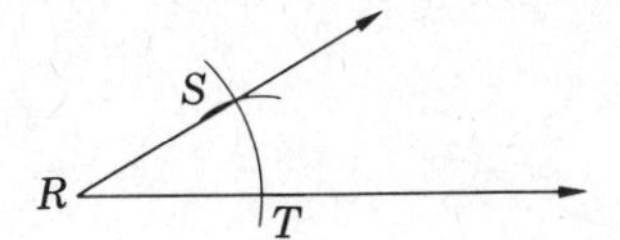

4 Draw ray $\overrightarrow{RS}$. $\angle SRT \cong \angle BAC$.

Proof

1 $\angle ABC \cong \Delta RST$ (SSS).

2 $\angle SRT \cong \angle BAC$ since they are corresponding angles of congruent triangles (Figure 10B.15).

FIGURE 10B.15

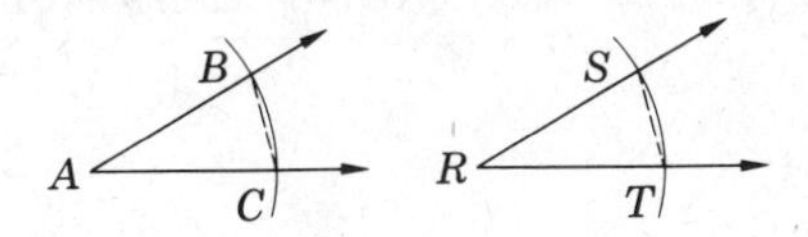

10B.5 Construction 5:

TO BISECT AN ANGLE

Example Given $\angle A$ (Figure 10B.16), how can you construct a ray with endpoint A that will bisect $\angle A$?

FIGURE 10B.16

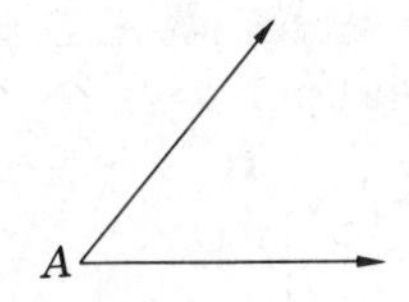

Solution

1 With A as center, draw an arc intersecting the sides of $\angle A$ at B and C (Figure 10B.17).

FIGURE 10B.17

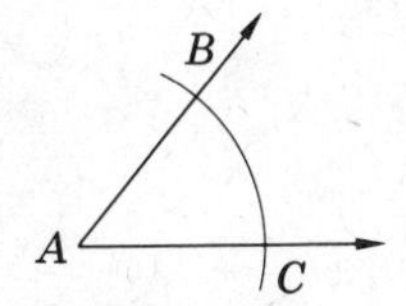

2 With B as center, and a radius greater than half of $\overline{BC}$, draw an arc in the interior of $\angle A$ (Figure 10B.18). Keeping the same radius and with C as center, draw an arc that intersects the first arc in point P.

FIGURE 10B.18

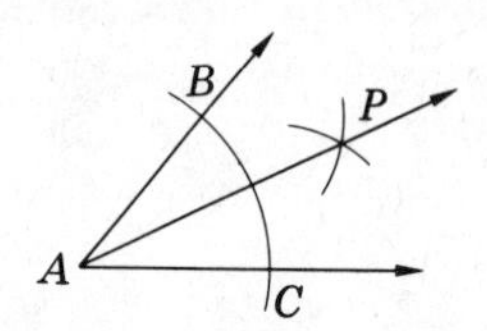

3 Draw the ray $\overrightarrow{AP}$. $\angle BAP \cong \angle PAC$. $\overrightarrow{AP}$ is called the *bisector of* $\angle BAC$.

Proof $\triangle ABP \cong \triangle ACP$ (SSS). Therefore, $\angle BAP \cong \angle PAC$, since they are corresponding angles of congruent triangles (Figure 10B.19).

FIGURE 10B.19

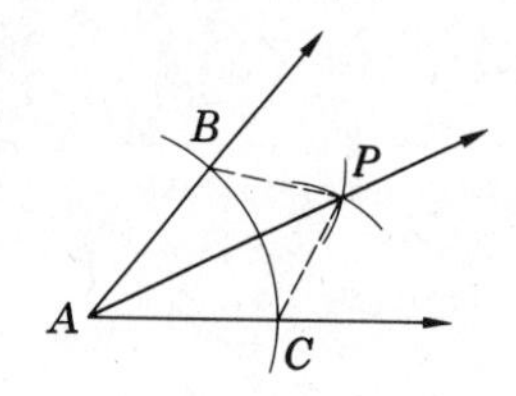

10B.6 Construction 6:

TO COPY A GIVEN TRIANGLE ON A GIVEN SIDE OF A GIVEN RAY

Example Given $\triangle ABC$ and a ray with endpoint at D (Figure 10B.20), how can you construct a $\triangle DEF$ congruent to $\triangle ABC$, with the side $\overline{DF}$ lying on the given ray?

FIGURE 10B.20

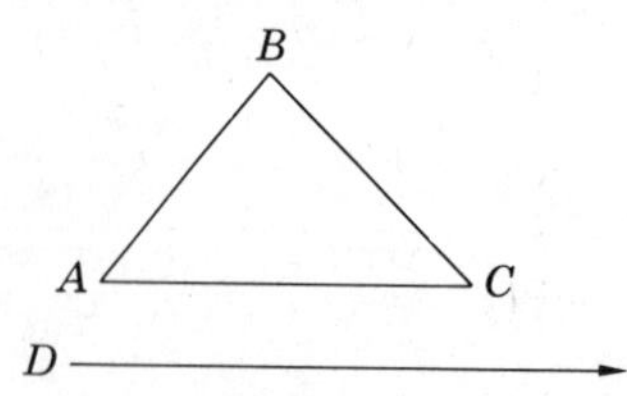

Solution See Figure 10B.21.

FIGURE 10B.21

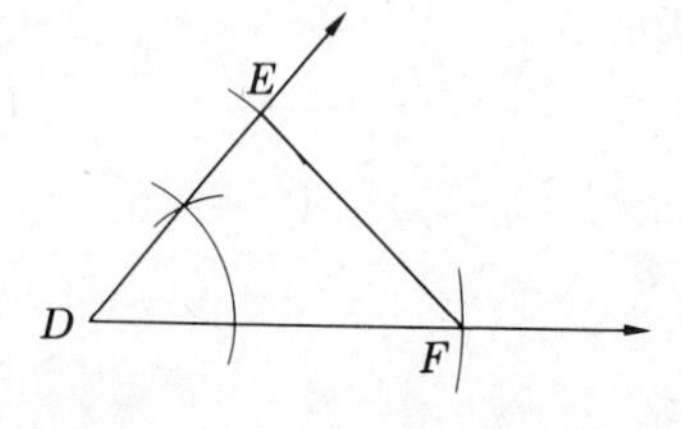

1. With D as center and AC as radius, draw an arc intersecting the given ray in F. $DF = AC$.
2. With D as vertex and $\overrightarrow{DF}$ as a side, construct $\angle D \cong \angle A$.
3. With D as center and AB as radius, draw an arc intersecting the other side of $\angle D$ at E. $DE = AB$. $\triangle DEF$ is the required triangle.

Proof $\triangle DEF \cong \triangle ABC$ (SAS).

EXERCISE SET 10B.1–10B.6

1. Our construction of an angle bisector was based on the SSS congruence property. Show a method of constructing an angle bisector that is based on the SAS congruence property.
2. Construct an angle of (a) 45°, (b) 60°, (c) 30°, (d) 15°, (e) 75°.
3. Construct a square, given the length of its diagonal.
4. Construct a square, given the length of its perimeter.
5. Given segments $\overline{AB}$ and $\overline{CD}$ and angle R as shown in the figure, construct $\triangle RST$ with $\overline{RS} \cong \overline{AB}$, and $\overline{RT} \cong \overline{CD}$. On which postulate is this construction based?

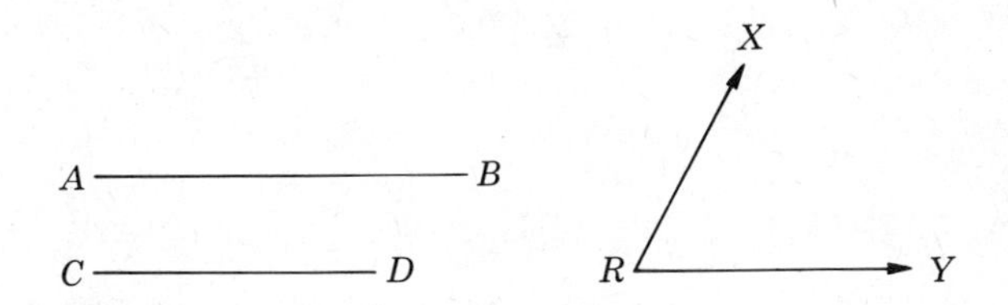

6. In $\triangle ABC$ and $\triangle A'B'C'$, $\overline{AB} \cong \overline{A'B'}$, $\overline{BC} \cong \overline{B'C'}$, and $\angle A \cong \angle A'$. Are the triangles necessarily congruent? Explain your answer.

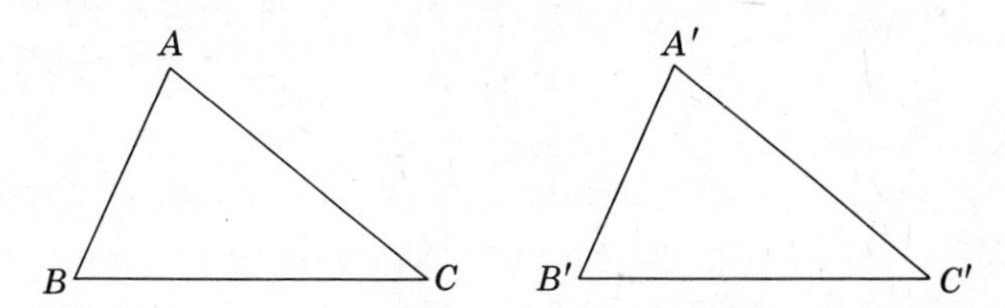

7. Construct an equilateral triangle the length of whose median is 1 inch.
8. We wish to construct a triangle with sides of lengths a, b, and c. What condition must be true of a, b, and c for the construction to be possible?
9. Since the metric system is based on powers of ten, you must be able to divide a given unit of length into ten equal parts if you wish to make home-made metric equipment. Given a cardboard strip $\overline{AB}$ measuring one meter, show how you would divide it into ten equal parts by a geometric construction. Each of the ten parts will then measure a *decimeter*.

Chapter 10C

MEASUREMENT

Objectives At the completion of this chapter, you should be able to:

Metric System

1. Name the basic units of length, capacity, and weight in the metric system.
2. Name and state the relationships among the prefixes used in the metric system.
3. Express metric measures in standard symbols.
4. Show the relationship between the metric system and the decimal numeration system.
5. Convert mentally from one metric measure to another.
6. Apply metric measures in the solution of problems.
7. State relationships among units of length, capacity, and weight.
8. Convert between American (English) and metric measures.
9. State some advantages of the metric system.

Measurement

1. Show the need for standard units of measure.
2. Explain why all measurements are approximate.
3. Define *perimeter*.
4. Determine the perimeter of a polygon, given its sides.

5 Define *area.*

6 Determine the area of a (a) rectangle, given its sides, (b) triangle, given its base and altitude (c) parallelogram, given its base and altitude, (d) trapezoid, given its altitude and bases, (e) geometric figure composed of a combination of the figures in (a) through (d).

7 Define *volume.*

8 Determine the volume of (a) a prism, given its altitude and base, and (b) a pyramid, given its altitude and base.

9 Define *circumference* in terms of limit.

10 Find the circumference of a circle given its radius.

11 Define the *area of a circle* in terms of limit.

12 Find the area of a circle, given its radius.

13 Determine the volume of:
a) A circular cylinder, given its altitude and base.
b) A circular cone, given its altitude and base.
c) A sphere, given its radius.

14 Find the surface area of a sphere, given its radius.

Prerequisites

You should:

1 Be familiar with the following geometric figures: square, rectangle, triangle, parallelogram, trapezoid, and circle.

2 Be able to use the decimal point to multiply and divide by powers of ten.

3 Be familiar with the American (English) units of measure.

INTRODUCTION

Although, today, we have many different kinds of measurements, the measurement of *length* was among the first measurements man needed. The early measures of length were based on parts of the human body. For instance, the width of the index finger was called a *digit.* The distance from the elbow to the end of the middle finger was called a *cubit.*

Since these units of length differed from one person to another, it became necessary to define the units more precisely. For instance, King Henry I of England decreed 850 years ago that a *yard* should be the distance from the end of *his* nose to the end of *his* thumb. In the sixteenth century a *rod* was the length of the left foot of 16 men lined up, as they left church on Sunday morning.

It was soon realized that using parts of the body for measuring may be convenient but is not reliable. A way had to be found by which a yard in England represented the same length as a yard in the United States. For these reasons certain units were agreed upon by large numbers of people. Such units are called *standard units.* The most common standard units of length are the inch, foot, yard, mile, and meter.

The only actual standard unit of length is the *meter.* The other units, such as the inch, foot, and yard, are specified by law in relation to the meter. The length of the meter was chosen to be one ten-millionth (0.0000001) of the meridional distance from the North Pole to the Equator.

Various methods have been used by the United States Bureau of Standards for maintaining a precise model of the standard meter. For many years the model was a platinum bar kept under carefully controlled atmospheric conditions. In 1960, a new definition of the meter was adopted. The meter is now defined as having a length that is 1,650,763.73 times the wavelength of orange light from krypton 86. This definition was adopted because it is easier to reproduce in scientific laboratories in different parts of the world, and because it provides a more precise model than the platinum bar.

In 1966, North America's most accurate rule was built by Ohio State University scientists. This ruler is a 500-meter line measured so accurately that its margin of error is less than the thickness of a sheet of paper. This remarkably accurate ruler, which is more than 1600 feet long, is used to calibrate instruments and other smaller measuring devices.

10C.1 Linear Measurement

Measurement of line segments is called *linear measurement.* When we *measure* a line segment $\overline{AB}$ (Figure 10C.1), we do three things:

FIGURE 10C.1

A ——————————— B

1. We first decide upon some segment, say $\overline{XY}$, to serve as a *unit* with which $\overline{AB}$ will be compared (Figure 10C.2(a)). By selecting $\overline{XY}$ as our unit, we are agreeing to consider its measure to be exactly 1 (Figure 10C.2(b)).

FIGURE 10C.2

(a) X ——— Y (b) X ——— Y (bracketed, labeled 1)

2. Then, starting at one end of $\overline{AB}$, we lay off a series of units $\overline{XY}$, end to end along $\overline{AB}$, until we reach (approximately) the

other end of the segment $\overline{AB}$ as shown in Figure 10C.3.

FIGURE 10C.3

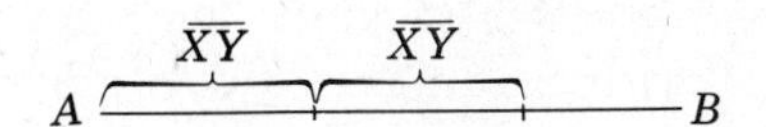

3 We then count the number of units $\overline{XY}$ needed to approximate the length of $\overline{AB}$. This number is the approximate measure of the length of $\overline{AB}$, expressed in terms of $\overline{XY}$.

a) If $\overline{AB}$ is "covered" by laying off the unit $\overline{XY}$ exactly twice, we say that the measure of $\overline{AB}$ is the number 2, and that the length of $\overline{AB}$ is exactly two $\overline{XY}$'s.

b) It is more likely, however, that $\overline{XY}$ will not fit into $\overline{AB}$ exactly a *whole* number of times (Figure 10C.4). In this case we can determine visually that the length of $\overline{AB}$ is, say, *nearer* to 3 units than to 2 units. So we say that, *to the nearest unit*, the length of $\overline{AB}$ is 3 units.

FIGURE 10C.4

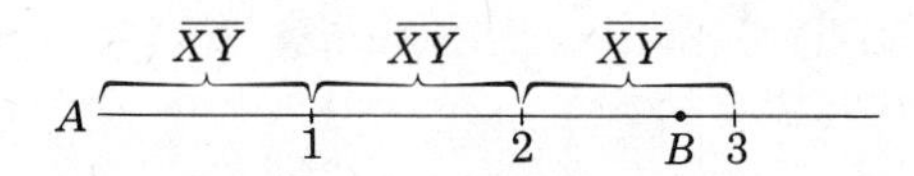

Comments

1 *Note that the* measure *of a segment is a number, while the* measurement *of a segment is a number* and *the unit used. For example, in the illustration above the* measure *of* $\overline{AB}$ *is approximately* 3; *the* measurement *of AB is approximately* 3$\overline{XY}$*'s.*

2 *To help estimate whether the measure of AB is nearer to* 3 *than to* 2, *we locate the midpoint of the unit (Figure 10C.5).*

FIGURE 10C.5

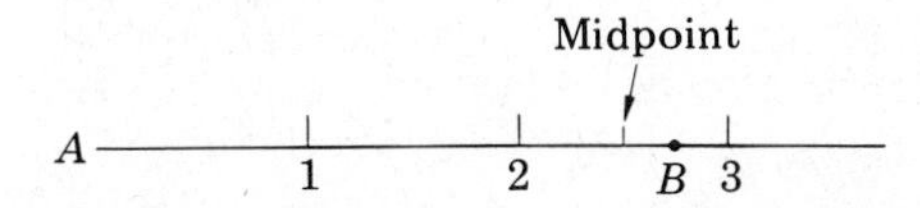

3 *Note that the size of the unit* $\overline{XY}$ *is completely arbitrary. We can measure a desk by using a pencil, eraser, or book as our unit of measure. The desk may measure "about 4 pencils long." A more abstract method of measuring a line segment is the use of another line segment as a unit of measure.*

THE APPROXIMATE NATURE OF MEASUREMENT

All measurements are approximate because unavoidable sources of error are always present.

Weather conditions affect the accuracy of the measuring instrument. The marks on the instrument are not exact and are not designated exactly. Neither are, say, the endpoints of the segment we wish to measure.

Carelessness, poor eyesight, and inexperience with the instrument are other sources of error that may creep into our measurements.

We can make finer distinctions between lengths by using a *smaller unit* of measurement.

THE RULER

A *ruler* is an instrument with a number line that has been subdivided into standard units, usually inches or centimeters. When we measure the edge of a desk with a ruler, we compare the length of the edge with the number line marked on the ruler.

Comments

1 *Comparing measurements of the same object made by different people and measures determined by instruments marked with varying degrees of precision, will help us see the approximate nature of measurement.*

2 *We have been talking only about* direct *measurement, where we apply a measuring instrument directly to the object whose measurement we wish to determine. But not all measurements can be obtained by such direct procedures. Some are obtained by* indirect measurement. *In indirect measurement we* calculate *certain quantities from other directly measured quantities.*

3 *It is the* line *at the edge of the ruler that is important. The width of the ruler only provides space for writing the numerals and for giving the instrument more "body."*

4 *A ruler is a convenient and clever instrument for measuring lengths because it combines all the requirements for measuring into one simple instrument:*

a) The unit of measurement is already indicated.
b) The units have already been "laid off," end to end.
c) The counting is done for us, since the units have been numbered.
d) The ruler contains a series of smaller subdivisions $\left(\frac{1}{2}\right.$ inch, $\frac{1}{4}$ inch, $\frac{1}{8}$ inch, etc.$\left.\right)$ for finer distinctions between lengths that are close together.

OTHER STANDARD UNITS OF MEASUREMENT

Scientific developments in recent years have brought about an extension of some of our already familiar standard units of measurement. For example, to facilitate the measurement of supersonic speeds, *Mach numbers* have been invented. These numbers have been defined in terms of the speed of sound. A plane traveling at, say, Mach 2, would be flying at twice the speed of sound, or about 1500 miles per hour.

The *light-year* was defined for better measurement of astronomic distances. One light-year is defined to be the distance light will travel in 1 year's time—a distance of about 5,870,000,000,000 miles. The North Star is a distance of about 47 light-years from the earth.

The *atomic mass unit* was invented for use with atomic weights. This unit is defined to be 0.00000000000000000000000000366 pound. Atomic nuclei are made up of *protons* and *neutrons*. The proton has a mass of 1.00758 atomic mass units, and the neutron has a mass of 1.00897 atomic mass units. The nucleus of an atom is surrounded by *electrons*, each of which has a mass of 0.00055 atomic mass unit.

If we should attempt to express these measurements in terms of miles and pounds, we shall quickly appreciate the convenience of having defined these new standard units of measurement.

EXERCISE SET 10C.1

1 Here are five segments:

(1) ______________ (2) ______

(3) ____________ (4) ____________________

(5) ____________________________________

Record the following data in the chart shown.

a) Estimate the length of each segment to the nearest inch.
b) Check your estimates with a ruler.
c) Measure each segment to the nearest $\frac{1}{4}$ inch; to the nearest $\frac{1}{8}$ inch; to the nearest $\frac{1}{16}$ inch.

	To Nearest Inch		*To Nearest $\frac{1}{4}$ Inch*	*To Nearest $\frac{1}{8}$ Inch*	*To Nearest $\frac{1}{16}$ Inch*
	Estimate	*Ruler*			
1					
2					
3					
4					
5					

10C.2 The Metric System

The *metric system* was born during the French Revolution and in the last two centuries it has spread to most of the world. Today, the United States is the only industrial country where the metric system is not yet generally used. However, the U. S. Metric Bill was signed into law on Dec. 23, 1975, putting the country on a ten-year program of converting to the metric system. But even today a large number of familiar products are sold in metric units: prescriptions are filled in milligrams, not ounces; camera lenses and film are described in metric units; electricity is measured in kilowatts, and water-meter readings are in cubic meters; scientists deal almost exclusively with metric weights and measures; all Olympic sports use metric measurement; the military has been using metric measurements for years because our NATO allies use them.

Some of the advantages of the metric system are its simplicity and its almost universal usage. Because it is a *decimal* system, like our numeration and monetary systems, conversion from one unit to another requires no more than moving a decimal point.

BASIC UNITS

The metric system uses three basic units:

Meter which measures *length*,
Liter which measures *volume* or *capacity*,
Gram which measures *weight.*

A *meter is a little more than 1 yard (Figure 10C.6).*

FIGURE 10C.6

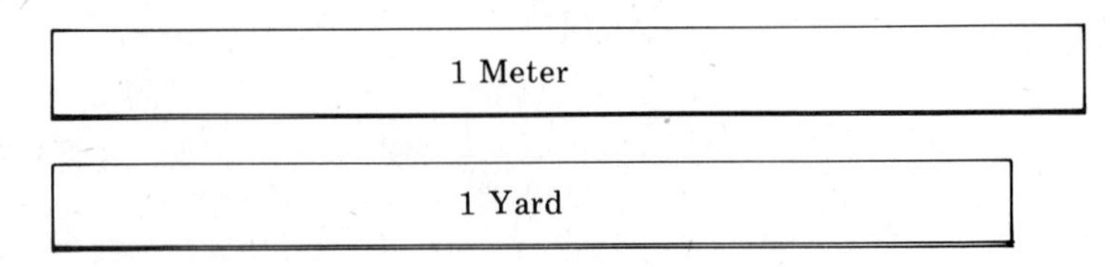

A *liter* is a little more than 1 quart (Figure 10C.7).

FIGURE 10C.7

A *gram* is about the weight of a paper clip (Figure 10C.8).

FIGURE 10C.8

The other units in the metric system are decimal parts or multiples of these three basic units. Certain *prefixes* are placed in front of each basic unit to indicate weights and measures smaller and larger than the basic units. Since the metric system is a decimal-based system, larger and smaller measurements are derived by multiplying and dividing a basic unit by 10 and its multiples.

The metric *prefixes* and their meanings are:

Milli means 0.001		*Deka* means 10	
Centi means 0.01		*Hecto* means 100	
Deci means 0.1		*Kilo* means 1000	

Meaning	1000	100	10	1	.1	.01	.001
Metric Prefixes	kilo	hecto	deka	basic unit	deci	centi	milli

When we combine these prefixes with the basic units, we obtain the table of metric measures shown in Table 10C.1.

In Figure 10C.9 the comparative sizes of centimeter to inch and kilogram to pound are shown.

FIGURE 10C.9

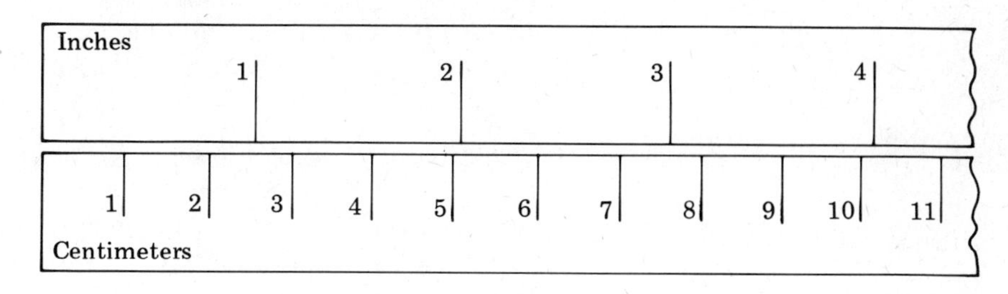

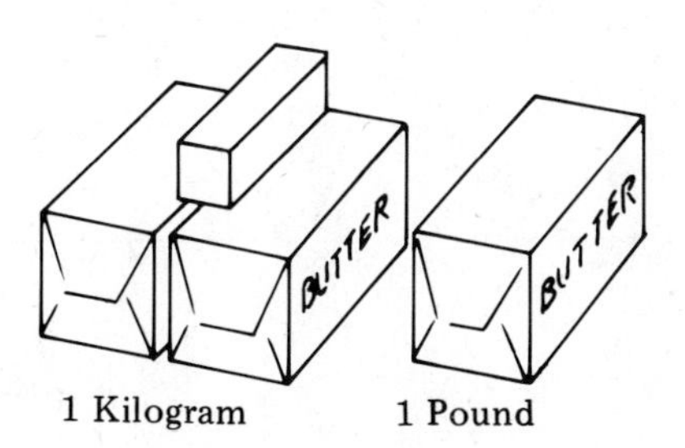

TABLE 10C.1
Metric Measures

Length		*Capacity*	
1 kilometer (km)	= 1000 meters	1 kiloliter (kl)	= 1000 liters
1 hectometer (hm)	= 100 meters	1 hectoliter (hl)	= 100 liters
1 dekameter (dkm)	= 10 meters	1 dekaliter (dkl)	= 10 liters
1 meter (m)	= 1 meter	1 liter (l)	= 1 liter
1 decimeter (dm)	= 0.1 meter	1 deciliter (dl)	= 0.1 liter
1 centimeter (cm)	= 0.01 meter	1 centiliter (cl)	= 0.01 liter
1 millimeter (mm)	= 0.001 meter	1 milliliter (ml)	= 0.001 liter

Weight	
1 kilogram (kg)	= 1000 grams
1 hectogram (hg)	= 100 grams
1 dekagram (dkg)	= 10 grams
1 gram (g)	= 1 gram
1 decimgram (dg)	= 0.1 gram
1 centigram (cg)	= 0.01 gram
1 milligram (mg)	= 0.001 gram

TABLE 10C.2
Approximate Equivalents

Metric-English		*English-Metric*	
1 kilometer	= 0.62 mile or 3280.8 feet	1 mile	= 1.6 kilometers
1 meter	= 39.37 inches	1 yard	= 0.914 meter
1 centimeter	= 0.39 inches	1 inch	= 2.54 centimeters
1 liter	= 1.057 quarts	1 foot	= 30.48 centimeters
1 kilogram	= 2.2 pounds	1 quart	= 0.95 liter
		1 pound	= 0.45 kilograms

Comments

1 *A* meter *is the basic unit of length in the metric system and is defined in terms of the wavelength of orange-red light emitted by a krypton-*86 *atom. A* kilogram *is the standard unit of mass in the metric system. There is but one standard for this unit and it is kept by the International Bureau of Weights and Measures near Paris. Every country has one or more duplicates of this standard. The United States duplicate is kept by the National Bureau of Standards in Washington.*

2 *The Celsius* (C) *or Centigrade scale for temperature is usually used with the metric system. (See Figure 10C.10.) Another temperature scale used with the metric system is called the* Kelvin Scale *where the starting or zero point is* absolute zero—*the lowest theoretical temperature that a gas can reach.*

FIGURE 10C.10

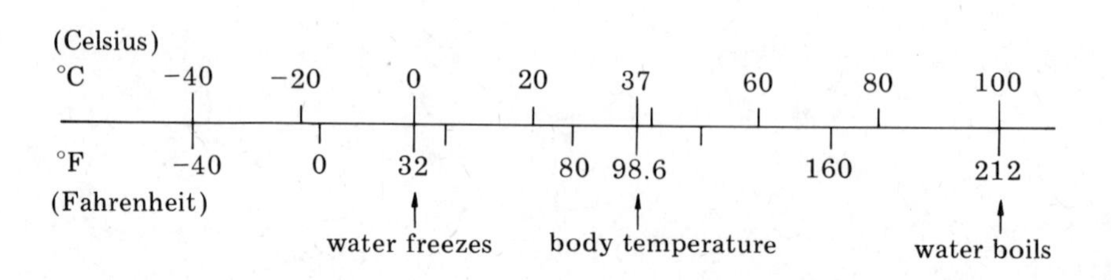

3 *Metric units have been designed to form a related system in which units of volume and weight are related to length:*

1 liter = 1000 cubic centimeters,
1 gram = weight of 1 cubic centimeter of water at 4° Centigrade.

4 *In the English units of measure, abbreviations for square and cubic units are usually written as sq. in., cu. ft., etc. In the metric system, abbreviations for square and cubic units are written with exponents: square meter is written m^2, cubic centimeter is written cm^3, etc.*

5 *Although for most people familiarity with the meter, liter, and gram is sufficient for their purposes, these units, by themselves, are insufficient for the engineer and scientist. The metric system has therefore been extended for accurate scientific work. This extended system was established by international agreement and is known as The International System of Units, SI. The system is built upon a foundation of seven basic units, plus two supplementary units.*

EXERCISE SET 10C.2

1 Which of each of the following pairs is larger?
a) A centimeter or an inch. b) A meter or a yard.
c) A kilometer or a mile. d) A liter or a quart.
e) A gram or an ounce. f) 400 grams or a pound.
g) A kilogram or 2 pounds. h) A milligram or .05 ounces.

2 Estimate, to the nearest centimeter, the length of the following line segments; then, verify your estimates with a ruler.

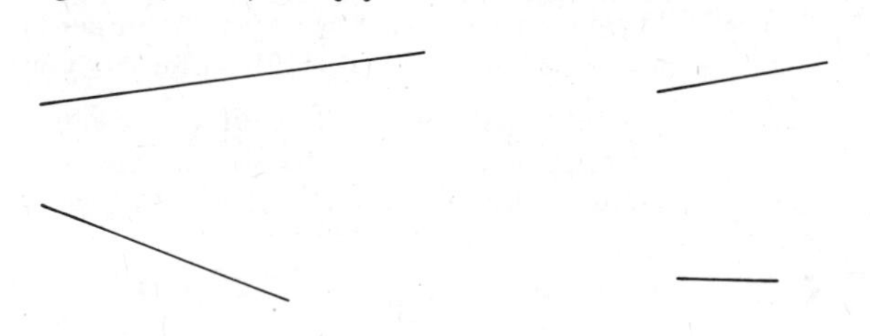

3 Draw line segments of the following lengths.
a) 5 cm b) 10 mm c) 1 dm d) 13.8 cm

4 Give a reasonable estimate in metric units for each of the following.
a) Your height. b) Your weight.
c) The weight of this book. d) The length of your index finger.
e) The circumference of the earth.
f) The weight of your watch.
g) The capacity of your hot water boiler.

5 Fill in the blanks.
a) 1 m = __dm = __cm = __mm b) 6 m = __ dm = __cm = __mm

c) 9 dm = __cm
d) 14 cm = __mm
e) 8 cm = __dm
f) 3 m 8 cm = __cm
g) 3 kg = __g = __mg
h) 2.3 g = __mg
i) 6 l = __dkl = __cl
j) 25 mm = __cm = __dm = __m
k) 36 cg = __dg = __g
l) 185 mm = __cg = __kg
m) 3250 m = __km

6 Do you get a smaller or larger number in your answer when:
a) You change a larger unit of measure to a smaller unit of measure?
b) You change a smaller unit of measure to a larger unit of measure?

7 Convert each of the following.
a) 10 centimeters to inches
b) 100 meters to feet
c) 20 meters to yards
d) 16 kilometers to miles
e) 1 meter to feet
f) 100 millimeters to inches
g) 100 yards to meters
h) 50 miles to kilometers
i) 1500 feet to kilometers
j) 150 pounds to kilograms
k) 3 quarts to liters
l) 16 gallons to liters
m) 3 ounces to milligrams

8 Fill in the blanks.
a) The height of a person 6 feet tall is about __meters.
b) A 3-inch nail is about __decimeters long.
c) The length of this page is about __centimeters.
d) A speed of 55 miles per hour is about __kilometers per hour.
e) The width of 35 mm film is about __inches.
f) A distance of 275 km is about __miles.

9 In the Olympic relay races there are events in the following distances: 400 meters, 800 meters, 1600 meters, 3200 meters, and 6,000 meters. Express each of these distances in yards.

10 Express the following distances in miles.
a) New York to San Francisco: 4118 km
b) New York to Boston: 306 km
c) New York to New Orleans: 1883 km
d) New York to Washington D. C.: 344 km

11 Assume that your car averages 18 km per liter of fuel. If the capacity of your fuel tank is 90 liters, how far can you drive on a full tank of fuel?

12 A pharmacist has 2 liters of a certain medicine with which to fill prescriptions. If he uses 30 ml for each prescription, how much medicine does he have left after filling 36 prescriptions?

10C.3 Square and Cubic Measurement

AREA: INFORMAL REMARKS

By the *area* of a closed region we mean a *number* that measures the "amount of surface" in the region. This number is called the *measure of the area* of the region.

We must be careful to distinguish between:

a) A *region*, which is a set of points, and its *area*, which is a number.

b) *Area*, which is a property of a region, and *perimeter*, which is a property of a boundary.

c) *Length*, which answers such questions as "How far?", and *area*, which answers such questions as "How much floor space?"

To determine the area of a region we follow the same process used to determine the length of a segment as shown in Table 10C.3.

TABLE 10C.3

To Determine Length of Segment	*To Determine Area of Region*
1 We chose a unit of length.	*1* We choose a unit of area.
2 We estimated the number of units needed to "cover" the segment.	*2* We estimate the number of units needed to "cover" the region.
3 This number of units is the *length* of the segment.	*3* This number of units is the *area* of the region.
4 More precise estimates of length can be gotten by using a smaller unit	*4* More precise estimates of area can be gotten by using a smaller unit.

Although our choice of shape and size for our *unit of area* is arbitrary, using a *square* unit has advantages over other shapes. For example, when circular units are used to cover a rectangular region, there are parts of the region left uncovered. Square units, however, cover the *entire* region (Figure 10C.11).

FIGURE 10C.11

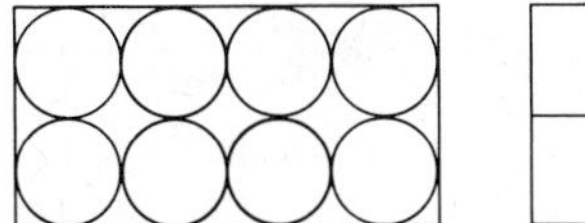

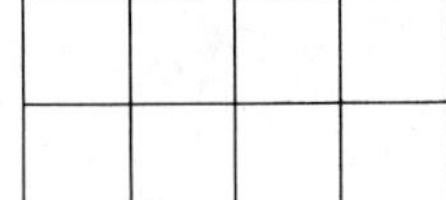

Comment

Using a unit in the shape of a right isosceles triangle would, like using squares, also completely cover a region. But squares are more convenient.

A *square inch* is the area of a region bounded by a square whose side is 1 inch. A *square centimeter* is the area of a region bounded by a square whose side is 1 centimeter. (See Figure 10C.12.)

FIGURE 10C.12

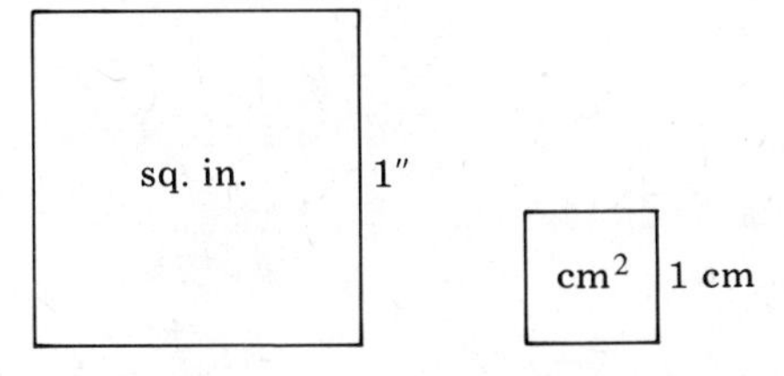

AREA THEOREMS

We shall now state a postulate that defines the *area of a square* by relating the concept of area to the concept of length.

POSTULATE 1 (*Unit Area Postulate*) The area of a square region is the square of the length of its side (Figure 10C.13).

FIGURE 10C.13

$A = s^2$ s

That is, the area of a square with side of length s is $A = s^2$.

From Postulate 1 we can (but won't in this book) derive the formulas for the areas of a rectangle, triangle, trapezoid, and parallelogram. We shall now state these formulas as theorems.

THEOREM 1 (*Area of a Rectangle*) The area of a rectangle is the product of its base and its altitude: $A = bh$. (See Figure 10C.14.)

FIGURE 10C.14

h $A = bh$

b

THEOREM 2 (*Area of a Triangle*) The area of a triangle is half the product of any base and the altitude to that base: $A = \frac{1}{2}bh$. (See Figure 10C.15.)

FIGURE 10C.15

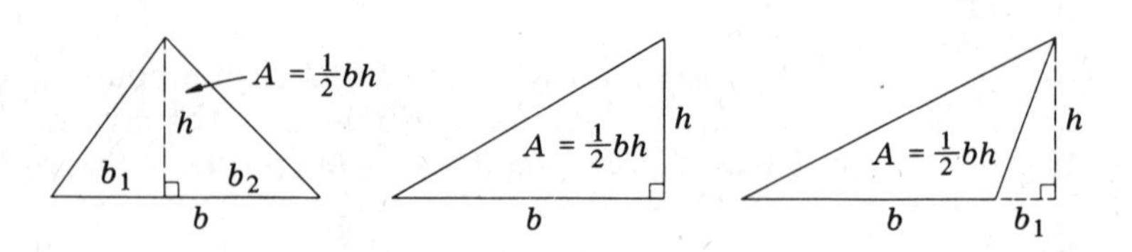

THEOREM 3 (*Area of a Trapezoid*) The area of a trapezoid is half the product of its altitude and the sum of its bases: $A = \frac{1}{2}h(b_1 + b_2)$. (See Figure 10C.16.)

FIGURE 10C.16

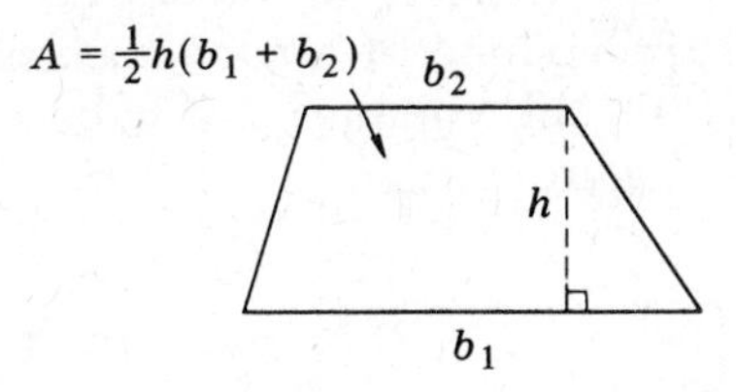

THEOREM 4 (*Area of a Parallelogram*) The area of a parallelogram is the product of any base and the corresponding altitude: $A = bh$. (See Figure 10C.17.)

FIGURE 10C.17

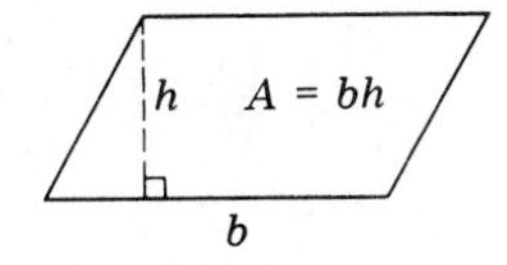

VOLUME: INFORMAL REMARKS

The *volume* of a solid is a *number* that measures the "amount of space" the solid occupies. The number is called the *measure of the volume* of the solid.

To determine the volume of a solid, we proceed the same way as when we determine length or area:

a) We choose a unit volume, or *cubic unit* (Figure 10C.18).
b) We estimate the number of these units needed to fill the solid (Figure 10C.19).
c) This *number* is the *volume* of the solid.
d) More precise estimates of volume can be obtained by using a smaller unit.

FIGURE 10C.18

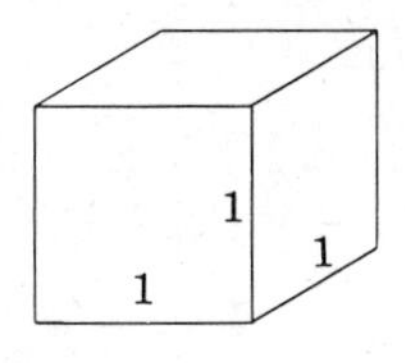

FIGURE 10C.19

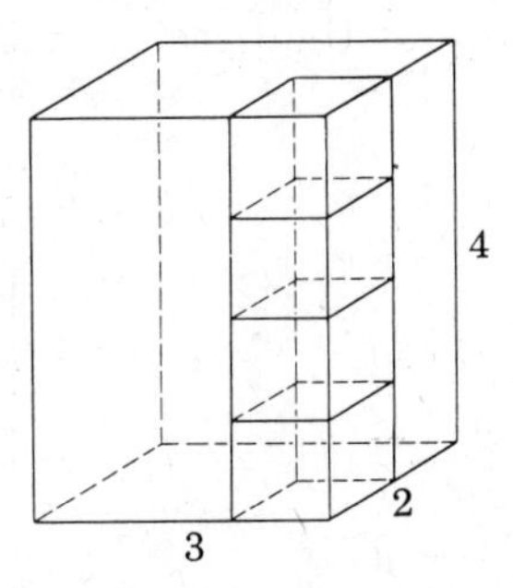

A *cubic inch* is the volume of a cube each side of which is 1 inch. A *cubic centimeter* is the volume of a cube each side of which is 1 centimeter. (See Figure 10C.20.)

FIGURE 10C.20

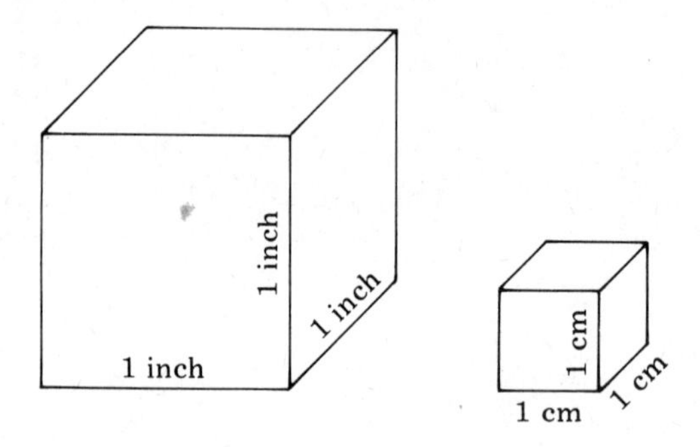

In our discussion of area we postulated the formula for the area of a square, from which we can derive the formulas for the areas of other polygonal regions. In the case of volume, we shall postulate the formula for the volume of a rectangular parallelepiped, from which we can derive the other volume formulas.

POSTULATE 2 (*Unit Volume Postulate*) The volume of a rectangular parallelepiped is the product of the altitude and the area of the base (Figure 10C.21).

FIGURE 10C.21

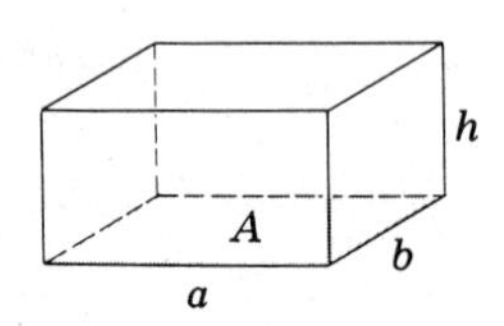

That is, $V = Ah = abh$.

VOLUME THEOREMS

THEOREM 5 (*The Volume of a Prism*) The volume of any prism is the product of the altitude and the area of the base: $V = Ah$. (See Figure 10C.22.)

FIGURE 10C.22

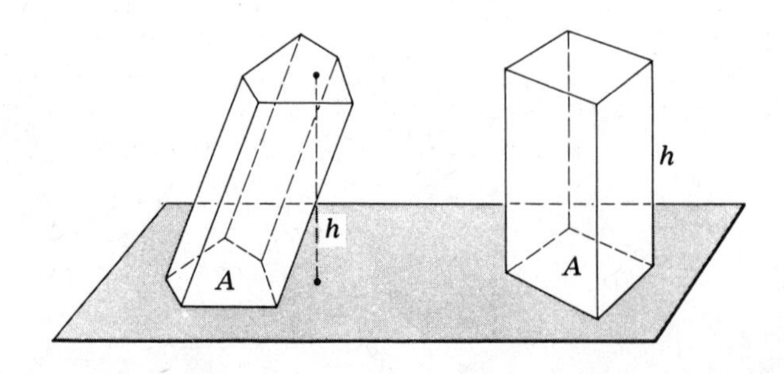

That is, if h is the altitude and A is the base area of the given prism, then its volume is $V = Ah$.

THEOREM 6 (*Volume of a Pyramid*) The volume of a pyramid is one third the product of its altitude and its base area: $V = \frac{1}{3}Ah$. (See Figure 10C.23.)

FIGURE 10C.23

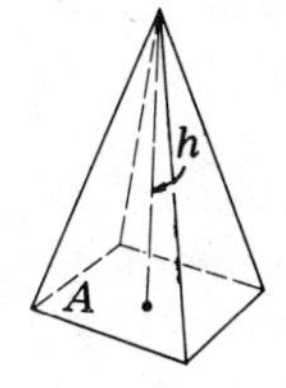

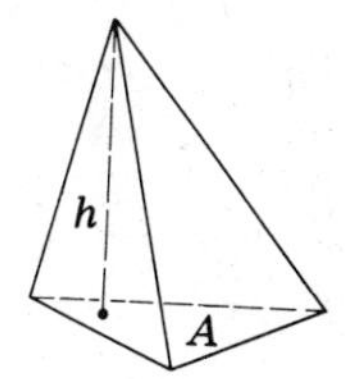

Comments

1 *A unit of* length *is a* segment *one unit long; a unit of* area *is a* square *one unit in each of two dimensions; a unit of* volume *is a* cube *one unit in each of three dimensions.*

2 *Theorem 6, in effect, tells us that the volume of a pyramid is one-third the volume of a prism with the same base and the same altitude as the pyramid.*

EXERCISE SET 10C.3

1 What is a square meter? A square yard? A square kilometer? A square mile?

2 a) If a rectangle is twice as long as another rectangle, but both have the same width, how do their areas compare?
b) If two rectangles have the same length, but the width of one is three times the width of the other, how do their areas compare?
c) One rectangle is twice as wide as another and three times as long. How do their areas compare?

3 What happens to the area of a triangle if:
a) Its altitude is doubled?
b) Its base is doubled?
c) Both its altitude and base are doubled?

4 How is the area of a square changed if its side is doubled? Halved? Tripled?

5 The altitude of a trapezoid is 5 inches and its area is 50 square inches. If the length of one base is 7 inches, what is the length of the other base?

6 How many bricks are needed to lay a walk 45 feet long and 12 feet wide if seven bricks are needed to cover a square foot?

7 A triangle and a parallelogram have equal areas and equal bases. How are their altitudes related?

8 The width of a rectangle is 5 inches less than its length. If the perimeter is 50 inches, find the area of the rectangle.

9 The length of a rectangle is two times its width. If the area is 98 square feet, find the perimeter.

10 A room is 18 feet long, 15 feet wide, and 8 feet high.
a) Find the total area of the walls and ceiling allowing 78 square feet for windows and doorway.
b) How many gallons of paint will be needed to cover the walls and ceiling if a gallon covers 400 square feet?

11 What is a cubic foot? A cubic yard? A cubic millimeter?

12 What happens to the volume of a rectangular solid:
a) If one of its measurements is doubled?
b) If two of its measurements are doubled?
c) If all three of its measurements are doubled?
d) If all three of its measurements are tripled?

13 A classroom measures 12 by 8 by 4 meters. How many cubic meters of air space are there for each pupil if there are 30 pupils in the class?

14 The volume of a box is $5\frac{1}{4}$ cubic inches. How long is the box if it is 2 inches wide and $\frac{3}{4}$ inch deep?

15 A toy is submerged in a rectangular water tank measuring 30 inches in length and 15 inches in width. What is the volume of the toy if it raises the water level 0.45 inch?

16 A sandbox measures 1.5 meters by 1.5 meters and is 0.3 meters deep. How many cubic meters of sand will it hold?

17 Find the volume of a prism whose altitude is 9 cm and whose base has an area of 72 square cm.

18 Find the volume of a pyramid whose altitude is 5 mm and whose base is an equilateral triangle with a side that measures 2 mm.

19 Given an irregularly shaped lump of rock, how would you approximate its volume?

10C.4 Measurement of Circular Figures

To find the perimeter of a triangle, we measure its three sides and add their lengths. We do the same thing to find the perimeter of a rectangle, pentagon, or any other polygon.

But how can we find the length, or circumference, of a circle? Can we measure the length of a curve the way we measure the length of a line segment? Can we use a ruler to measure the length of a circle?

We can, of course, take a tape measure and bend it into the shape of the circle. But this is not always convenient, accurate, or even possible. If we want to measure the length of a circular object, such as a disc or wheel, we might roll the object along the edge of a ruler and determine the distance covered by one complete revolution of the wheel. This distance will be the same as the circumference of the wheel. But this method, too, is not practical and seldom possible.

There is, however, a way of determining the circumference of a circle without bending the measuring instrument or rolling the object to be measured. We can determine the circumference merely by measuring a line segment! What we are saying is that it is possible to measure a curve by using a ruler! What makes this possible is the fascinating relation that exists between a particular line segment connected with a circle and the circumference of that circle.

In exploring this method of determining the circumference of a circle, we encounter a new and fascinating number—a number that has intrigued people for thousands of years. Although this number was first used by the ancient Greeks in connection with circles and is still best known in this connection, it has been found equally useful in situations that have nothing to do with circles or even with geometry. It is used, for instance, in public opinion polls to predict the winner of a forthcoming election, by insurance companies to determine the premium you pay on a life insurance policy, by a manufacturer to test the quality of his product, and by a scientist to understand the laws of heredity. We are talking, of course, about the number π.

In trying to determine the area of a circular region, we again face the problem of measuring something round with something straight—in this case, measuring a circular region with a square unit. And once again we find that this new, magic number comes to the rescue.

CIRCUMFERENCE OF A CIRCLE

We can approximate the circumference C of a circle by inscribing in it a regular polygon with a large number of sides as shown in Figure 10C.24. The perimeter p of such a polygon will be a good approximation of the circumference of the circle. In fact, we can make p as close as we wish to C by making the number of sides n large enough. We define the circumference as that number which the perimeter approaches as the number of sides gets larger and larger. This situation is described by the following definition.

FIGURE 10C.24

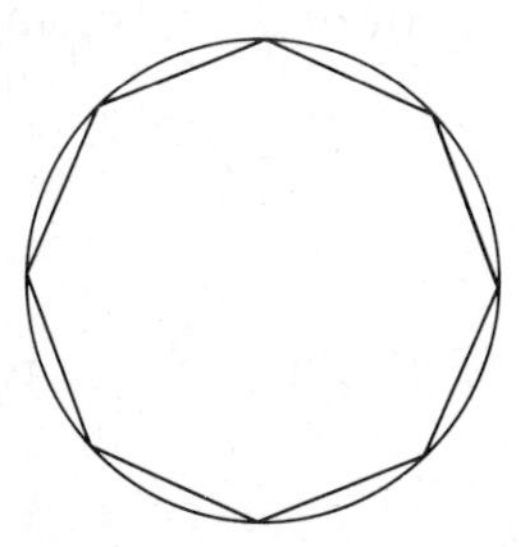

DEFINITION 1 The *circumference* of a circle is the limit of the perimeters of the inscribed regular polygons.

An important and interesting property of circles that we can prove is that *the ratio of the circumference C of any circle to its diameter d is a constant irrational number called* π. That is,

$$\frac{C}{d} = \pi; \quad \text{or} \quad C = \pi d, \quad \text{or} \quad C = 2\pi r$$

where r is the length of the radius of the circle. Since π is an irrational number, its decimal representation, 3.141592..., is infinite and nonrepeating. The numbers 3.14 and $\frac{22}{7}$ are often used as approximations of π.

THEOREM 7 The circumference of a circle of radius r and diameter d is $C = 2\pi r$ or $C = \pi d$.

AREA OF A CIRCLE

We approximated the circumference of a circle by finding the perimeter of an inscribed regular polygon with a very large number of sides. We can approximate the *area* of the circle by finding the *area* of such a polygon.

If the polygon has n sides, its area may be found by subdividing it into n congruent triangles as shown in Figure 10C.25 and then finding the sum of the areas of the triangles.

FIGURE 10C.25

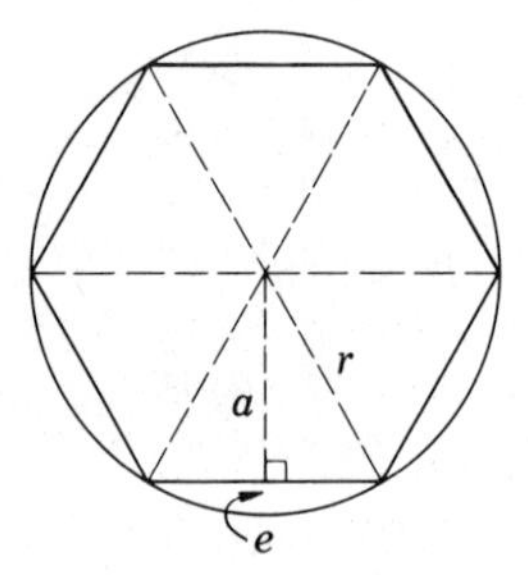

Since each triangle has an area of $\frac{1}{2}ae$, and since there are n such triangles in the polygon, the area of the polygon is

$$n \cdot \frac{1}{2}ae.$$

Since $n \cdot e$ is the perimeter p of the polygon, the area becomes $\frac{1}{2}pa$. That is,

$$A = \frac{1}{2}\text{ (perimeter of the polygon)} \times a.$$

As the number of sides of the polygon becomes larger and larger, the perimeter of the polygon approaches the circumference C of the circle, and the length of a approaches the length of the radius r. That is, the area of the circle becomes

$$A = \frac{1}{2}r \cdot C.$$

Since $C = 2\pi r$,

$$A = \frac{1}{2}r\,(2\pi r), \quad \text{or} \quad A = \pi r^2.$$

This discussion leads to the following definition and theorem.

DEFINITION 2 The *area of a circle* is the limit of the areas of the inscribed regular polygons.

THEOREM 8 The area of a circle of radius r is $A = \pi r^2$.

CYLINDER

DEFINITION 3 The area of the curved surface of a right circular cylinder is called its *lateral surface area* (Figure 10C.26).

FIGURE 10C.26

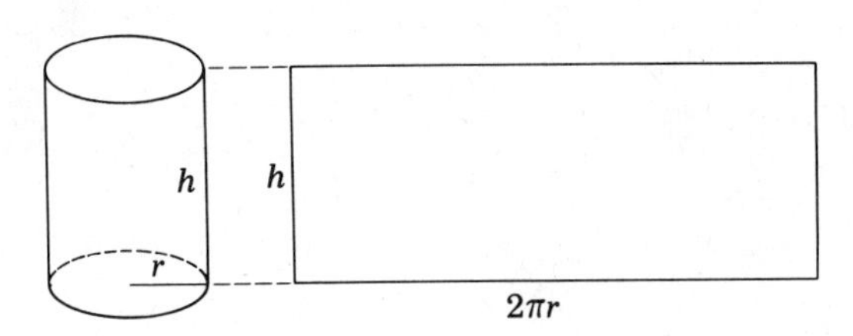

Think of the surface of the cylinder when it is flattened out. It becomes a rectangle with height h and base equal to the circumference of the cylinder. The area of this rectangle is the product of its height h and its base $2\pi r$; i.e., $2\pi rh$, or πdh. This discussion leads to the following theorem.

THEOREM 9 The lateral surface area of a right circular cylinder is $S = 2\pi rh$, where r is the radius of the base and h is the height of the cylinder (Figure 10C.26).

THEOREM 10 The volume of a circular cylinder is the product of its altitude and the area of its base (Figure 10C.27). That is, $V = Bh$, or $V = \pi r^2 h$.

FIGURE 10C.27

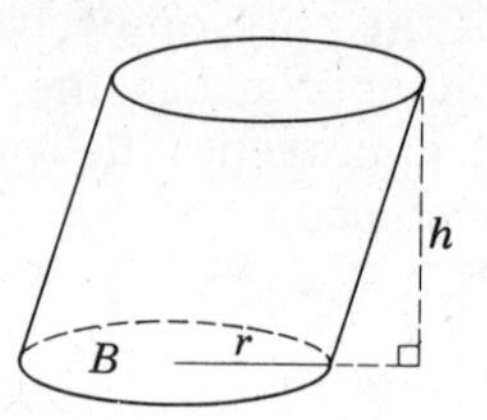

CONE

THEOREM 11 The lateral area of a right circular cone is equal to half the product of the circumference of the base and the slant height (Figure 10C.28). That is, $S = \pi rs$.

FIGURE 10C.28

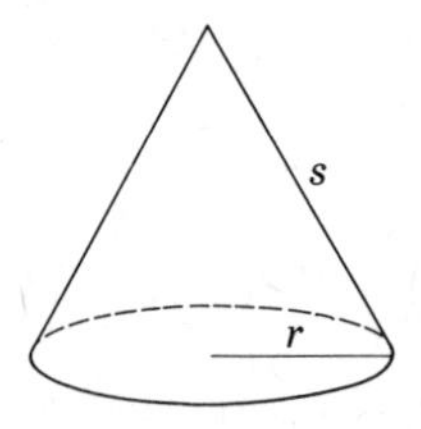

THEOREM 12 The volume of a circular cone is one-third the product of its altitude and the area of its base (Figure 10C.29). That is, $V = \frac{1}{3}Bh$, or $V = \frac{1}{3}\pi r^2 h$.

FIGURE 10C.29

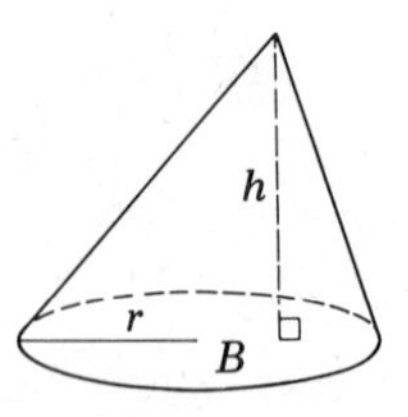

SPHERE

THEOREM 13 The volume of a sphere of radius r is $V = \frac{4}{3}\pi r^3$ (Figure 10C.30).

FIGURE 10C.30

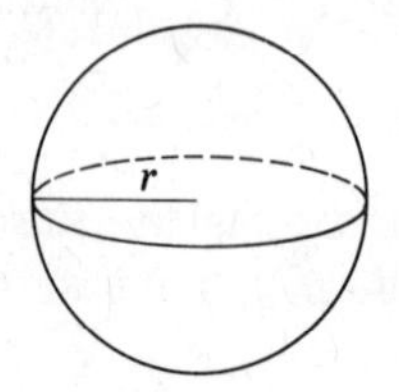

THEOREM 14 The surface area of a sphere of radius r is $S = 4\pi r^2$.

Comments

1 *The surface area of a sphere is four times the area of the plane region inside one of its great circles.*

2 *Unlike the surface of a cylinder, the surface of a sphere cannot be flattened out onto a plane without stretching.*

EXERCISE SET 10C.4

1 Find the missing information for each circle.

	Radius	*Diameter*	*Circumference*	*Area*
a)	5 mm			
b)		8 ft		
c)				π sq ft
d)			44 yd	
e)	$1\frac{1}{4}$ in.			
f)			15 m	
g)		2.6 cm		
h)			150 yd	

2 What is the radius of a circle whose circumference is π?

3 A circular rug, 12 feet in diameter, costs $52. How much should a similar rug with a 6-foot diameter cost?

4 A tire on a car has a diameter of 26 inches. If it revolves at 10 revolutions per second, what is the approximate speed of the car in miles per hour?

5 What distance does a wheel with a 15-inch-diameter roll if it turns 150°?

6 The minute hand of a wrist watch is $\frac{1}{2}$ inch long. How many inches does the tip of the hand travel in 25 minutes?

7 The diameter of a wagon wheel is 20 inches. If the wheel makes 250 revolutions as it moves from point A to point B, what is the distance from A to B?

8 How much farther will a 30-inch bike travel in two revolutions of the wheel than a 22-inch bike?

9 a) How many circular pieces of metal, each having a 3-inch diameter, can be cut from a rectangular sheet measuring 36 by 42 inches?
b) How many square inches of metal are wasted?

10 If the area of the square shown below is 49 square centimeters, find the total area of the four shaded surfaces.

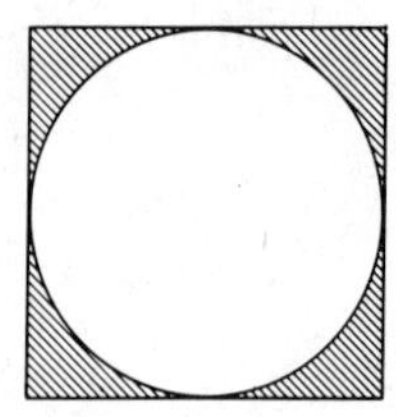

11 a) How many gallons of paint will be required for painting a cylindrical tank measuring 12 feet in diameter and 25 feet in height, if a gallon of paint covers 400 square feet?
b) At \$6.50 a gallon, what will be the cost of applying two coats of paint?

12 How much metal is needed to make a fruit juice can measuring 4 inches in diameter and 8 inches in height?

13 Compare the volumes of two cans of dessert, each 10 inches high, one can having a 3-inch radius and the other a 6-inch radius.

14 What is the effect on the surface of a sphere if you double its radius?

15 Find the weight of a steel ball with a 9-inch diameter if a cubic foot of steel weighs 490 pounds?

16 Find the altitude of a right circular cone whose volume is 36π, if the radius of its base is 4.

17 Find the diameter of a sphere whose surface area is equal to its volume.

18 What is the ratio of the radii of two spheres if the volume of one sphere is three times the volume of the other sphere?

Chapter 10D

GEOMETRY THROUGH TRANSFORMATIONS

Objectives

At the completion of this chapter, you should be able to:

1. Define a *geometric transformation*.
2. Define a *rigid motion* and state its properties.
3. Define a *translation*, a *rotation*, and a *reflection*, and state the properties of each.
4. Show how a given translation or a given rotation can be accomplished by the successive application of reflections.
5. Define *congruence* in terms of rigid motions.
6. Define *similarity* in terms of rigid motions.
7. Define *line symmetry* and *point symmetry* and state the properties of each type of symmetry.
8. Recognize figures with line symmetry and locate their axes of symmetry.
9. Recognize figures with point symmetry and locate their centers of symmetry.
10. Define *orientation* and recognize which transformations preserve and which transformations reverse orientation.

Prerequisites You should:

1. Understand the concepts of *congruence* and *similarity.*
2. Be familiar with the properties of *closure*, *associativity*, *identity*, and *inverses.*

INTRODUCTION

An interesting and valuable approach to the study of geometry is through *transformations.* A geometric transformation, as we shall see, is an operation that assigns to each point P in the plane a point P' which may or may not be P itself. We are interested particularly in those transformations that slide or rotate a figure to form an *identical* figure, or that reflect a figure to form a *mirror image* of itself, or that alter the size of a figure to form a larger or smaller figure of the *same shape.* A study of these transformations provides us with a fresh way of looking at such basic geometric concepts as congruence, similarity, and symmetry.

10D.1 Rigid Motions

Imagine a block of wood resting on a table. As we slide the block along the table top, every point of the block changes position in relation to the points of the table top. Let P be any point on the table top that was initially occupied by a particular point of the block, and let P' be the new point on the table top to which the point of the block has been moved. Sliding the block across the table associates the initial point P with the terminal point P'. This association of each point of the table top with another point of the table top is called a *transformation* of the table top. Note that the transformation is on the set of points of the table top to the same set of points of the table top. Because no two points of the table top are carried to the same new position, we shall call such a transformation a *one-to-one transformation.* So by a *transformation* we mean a rule, operation, or function that assigns to every point P of the plane a unique point P' called the *image* of P.

Now, instead of sliding a block across a table top, let us imagine that the table top is a plane and that the plane itself is in motion. Such motion will carry any point P to a new position P'. If the plane moves as if it were made of stiff cardboard, then the length of any segment in the plane is preserved by the motion of the plane. This means that every segment before and after the transformation remains the same. We call such a distance-preserving transformation a *rigid motion* or an *isometry* ("same measure").

DEFINITION 1 A *rigid motion*, or an *isometry*, is a transformation that preserves distances.

It can be proved that there are only three possible kinds of rigid motions: *translations*, *rotations*, and *reflections*. Any other rigid motion is the result of the successive application of a finite number of translations, rotations, and reflections.

TRANSLATIONS

Intuitively, a *translation* suggests the sliding of a rigid object, without twists, turns, or flips, from one position to another.

DEFINITION 2 A *translation* of the plane is a transformation that moves every point of the plane the same distance in the same direction.

Figure 10D.1 shows a translation of $\overline{AB}$, one inch, in a direction parallel to line l.

FIGURE 10D.1

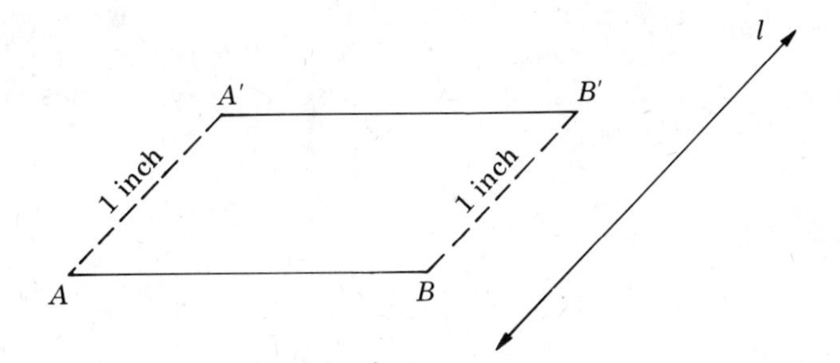

ROTATIONS

DEFINITION 3 A *rotation* of the plane about a fixed point O is a transformation that moves every ray $\overrightarrow{OP}$ into a ray $\overrightarrow{OP'}$ through a fixed angle θ such that $\overline{OP} \cong \overline{OP'}$ (Figure 10D.2).

FIGURE 10D.2

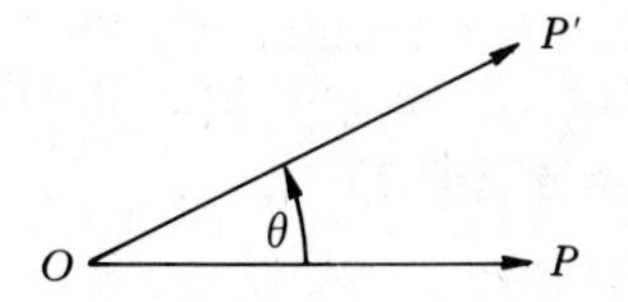

Comments

1 *There can be either a counterclockwise rotation or a clockwise rotation. In both cases the definition remains the same.*

2 *A rotation with a* 0° *angle or with any multiple of* 360° *transforms every point into itself. This transformation is called the* identity transformation.

3 *If the angle of rotation is* 180°, *then O is the midpoint of PP′ (Figure 10D.3). Such a rotation is called a* reflection in *O*.

FIGURE 10D.3

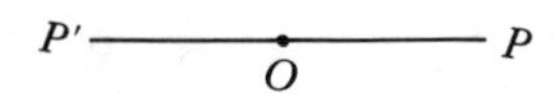

REFLECTIONS

A *reflection* may be described intuitively as a flipping of the plane about a fixed line. You can obtain a reflection of a figure in a plane by drawing the figure on a piece of tissue paper, folding the paper over the figure so that it is completely covered, and then tracing the figure on the folded-over part of the paper. When you unfold the paper, it will resemble Figure 10D.4. The tracing is called the *reflection image* of the original figure with respect to the line represented by the fold in the paper. The line is called the *line of reflection*. A more precise way of describing a reflection is as follows:

FIGURE 10D.4

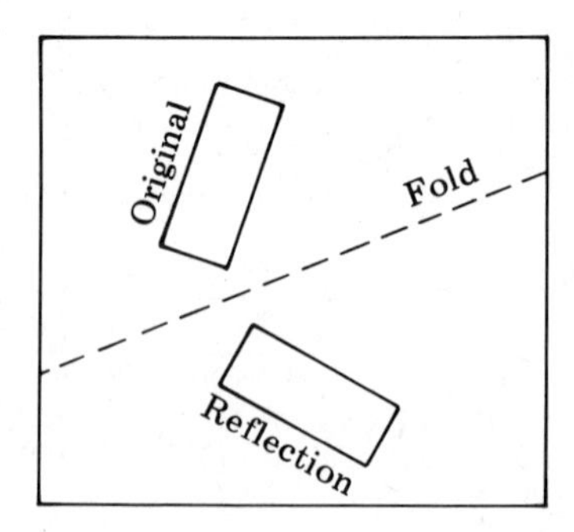

DEFINITION 4 A *line reflection* in l is a transformation which moves each point P to a point P' on the other side of l so that l is the perpendicular bisector of $\overline{PP'}$ (Figure 10D.5).

The reflection of $\triangle ABC$ in l is shown in Figure 10D.6.

FIGURE 10D.5

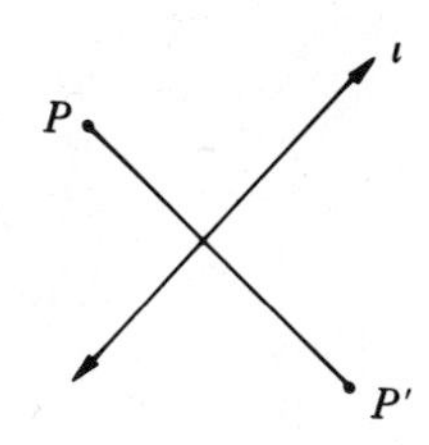

FIGURE 10D.6

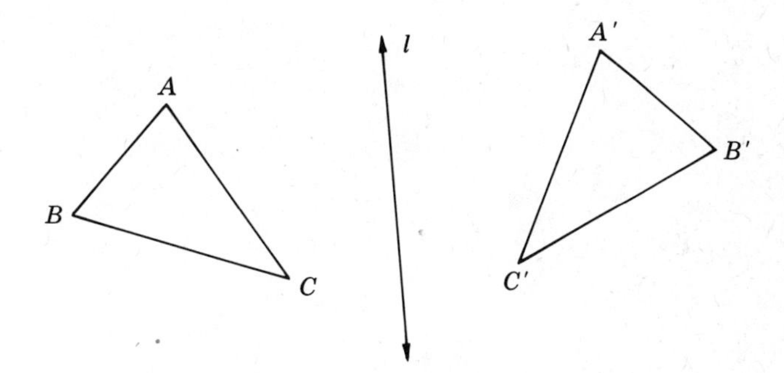

Comments

1. *When you look into a mirror, you see a* reflection image *of yourself.*
2. *If P lies on l, then the reflection of P is P itself; that is, P and P' are the same point.*
3. *Note that in a reflection one figure cannot be superimposed on the other without lifting one of the figures out of the plane.*

10D.2 Orientation

In the reflection shown in Figure 10D.6, note that as we move from point A to point B to point C of $\triangle ABC$ we move in a counterclockwise direction. But if we move around the corresponding vertices of the *reflection* of $\triangle ABC$, i.e., if we move from A' to B' to C' of $\triangle A'B'C'$, we move in a clockwise direction. We describe this change of direction by saying that under a reflection *the orientation of the triangle is reversed.*

In the translation shown in Figure 10D.7 note that the orientation of the triangle is preserved since the path from A to B to C and the path from A' to B' to C' are both counterclockwise.

FIGURE 10D.7

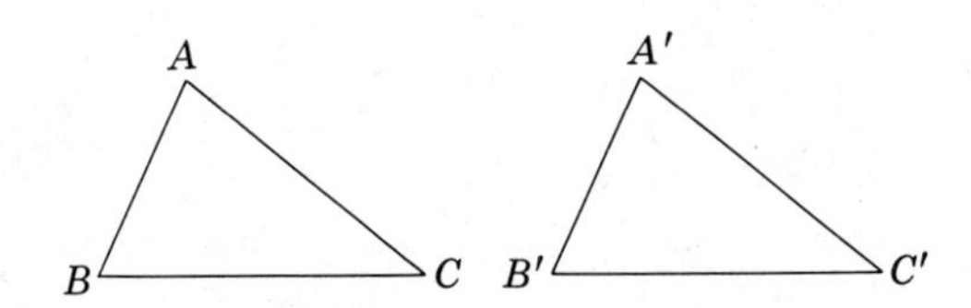

In the rotation shown in Figure 10D.8, again note that the orientation of the triangle is preserved.

FIGURE 10D.8

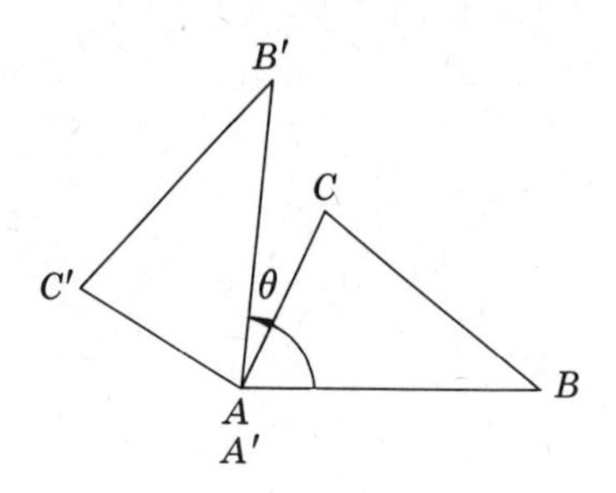

In general, translations and rotations preserve orientations while reflections reverse orientations.

RIGID MOTIONS AS REFLECTIONS

If T_1 is a rigid motion that moves the point P to P' and if T_2 is a rigid motion that moves P' to P'', then the combined effect of performing T_1 and T_2 in succession is a rigid motion that moves P to P''. We call the resulting rigid motion the *product* of T_1 and T_2 and denote it by $T_1 T_2$.

A closer look at a translation and a rotation will show that each can be accomplished by the successive application of a finite number of reflections. Let us see why this is so.

The product of a reflection in line l_1 and a reflection in l_2, when l_1 is parallel to l_2, is actually a translation through twice the distance from l_1 to l_2 in a direction perpendicular to the two lines as shown in Figure 10D.9.

FIGURE 10D.9

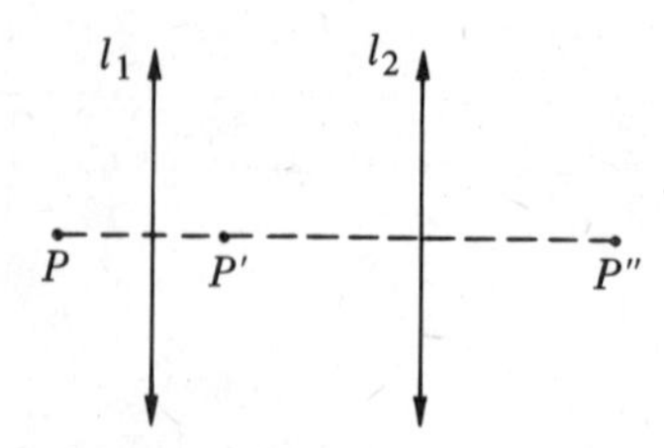

Conversely, if we begin with a translation, we can always find a sequence of reflections in parallel lines, and this sequence is equivalent to the translation.

The product of a reflection in line l_1 and a reflection in line l_2, when l_1 and l_2 intersect, is actually a rotation about their point of intersection, O, through twice the angle formed by the intersection (Figure 10D.10).

FIGURE 10D.10

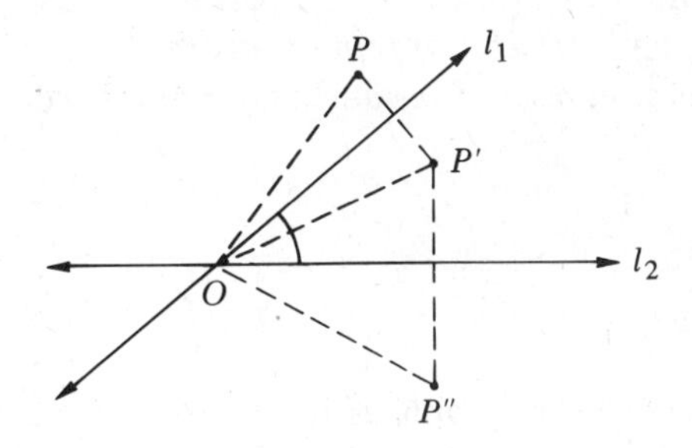

If l_1 is perpendicular to l_2, then the product is a rotation of 180°, or a reflection in the point O (Figure 10D.11).

FIGURE 10D.11

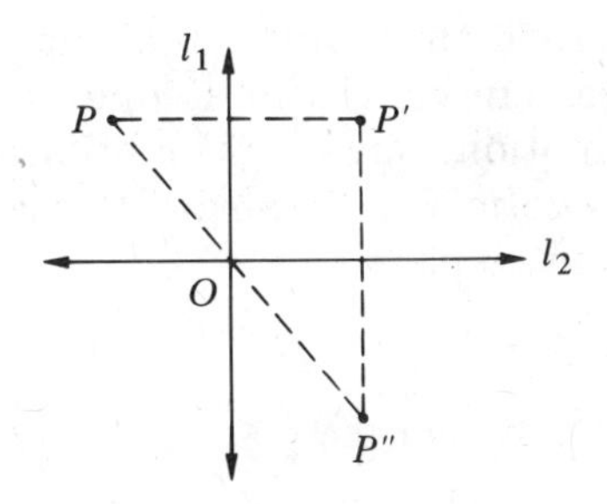

Conversely, if we start with a rotation, we can always find a sequence of reflections in intersecting lines, and this sequence is equivalent to the rotation.

It can be proved that any translation or rotation can be accomplished by a sequence of at most three reflections.

Comment

If we define "multiplication" of two rigid motions T_1 and T_2 as the product $T_1 T_2$ defined above, then it can be shown that the set of rigid motions with respect to the operation of multiplication is closed, is associative, contains an identity, and contains an inverse for each element in the set. A set that satisfies these four properties with respect to a given operation is called a group. *The set of rigid motions is, therefore, a* group.

Space does not allow a further elaboration of the concept of a group, but we should note that it is one of the most frequently found mathematical structures, not only in mathematics but in the physical world as well. Aside from the set of rigid motions, the set of nonzero real numbers under multiplication forms a group as does the set of integers under addition. Through the theory of groups, it has been shown that there are exactly 17 *basic wallpaper designs which are repetitive. In other words, there are only* 17 *basic designs that contain a pattern which repeats in two non-parallel directions across a plane surface.*

10D.3 Congruence

We can now define *congruence* of geometric figures in terms of rigid motions.

DEFINITION Two figures A and B are *congruent* if and only if they are related to each other by a rigid motion.

That is, there is a rigid motion such that the image of A is B. In other words, two figures are congruent if one can be obtained from the other by a rigid motion. Since rigid motions preserve angle measure and distance, this definition of congruence agrees with the definition of congruence given on page 243.

10D.4 Similarity

The concept of *similarity* of geometric figures can be viewed as a transformation that preserves shape but not size. Let P be any point in the plane, and let P' be its image. Under this new transformation, the distance from P' to a point C called the *center* is k times the distance from C to P. In Figure 10D.12 we see a transformation with center C and $k = 2$ for the points P and Q. Here,

$$CP' = 2(CP) \text{ and } CQ' = 2(CQ).$$

We call such a transformation a *size transformation.*

FIGURE 10D.12

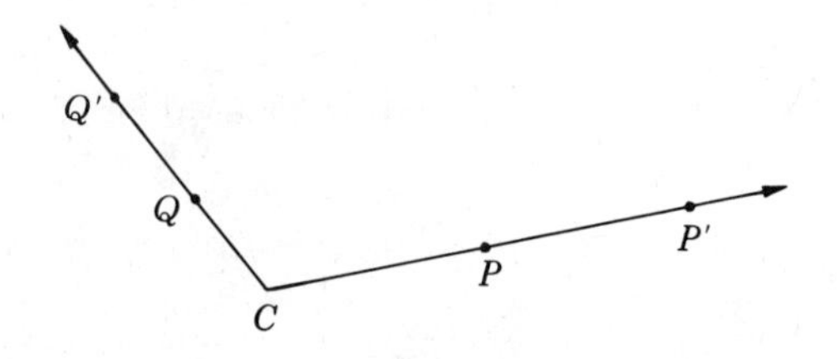

Figure 10D.13 shows two triangles under a size transformation with center O and $k = 3$. Figure 10D.14 shows two triangles under a size transformation with center O and $k = \frac{1}{2}$.

FIGURE 10D.13 *FIGURE 10D.14*

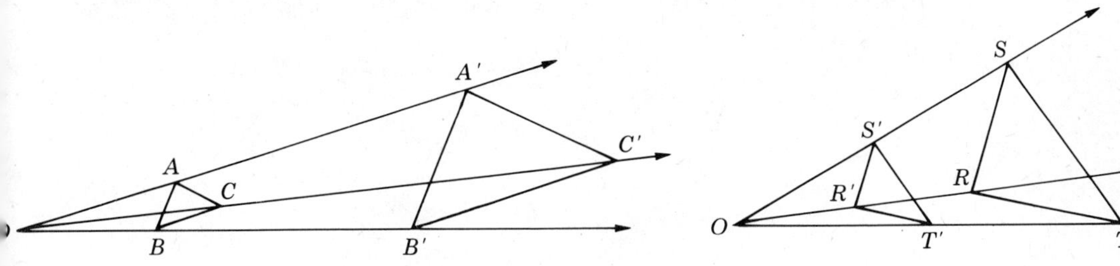

If $k > 1$, the transformation is called an *expansion*. If $k < 1$, the transformation is called a *contraction*. If $k = 1$, then each point is its own image. Figure 10D.13 illustrates an *expansion*; Figure 10D.14 illustrates a *contraction*.

Since a size transformation multiplies the length of each segment in a figure by the magnitude k, it follows that, under a size transformation, the corresponding segments of a figure and its image are proportional. Like a rigid motion, a size transformation preserves angles and angle measure. A composite of rigid motions and size transformations is called a *similarity transformation*.

DEFINITION Two figures A and B are *similar* if and only if they are related to each other by a similarity transformation.

From this definition of similarity it follows that (1) corresponding angles of similar figures are congruent, and (2) corresponding segments of similar figures are proportional. These conclusions agree with the definition of similarity given on page 263.

10D.5 Symmetry

Nature and man-made objects abound with examples of symmetry: leaves, butterflies, flowers, crystals, wallpaper designs, vases, architecture, living things, works of art (Figure 10D.15 on page 306). We shall now examine more closely the meaning of symmetry.

FIGURE 10D.15

Snow crystals from *Hornung's Handbook of Designs and Devices*, Dover Publications, Inc., New York, 1959, p. 142.

LINE SYMMETRY

A reflection about a line sometimes transforms a figure into itself as in the case of the circle and the isosceles triangle in Figure 10D.16. Observe that a line l can be drawn through each of the figures in such a way that every point P of the figure on one side of l has a *symmetric* point P' of the figure on the opposite side of l. The points P and P' are *symmetric about* l if l is the perpendicular bisector of $\overline{PP'}$. The line l is called the *line* (or axis) of *symmetry.* Also observe that if each figure in Figure 10D.16 is folded along the line of symmetry the two halves will coincide.

FIGURE 10D.16

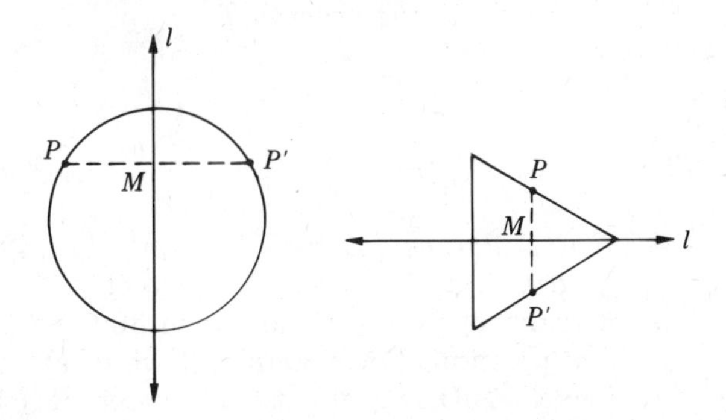

DEFINITION 1 A figure F is *symmetric about a line* l if for every point P of F there exists a corresponding point P' of F (not necessarily distinct) such that l is the perpendicular bisector of $\overline{PP'}$. The line l is called a *line of symmetry.*

Plane figures can have one or more lines of symmetry. The isosceles triangle in Figure 10D.16 has only one line of symmetry, while the circle has infinitely many lines of symmetry—every line that passes through its center. A square has four lines of symmetry as shown in Figure 10D.17.

FIGURE 10D.17

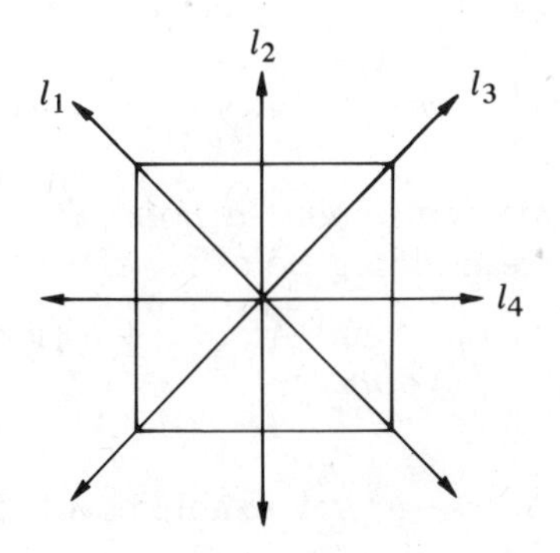

Comments

1 *We must not confuse* line symmetry *with* line reflection. *Only some figures are symmetric, but* any *figure may be reflected. Symmetry is a property that a figure may or may not have, while a reflection is an operation that we may apply to any figure we like.*

2 *We can also define line symmetry as follows: figure F is* symmetric about a line *if F is its own image with respect to that line.*

3 *If, in Figure 10D.16, we place a mirror on line l perpendicular to the plane of P, P′, and l, then the* mirror image *of P is its symmetric point P′; one half of the figure is the* mirror image *of the other half.*

POINT SYMMETRY

Geometric figures can have not only line symmetry but also *point symmetry*. A figure has *point symmetry* if it can be rotated about a point P through an angle less than $360°$ so that it coincides with its original position. For example, if we rotate rectangle $ABCD$, shown in Figure 10D.18, about its center P (the intersection of its diagonals) through an angle of $180°$, the new position of the rectangle will coincide with its original position. Therefore we say that a rectangle has *point symmetry*. We can describe *point symmetry* more precisely by the following definition.

FIGURE 10D.18

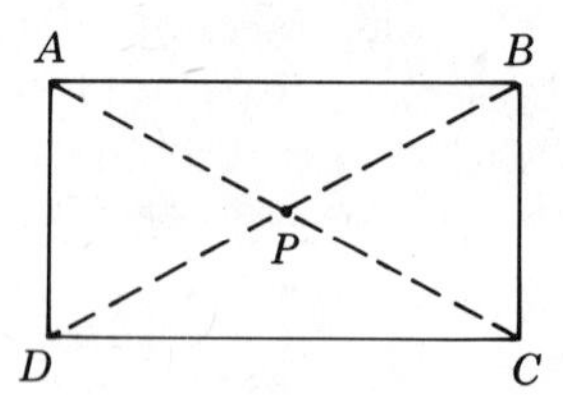

DEFINITION 2 A figure F is *symmetric about a point* P if for each point A of F there exists a corresponding point A' of F (not necessarily distinct) such that $\overline{AA'}$ lies on P and $\overline{AP} \cong \overline{PA'}$ (Figure 10D.19). The point P is called a *center of symmetry*.

Figure 10D.20 shows several examples of figures that have point symmetry.

FIGURE 10D.19

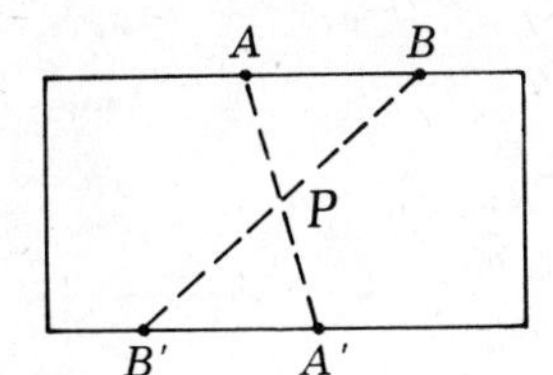

FIGURE 10D.20

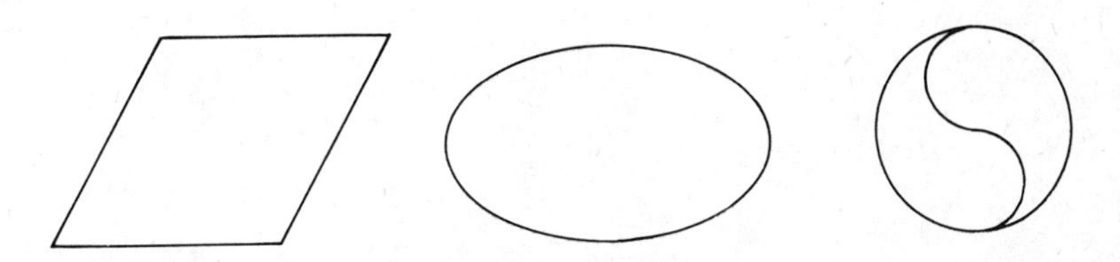

Some figures with point symmetry can be rotated about a point through more than one angle so that they coincide with their original positions. For example, equilateral $\triangle ABC$ in Figure 10D.21 can be rotated about its centroid P (the point of intersection of the medians of the triangle) through either 120° or 240° to make it coincide with its original position.

FIGURE 10D.21

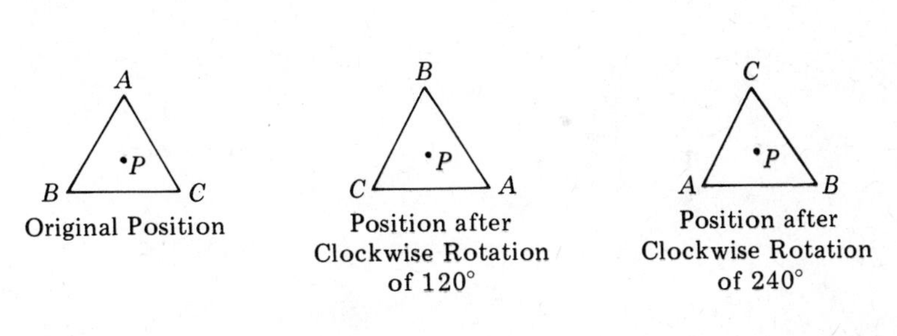

EXERCISE SET 10D.1–10D.5

1 Draw the letter T on a sheet of tissue paper. Find its reflection image with respect to:
a) A horizontal fold in the paper.
b) A vertical fold in the paper.
c) A slanted fold in the paper.

2 Find the reflecting line so that point A is the image of point B in the figure.

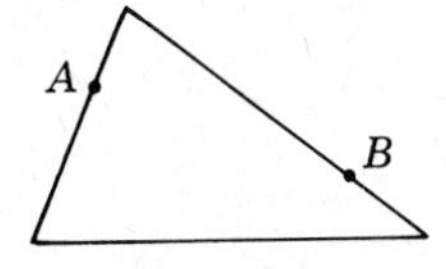

3 Given: isosceles $\triangle ABC$ with $\overline{AB} \cong \overline{AC}$ as shown in the figure. If the triangle is reflected about $\overleftrightarrow{AD}$, then what is the image of each of the following?

a) A b) B c) C d) $\overline{AC}$ e) $\overline{AB}$ f) $\overline{BC}$ g) $\overline{AD}$

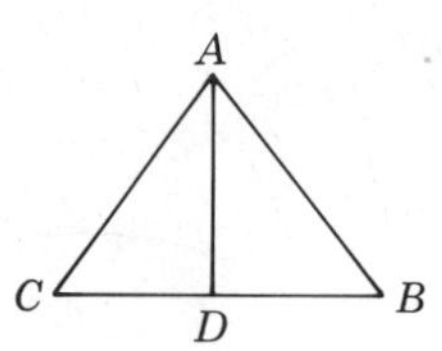

4 In $\triangle ABC$ shown in Exercise 3, what is the image of angle C? What conclusion about the two angles can you draw from this relationship?

5 Which rigid motions, if any, can move figure 1 into figure 2 in each of the following situations?

a)

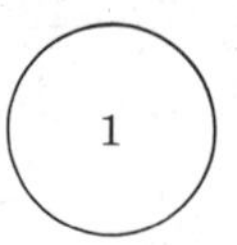

b)

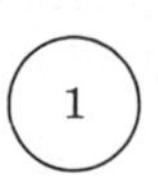

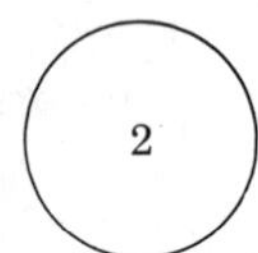

c)

d)

6 a) A sequence of two translations is equivalent to which isometry?
b) A sequence of two rotations with the same center is equivalent to which isometry?
c) A sequence of two reflections is equivalent to which isometry?

7 How many points of the plane does
a) A translation leave fixed?
b) A rotation leave fixed?
c) A line reflection leave fixed?

8 Draw any $\triangle ABC$ and then transform it to $\triangle A'B'C'$ through a rotation about point A through any convenient angle. Show how you could have achieved the same transformation through a sequence of reflections.

9 Which letters of the alphabet, when capitalized, have line symmetry?

10 Which letters of the alphabet, when capitalized, have point symmetry? Which have both line and point symmetry?

11 Which of the numerals 0 through 9 have line symmetry? Which have point symmetry? Which have both line and point symmetry?

12 Does a scalene triangle have a line of symmetry? A center of symmetry?

13 a) Draw an isosceles triangle, an equilateral triangle, a square, and a rectangle. Then draw all the lines of symmetry of each figure.
b) Draw a quadrilateral that has exactly one line of symmetry.

14 Which of the following figures have line symmetry? Draw their line(s) of symmetry.

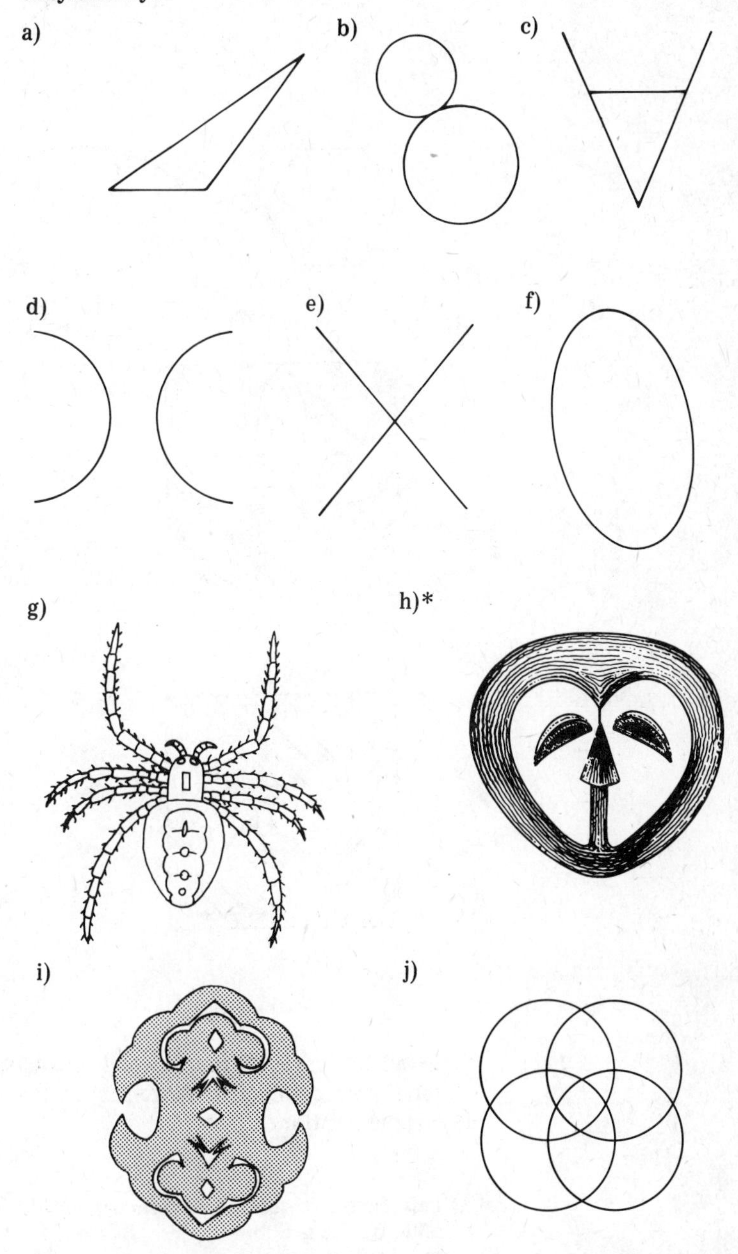

*Kwele mask, Congo Republic. The heart-shaped face is characteristic of this region. This figure is reprinted from *Africa Counts* by Claudia Zaslavsky, Prindle, Weber & Schmidt, Inc., Boston, 1973, p. 175.

15 Show the line(s) of symmetry of (a) a parabola, (b) a regular pentagon.

16 Which of the following figures have point symmetry? Locate their center of symmetry.

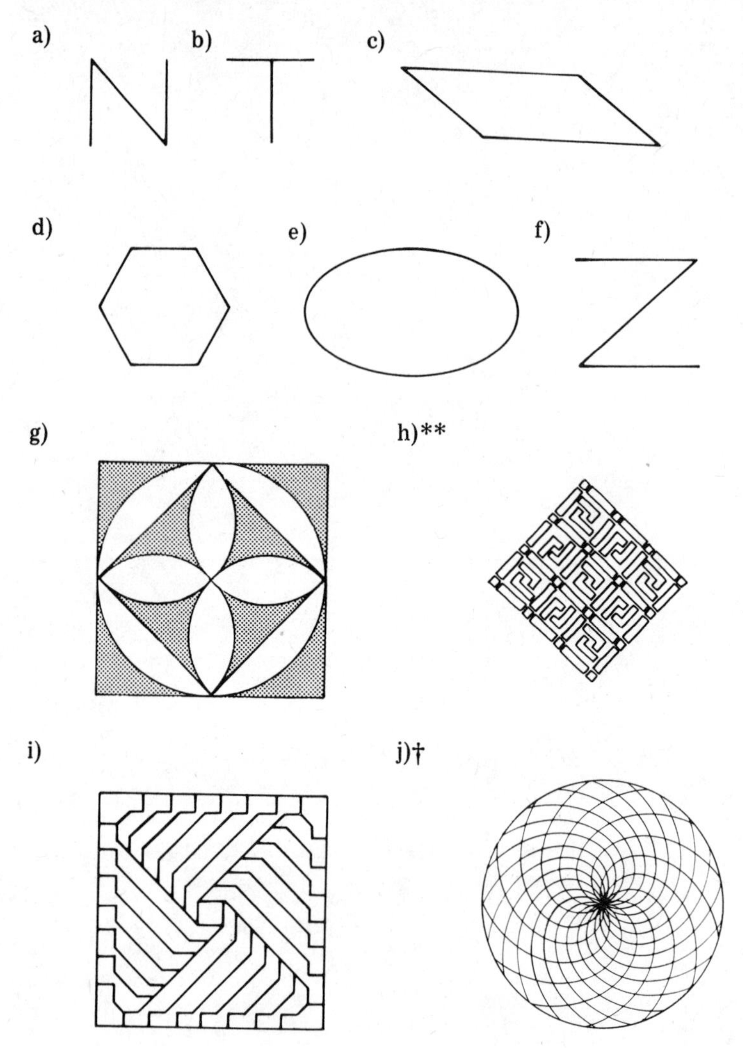

17 Name all the angles through which a square can be rotated about its center (intersection of its diagonals) so that the square coincides with its original position.

**This figure is reprinted from *Africa Counts* by Claudia Zaslavsky, Prindle, Weber & Schmidt, Inc., Boston, 1973, p. 192.

†This drawing illustrates a seed head of a sunflower, which is often used to illustrate the Fibonacci sequence as it occurs in nature. (The Fibonacci sequence is the sequence of numbers1,1, 2, 3, 5, 8, 13, 21, 34, ... in which each term after the second is the sum of the two preceding terms.)

18 Through how many angles less than 360° can a regular hexagon be rotated to coincide with its original position? Name them.

19 Bearing in mind the definition of *line symmetry* as applied to plane figures, define *plane symmetry* for three-dimensional figures.

20 Describe point and line symmetry for three-dimensional figures. Give two examples of each kind of symmetry.

Chapter 10E

COORDINATE GEOMETRY

Objectives

At the completion of this chapter, you should be able to:

1. Plot points with given *coordinates.*
2. Name the coordinates of a given point.
3. Determine the *quadrant* in which a point with given coordinates lies.
4. Find the distance between two points.
5. Find the *midpoint* of a segment joining two points.
6. Find the equations of lines:
   a) That pass through the origin.
   b) That are parallel to the x-axis.
   c) That are parallel to the y-axis.
   d) That have a given slope and pass through a given point.
7. Find the *slope* and *y-intercept* of a line passing through two given points.
8. Find the x- and *y-intercepts* of a line represented by a given equation.

Prerequisites

You should:

1 Be familiar with the properties of similar triangles.

2 Be able to apply the Pythagorean Theorem to find the distance between two points.

3 Know what is meant by *absolute value.*

4 Be familiar with linear equations, both integeral and fractional.

INTRODUCTION

The basic idea of *coordinate geometry* is so simple that one wonders how such an idea can be regarded, as it is, as one of the greatest contributions of all time to mathematics.

Draw any two intersecting lines in a plane as shown in Figure 10E.1. To keep things simple, draw them perpendicular to each other so that one line runs north–south and the other runs east–west. We shall refer to these two lines as the *axes* and call the point of intersection of the axes the *origin.*

FIGURE 10E.1

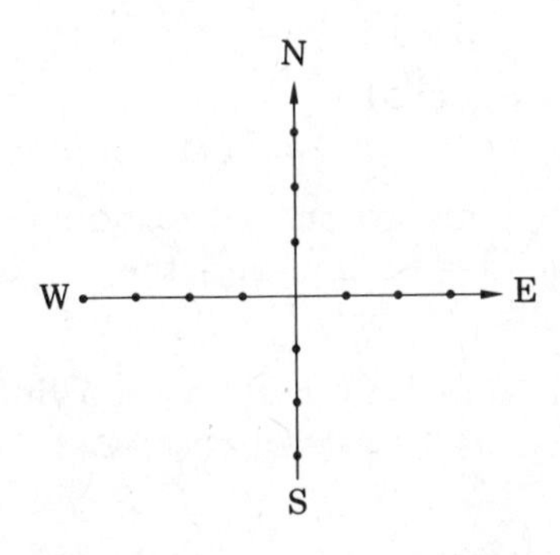

Let us further imagine that a city is laid out by this plan, with the avenues numbered consecutively along the north–south axis and the streets numbered consecutively along the east–west axis. It is obvious that on such a map we can locate instantly any location in the city. We only need to give the pair of numbers that measure its east–west and north–south distances from the axes. These two numbers are called the *coordinates* of the point with respect to the axes.

We can extend the coordinate system to any number of dimensions. For the plane we need *two* coordinates. For three-dimensional space we need *three* coordinates. For example, any position on the earth can be specified by its latitude, longitude, and height above sea level. For n-dimensional space we need n coordinates.

The method of coordinate geometry is so powerful that high school students today can use it to solve problems that would have baffled the greatest geometers of antiquity. The method is powerful because it permits us to bring to bear on the solution of a geometrical problem our knowledge of algebra and the properties of the real number system. We can, thereby, transform a geometrical problem into a corresponding, and often simpler, algebraic problem.

Although coordinates were used by the ancient Egyptians and Romans in surveying and by the Greeks in map making, most historians agree that the decisive contributions to the creation of coordinate geometry were made by René Descartes (1596–1650) and Pierre de Fermat (1601–1665).

According to one legend, the idea of coordinate geometry first came to Descartes as he was watching a fly crawl on the ceiling of his room. He surmised that the path of the fly might be described in terms of the distances from the two adjacent walls.

Descartes began with a curve and used coordinate geometry to find the equation that corresponds to that curve. Fermat started with an equation and then studied its corresponding curve. These are the two fundamental ideas of coordinate geometry.

10E.1 Coordinate Systems in a Plane

We can establish a one-to-one correspondence between the set of points of the Euclidean plane and the set of ordered pairs of real numbers in the following way:

We first set up a coordinate system on a line X (Figure 10E.2). We call this line the *x-axis* (or axis of *abscissas*).

FIGURE 10E.2

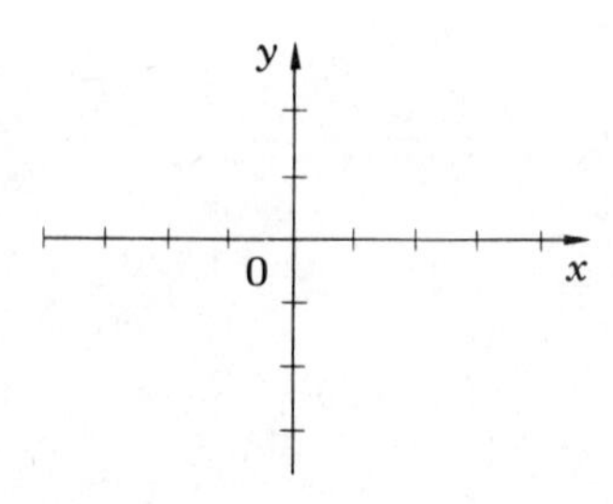

We then set up a coordinate system on a line Y perpendicular to the x-axis through the point with coordinate 0. We make the zero point on Y the same as the zero point on X. We call this line the *y-axis* (or axis of *ordinates*).

The point where the x-axis intersects the y-axis is called the *origin*. We call the x-axis and the y-axis the *coordinate axes for the plane.*

DEFINITIONS

1 The x *coordinate* of a point P is the coordinate of the foot of the perpendicular from P to the x-axis.

2 The y *coordinate* of P is the coordinate of the foot of the perpendicular from P to the y-axis.

3 If P has an x coordinate of a and a y coordinate of b, then we write $P(a, b)$.

If P_1 is on the x-axis, then P_1 has coordinates $(x, 0)$; if P_2 is on the y-axis, then P_2 has coordinates $(0, y)$ (Figure 10E.3).

FIGURE 10E.3

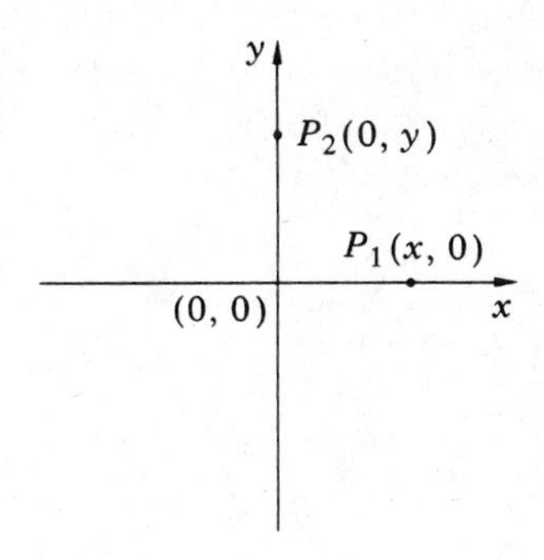

The origin has coordinates $(0, 0)$ (Figure 10E.4).

FIGURE 10E.4

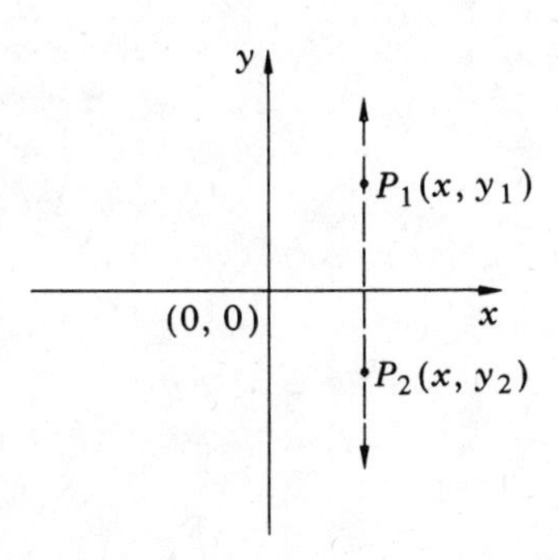

Any two points with the same first coordinate are on the same line parallel to the y-axis (Figure 10E.4).

Any two points with the same second coordinate are on the same line parallel to the x-axis (Figure 10E.5).

FIGURE 10E.5

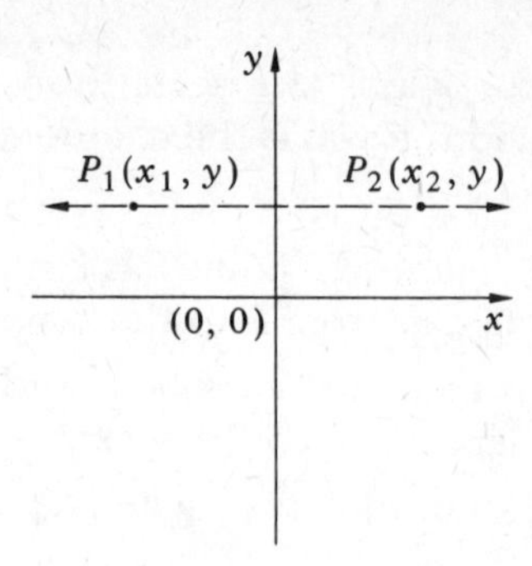

Comments

1 *In drawing the coordinate axes, we usually put an arrowhead on the x-axis and on the y-axis to specify the positive direction on each axis. This is particularly useful when dealing with* directed segments *and* directed distances.

2 *Although we chose the coordinate axes to be perpendicular, this is not essential. For some purposes it is just as easy to use oblique axes, as in Figure 10E.6. Starting from the origin O, we reach the point P(x, y) by moving a distance x along the x-axis, and then a distance y along a line parallel to the y-axis.*

FIGURE 10E.6

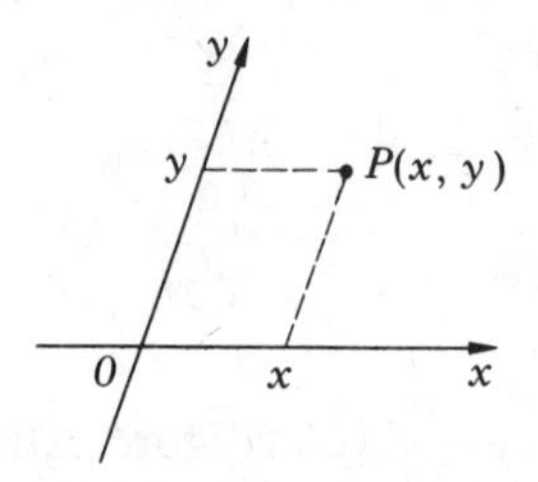

3 *The unit of length used on the y-axis need not be the same as the unit of length used on the x-axis if the scales measure different kinds of things. But in geometry, the scales need to be the same for both axes in order not to distort geometric figures. For instance, a circle will look like an ellipse if different scales are used.*

4 *For the sake of convenience, we shall refer freely to points by their coordinates and to coordinates by their corresponding points. For example, we shall speak of "the point* $(1, {}^{-}4)$*" and "*$P = ({}^{-}2, 5)$*."*

The coordinate axes separate the rest of the plane into four regions called *quadrants.* These quadrants are numbered I, II, III, and IV, as shown in Figure 10E.7.

FIGURE 10E.7

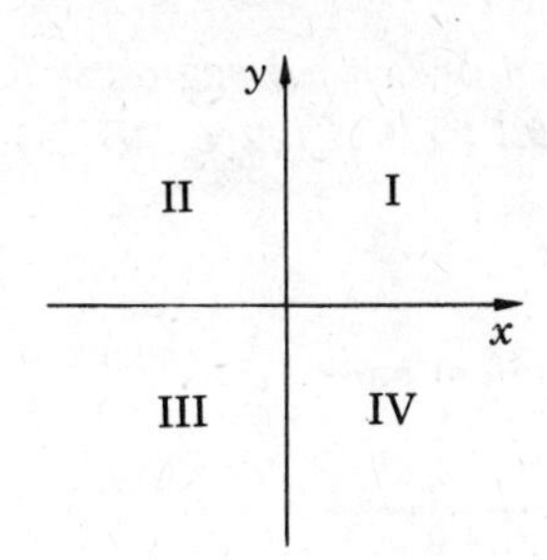

10E.2 The Slope of a Line

DEFINITION If $P_1 = (x_1, y_1)$ and $P_2 = (x_2, y_2)$ and $\overline{P_1P_2}$ is nonvertical, then the *slope* of $\overline{P_1P_2}$ is the number

$$m = \frac{y_2 - y_1}{x_2 - x_1} \qquad \text{(Figure 10E.8)}.$$

FIGURE 10E.8

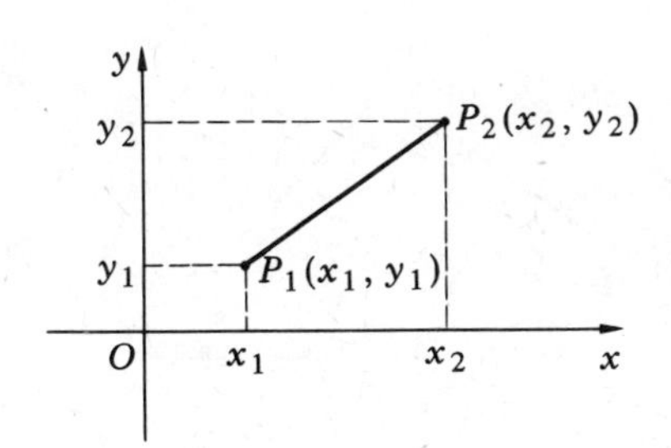

Comments

The following are immediate consequences of the definition of slope:

1 *The slope of $\overline{P_1P_2}$ is the same as the slope of $\overline{P_2P_1}$, since*

$$\frac{y_2 - y_1}{x_2 - x_1} = \frac{-(y_1 - y_2)}{-(x_1 - x_2)} = \frac{y_1 - y_2}{x_1 - x_2}.$$

But the order in which y_2 and y_1 are written in the formula must be the same as that of x_2 and x_1. For instance,

$$m = \frac{(y_2 - y_1)}{(x_2 - x_1)}$$

cannot be written as

$$m = \frac{(y_2 - y_1)}{(x_1 - x_2)}.$$

2 *The slope of a* vertical *segment is undefined, since the denominator, $x_2 - x_1$, is equal to* 0.

3 *The slope of a horizontal segment is* 0, *since the numerator,* $y_2 - y_1$, *is equal to* 0 (*Figure 10E.9*).

FIGURE 10E.9

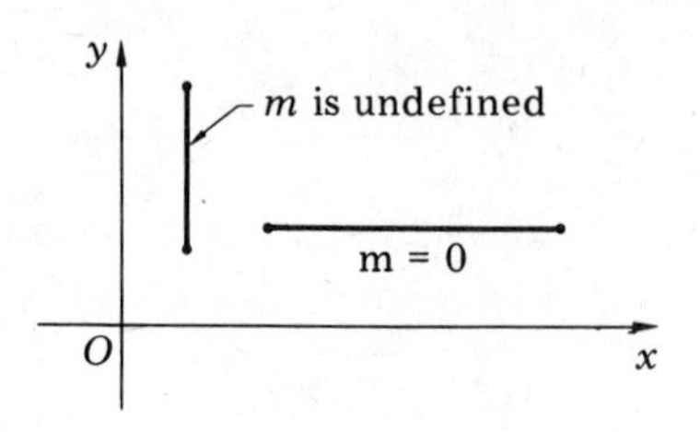

4 *If a segment rises from left to right, its slope is positive, since the numerator and denominator are both positive as shown in Figure 10E.10. If the segment drops from left to right, its slope is negative, since the numerator is negative while the denominator remains positive.*

FIGURE 10E.10

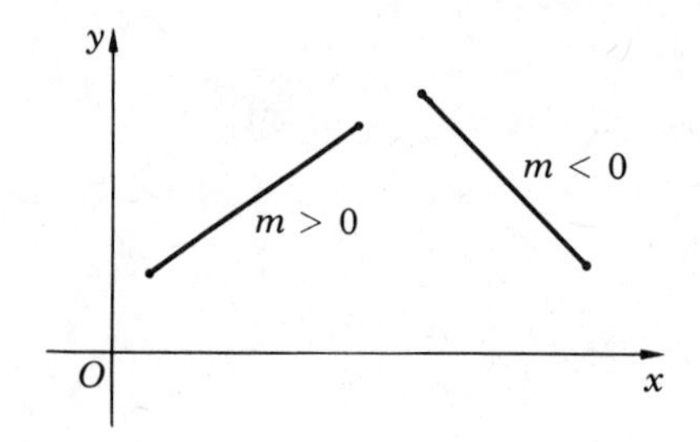

10E.3 The Distance Formula

THEOREM 1 (*The Distance Formula*) The distance between the points (x_1, y_1) and (x_2, y_2) is

$$d = \sqrt{(x_2 - x_1)^2 + (y_2 - y_1)^2}.$$

Proof If $\overline{P_1P_2}$ is horizontal or vertical, the formula is easily verified. If $\overline{P_1P_2}$ is neither horizontal nor vertical, then the proof is as follows:

1 Let A_1 and A_2 be the feet of the perpendiculars from $P_1(x_1, y_1)$ and $P_2(x_2, y_2)$ to the x-axis, and let B_1 and B_2 be the feet of the perpendiculars from P_1 and P_2 to the y-axis (Figure 10E.11). Also, let C be the point of intersection of the horizontal line through P_1 and the vertical line through P_2.

FIGURE 10E.11

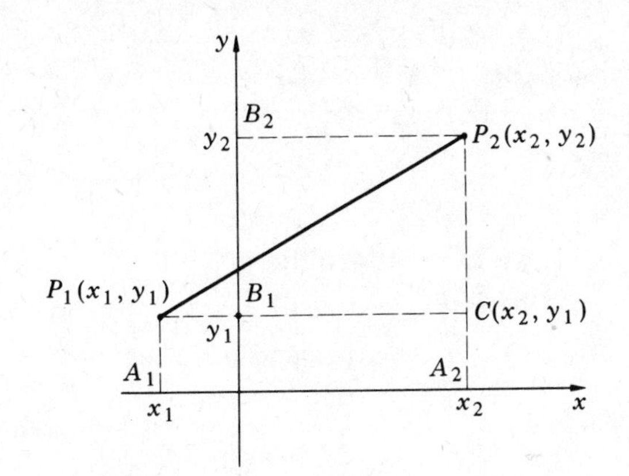

2 Then $(P_1P_2)^2 = (P_1C)^2 + (CP_2)^2$, by the Pythagorean Theorem.

3 Since $P_1CA_2A_1$ is a rectangle,

$P_1C = A_1A_2 = |x_2 - x_1|$ and

$CP_2 = B_1B_2 = |y_2 - y_1|$.

4 Therefore, $(P_1P_2)^2 = |x_2 - x_1|^2 + |y_2 - y_1|^2$.

5 But $(P_1P_2)^2 = (x_2 - x_1)^2 + (y_2 - y_1)^2$, since the square of a number is the same as the square of its absolute value.

6 Since $P_1P_2 \geqslant 0$, we get

$$P_1P_2 = \sqrt{(x_2 - x_1)^2 + (y_2 - y_1)^2}.$$

Comment *In using the Distance Formula it makes no difference which point is considered to be (x_1, y_1), since*

$(x_1 - x_2)^2 = (x_2 - x_1)^2$ and

$(y_1 - y_2)^2 = (y_2 - y_1)^2$.

10E.4 The Midpoint Formula

THEOREM 2 (*The Midpoint Formula*) Let $P_1 = (x_1, y_1)$ and $P_2 = (x_2, y_2)$. The midpoint of $\overline{P_1P_2}$ is the point

$$P = \left(\frac{x_1 + x_2}{2}, \frac{y_1 + y_2}{2}\right).$$

Proof 1 Consider a segment $\overline{P_1P_2}$ on the x-axis, with the coordinates as shown in Figure 10E.12 and with $x_1 < x_2$.

FIGURE 10E.12

P_1 P P_2

O x_1 x x_2

a) If P is the midpoint of $\overline{P_1P_2}$, then $P_1P = PP_2$.

b) $P_1P = |x - x_1| = x - x_1$, and $PP_2 = |x_2 - x| = x_2 - x$.

c) From parts a and b we get

$$x - x_1 = x_2 - x \quad \text{or}$$

$$x = \frac{x_1 + x_2}{2}.$$

d) Similarly, if $\overline{P_1P_2}$ is on the y-axis, we get

$$y = \frac{y_1 + y_2}{2}$$

for the midpoint.

2 To get the midpoint of an oblique segment, we note (Figure 10E.13) that

a) If P is the midpoint of $\overline{P_1P_2}$, then A is the midpoint of $\overline{A_1A_2}$ because $\Delta P_1PB \sim \Delta P_1P_2C$.

b) Therefore,

$$x = \frac{(x_1 + x_2)}{2}$$

and, in a similar way, we get

$$y = \frac{(y_1 + y_2)}{2}.$$

That is,

$$P = \left(\frac{x_1 + x_2}{2}, \frac{y_1 + y_2}{2}\right).$$

FIGURE 10E.13

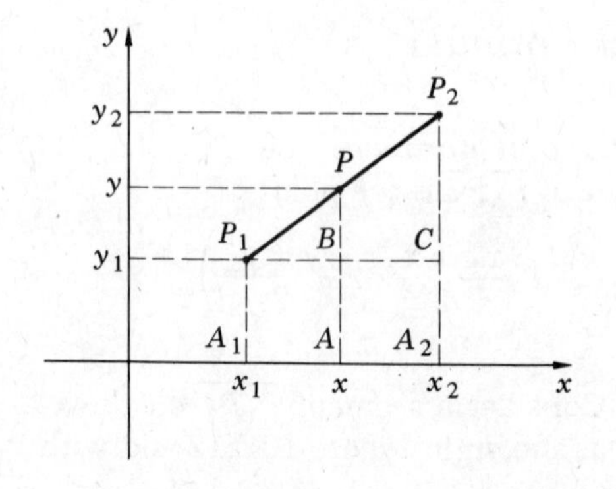

10E.5 Equations of a Straight Line

We can think of a curve such as a circle or a straight line as a set of points that satisfy certain conditions. These conditions may be stated in words, or they may be stated as equations or as inequalities. The set of all points of the plane that satisfy a given condition is called the *graph of the condition.*

For instance, the graph of the condition that the set of points in a plane be at a given distance (r) from a point (P) in the plane is a circle (Figure 10E.14).

FIGURE 10E.14

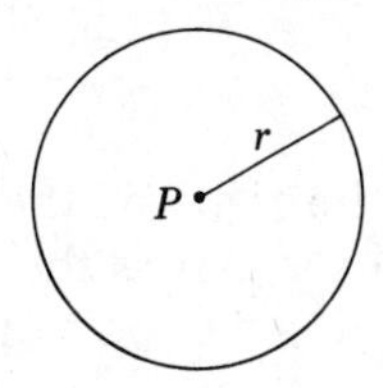

The graph of the condition $x = 2$ is the vertical line 2 units to the right of the y-axis (Figure 10E.15).

FIGURE 10E.15

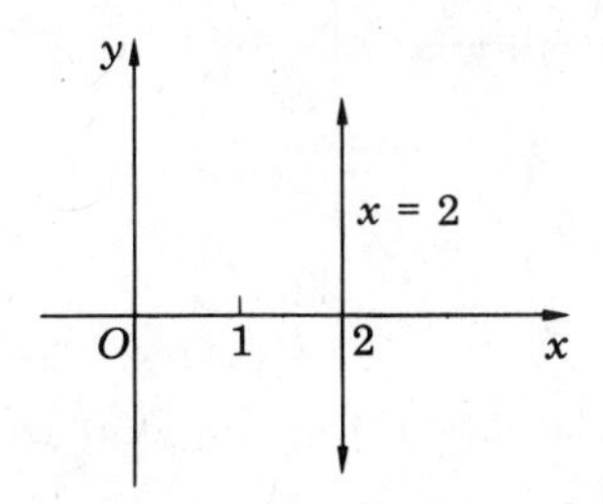

The graph of the condition $y > 1$ is the set of points more than 1 unit above the x-axis (Figure 10E.16).

FIGURE 10E.16

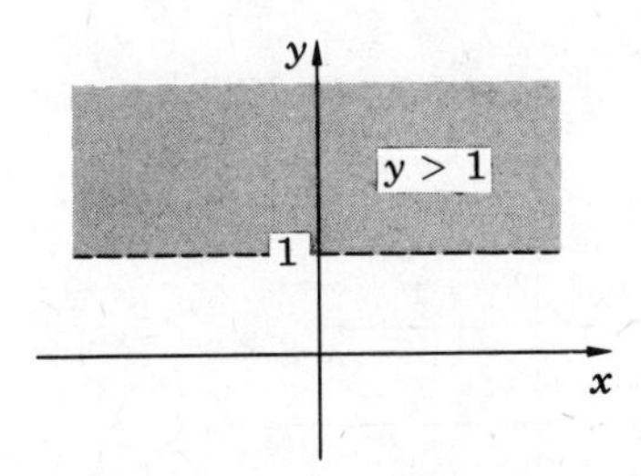

In the graph-equation relationship we can start with an equation and find its graph, or we can start with a graph and find its equation. A graph and equation correspond (1) if the coordinates of each point in the graph satisfy the equation, and (2) if each point whose coordinates satisfy the equation lies on the graph.

We shall start with the simplest graph, the straight line, and investigate its relationship to its corresponding equation. We shall see that the equation of an oblique straight line may be expressed in any one of several forms, each more appropriate for certain purposes.

EQUATIONS OF LINES PARALLEL TO THE AXES

THEOREM 3 The equation of the straight line parallel to the y-axis and a units from it is $x = a$.

If a is positive, the line is to the right of the y-axis; if a is negative, then the line is to the left of the y-axis as shown in Figure 10E.17. In either case, every point on the line will be a units from the y-axis.

FIGURE 10E.17

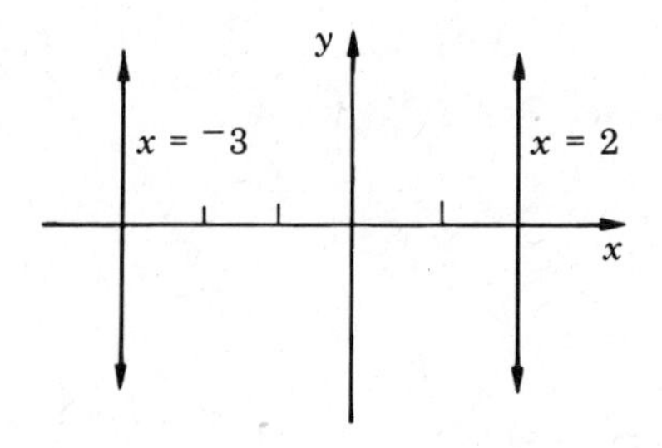

THEOREM 4 The equation of the straight line parallel to the x-axis and b units from it is $y = b$.

If b is positive, the line is above the x-axis; if b is negative, then the line is below the x-axis as shown in Figure 10E.18. In either case, every point on the line will be b units from the x-axis.

FIGURE 10E.18

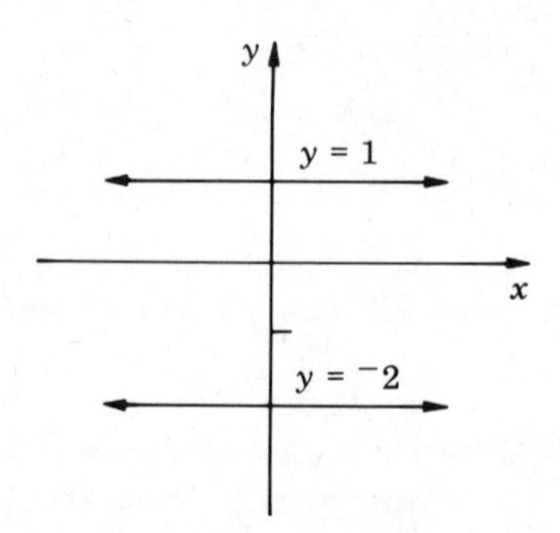

EQUATIONS OF OBLIQUE LINES

Point-slope Form of the Equation Consider the line l, which passes through the point (x_1, y_1), and has slope m (Figure 10E.19). For every point (x, y) of l, the equation

$$\frac{y - y_1}{x - x_1} = m \quad \text{or}$$

$$y - y_1 = m(x - x_1)$$

is satisfied. This equation is called the *point-slope form* of the equation of the line.

FIGURE 10E.19

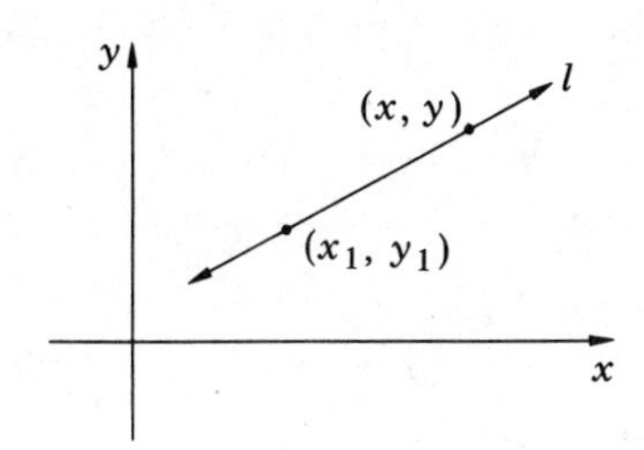

THEOREM 5 The equation of the straight line passing through the point (x_1, y_1) with slope m is

$$y - y_1 = m(x - x_1).$$

Slope-intercept Form of the Equation Every oblique line intersects the y-axis in some point whose x coordinate is 0. The y-coordinate of this point is called the *y-intercept* of the line and is often denoted by b. For every point (x, y) of l (Figure 10E.20), we get

$$\frac{y - b}{x - 0} = m \quad \text{or} \quad y - b = m(x - 0) \text{ or}$$

$$y - b = mx \quad \text{or} \quad y = mx + b.$$

This equation is called the *slope-intercept form* of the equation of the line.

FIGURE 10E.20

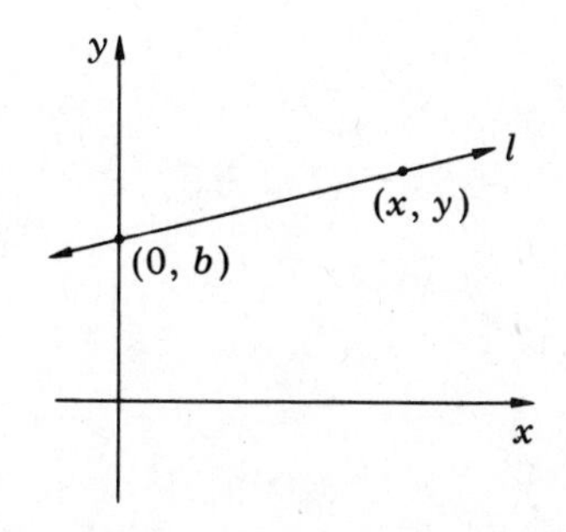

DEFINITIONS

1 The point where the line crosses the y-axis is called the *y-intercept.*

2 The point where the line crosses the x-axis is called the *x-intercept.*

THEOREM 6 The equation of the straight line with slope m and y-intercept b is

$$y = mx + b.$$

Intercept Form of the Equation If the x- and y-intercepts of a line are a and b, respectively, with $a \neq 0$ and $b \neq 0$, the line passes through the points $(a, 0)$ and $(0, b)$ and has the slope $m = -\frac{b}{a}$ (Figure 10E.21). Using the point-slope form, the equation of the line is

$$y - 0 = -\frac{b}{a}(x - a) \quad \text{or}$$

$$ay = {}^{-}bx + ab \quad \text{or} \quad bx + ay = ab.$$

FIGURE 10E.21

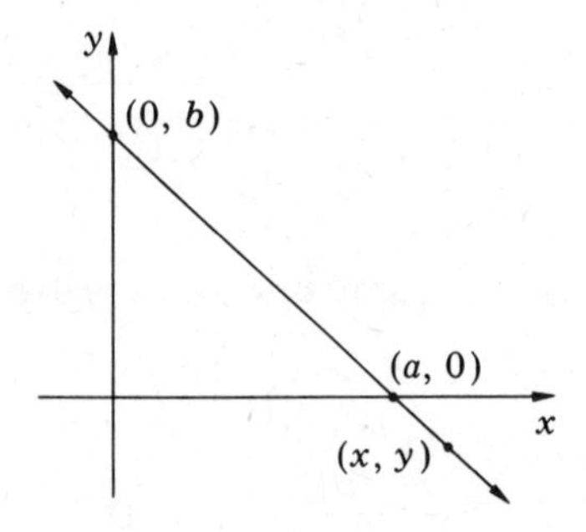

Dividing both members of the last equation by ab, we get

$$\frac{x}{a} + \frac{y}{b} = 1.$$

This equation is called the *intercept form* of the equation of the line.

THEOREM 7 The equation of the straight line with x-intercept a and y-intercept b is

$$\frac{x}{a} + \frac{y}{b} = 1.$$

FIRST-DEGREE EQUATIONS AND THEIR GRAPHS

DEFINITION An equation of the form

$$Ax + By + C = 0$$

is called a *first-degree equation* in x and y if A, B, and C are real numbers and A and B are not both zero.

A fundamental relationship between algebra and geometry is expressed by the following theorem.

THEOREM 8 All straight lines have first-degree equations in two unknowns, and, conversely, all first-degree equations in two unknowns represent straight lines.

Comments

1. *Because of the relationship established in Theorem 8, equations of the first degree are called* linear equations, *and the equation* $Ax + By + C = 0$ *is called the* general linear equation.
2. *Two different equations may represent the same line. For example, the two equations* $y = x$ *and* $3y - 3x = 0$ *represent the same line. We call such equations* equivalent equations.

EXERCISE SET 10E.1–10E.5

1. Plot the following points on a sheet of graph paper.
   a) (1, 3) b) ($^-2$, 1) c) (3, $^-5$) d) ($^-2$, $^-4$)
2. What are the coordinates of P_1, P_2, P_3, and P_4 as shown in the figure?

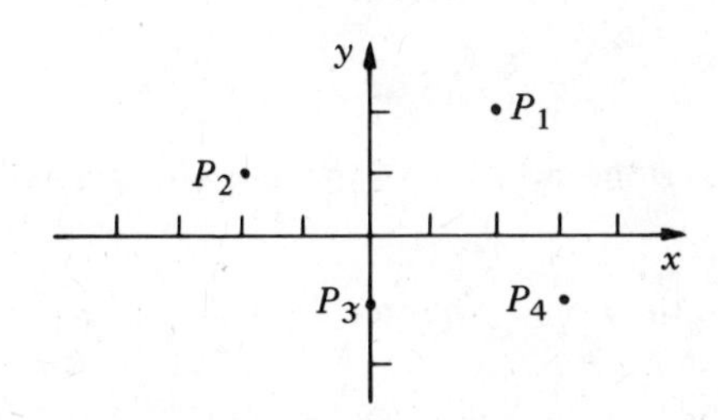

3. a) The point (0, y) lies on ____.
   b) The point (x, 0) lies on ____.
   c) The point ($^-2$, y) lies on ____.
   d) The point (x, 3) lies on ____.
4. In which quadrant is $P(x, y)$ located if:
   a) $x < 0$ and $y > 0$? b) $x < 0$ and $y < 0$? c) $x > 0$ and $y > 0$?
   d) $x > 0$ and $y < 0$?
5. If a line, different from and not parallel to either axis, does not pass through the origin, in how many quadrants does it lie?
6. A straight line passes through the points (0, 0) and (2, 2). What acute angle does it make with the positive x-axis?
7. A line passes through the points (2, 3) and (0, $^-2$). Give the coordinates of any other point of the line.

8 Does the line through the points ($^-2$, $^-2$) and (2, 3) pass through the origin? Give a reason for your answer.

9 What must be the value of r if:
a) The line through (0, $^-2$) and (3, r) is horizontal?
b) The line through (r, y_1) and (x_2, y_2) is vertical?
c) The line through ($^-3$, r) and (2, $^-3$) is horizontal?
d) The line through ($^-2$, 0) and (3, r) is to have a slope of $\frac{2}{3}$?

10 Find the distance between each of the following pairs of points.
a) (4, 5) and (3, 1) b) (2, 0) and ($^-3$, 3) c) (2, 5) and (4, $^-1$)
d) (2, 5) and ($^-5$, $^-1$)

11 Find the midpoint of the segment joining each of the following pairs of points?
a) (2, 3) and (6, 4) b) (1, 2) and ($^-3$, 5) c) (0, $^-5$) and ($^-5$, 2)

12 The vertices of a triangle are $A(^-2, 0)$, $B(2, 1)$, and $C(1, 5)$.
a) Show that ΔABC is a right triangle, and name the vertex of the right angle.
b) Find the perimeter of the triangle.

13 The vertices of a triangle are $A(0, 0)$, $B(6, 2)$, and $C(3, ^-2)$. Find:
a) The slopes of its sides.
b) The slopes of its altitudes.
c) The lengths of its medians.

14 Find an equation of the line that passes through:
a) The origin and (2, $^-3$). b) (6, $^-3$) and (2, 7).
c) ($^-2$, $^-5$) and ($^-2$, 4). d) (1, 3) and ($^-4$, 3).

15 Draw the lines in Exercise 14 and find the slope of each line.

16 Draw the line represented by each of the following equations.
a) $x = {^-5}$ b) $y = 3$ c) $3x + 2y = 5$ d) $^-2x - y = 1$
e) $y = \frac{3}{4}x + 1$

17 What is the equation of the line with a slope of $\frac{^-2}{3}$ and passing through (3, $^-2$)?

18 Find the x- and y-intercepts of the line represented by:
a) $2x + y = {^-8}$. b) $x - 3y = 6$.

19 Prove that the points (0, $^-8$), ($^-4$, 0), (2, $^-12$) are collinear.

20 Use coordinate geometry to prove that the diagonals of a rectangle are congruent.

PUZZLES FOR CHAPTER 10

1 In the accompanying figure is a set of six squares made with 17 line segments. How can you change the six squares into three squares by removing 5 line segments?

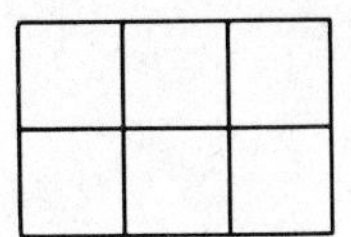

2 Without taking your pencil off the paper, draw 4 line segments that pass through every point.

3 Without taking your pencil off the paper, draw 6 line segments that pass through every point.

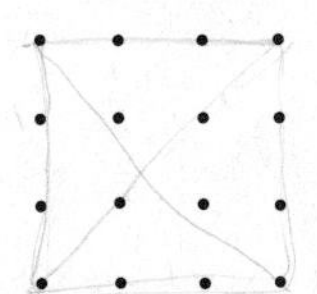

4 Starting at point P, draw the figure shown without lifting your pencil from the paper, without tracing any part of the figure more than once, and without crossing any part of the figure.

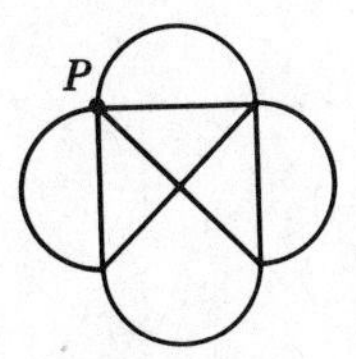

5 Connect like numerals by lines that do not cross each other or any other lines in the figures shown.

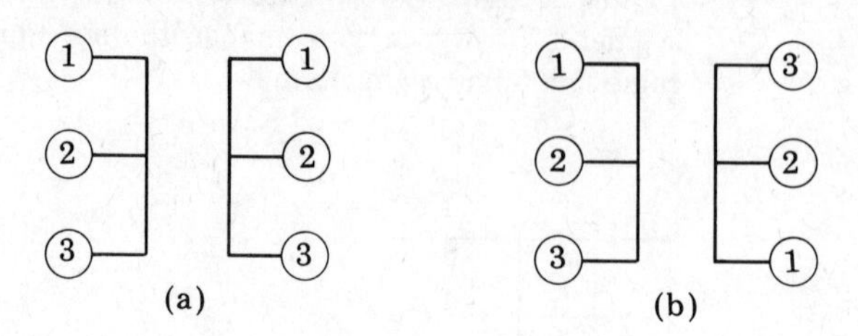

6 What is the total number of triangles in the figure?

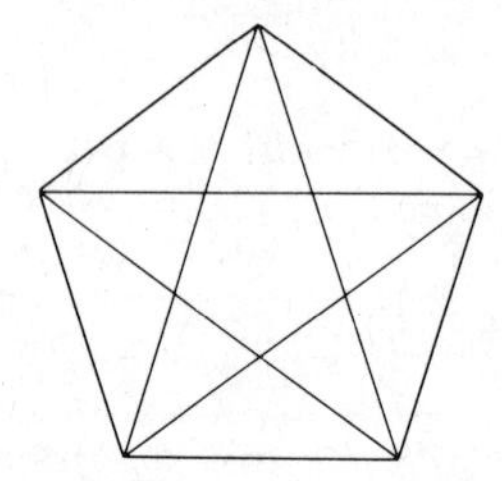

7 You are given six toothpicks of equal length. Arrange them into four equilateral triangles, the sides of each triangle being the length of a toothpick.

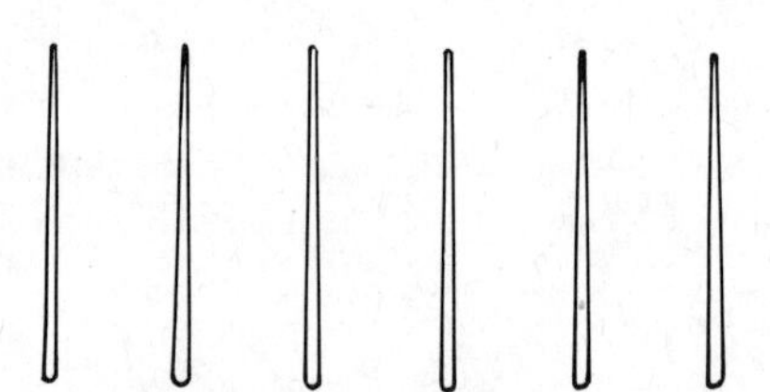

8 You are given eight segments, four of which are half the length of the other four. Can you form three congruent squares with these eight segments?

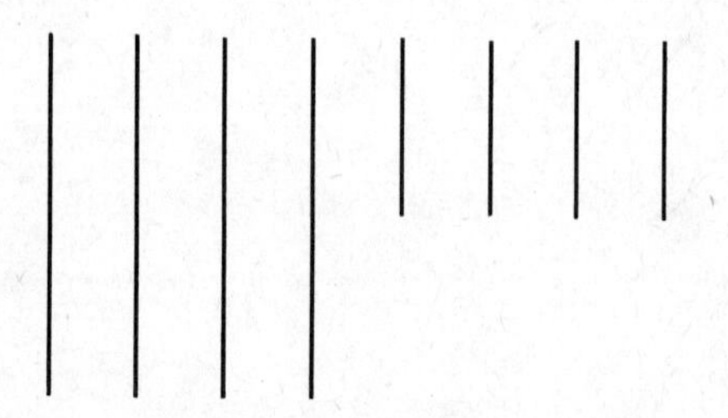

9 Can you arrange four coins, all the same size, on a table so that they are all the same distance apart?

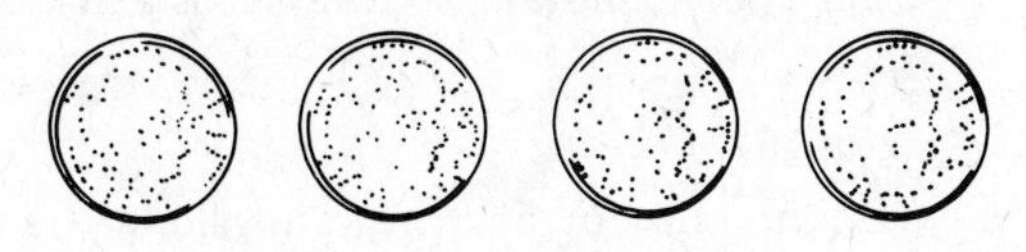

10 Given a circle but not the location of its center, how can you locate the center of the circle?

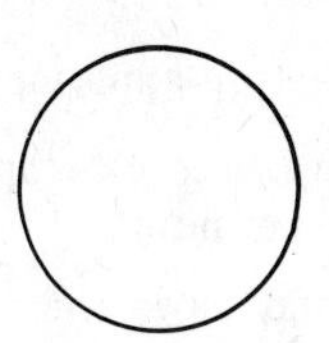

BASICS REVISITED FOR CHAPTER 10

1 Express 275% as a decimal.

2 Subtract $.14\frac{2}{7}$ from .4 in two different ways.

3 Estimate the square root of 159 to the nearest whole number.

4 $\frac{9}{16} \div 1\frac{2}{9} = ?$ 5 $\left(1\frac{3}{8} - \frac{15}{16}\right) \times \frac{3}{4} = ?$

6 Estimate the product of 403 and 801.

7 4.706 - .87 = ?

8 What percent of $2.25 is $.50?

9 Arrange in order of size:

2.41, 6, .09, .1

10 29.8 is 4% of what number?

11 Find 5.7% of 25.

12 If you earn $\$3.42\frac{1}{2}$ an hour, how much do you earn for 8 hours of work?

13 Which is larger and by how much: .1 or .08867?

14 Find the quotient of 32.05 and .14, rounded to 2 places.

15 12 is $16\frac{2}{3}$% of what number?

16 How many 324s are there in 266328?

17 Find the difference between 3500 and 2696.

18 Find the greatest common divisor of 84, 72, and 108.

19 Using the short method, multiply 1.203 by (a) 10, (b) 100, (c) 1000.

20 Using a ruler (showing sixteenths) as a number line, show that

$$\frac{3}{8} \div \frac{1}{16} = 6$$

21 Gasoline sells for 56.9 cents a gallon. If a car averages 11.7 miles per gallon, what will be the cost of gasoline for a 300-mile trip?

22 After $\frac{5}{6}$ of Jane's vacation time went by, she had 4 more days left. How many days was Jane on vacation?

23 If a driver's reaction time before applying his brakes is $\frac{3}{4}$ second, how many feet will his car move before he applies the brakes if the car is moving at a rate of 65 miles per hour?

24 If a person should not spend more than 25% of his income on rent, what is the highest monthly rent he can afford to pay if he earns $13,500 a year?

25 How much should a man invest at 5.8% in order to earn $1,000 a year on his investment?

EXTENDED STUDY

10.1 The History of Geometry

GEOMETRY OF THE EGYPTIANS

Geometry had its beginnings in measurement. The ancient Egyptians were concerned with land measures because each year they had to restore washed-out property boundaries caused by the annual floods of the Nile River. As a result, they became very proficient in dealing with these practical aspects of geometry. They undoubtedly knew some of the fundamental formulas for measurement. In the construction of their pyramids, the ancient Egyptians showed such remarkable accuracy that to this day we do not fully know how they achieved it.

GEOMETRY OF THE GREEKS

The Greeks moved away from this practical concept of geometry and, instead, developed geometry as a deductive system. In about 300 B.C., Euclid was the first to organize and systematize geometry into the deductive science which we still study today. Other important Greek mathematicians were Thales (600 B.C.), who was the first to prove that any circle is bisected by its diameter; Pythagoras (540 B.C.), who is believed to have proved the Pythagorean Theorem; Aristotle (384-322 B.C.), who developed the logic that we still use today; and Archimedes (287-212 B.C.), one of the greatest mathematicians of all time, who developed some of the methods of the integral calculus and who also found a method of computing π.

ANALYTIC GEOMETRY

In the seventeenth century, René Descartes (1596-1650) "algebratized" geometry by his invention of *analytic geometry.* What he invented was a method of combining algebra with geometry to produce what is considered one of the most useful and revolutionary inventions of the human mind.

His method makes use of two intersecting straight lines on a plane and a way of locating points on the plane with reference to these lines. Analytic geometry was the first major advancement from classic Greek geometry and became a major factor in the development of the calculus and modern mathematics.

Pierre de Fermat (1601-1665) discovered analytic geometry independently of Descartes.

PROJECTIVE GEOMETRY

Out of the experiments with perspective by the seventeenth century artists grew an entirely new kind of geometry, known as *projective geometry.* In this geometry the basic concepts are *projection* and *section.*

A *projection* may be thought of as a set of lines of light from the eye to the points of an object as shown in Figure 10.1. A *section* is the pattern formed by the intersection of these lines with a glass screen placed between the eye and the object.

FIGURE 10.1

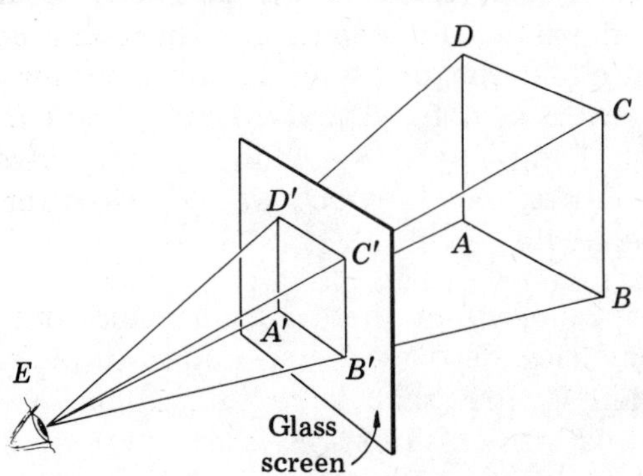

A main goal of projective geometry is to find properties of geometric figures that remain invariant for any section of any projection of those figures. These properties have been shown to involve collinearity of points, concurrence of lines, cross-ratio, and the principle of duality according to which we can interchange "point" and "line" in a theorem and obtain a valid statement.

In contrast to projective geometry, Euclidean geometry seeks to study properties of geometric figures that remain invariant under rigid motions. These properties involve equality of lengths, angles, and areas.

Gérard Desargues (1593-1662) and Blaise Pascal (1623-1662) made important contributions to projective geometry.

NON-EUCLIDEAN GEOMETRY

For two thousand years most mathematicians accepted as a "self-evident" truth Euclid's postulate which implied that there exists only one line parallel to a given line through a given point not on the line. They also assumed that the geometry of Euclid was the only geometry possible.

But in the last 200 years mathematicians have successfully challenged these beliefs. They have shown that it is possible, for example, to build consistent and useful geometries with a postulate that contradicts Euclid's parallel postulate. These geometries are called *non-Euclidean geometries.*

In the early part of the nineteenth century, three mathematicians, working independently, developed a geometry with a postulate that assumed that *more* than one parallel can be drawn to a given line through a given point not on the line. This geometry, later named hyperbolic geometry, was developed by N. I. Lobachevsky (1793-1856), Janos Bolyai (1802-1860), and K. F. Gauss (1777-1855).

Later, G.F.B. Riemann (1826-1866) assumed that *no* line can be drawn parallel to a given line through a point not on the line. This resulted in the creation of another new and consistent geometry which differed from that of Euclid and from that of Lobachevsky, Bolyai, and Gauss. Riemann, in fact, created a whole class of new geometries, one of which was the basis for Einstein's general theory of relativity.

The discovery of these equally valid and consistent non-Euclidean geometries destroyed the notion that Euclid's axioms were self-evident truths that were immutable. We now regard axioms, or postulates, as *assumptions* that need only be consistent. (For a more detailed discussion of non-Euclidean geometries, see pages 349-361.)

TOPOLOGY

Since the middle of the nineteenth century, geometry has taken a new and exciting turn through the development and spectacular growth of a branch of mathematics called *topology*. Some of the men who made important contributions to this subject are Leonard Euler (1707-1783), A. F. Möbius (1790-1868), Bernhard Riemann (1826-1866), Henri Poincaré (1854-1912), and Felix Hausdorff (1868-1942).

The object of topology is to study those properties of geometric figures that remain invariant, that do not change, even after the size and shape of these figures have been altered. Such deformations would occur if we imagine our figures to be drawn on a rubber sheet and then allow them to be distorted by stretching the rubber sheet in any way we wish without tearing it or folding it back onto itself.

For example, suppose we draw a triangle on a rubber sheet and mark a dot inside the triangle (Figure 10.2). No matter how we stretch or bend the rubber sheet, the dot is still *inside* the triangle (Figure 10.3). This "inside" position of the dot remains invariant even though the triangle was subjected to so drastic a deformation that it lost its original shape and size.

FIGURE 10.2

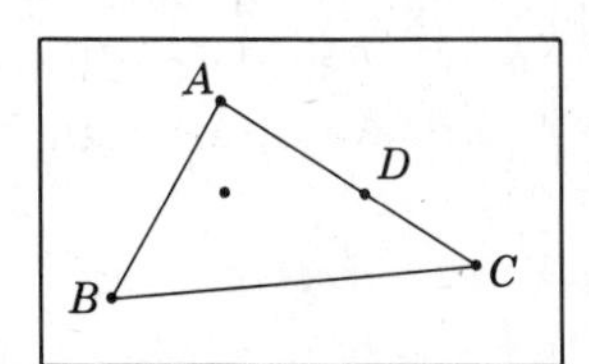

FIGURE 10.3

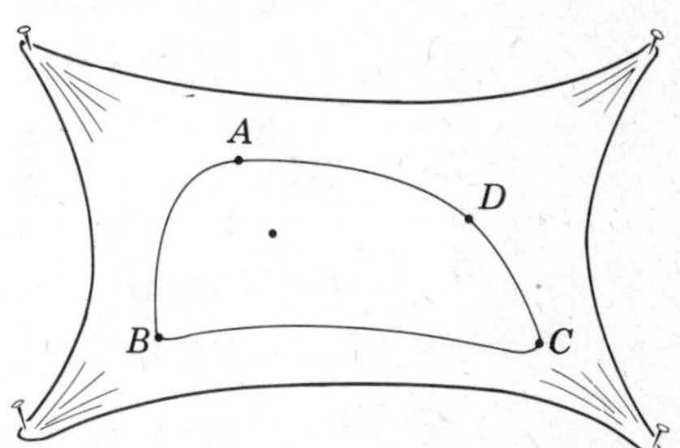

Certain other properties will remain invariant under this deformation. For instance, the *order* of the points A, B, C, and D remains the same. No matter how we stretch the rubber sheet, the path from A to B to C to D to A remains a path that does not cross itself. No matter where we start on this path, we will return to the starting point without crossing the path.

Topology has been used to solve some fascinating puzzles and problems, such as the problem of connecting like numerals by lines that do not cross each other or any other lines in the figure (Figure 10.4).

FIGURE 10.4

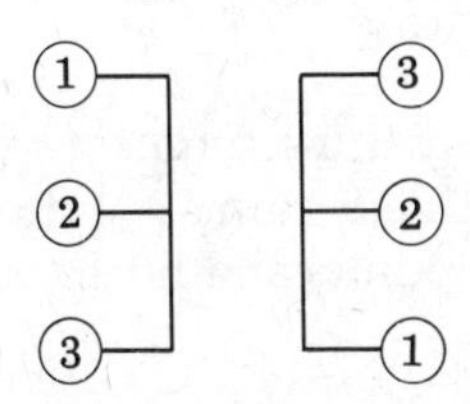

A famous problem that substantially led to the beginnings of topology was the Koenigsberg bridge problem. This relates to the Preger River, which ran through the town of Koenigsberg and was crossed in part by seven bridges (Figure 10.5). The problem was how to start out from any point in the town and cross each bridge once, and only once, in returning to the starting point. Euler became interested in this problem and proved that it was impossible to cross each bridge only once and return to the starting point. Euler's solution led to the founding of topology.

FIGURE 10.5

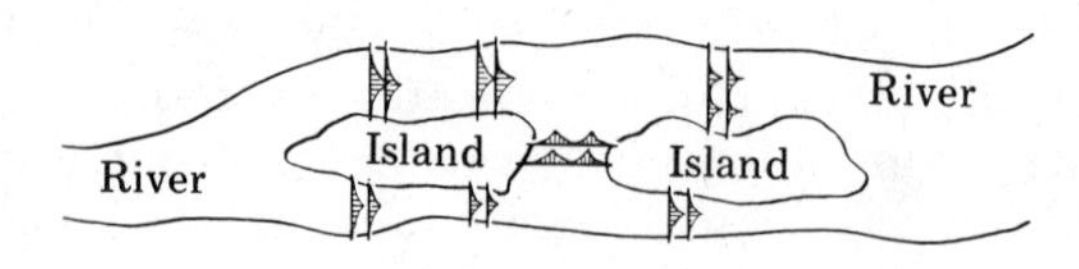

A famous topological problem, known as the four-color conjecture, asks us to prove or disprove that four different colors are sufficient to color any map on a plane or sphere so that countries with a common border are colored differently. After a century of effort, this conjecture was finally proved in 1976 by the use of a computer.

Although topology can be a "fun" subject and can be used to solve some fascinating problems, it is being applied more and more to many practical problems and serves to illuminate other branches of mathematics.

10.2 Finite Geometries

Euclidean geometry is the most famous example of an abstract mathematical system, but it cannot be "seen" easily as a *whole* because of its many postulates and theorems. For a clearer grasp of the structure and workings of such a system we need a smaller and less complex example—based on only a few postulates that yield only a few theorems—that exposes more clearly the interrelation of its various parts. We have just such an example in the *finite geometries.*

A finite geometry is a miniature geometry which is based on a small set of postulates, undefined terms, and undefined relations and which limits the set of all points and lines to a finite number.

CRITERIA FOR POSTULATES

The postulates in a finite geometry must meet the same criteria as the postulates of any other mathematical system. They must be *consistent;* that is, no two postulates can contradict each other. It is desirable that they be *independent*, that is, that no postulate is a logical consequence of, or deducible from, the remaining postulates. It may also be desirable that the postulates be *complete*, that is, that it is always possible to prove every statement that can be made about the undefined terms of the system from the postulates.

SEVEN-POINT GEOMETRY

Seven-point geometry is an example of a finite geometry. It consists of a set of undefined elements called "points," certain undefined subsets of points called "lines," the undefined relations "contain" and "lies on," and a set of seven postulates.

Although we have given specific names to our undefined terms, it is essential that we not read Euclidean or any other meanings into them. We must not confuse these "points" and "lines" with Euclidean points and lines. Our present points and lines will be completely defined by our seven postulates. From them it will soon be seen, for example, that our "line," unlike the Euclidean line, is not continuous and consists of a finite number of points.

POSTULATES FOR SEVEN-POINT GEOMETRY

The seven postulates for our geometry are:

POSTULATE 1 If P_1 and P_2 are any two distinct points, there is at least one line containing both P_1 and P_2.

POSTULATE 2 If P_1 and P_2 are any two distinct points, there is at most one line containing both P_1 and P_2.

POSTULATE 3 If l_1 and l_2 are any two distinct lines, there is at least one point which lies on both l_1 and l_2.

POSTULATE 4 Every line contains at least three points.

POSTULATE 5 Every line contains at most three points.

POSTULATE 6 If l is any line, there is at least one point which does not lie on l.

POSTULATE 7 There exists at least one line.

Note that Postulates 1, 2, 4, 6, and 7 are also properties of Euclidean plane geometry. However, Postulate 3 contradicts Euclid's parallel postulate, since it asserts that any two lines intersect, and Postulate 5 is not true in Euclidean geometry.

MODELS FOR SEVEN-POINT GEOMETRY

Can we find some interpretation or model for the postulates of our seven-point geometry? If in the arrangement

$$\begin{matrix} P_1 & P_2 & P_3 & P_4 & P_5 & P_6 & P_7 \\ P_2 & P_3 & P_4 & P_5 & P_6 & P_7 & P_1 \\ P_4 & P_5 & P_6 & P_7 & P_1 & P_2 & P_3 \end{matrix}$$

we let the *P*s represent the "points" in our geometry and we let the vertical columns represent the "lines," then Figure 10.6 can be a geometric model for our seven-point geometry.

FIGURE 10.6

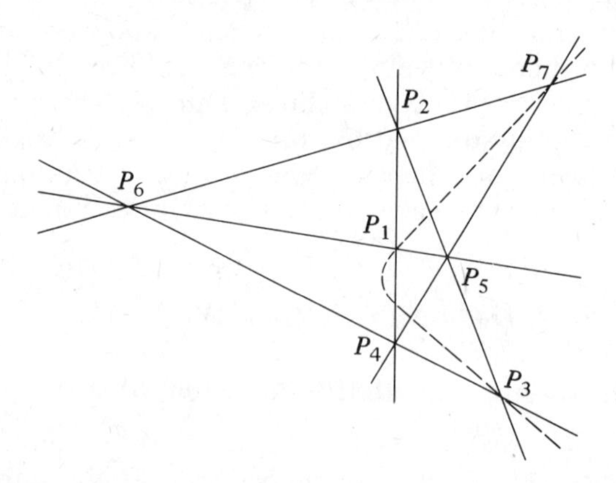

If we go back to the seven postulates and check them against this model, we will see that the model satisfies each of them. Note that since the postulates do not say that a line must be "straight," it may be curved like line $P_7P_1P_3$.

Because we have exhibited at least one model that satisfies all our postulates, we can conclude that the seven postulates are consistent.

We can construct a different geometric diagram (Figure 10.7) that satisfies all our postulates, again letting the *P*s represent "points" and the vertical columns represent "lines":

$$\begin{matrix} P_1 & P_1 & P_2 & P_1 & P_3 & P_2 & P_3 \\ P_2 & P_4 & P_4 & P_6 & P_5 & P_5 & P_4 \\ P_3 & P_5 & P_6 & P_7 & P_6 & P_7 & P_7. \end{matrix}$$

FIGURE 10.7

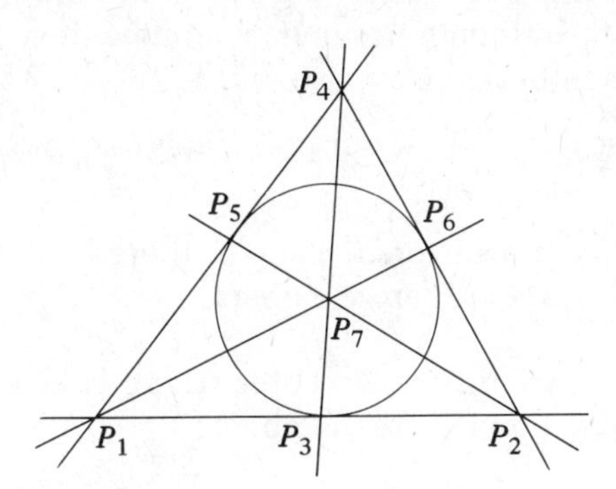

Here, again, inspection of the model will show that it satisfies each of the seven postulates.

Geometric diagrams are not the only models we can use to represent our postulates. We can translate the postulates of seven-point geometry into statements about numbers. For instance, let us adopt these definitions:

1 The seven "points" correspond to the seven numbers 1, 2, 3, 4, 5, 6, and 7.

2 The seven "lines" correspond to the seven numbers 123, 145, 246, 167, 356, 257, and 347.

3 A point is "on" a line if and only if the digit corresponding to the point is a digit of the numeral corresponding to the line. If a point and a line have this relation, then the line is said to "contain" the point.

With these definitions, every statement of the seven-point geometry becomes a statement about certain numbers. Inspection of this model will show that it, too, satisfies all seven postulates.

We can interpret the postulates of our seven-point geometry in still another way. We can relate them to people and committees. To create this new model, we adopt these definitions:

1 The seven "points" correspond to the seven people *A*, *B*, *C*, *D*, *E*, *F*, and *G*. "People" and "members" will be used interchangeably.

2 The seven "lines" correspond to seven committees, as indicated by the following seven columns:

A A B A C B C
B D D F E E D
C E F G F G G.

3 A line "contains" a point if and only if a person is a member of the committee. If a point and a line have this relation, then the point is also said to "lie on" the line.

With these definitions, our postulates become:

POSTULATE 1 If A and B are different people, there is at least one committee of which both A and B are members.

POSTULATE 2 If A and B are different people, there is at most one committee on which both A and B are members.

POSTULATE 3 Any two committees have at least one member in common.

POSTULATE 4 Every committee contains at least three members.

POSTULATE 5 Every committee contains at most three members.

POSTULATE 6 For any committee, there is at least one person who is not a member of it.

POSTULATE 7 There exists at least one committee.

An inspection of the membership of each committee will again show that all seven postulates of our geometry are satisfied. The isomorphism between the point—line model and the people—committee model becomes apparent when we rename the points with people as shown in Figure 10.8.

FIGURE 10.8

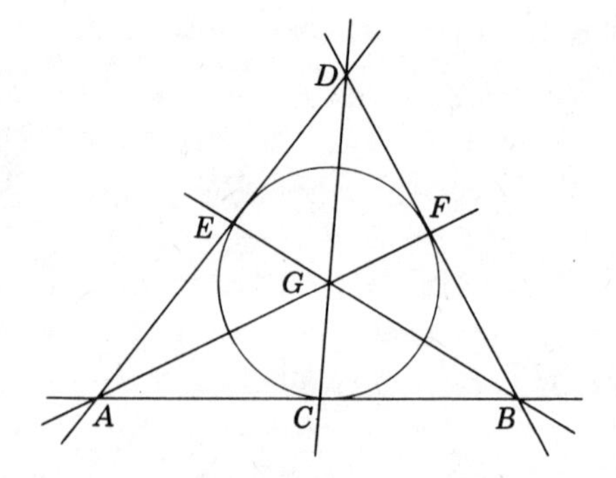

THEOREMS OF SEVEN-POINT GEOMETRY

Now that we have described the postulates of seven-point geometry, we shall see what *theorems* can be deduced from them. To prove theorems in this system, we do precisely what we do in other mathematical systems—but in this system we have only seven postulates and four undefined terms to consider.

THEOREM 1 If l_1 and l_2 are any two distinct lines, there is at most one point which lies on both l_1 and l_2 (Figure 10.9).

Proof Assume the contrary, that is, that two distinct lines have two points in common. This contradicts Postulate 2. Therefore, two distinct lines have at most one point in common.

FIGURE 10.9

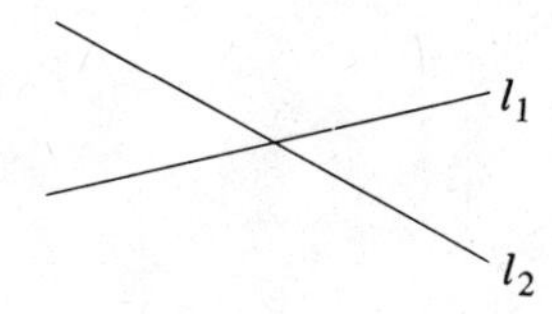

THEOREM 2 There exists at least one point.

Proof By Postulate 7 there exists at least one line, and by Postulate 4 every line contains at least three points. Therefore, there exists at least one point.

THEOREM 3 If P is any point, there is at least one line which does not pass through P.

Proof

1. By Postulate 7 there exists at least one line l.
2. If this line does not pass through P, our proof is complete.
3. Suppose l passes through P (Figure 10.10). Then, by Postulate 4, l contains at least one point P_1 besides P.

FIGURE 10.10

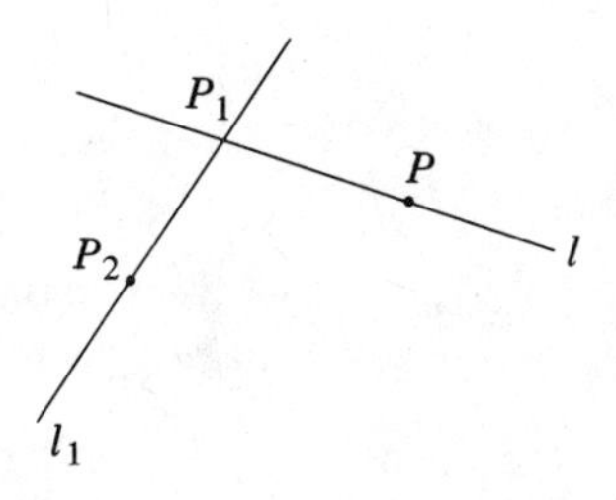

4. By Postulate 6, there is at least one point P_2 which does not lie on l.
5. By Postulates 1 and 2, there is a unique line l_1 which contains P_1 and P_2.
6. l and l_1 are distinct, since l_1 contains P_2 and l does not.
7. By Postulate 3 and Theorem 1, l and l_1 have exactly one point in common, P_1.

8 Therefore, P, which lies on l, cannot also lie on l_1. In other words, l_1 is a line that does not pass through P.

THEOREM 4 Every point lies on at least three lines.

Proof

1 By Theorem 3, there is at least one line l that does not pass through a given point (Figure 10.11).

FIGURE 10.11

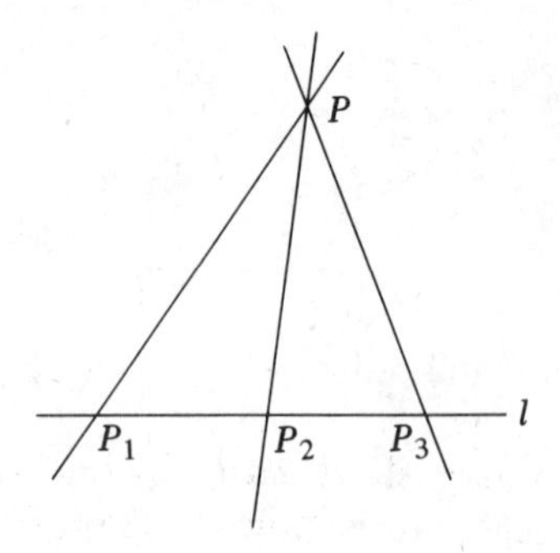

2 By Postulate 4, l contains at least three points, P_1, P_2, and P_3.

3 By Postulates 1 and 2, each of these points determines with P a unique line.

4 These lines are distinct, since if two of them coincided, that line would have two points in common with l. This would contradict Postulate 3 and Theorem 1.

5 Therefore there are at least three lines passing through every point P.

DUALITY

By proving these four theorems we have established a very interesting and important property of seven-point geometry—that it has *duality.*

According to the principle of *duality*, any true statement about lines and points is also true when "line" and "point" are interchanged, and when a corresponding change is also made in the connective used. For instance, *all lines do not pass through the same point* implies, by duality, that *all points do not lie on the same line.*

If we examine the theorems and postulates of our seven-point geometry, we find that

1 Postulate 1 is the dual of Postulate 3.

2 Theorem 1 is the dual of Postulate 2.

3 Theorem 2 is the dual of Postulate 7.

4 Theorem 3 is the dual of Postulate 6.

5 Theorem 4 is the dual of Postulate 4.

6 The dual of Postulate 5 is the theorem that every point lies on at most three lines.

This proves that our seven-point geometry has duality.

In general, the importance of this property lies in the fact that in any system that has duality, each proof actually establishes two theorems, the theorem we set out to prove and its dual.

The following are two other theorems in this geometry.

THEOREM 5 Seven-point geometry contains exactly seven points.

THEOREM 6 Seven-point geometry contains exactly seven lines.*

Note that Theorems 5 and 6 are duals of each other; therefore one theorem implies the other.

Theorems 5 and 6 have been generalized to yield information about the number of points and lines contained in any finite projective geometry:

1 If a line contains exactly n points, then the system contains exactly $n^2 - n + 1$ points and $n^2 - n + 1$ lines.

2 If a line contains exactly n points, then exactly n lines pass through every point.

For example, in a finite projective geometry where a line contains 4 points,

1 Exactly 4 lines will pass through every point.

2 The system will contain $n^2 - n + 1$ points and lines, or 13 points and 13 lines.

The study of finite geometries goes back to 1892, when G. Fano first considered a three-dimensional geometry containing 15 points and 35 lines, each plane containing 7 points and 7 lines. About 15 years later, O. Veblen and J. W. Young developed seven-point geometry in their book *Projective Geometry*, Vol. 1, Ginn & Company, Boston, 1910.

10.3 The Famous Construction Problems of Antiquity

During their study of geometry, the Greeks tried to solve all construction problems by the use of only two tools—the straightedge

*For a proof of Theorems 5 and 6, see B. E. Meserve, *Fundamental Concepts of Geometry*, Reading, Mass.: Addison-Wesley Publishing Company, Inc., 1955, p. 16.

and compass. This was a kind of game with self-imposed restrictions. You could use these two tools only in a special way: With the straightedge you were permitted to draw a line segment of indefinite length through any two given distinct points; with the compass you were permitted to draw a circle with any given point as center and passing through any given second point.

But the Greeks soon learned that even though they could construct many complicated curves with these two tools alone, not all problems could be solved this way. Three of the most famous of these "failures" were the problems of how (1) to trisect any angle, (2) to construct a square whose area is equal to that of a given circle, and (3) to construct a cube whose volume is twice that of a given cube.

In their fruitless efforts to find solutions to these problems, the Greeks discovered the conic sections, cubic curves, and even some transcendental curves.

It was not until more than 2000 years later that the *reason* for the failure of the Greeks to solve these seemingly easy construction problems was established. The reason turned out to be that these problems are *impossible* to solve by the use of a straightedge and a compass alone. This impossibility was established by algebraic means.

TRISECTION OF AN ANGLE

Given *any* angle $\angle BAC$ as shown in Figure 10.12, the problem is to construct, by straightedge and compass, two rays $\overrightarrow{AD}$ and $\overrightarrow{AE}$ so that

$$\angle BAD \cong \angle DAE \cong \angle EAC.$$

The impossibility of this construction rests on a theorem which states: From a given unit length, it is impossible to construct with Euclidean tools a segment the magnitude of whose length is a root of a cubic equation with rational coefficients but no rational roots.

FIGURE 10.12

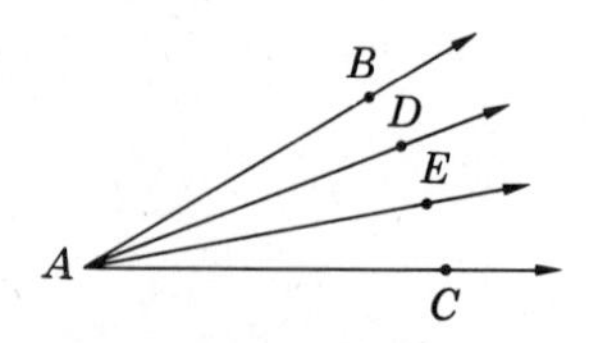

If we can prove that any *one* angle, say an angle of 60°, cannot be trisected, then we will have established the nonexistence of a *general* method of trisecting an angle.

The problem of trisecting an angle of 60° can be shown to be equivalent to the problem of constructing a segment the magnitude of whose length is a root of the cubic equation

$$8x^3 - 6x - 1 = 0$$

that has no rational root.

Therefore, by the theorem just stated, this construction is impossible. Despite this proof, a new supply of "solutions" are offered each year, a recently publicized one in March 1969 (*New York Times*).

To say that the trisection of an angle is impossible is not to say that *no* angle can be trisected by straightedge and compass. Special angles such as 90°, 45°, or 135° can be so trisected.

If we remove the restriction of having to use *only* a straightedge and compass, the problem becomes solvable. A number of instruments have been devised for trisecting an angle.

SQUARING THE CIRCLE

Given a circle, the problem is to construct, by straightedge and compass, a square whose area is the same as that of the circle (Figure 10.13). That is, we wish $s^2 = \pi r^2$, or $s = r\sqrt{\pi}$. We could construct s if we could construct a segment whose length is π units.

FIGURE 10.13

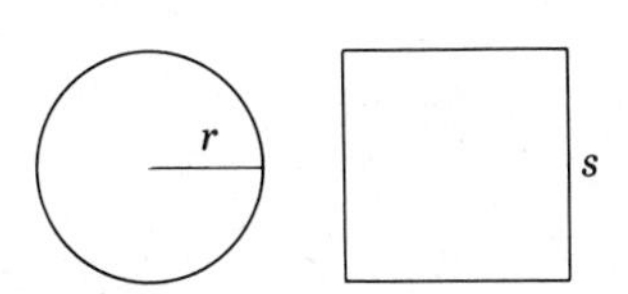

The impossibility of this construction rests on a theorem which states that the magnitude of any length constructible with Euclidean tools from *a given unit length* is an algebraic number.

A number r is an *algebraic* number if it satisfies an equation of the form

$$a_0x^n + a_1x^{n-1} + \cdots + a_{n-1}x + a_n = 0,$$

where the a's are all rational numbers and n is a positive integer. Otherwise r is *transcendental.* In other words, a number that cannot be the root of an algebraic equation with rational coefficients is called a *transcendental number.*

About a hundred years ago (1882), F. Lindemann (1852-1939) showed that the number π is transcendental. Therefore, a segment of length π cannot be constructed from a given unit segment by straightedge and compass alone, and our problem becomes inconstructible.

DUPLICATION OF THE CUBE

Given a segment of length a, the problem is to construct, by straightedge and compass, a segment of length b such that a cube of edge b has twice the volume of a cube of edge a (Figure 10.14). That is, we wish $b^3 = 2a^3$. If a is unit length, then $b^3 = 2$ or $b^3 - 2 = 0$.

Since this equation has no rational root, b cannot be constructed from segment a with straightedge and compass.

FIGURE 10.14

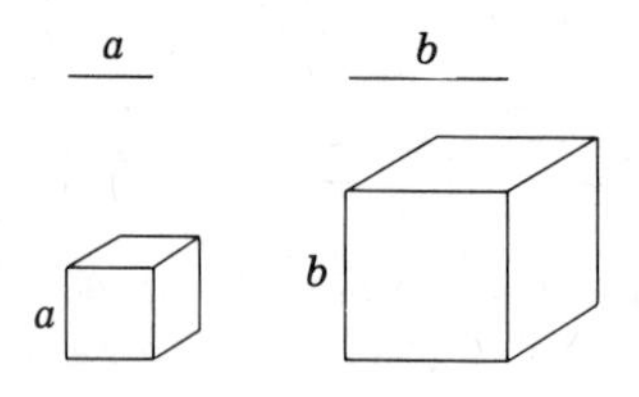

10.4 *The History of* π

Interest in computing the value of π goes back 2200 years when Archimedes (287-212 B.C.), by using regular inscribed and circumscribed polygons, computed the value of π to be between $\frac{223}{71}$ and $\frac{22}{7}$ or, to two decimal places, as 3.14. This improves upon the value of 3 given in the Bible (I Kings 7:23) which describes a basin as 10 cubits from one brim to the other, "and a line of 30 cubits did compass it round about."

Archimedes inscribed and circumscribed regular polygons about a circle, determined their perimeters, and then used these values as upper and lower bounds to the length of the circumference. He kept doubling the number of sides until the perimeters of polygons of 96 sides were obtained. With these perimeters he established the value of π to be $3\frac{10}{71} < \pi < 3\frac{1}{7}$.

Archimedes' method, known as the *classical method* of computing π, was the one used for hundreds of years after Archimedes. This method can yield approximations for π to as many decimal places as we wish. In 1579, Francois Viète used polygons with 393,216 sides to find π correct to nine decimal places.

In 1610, Ludolph van Ceulen worked on π almost to the day of his death at the age of 70, and computed π to 35 decimal places by using regular polygons with 2^{62} sides. Before he died, he requested that the 35 digits of π be engraved on his tombstone. This was actually done. The tombstone seems now to be lost.

About a hundred years after Archimedes, Ptolemy computed the value of π as 3.1416 from the lengths of the chords of a circle subtended by central angles of 1°. If the length of the chord of the central angle is multiplied by 360, and the result divided by the length of the diameter of the circle, this value for π is obtained.

ALGEBRAIC METHODS FOR COMPUTING π

In the seventeenth century the classical method of Archimedes was abandoned and the algebraic methods of convergent infinite series, products, and continued fractions became the method for computing the value of π. For instance,

1 John Wallis (1616-1703) expressed the value of π this way:

$$\frac{\pi}{2} = \frac{2}{1} \times \frac{2}{3} \times \frac{4}{3} \times \frac{4}{5} \times \frac{6}{5} \times \frac{6}{7} \times \frac{8}{7} \times \frac{8}{9} \times \cdots.$$

2 Lord Brouncker (1620-1684) expressed the value of π this way:

$$\pi = \cfrac{4}{1 + \cfrac{1^2}{2 + \cfrac{3^2}{2 + \cfrac{5^2}{2 + 7^2 \ldots}}}}$$

3 The simplest and most elegant expression for the value of π was given by G. W. von Leibniz (1646-1716):

$$\frac{\pi}{4} = 1 - \frac{1}{3} + \frac{1}{5} - \frac{1}{7} + \frac{1}{9} - \frac{1}{11} + \frac{1}{13} - \frac{1}{15} + \cdots.$$

About 100 years ago, William Shanks computed π to 707 places. In 1945 his result was proven wrong starting with the 528th place!

In 1961, the value of π was calculated to 100,265 decimal places by an IBM 7090 computer. The job was completed in 8.7 hours.

In 1967, π was computed on a CDC 6600 computer to 500,000 decimal places by Jean Guilloud in Paris.

WHY THIS UNUSUAL INTEREST IN π?

The preoccupation with calculating π to so many decimal places is all the more remarkable in light of at least one authoritative opinion that if π, expressed to only 10 decimal places, were used to compute the circumference of the earth, the result would be correct to inches!

There are several reasons, however, for this astonishing interest. It was suggested earlier that the appearance of π in a context different from geometry has contributed to this interest. For example, the equation

$$y = \frac{N}{\sigma\sqrt{2\pi}}\, e^{-x^2/2\sigma^2}$$

defines the normal-distribution curve, which is most essential in inferential statistics. It is this curve that enables us to make predictions about entire populations on the basis of small samples.

Another example of the appearance of π in a nongeometric context is in the famous Buffon Needle Problem. In 1760, G. L. Leclerc, Comte de Buffon, investigated the probability of dropping at random a uniform needle of length l onto a plane ruled with parallel lines d distance apart ($l < d$) in such a way that the needle will touch one of the ruled lines (Figure 10.15). He found this probability to be $\frac{2l}{\pi d}$.

FIGURE 10.15

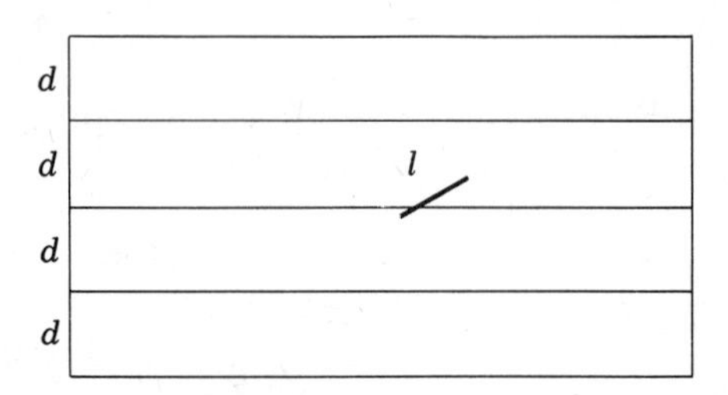

If the length of the needle is half the distance between the parallel lines, that is,

$$l = \tfrac{1}{2}d,$$

then the probability of success becomes $\frac{1}{\pi}$. This provides a good laboratory-type procedure for obtaining an approximation of π.

A second reason for the exceptional interest in calculating π to a large number of decimal places is the desire to test the "regularity" or "normalcy" of π. By "normalcy" we mean the following: A real number is said to be *simply normal* if in its infinite decimal expression all digits occur with equal frequency. It is said to be *normal* if all blocks of digits of the same length occur with equal frequency.

Starting in 1949, computers were used to find out by actual count whether π is *normal* or *simply normal.* The results seem to indicate that π is *normal.*

10.5 Non-Euclidean Geometries

We have seen how the rejection of Euclid's fifth postulate led to the development of non-Euclidean geometries. We now wish to go more deeply into the ideas and implications of these geometries.

EUCLID'S FIFTH POSTULATE

Let us take a closer look at the fifth postulate. It asserts that if two lines r and s are intersected by a third line l so as to make the sum of angles 1 and 2 less than 180°, then the lines r and s *meet* on the same side of line l as are angles 1 and 2 (Figure 10.16). This assertion implies the existence of parallel lines.

FIGURE 10.16

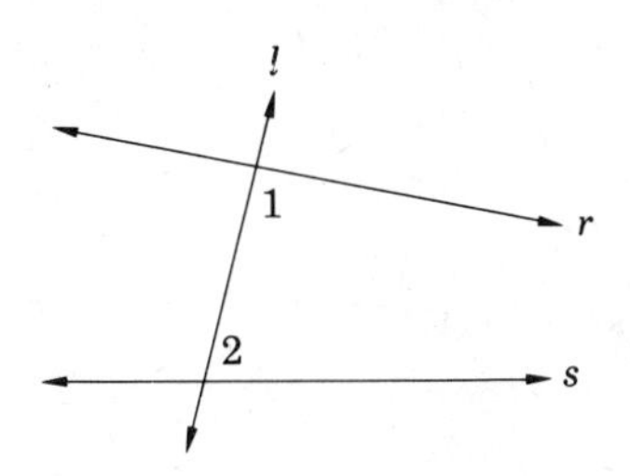

An equivalent axiom, known as *Playfair's axiom*, expresses Euclid's fifth postulate this way:

PLAYFAIR'S AXIOM Through a point not on a given straight line there is one and only one straight line that does not intersect the given straight line.

This says that, given line s and point P not on s, there is one and only one line r in the plane of P and s which passes through P and does not meet s (Figure 10.17).

FIGURE 10.17

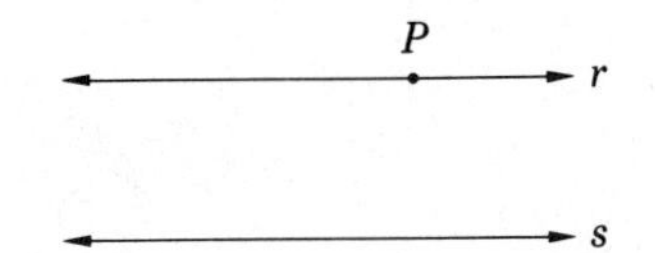

We say that Playfair's axiom and Euclid's fifth postulate are equivalent because from Euclid's postulates we can derive Playfair's axiom, and from Euclid's postulates with the Playfair axiom replacing Euclid's fifth postulate we can derive Euclid's fifth postulate.

The fifth postulate aroused controversy almost from the beginning. Postulates were supposed to be "self-evident truths," but the fifth is more complex than any of the others and is not quite as self-evident as, say, the fourth, which says that "all right angles are equal to one another." The fifth, unlike the other nine postulates, makes an assertion about lines that are "produced indefinitely." This transcends our experience, is not subject to experimental verification, and is, therefore, unlike the other postulates that deal only with finite segments of lines. Besides, the *converse* of the fifth postulate can be proved from the other four postulates and axioms, so why cannot the fifth be proved in a similar way? (Euclid's fifth postulate is the converse of Proposition 17 of Book I of the *Elements.*)

So for 2000 years mathematicians tried to *prove* that the fifth postulate is a necessary *consequence* of the other nine postulates. But all their attempts resulted in failure. Perhaps the most famous of these attempts was the one made by Girolamo Saccheri (1667-1733), who used the indirect method of proof to try to show that the fifth postulate is a consequence of the other nine.

The fifth postulate asserts the existence of one and only one line parallel to a given line and passing through a given point not on the given line. There are two alternatives to this assertion:

1 There is *no* line parallel to the given line and passing through a given point not on the given line.

2 There is *more* than one such line.

Saccheri's plan was to show that by assuming either of the two alternatives he would end up with a contradiction. Therefore, the only other possibility, Euclid's fifth postulate, must be true. Saccheri was able to show that the assumption that there are *no* parallels leads to a contradiction. He was not able to show a contradiction by assuming that there is *more* than one parallel.

Saccheri's efforts centered around a special quadrilateral, now called a *Saccheri quadrilateral,* in which two equal sides are both perpendicular to a third side. Let $ABCD$ in Figure 10.18 be a Saccheri quadrilateral with $AD = BC$ and angles A and B right angles. $\overline{AD}$ and $\overline{BC}$ are called the *legs* of the quadrilateral; $\overline{AB}$ is called the *base* of the quadrilateral; $\overline{DC}$ is called the *summit;* and angles C and D are called the *summit angles.*

FIGURE 10.18

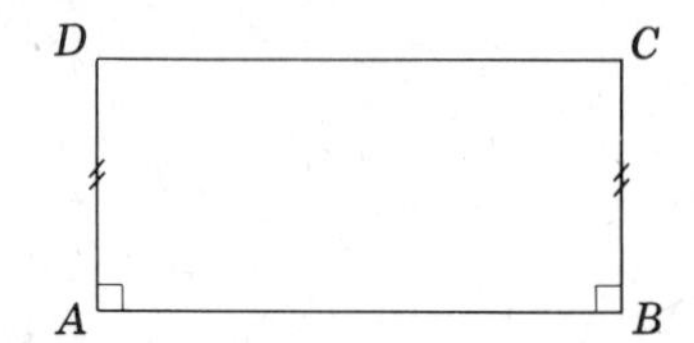

After proving that $\angle C \cong \angle D$,* Saccheri considered three different hypotheses concerning these two summit angles:

1 The summit angles are right angles; that is, $m\angle C = m\angle D = 90$.

2 The summit angles are obtuse angles; that is, $m\angle C = m\angle D > 90$.

3 The summit angles are acute angles; that is, $m\angle C = m\angle D < 90$.

Saccheri was able to prove that the assumption that the summit angles are obtuse leads to a contradiction. When he assumed the summit angles to be acute, he ended up with several "strange" theorems, such as

1 The sum of the measures of the angles of any triangle is less than 180.

2 Two triangles are congruent if the three angles of one are congruent to the three angles of the other.

He concluded that this assumption, too, leads to a contradiction.

Believing that he proved that hypothesis 2 and hypothesis 3 lead to contradictions, Saccheri concluded that hypothesis 1, which is equivalent to Euclid's parallel postulate, must be true.

Although Saccheri believed he proved that the hypothesis that the summit angles are acute leads to a contradiction, he actually did not. He only proved a number of theorems in the non-Euclidean geometry that were developed a century later by Bolyai and Lobachevsky.

Later attempts by other mathematicians to prove Euclid's fifth postulate—J. H. Lambert (1728-1777), A. M. Legendre (1752-1833), and many others—were equally unsuccessful.

The inability of mathematicians to prove the fifth postulate as a consequence of Euclid's other nine postulates finally convinced them that their failure was due to the fact that the fifth is *independent* of the other nine.

This meant that it is possible to construct a consistent geometry from a set of postulates in which the parallel postulate was replaced with a contradictory postulate. Such a geometry is called a *non-Euclidean geometry*.

*Join D to B and C to A. $\triangle ABD \cong \triangle ABC$, making $BD = AC$. Therefore, $\triangle ACD \cong \triangle BDC$ and $\angle ADC \cong \angle BCD$.

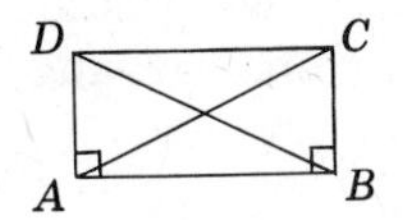

Why is the parallel postulate so important? What role does it play in geometry? It is no exaggeration to say that this single postulate leads to some of the most basic, useful, and important results in Euclidean geometry. It leads to:

1 The conclusion that the sum of the measures of the angles of a triangle is 180.

2 The existence of rectangles and squares.

3 The existence of similar figures.

4 The Pythagorean relation.

If we change just this one postulate but keep all the other postulates, then we are creating a new geometry in which *none* of these results hold, as we shall soon see.

It was not until the nineteenth century, after centuries of failure to prove the parallel postulate, that mathematicians began to develop geometries based on a contradiction of Euclid's fifth postulate. By 1830, J. Bolyai (1802-1860), N. I. Lobachevsky (1793-1856), and K. F. Gauss (1777-1855) independently developed a geometry, called *hyperbolic geometry*, in which they assumed that there is *more* than one line parallel to a given line through an external point.

In 1854, G. F. B. Riemann (1826-1866) introduced a different non-Euclidean geometry, called *elliptic geometry*, in which he assumed that there are *no* parallel lines.

Euclidean, hyperbolic, and elliptic geometries have very different properties, and we shall now see what some of these properties are.

HYPERBOLIC GEOMETRY

In *hyperbolic geometry*, also called *Lobachevskian geometry*, all the Euclidean postulates are assumed except the parallel postulate, which is replaced with another that asserts that there is *more than one* line parallel to a given line through an external point.

Two consequences immediately follow.

1 Every theorem of Euclidean geometry can also be a theorem of hyperbolic geometry as long as it does not depend on Euclid's parallel postulate.

2 There are infinitely many parallels to a given line through an external point. The reason for this conclusion is as follows:

a) Suppose there are two distinct lines l_1 and l_2, both parallel to line m and passing through point P (Figure 10.19).

b) Then *any* line $\overleftrightarrow{PC}$, where C is in the interior of $\angle APB$, cannot intersect m because to do so $\overleftrightarrow{PC}$ would have to cross l_1 or l_2 and therefore meet l_1 or l_2 in a second point. But since two lines can intersect in at most one point, it follows that all lines through P which lie in the interior of $\angle APB$ will not intersect m.

c) Therefore, the assumption that there are at least two parallels to m passing through an external point P implies that there are infinitely many parallels to m passing through P.

FIGURE 10.19

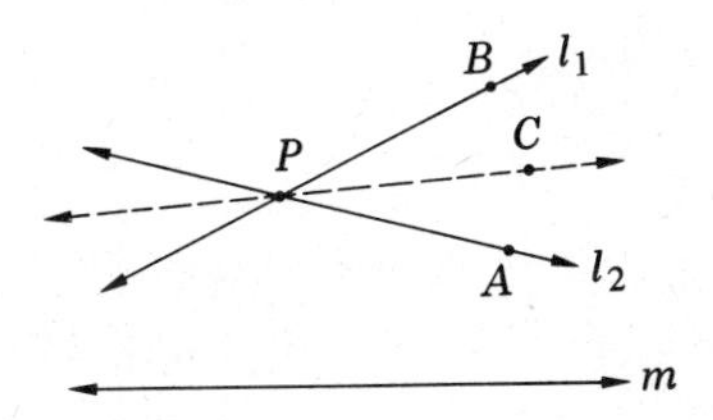

A MODEL FOR HYPERBOLIC GEOMETRY

We use the interior of a Euclidean circle to serve as a model for hyperbolic geometry. The points of the circle itself are not included in this geometry.

The "points" in this geometry are the points *inside* the circle, and the "lines" are chords without endpoints. "Segments" are represented by Euclidean segments whose endpoints are inside the circle. The Lobachevskian "plane" is represented by the interior of the circle (Figure 10.20).

FIGURE 10.20

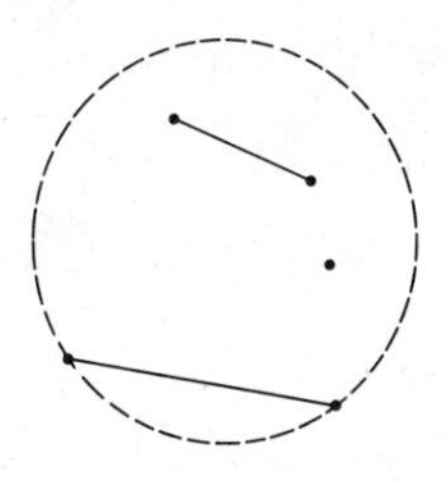

From the model in Figure 10.21 we can see that it satisfies the "more-than-one-parallel" postulate of hyperbolic geometry, since neither m nor n intersects line l, making both m and n parallel to l. It is also apparent that lines parallel to the same line are not necessarily parallel to each other. But the other postulates must be verified in the model, and this requires definitions of measures of line segments and of angles.

FIGURE 10.21

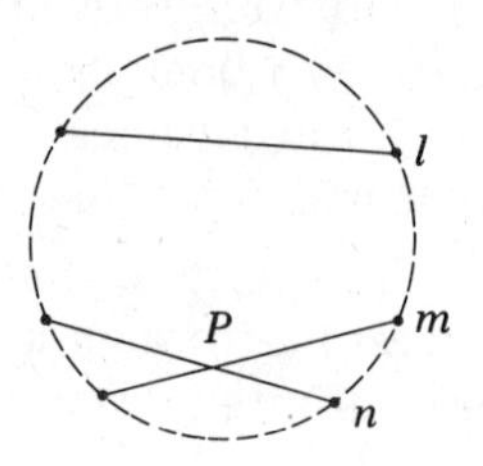

THE SUM OF THE ANGLE MEASURES OF A HYPERBOLIC TRIANGLE

Before discussing a theorem about the sum of the angle measures of a hyperbolic triangle, we shall state but not prove another theorem in hyperbolic geometry:

THEOREM 7 Let n be a line, A a point not on n, and B a point on n; let a side of $\overleftrightarrow{AB}$ be given (Figure 10.22). Then there exists a point C of n, on the given side of $\overleftrightarrow{AB}$ such that the measure of $\angle ACB$ is as small as we please.

FIGURE 10.22

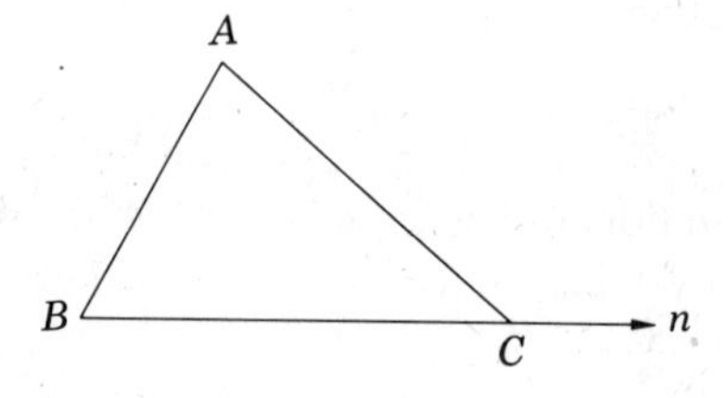

An important theorem in hyperbolic geometry states that the sum of the measures of the angles of every triangle is less than 180.

Proof

1 Let n be a line and A a point not on n. Also, let $\overleftrightarrow{AB}$ be perpendicular to n at B and let m be perpendicular to $\overleftrightarrow{AB}$ at A (Figure 10.23).

FIGURE 10.23

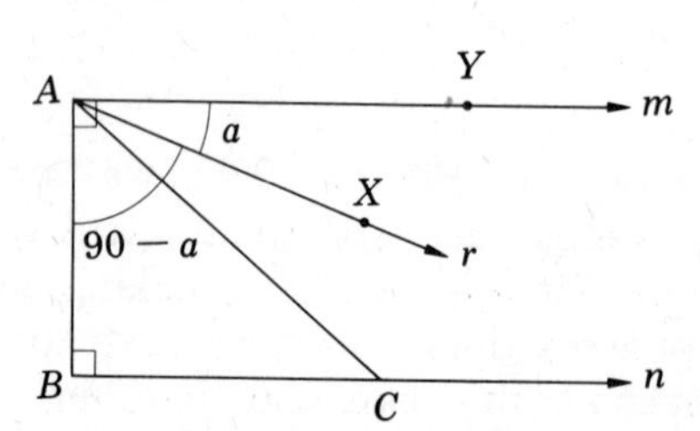

2 By the Lobachevskian parallel postulate, there is another line r through A parallel to n.

3 One of the angles that r makes with $\overleftrightarrow{AB}$ must be acute. Let X be a point of r such that $\angle BAX$ is acute.

4 Let Y be a point of m on the same side of $\overleftrightarrow{AB}$ as X. Let $m\angle XAY = a$. Then $m\angle BAX = 90 - a$.

5 By the theorem stated above, let C be a point of n, on the side of $\overleftrightarrow{AB}$ containing X, such that $m\angle ACB < a$.

6 In $\triangle ABC$, $m\angle ABC = 90$, $m\angle ACB < a$, and $m\angle BAC < m\angle BAX$, or $m\angle BAC < 90 - a$.

7 By adding, we obtain

$$m\angle ABC + m\angle ACB + m\angle BAC < 90 + a + 90 - a \quad \text{or}$$

$$m\angle ABC + m\angle ACB + m\angle BAC < 180.$$

That is, the sum of the measures of the angles of $\triangle ABC$ is less than 180.

8 What we have shown, so far, is that there exists a triangle ($\triangle ABC$) whose angle sum is less than 180°. But by another theorem we know that if one triangle has an angle sum that is less than 180°, then *every* triangle has an angle sum less than 180°.

9 Therefore, in hyperbolic geometry, the sum of the measures of the angles of every triangle is less than 180.

PROPERTIES OF HYPERBOLIC GEOMETRY

Unlike the Euclidean property, in hyperbolic geometry the angle sum of a triangle is *not* constant. Here the angle sum can be any value between 0° and 180°.

As a consequence of this theorem, the following properties hold in hyperbolic geometry:

1 The angle sum of every quadrilateral is less than 360°.

2 No two parallel lines are everywhere equidistant.

3 No parallelograms, rectangles, or squares exist.

4 If two triangles have their corresponding angles congruent, then the triangles are congruent. In other words, no triangles are similar unless they are congruent. This property suggests that in hyperbolic geometry exact scale models are impossible.

5 The Pythagorean relation does not hold.

AREA IN HYPERBOLIC GEOMETRY

Since squares do not exist in hyperbolic geometry, we can no longer define area in terms of square units. We must, therefore, formulate a new definition of area for this geometry.

Any definition of area must satisfy three properties:

1. To each region there must correspond a uniquely determined positive real number, called its *area*.
2. Congruent regions must have equal areas.
3. If a region R is split into any finite number of subregions $R_1, R_2, \ldots, R_n$, then the area of R is the sum of the areas of $R_1, R_2, \ldots, R_n$.

A function that assigns to each region a specific real number in such a way that properties 1, 2, 3 are satisfied is called an *area function* or *area measure.*

We need one more definition before we can define area in hyperbolic geometry. For the sake of simplicity, we shall consider only the area of a triangle.

DEFINITION The amount by which the sum of the angles of a triangle differs from 180° is called the *defect* of the triangle.

That is, if d is the defect of $\triangle ABC$, then

$$d = 180 - (\angle A + \angle B + \angle C) \qquad \text{(Figure 10.24).}$$

Note that in this definition, $\angle A$, $\angle B$, and $\angle C$ represent the *degree measures* of the angles. Therefore, the defect of a triangle is a *real number* rather than a number of degrees.

FIGURE 10.24

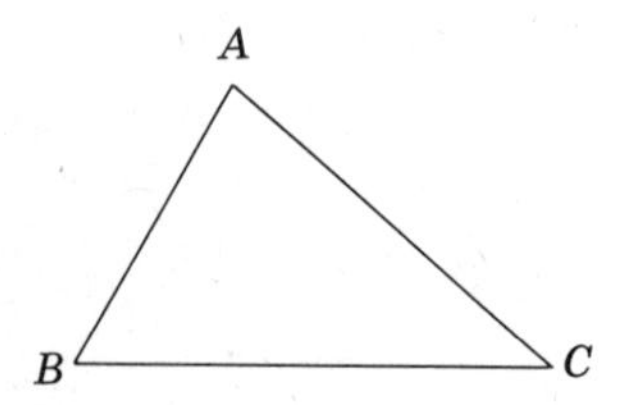

It can be shown that the defect of a triangle satisfies the three properties of an area function. We may therefore regard the defect as an area measure for triangles in hyperbolic geometry. The closer that the sum of the angle measures is to 180, the smaller the area.

CONGRUENCE IN HYPERBOLIC GEOMETRY

The congruence theorems in Euclidean geometry apply equally well to hyperbolic geometry since their proofs do not depend on Euclid's parallel postulate.

There is, however, an additional method for proving triangles congruent in hyperbolic geometry. This method is based on property 4 on page 355; that is, two triangles in the hyperbolic plane are congruent if the three angles of one triangle are, respectively, congruent to the three angles of the other triangle.

ELLIPTIC GEOMETRY

We have already seen that in the selection of an axiom of parallels there are three choices open to us: Through a point not on a given line there is

1 One and only one parallel.

2 More than one parallel.

3 No parallel.

The first choice leads to Euclidean geometry in which the angle sum of a triangle is 180°.

The second choice leads to hyperbolic geometry in which the angle sum of a triangle is less than 180°.

The third choice leads to the elliptic geometry of Riemann in which the angle sum of a triangle is greater than 180°.

Unlike hyperbolic geometry where all Euclidean postulates except the parallel postulate are retained, elliptic geometry requires abandoning not only Euclid's parallel postulate but other postulates as well. Since there are no parallels in elliptic geometry, any postulate that implies the existence of parallel lines has to be abandoned.

Elliptic geometry is incompatible with Euclid's assumption that there exist straight lines of infinite extent. Therefore, in this geometry a distinction is made between "infinite" and "boundless." If we move on a circle, we can keep going around the circle endlessly, yet its length is finite. Lines may be finite though *boundless.* Riemann therefore replaced the Euclidean assumption of the infinitude of a straight line by the postulate that a straight line is *finite and boundless.*

Since the assumption of the infinitude of a straight line is related to the axioms of betweenness, these axioms, too, have to be abandoned.

If two lines cannot be parallel in elliptic geometry, how many points of intersection will they have?

Two answers have been given. The first is that any two lines intersect in exactly *one* point, as in the Euclidean plane. This leads to an elliptic plane called the Cayley-Klein plane, first suggested by Arthur Cayley (1821-1895) and Felix Klein (1849-1925). The geometry of the Cayley-Klein plane is sometimes referred to as *single elliptic geometry.*

The second answer is that two lines intersect in exactly *two* points. This leads to an elliptic plane called the Riemann plane. The geometry of this plane is sometimes referred to as *double elliptic geometry.*

A MODEL FOR ELLIPTIC GEOMETRY

A model of a Riemann plane is the geometry on a sphere. The points on a sphere correspond to the "points" in elliptic geometry. The "lines" of this geometry correspond to the great circles of the sphere. "Segments" are represented by arcs of great circles. "Distance" between two points is represented by the length of the shortest arc of a great circle joining the two points. "Angles" are represented by the spherical angles formed by two great circles, and the measure of an angle is represented by the measure of a spherical angle.

Every true statement about a sphere can be converted into a true statement about a Riemann plane by replacing the term "great circle" with the term "straight line." This is interesting when you consider that elliptic geometry and Euclidean geometry contradict each other in some important ways.

PROPERTIES OF ELLIPTIC GEOMETRY

1. We have seen that there are no parallel lines in this geometry and that any two lines in the Riemann plane meet in exactly two points. These points are opposite ends of a diameter (Figure 10.25).

FIGURE 10.25

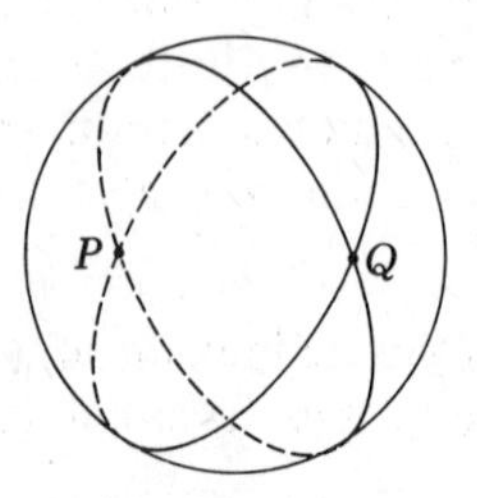

2 Two points do not necessarily determine a unique line. For example, points P and Q, shown in Figure 10.25, lie on infinitely many great circles.

3 All the lines perpendicular to a given line meet at two particular points associated with the given line and known as its *poles* (Figure 10.26). This means that the perpendicular to a line from an external point is not necessarily unique.

FIGURE 10.26

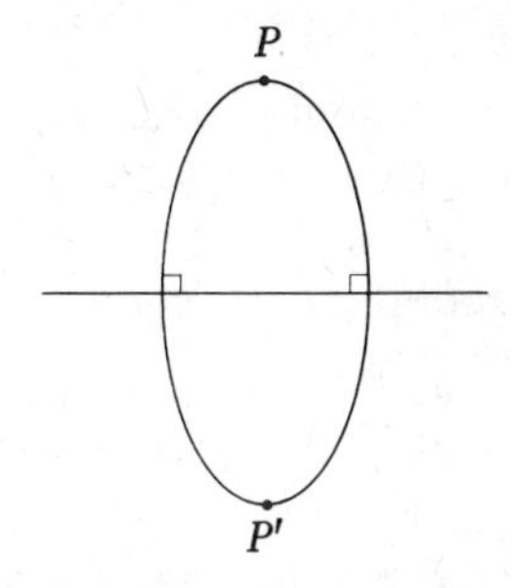

4 Although the "lines" in this geometry are boundless in the sense that they never come to an end at any point, they are nevertheless finite in extent. If the radius of our model sphere is 1, then the maximum possible distance between any two points is π.

5 We said earlier that some of the order properties of Euclidean geometry have to be abandoned in elliptic geometry. For example, if $\overleftrightarrow{AC}$ is a straight line in Euclidean geometry (Figure 10.27) and if B is between A and C, then A is not between B and C, and C is not between A and B. However, on the great circle ABC (Figure 10.28), not only is B between A and C, but A is between B and C, and C is between A and B.

FIGURE 10.27

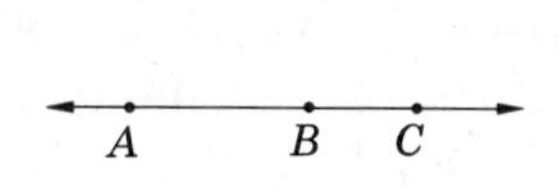

FIGURE 10.28

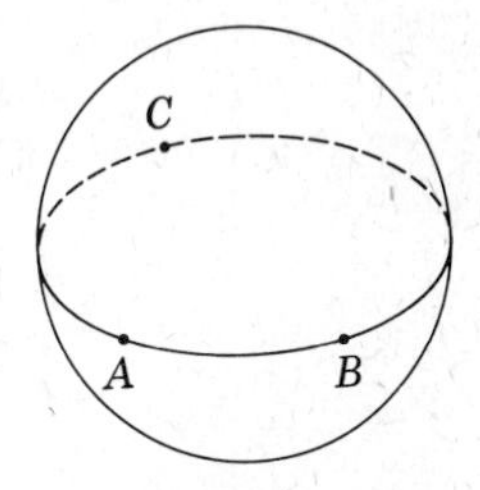

6 We have already said that the sum of the measures of the angles of a triangle in this geometry is greater than 180. Therefore, it is possible for a triangle to contain more than one right angle. In Figure 10.29, if P is the North Pole and A and B are points on the equator that are 90° of longitude apart, then $\triangle ABP$ has three right angles.

FIGURE 10.29

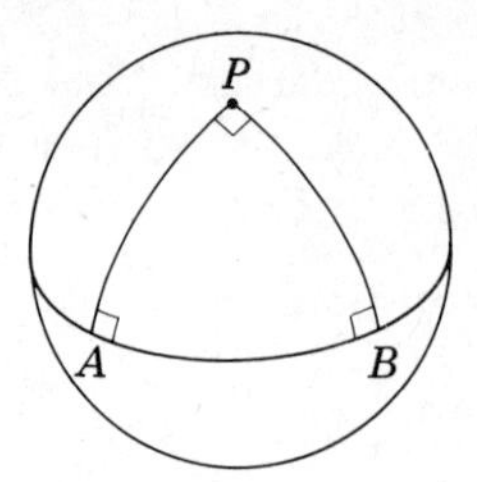

7 Since there exist triangles with more than one right angle, the Pythagorean relation does not hold in this geometry.

8 If the sum of the angle measures of a triangle is always greater than 180, then the sum of the angle measures of a quadrilateral will always be greater than 360. This means that we can have no rectangles or squares in this geometry.

9 The amount by which the sum of the angles of a triangle exceeds 180 is called the *excess* of the triangle. The excess of a triangle is a measure of its area, just as in hyperbolic geometry the *defect* of the triangle is a measure of its area.

10 Two similar triangles in this geometry, as in hyperbolic geometry, are necessarily congruent. This means that two figures cannot have exactly the same shape unless they also have exactly the same size.

CONSISTENCY OF NON-EUCLIDEAN GEOMETRIES

We said that the non-Euclidean geometries are "consistent." How do we know that they are consistent? Can we *prove* that they are?

The way consistency is established is to relate the non-Euclidean geometries to their "models," which satisfy all the axioms of Euclidean geometry except the parallel postulate.

The argument then proceeds as follows. Hyperbolic geometry has a model in Euclidean geometry—the interior of a Euclidean circle. If Euclidean geometry is consistent, then hyperbolic geometry must also be consistent. Similarly, elliptic geometry has a model on a Euclidean sphere. If elliptic "lines" contain a contradiction, then there must be a corresponding contradiction for Euclidean great circles.

What we are really saying is that the consistency of one geometric system implies the consistency of another. Either *both* are consistent or *both* are inconsistent.

We should note that such a "proof" of consistency is not absolute but *relative*. Consistency in one system is contingent on consistency in another system. This means that we cannot really be sure that the non-Euclidean geometries are consistent. We can only be sure that they are *as consistent* as Euclidean geometry. To be able to say even this much, we must first assume the consistency of the theory of logical inference used in developing these geometries.

SOME GENERAL COMMENTS

The creation of non-Euclidean geometries had profound implications not only for geometry but for all mathematics. It demolished the notion that Euclid's axioms were immutable and that axioms are "self-evident truths." It liberated mathematics from dependence on the physical world and revealed most dramatically the abstract nature of mathematics. It established mathematics as a purely abstract science and led to the development of such abstract concepts as complex numbers, n-dimensional spaces, postulate sets and their properties (axiomatics), and symbolic logic.

We should note that even though Euclidean, hyperbolic, and elliptic geometries have different properties, they almost coincide if we confine ourselves to a small portion of the plane. The geometry of a sphere in the neighborhood of a point P nearly coincides with the geometry of the point P on the plane tangent to the sphere at P. Very small triangles in any of these non-Euclidean geometries have angle sums that almost equal $180°$. So even if astronomical space may be best described as non-Euclidean, physical space in our immediate environment may be described as Euclidean.

10.6 Ideal Points

In Euclidean geometry, two lines in a plane either intersect or are parallel. Therefore, any complete discussion involving two lines in a plane requires us to draw conclusions first, say, when the lines intersect and then when they are parallel. Such a procedure is not as convenient as it would be if both cases could be treated by one formulation. If we could modify our geometry so that *every* pair of lines intersect, this inconvenience would be eliminated.

We can so modify our geometry if we add to the ordinary points on each line just one more point called an *ideal point*, or a *point at infinity*. This point is considered to belong to all the lines parallel to a given line and to no other lines. By postulating the existence of *ideal points*, we have modified our geometry so that every pair of lines in the plane intersect in one point. If the lines are not parallel, they intersect in an "ordinary" point. If the lines are parallel, they intersect in an *ideal point*.

Our willingness to lessen our difficulties in what must appear to be an arbitrary fashion should not surprise us. We often do this in mathematics. When we found the system of natural numbers inadequate, didn't we arbitrarily decide to enlarge it by the addition of new numbers? Isn't this the way all our number systems were created? Now we find the points in the Euclidean plane inadequate, so we decide to enlarge it by the addition of ideal points. The only restriction on our freedom to do this is that the properties we choose to assign to these ideal points must be consistent in their relations to ordinary points and to each other.

From the standpoint of intuition, the existence of ideal points should not be difficult to accept.

Think of a straight railroad track on a level plane (Figure 10.30). Although we know the tracks to be parallel, they still appear to converge in a common point at the horizon. This common point at the horizon is what we defined to be an *ideal point*, or a *point at infinity.* All lines running in the same direction as these two tracks pass through the same ideal point. Every direction in the plane has a different ideal point.

FIGURE 10.30

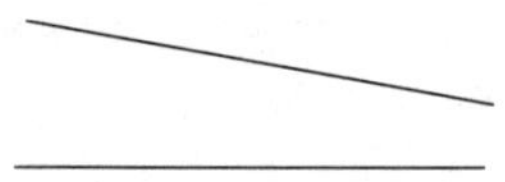

Drawing in another way on our intuition, think of a line l, a point P not on l, and a line m that passes through P and intersects l at A (Figure 10.31). Rotate m clockwise about P. As m approaches the parallel position m' as a limit, the point of intersection A will recede to infinity. This suggests the idea that parallel lines meet at a *point at infinity.*

FIGURE 10.31

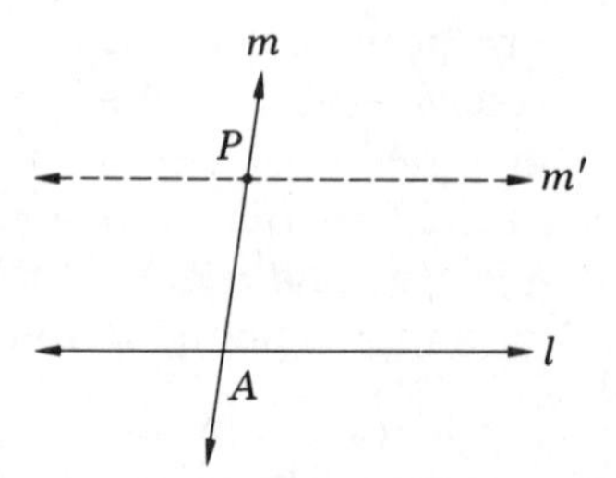

Since a line recedes to infinity in two directions, should we not add *two* ideal points to each line—one for each direction? Let us see what happens if we do this.

Assume that every line l has two ideal points $P_{-\infty}$ and P_{∞}, and let A be a point not on l (Figure 10.32). Then a line parallel to l and passing through A would intersect l at two points, $P_{-\infty}$ and P_{∞}. In fact, since l would share the two points $P_{-\infty}$ and P_{∞} with *every* line parallel to it, infinitely many lines would pass through these two points. But this contradicts the property that through any two distinct points one and only one line may be drawn. Therefore, every line can have one and only one ideal point.

FIGURE 10.32

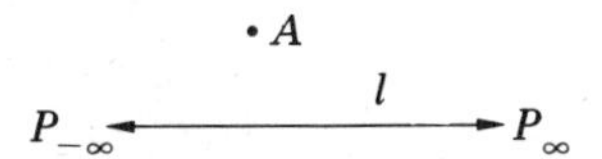

This means that if you look down a straight road in each direction, say east and west, the two horizons are at the same point. It also means that no matter in which direction you travel on a straight road, east *or* west, you will ultimately arrive at the same point, its ideal point. This situation can be visualized if we think of a straight line as a huge circle with its ends connected at the ideal point as shown in Figure 10.33.

FIGURE 10.33

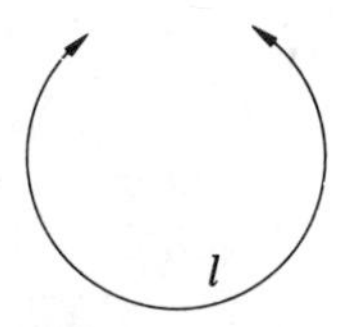

Just as we added to the ordinary points on each line a single ideal point, we shall now add to the ordinary lines in a plane a single *ideal line.* This *ideal line* contains all the ideal points in the plane, but it contains no ordinary points.

By adding ideal points and an ideal line to the Euclidean plane, we have created the *real projective plane* and a new kind of geometry, *projective geometry.* In this geometry, we have simplified certain incidence relations by removing the exceptions caused by parallelism. Thus, in projective geometry, any two distinct lines always meet in a unique point, without exception.

The concepts of an ideal point and an ideal line are not new. They go back to Johannes Kepler (1571-1630), who postulated their existence in connection with his *principle of continuity.* He thought of a line as closed at infinity, and two parallel lines as intersecting at infinity. Although these ideas did not receive wide and immediate acceptance when Kepler first suggested them, they eventually led to the creation of *projective geometry.*

Chapter 11

PROBABILITY

Objectives

At the completion of this chapter, you should be able to:

1. Define and illustrate *sample space*, *sample point*, *event*, *outcome*.
2. Define *probability* of an event E and its *complement* $\overline{E}$.
3. Solve problems involving the calculation of probabilities and their complements.

4. Define *odds for* and *odds against.*
5. Solve problems involving the calculation of odds.
6. Define *mathematical expectation.*
7. Solve problems involving the calculation of mathematical expectation.
8. State and illustrate the Fundamental Counting Principle.
9. Solve problems involving the Fundamental Counting Principle.
10. Define and give examples of a *permutation;* of a *combination.*
11. Distinguish between a permutation and a combination.
12. Solve problems involving the calculation of permutations and combinations.
13. State what is meant by *mutually exclusive events* and identify such events.
14. State the *addition property* for any two events and for mutually exclusive events.
15. Solve problems involving the application of the addition property.
16. State what is meant by *independent events* and identify such events.
17. State the *multiplication property* for two or more independent events.
18. Solve problems involving the application of the multiplication property.
19. Explain the meaning of *conditional probability.*
20. Solve problems involving the calculation of conditional probabilities.

Prerequisites You should understand the meaning of ratio.

INTRODUCTION

If two card players are forced to quit before their game is finished, how should the pot be divided?

In the middle of the seventeenth century, a French gambler and member of the aristocracy, the Chevalier de Méré, asked this question of the devoutly religious mathematician, Blaise Pascal (1623-1662), who in turn discussed it with another mathematician, Pierre de Fermat (1601-1665). In the correspondence that followed, the *theory of probability* was born. The two mathematicians decided that the player with the greater "probability" of winning should receive the greater share, and they proceeded to formulate the basic methods of determining each player's probability of winning.

Today, the theory of probability has far greater depth and significance than it did when it was first originated. Today, it has extremely important applications in such diverse areas as determining life expectancy tables on which life insurance premiums are based; in predicting the weather; in checking the reliability of equipment; in predicting elections; in evaluating the significance of medical research; in determining the popularity of television programs; in the study of military strategy; and in the study of physics, genetics, economics, and psychology.

Actually, the first book on the subject of probability was written by the Italian mathematician Girolamo Cardano (1501-1576) who wrote *The Book on Games of Chance* in which he discusses the probabilities of winning in various games of chance and offers advice on how to cheat and how to detect cheating. But whereas Cardano solved just a few problems of probability, Pascal aimed to create a whole new science—"the mathematics of chance."

The mathematics of probability becomes operative when we are called upon to make decisions or judgments on the basis of incomplete or uncertain information. Every day we have occasions to make such judgments: whether to wear a raincoat, as members of a jury, in a card game, in making an investment, or in selecting a winning team.

It is interesting to observe that we are able to predict the precise moment of an eclipse of the moon, and the precise second when a space ship will land on the moon, but we are unable to predict the outcome of tossing a coin or of throwing a pair of dice. In the case of an eclipse or of a moon landing, we have all the information needed to make a prediction. In the case of the coin or dice, we lack the necessary information, and so we ascribe their behavior to "chance." But even in the case of chance, there exists a certain regularity, and this regularity has been formulated in the study of *probability*. In this chapter we shall define precisely the meaning of probability and how we can use its ideas to measure the likelihood of an occurrence.

11.1 Probability in the Context of Experiments

We shall find it useful to discuss *probability* in the context of "experiments" and their outcomes. We shall call any set of possible outcomes an "event." Familiar experiments that we shall often refer to involve standard decks of 52 well-shuffled cards, coins, "honest" dice, and boxes containing objects such as checkers or discs of different colors. In these experiments we shall be interested in determining the chance that some event from among a set of *equally likely events* will occur or will not occur. The measure of this chance occurrence is called its *probability*.

Consider a coin tossed at random, where by "random" we mean that the coin is allowed to fall freely, without any interference. If we disregard the possibility of its landing on its edge, the coin can land in only one of two possible ways: either heads or tails. That is, there is one chance in two of getting a head, and there is one chance in two of getting a tail. We describe these "chances" by saying that the *probability* of obtaining a head is one out of two, or $\frac{1}{2}$, and the probability of obtaining a tail is likewise $\frac{1}{2}$.

As another example, suppose that a box contains 3 red discs and 5 blue discs. If we withdraw one of these 8 discs from the box, at random, what is the probability that the disc we withdraw will be a red one? Since there are 3 red discs in a box containing 8 discs altogether, and since we assume that each of the 8 discs is *equally likely to be drawn*, then we have 3 ways out of 8 possible ways to draw a red disc. We therefore say that the probability of drawing a red disc is $\frac{3}{8}$. By the same reasoning, the probability of drawing a blue disc is $\frac{5}{8}$.

11.2 Sample Space

A set of possible outcomes of an experiment is called a *sample space* of the experiment if every outcome is equally likely and exactly one outcome must occur. When we toss a coin, the possible outcomes are a head (H) or a tail (T). A sample space of this experiment is, therefore, the set $\{H, T\}$. Every element of a sample space is called a *sample point*. The sample space $\{H, T\}$ has two sample points: H and T.

If a nickel and dime are tossed, the possible outcomes are:

Nickel	*Dime*
H	H
H	T
T	H
T	T

A sample space of this experiment is, therefore, $\{HH, HT, TH, TT\}$ with four sample points.

An *event* may be regarded as a subset of a sample space.

11.3 The Probability of an Event

DEFINITION If, of n equally likely outcomes, m are favorable to the happening of an event E, then the *probability* of E happening is

$$P(E) = \frac{m}{n}.$$

That is, *probability* is a *ratio* of the number of successful outcomes to the number of possible outcomes.

Example 1 If a box contains 10 slips of paper on which are written the names of 7 men and 3 women, the probability of drawing at random a man's name is $P\,(\text{man}) = \frac{7}{10}$, and the probability of drawing a woman's name is $P\,(\text{woman}) = \frac{3}{10}$.

We shall use the symbol $\overline{E}$ to mean "not E"; that is, the event E *not* happening. We refer to $\overline{E}$ as the *complement of E*. The probability of event E not happening is denoted by $P(\overline{E})$.

If there are n possible outcomes for an event E of which m are favorable to the happening of the event, then there are $n - m$ ways of the event *not* happening. It therefore follows from the definition of probability that

$$P(\overline{E}) = \frac{n-m}{n} = 1 - \frac{m}{n} = 1 - P(E).$$

Therefore, $P(\overline{E}) + P(E) = 1$.

That is, *the sum of the probabilities of an event E and its complement $\overline{E}$ is always equal to 1.*

Example 2 In Example 1, the probability of selecting a man's name *and* the probability of *not* selecting a man's name is

$$P\,(\text{man}) + P\,(\overline{\text{man}}) = \frac{7}{10} + \frac{3}{10} = 1.$$

Example 3 If there are 4 black checkers and 9 red checkers in a box, what is the probability that a checker removed from the box at random will be:

a) a red checker? *Answer:* $P(R) = \frac{9}{13}$

b) a black checker? *Answer:* $P(B) = \frac{4}{13}$

c) not a red checker? *Answer:* $P(\overline{R}) = 1 - \frac{9}{13} = \frac{4}{13}$

d) a white checker? *Answer:* $P(W) = \frac{0}{13} = 0$

e) either a black or a red checker? *Answer:* $P(B \text{ or } R) = \frac{13}{13} = 1$

Note that the solutions to parts d and e illustrate that the probability of an event that cannot happen is 0, while the probability of an event that is certain to happen is 1.

Example 4 If two coins are tossed, what are the probabilities of getting (a) 2 heads, (b) 1 head and 1 tail, (c) 2 tails?

Answer We note that there are four different, but equally likely, ways in which the two coins can fall:

$$HH, \quad HT, \quad TH, \quad \text{and} \quad TT.$$

Of these possible outcomes, there is only 1 way of obtaining 2 heads, 2 ways of obtaining 1 head and 1 tail, and only 1 way of obtaining 2 tails. Therefore, the answers are (a) $\frac{1}{4}$, (b) $\frac{2}{4} = \frac{1}{2}$, (c) $\frac{1}{4}$.

Note that it does not matter whether the two coins are tossed simultaneously or consecutively since each coin falls independently of the other. In either case the result is the same.

Comments

1 *"Probability" has meaning only as it relates to a large number of trials of an experiment under the same conditions each time. The statement that "the probability that a coin will come up heads is* $\frac{1}{2}$*" means that in a large number of trials we expect to get heads about half of the time.*

If a coin is tossed ten times, it can happen that a head will come up nine of those ten times. However, the more times the coin is flipped, the better will the ratio of the number of heads showing to the total number of tosses approximate the ratio $\frac{1}{2}$.

2 *The definition of* probability *does not tell us how to determine whether the possible outcomes of an experiment are equally likely. That must be settled by nonmathematical means such as the physics of the situation, past experience, or even faith or intuition.*

3 *Can we say that the probability of flying safely from*

Philadelphia to Tucson is $\frac{1}{2}$ *because there are two possible outcomes, flying safely and not flying safely, and of these two, only one is favorable? The answer is no because the two possible outcomes, flying safely and not flying safely,* are not equally likely.

4 *The definition of* probability *given here is sometimes called* a priori probability *because it can be determined before or even without experimentation. But suppose we wish to determine the probability that a man of age thirty will be alive at age fifty. How does our definition of*

$$\frac{\text{favorable outcomes}}{\text{total outcomes}}$$

apply here? In this kind of situation, insurance companies, for whom life expectancy probabilities are of vital importance, use the concept of relative frequency *as their definition of probability. Over a period of many years, they collected a great deal of data about the birth and death records of people. Suppose they found that of* 100,000 *men alive at age thirty,* 75,000 *were still alive at age fifty. They then took the ratio*

$$\frac{75{,}000}{100{,}000} \quad \text{or} \quad 0.75$$

as the probability *that a man of age thirty will be alive at age fifty. This ratio is called the* statistical probability. *This definition of probability says, in effect, that the probable outcome is that which usually happens as reflected in a large collection of data. It is possible for the statistical probability of an event to change as more data is accumulated or, as in the present example, if medical advances succeed in expanding people's life spans. A priori probability, however, is not subject to change; the probability of tossing a head with one throw of a coin will always remain* $\frac{1}{2}$.

EXERCISE SET 11.1–11.3

1. What is meant by the *probability* of an event?
2. The probability of event E occurring is one in a billion.
   a) Is this event impossible? b) Is this event improbable?
3. A box contains 5 blue discs, 3 yellow discs, and 9 brown discs. If one disc is drawn from the box, what is the probability that it is:
   a) A blue disc? b) A yellow disc? c) A brown disc?
   d) A blue or a brown disc? e) An orange disc?
   f) A blue, yellow, or brown disc?

4. In a regular deck of 52 cards, what is the probability of obtaining on a single draw:
   a) The ace of spades? b) A number 5 card? c) A black card?
5. What is the probability of obtaining on a single toss of a die:
   a) An even number? b) An odd number?
   c) A number less than 2? d) A number greater than 5?
   e) The number 6? f) A number different from 6?
   g) A prime number? h) A composite number?
6. If the probability of winning a prize is 0.02, what is the probability of not winning a prize?
7. Tabulate a sample space for an experiment involving the tossing of a penny three times.
8. List a sample space for each of the following experiments.
   a) Three coins are tossed simultaneously.
   b) A coin is tossed, then a die is tossed.
   c) A box contains 2 white checkers and 5 black checkers. You draw one checker from the box at a time, replace the checker in the box, then draw a second checker. (*Suggestion* Call the white checkers W_1 and W_2 and the black checkers B_1, B_2, B_3, B_4, B_5.)
9. Three coins are tossed. What is the probability of getting:
   a) 3 heads? b) 1 head and 2 tails?
   c) 2 tails and 1 head? d) More than 1 head?
   e) No heads? f) No tails?
10. If three coins are tossed, what is the probability of not throwing 2 heads and 1 tail?
11. Suppose you have two sets of five tickets, each set numbered 1 through 5, and one card is drawn from each set. What is the probability that the sum of the numbers is:
    a) 9? b) 10? c) Greater than 5? d) Less than 5?
12. What, if anything, is wrong with the following reasoning.
    The probability that you will earn a million dollars next year is $\frac{1}{2}$ since there are two possible outcomes: you will either earn or not earn a million dollars next year, and only one of these possibilities is favorable.

11.4 Odds

We often speak of the *odds* in favor of, or against, some event occurring. We may hear that, in a boxing match about to take place, the odds in favor of the champion to retain his title are "5 to 1". What does this statement mean?

If a box contains 1 red marble and 5 green marbles, then the probability of drawing a green marble is $\frac{5}{6}$, and the probability of *not* drawing a green marble is $\frac{1}{6}$. That is,

$$P(E) = \frac{5}{6} \quad \text{and} \quad P(\overline{E}) = \frac{1}{6}.$$

The *odds in favor* of drawing a green marble is *the ratio of the probability that a green marble will be drawn to the probability that a green marble will not be drawn.* Thus, the *odds in favor* of drawing a green marble are the ratio

$$\frac{P(E)}{P(\overline{E})} = \frac{\frac{5}{6}}{\frac{1}{6}} = \frac{5}{1}, \text{ or 5 to 1.}$$

The *odds against* drawing a green marble are the ratio

$$\frac{P(\overline{E})}{P(E)} = \frac{\frac{1}{6}}{\frac{5}{6}} = \frac{1}{5}, \text{ or 1 to 5.}$$

Thus, the statement, "the odds in favor of the champion to retain his title are 5 to 1," means that the probability that the champion will win is five times as great as the probability that he will lose.

DEFINITION If, of n equally likely outcomes of an event E, m are favorable, then the *odds in favor* of E are

$$\frac{P(E)}{P(\overline{E})} = \frac{\frac{m}{n}}{\frac{n-m}{n}} = \frac{m}{n-m},$$

and the *odds against* E are

$$\frac{P(\overline{E})}{P(E)} = \frac{\frac{n-m}{n}}{\frac{m}{n}} = \frac{n-m}{m}.$$

Since m is the number of favorable ways E can occur and $n - m$ is the number of unfavorable ways, the definition just given is equivalent to saying that

$$\textit{odds for} \text{ an event} = \frac{\text{number of favorable outcomes}}{\text{number of unfavorable outcomes}}, \text{ and}$$

$$\textit{odds against} \text{ an event} = \frac{\text{number of unfavorable outcomes}}{\text{number of favorable outcomes}}.$$

Note that the *odds for* ratio is the *reciprocal* of the *odds against* ratio. Also, since $P(E) + P(\overline{E}) = 1$, we can say that if the probability of an event is p, then the *odds for* its occurrence are p to $1 - p$, and the *odds against* its occurrence are $1 - p$ to p.

Example A box contains 15 slips of paper on which are written the names of 12 boys and 3 girls.

a) What are the odds favoring the random selection of a boy's name?

b) What are the odds against the random selection of a boy's name?

Solution a) *odds for* $= \dfrac{\text{favorable outcomes}}{\text{unfavorable outcomes}} = \dfrac{12}{3} = \dfrac{4}{1}$, or 4 to 1.

b) *odds against* $= \dfrac{\text{unfavorable outcomes}}{\text{favorable outcomes}} = \dfrac{3}{12} = \dfrac{1}{4}$, or 1 to 4.

11.5 Mathematical Expectation

Closely related to *odds* is the notion of *mathematical expectation.* Suppose a man stands to win \$12 if he throws a 4 with a single throw of a die. Then we say that his *mathematical expectation* is the product of the probability that he will throw a 4 and the \$12 he stands to win; that is, $\frac{1}{6}(\$12) = \$2$. This mathematical expectation of \$2 is his expected average winnings per throw over a long sequence of trials.

Now suppose that the man will win \$6 if he rolls a 3, 4, 5, or 6 with a single die, but he will pay \$9 if he rolls a 1 or 2. In this case, his mathematical expectation is the sum of the expectations of each outcome:

$$\frac{1}{6}(\$6) + \frac{1}{6}(\$6) + \frac{1}{6}(\$6) + \frac{1}{6}(\$6) + \frac{1}{6}(\$^-9) + \frac{1}{6}(\$^-9) = 4 - 3 = \$1.$$

In a *fair* game, the mathematical expectation of each of the players should be zero. In both of the above examples, our man had an advantage.

DEFINITION If $w_1, w_2, \ldots, w_n$ are winnings for a player, and $p_1, p_2, \ldots, p_n$ are the corresponding probabilities for each of these winnings, then the *mathematical expectancy* for the player is

$$w_1p_1 + w_2p_2 + \cdots + w_np_n.$$

EXERCISE SET 11.4, 11.5

1 What are the odds in favor of obtaining:
a) A head and a tail in a single toss of two coins?
b) Two tails in a single toss of two coins?
c) Two heads in a single toss of three coins?
d) A king in a single draw from a deck of 52 cards?
e) A jack or queen in a single draw from a deck of 52 cards?

2 What are the odds against obtaining:
a) Two tails in a single toss of two coins?
b) A queen in a single draw from a deck of 52 cards?
c) Two aces in two draws from a deck of 52 cards if the first card is replaced before the second is drawn? If there is no replacement?

3 If a single die is tossed:
a) What are the odds in favor of rolling an even number? an odd number? a prime number? a composite number?
b) What are the odds that neither a prime nor a composite number will be face up?

4 What are the odds against an event whose probability is $\frac{3}{8}$?

5 If two dice are rolled, find the odds against a sum:
a) Of 5. b) Of 7 or 11. c) Less than 5. d) Greater than 5.

6 A player gets \$2 if a coin comes up heads in a single throw. What is his mathematical expectation?

7 A player receives \$3 if a coin, in a single throw, comes up heads, and pays \$3 if the coin comes up tails. Find his mathematical expectation.

8 If a player is paid the number of dollars equal to the number showing on the face of a die after a single throw, what is his mathematical expectation?

9 A man wins \$5 if he throws a total of 7 or a total of 11 with a pair of dice, and loses \$2 if he does not. Find his mathematical expectation.

10 A student is promised \$5 if he gets a B in a course, \$10 if he gets an A, and nothing if he receives any other grade. What is his mathematical expectation if the probability for him receiving a B is 0.34 and the probability for an A is 0.16?

11 A thousand raffle tickets are to be sold with the winner receiving a car worth \$4000. How much should you pay for a ticket if the raffle is fair?

11.6 *The Fundamental Counting Principle*

Before we can predict the likelihood that an event will occur, we have to know how to calculate the number of different ways the event *can* occur. The probability problems we encountered so far involved small numbers so it was easy to determine sample spaces and sample points. In real-life situations the numbers may not be that small. For example, the sample space for all possible 13-card bridge hands, which can be dealt with an ordinary deck of 52 cards, contains more than 635 *billion* possibilities. Obviously, we cannot determine this number of possibilities by listing them. Even if we could, how can we be sure that we didn't repeat or overlook some of the possibilities. It is clear, therefore, that to compute probabilities we need efficient and systematic ways of determining the number of possible outcomes in the experiment. One such way is provided by a basic mathematical principle called the *Fundamental Counting Principle.*

A man has three shirts (s_1, s_2, s_3) and two ties (t_1, t_2). How many dressing outfits can he make with his shirts and ties?

To examine all the possibilities, it is helpful to refer to a *tree diagram* like that in Figure 11.1. The diagram shows that we start with three possibilities: the man can wear s_1, s_2, or s_3. Each of these three possibilities grows into two branches, since each shirt can go with either of the two ties.

FIGURE 11.1

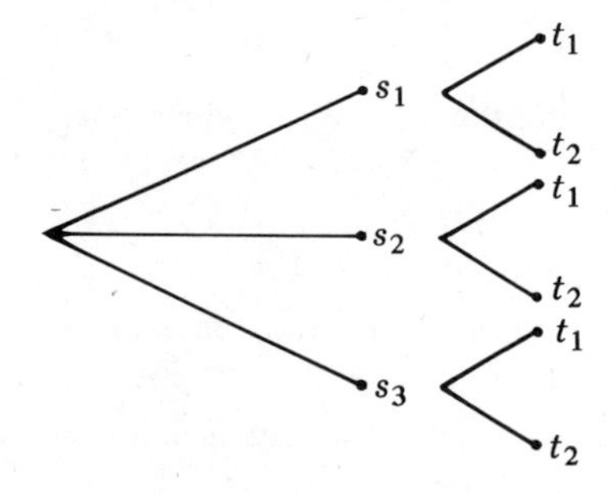

Counting from left to right in the diagram, we find there are altogether 6 different paths along the branches of the tree; that is, the sample space has 6 sample points:

$$\{(s_1, t_1), (s_1, t_2), (s_2, t_1), (s_2, t_2), (s_3, t_1), (s_3, t_2)\}.$$

So, the 3 shirts and the 2 ties produce $3 \times 2 = 6$ possible outfits. If we reverse the order by starting with the ties and branching out to the shirts, we will come out with the same result: 6 possible outfits.

Now let us consider by means of a tree diagram, shown in Figure 11.2, all the possible outcomes when 3 coins are tossed. Starting at the left of the diagram, we see that tossing the first coin results in two possible outcomes: heads (H) or tails (T).

FIGURE 11.2

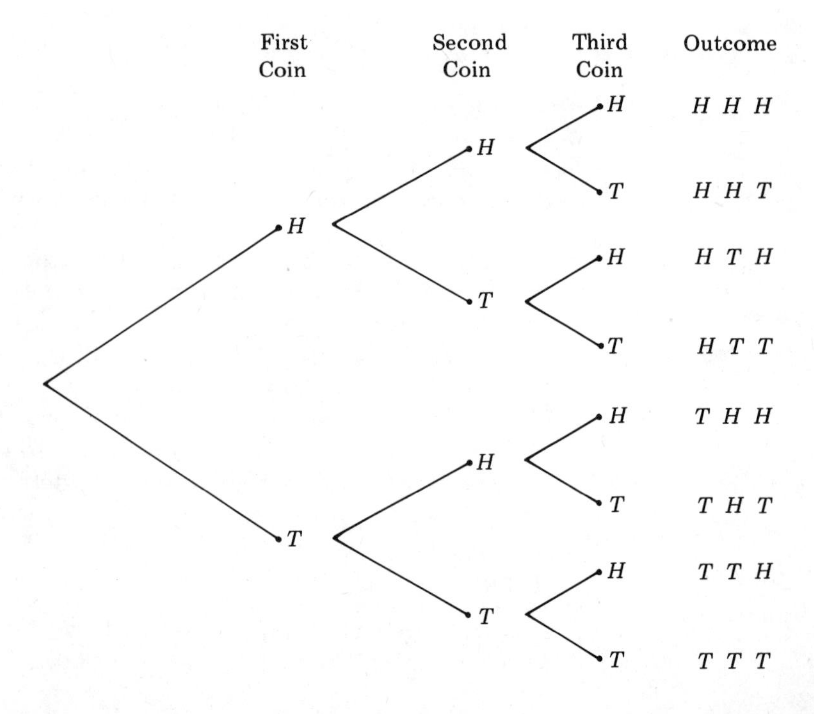

Each of these 2 possibilities can then be associated with each of the 2 possible outcomes which result from tossing the second coin. So tossing the first and second coins results in $2 \times 2 = 4$ possible outcomes.

Each of these 4 possibilities can, in turn, be associated with each of the two possibilities resulting from tossing the third coin. Tossing the three coins results, therefore, in $2 \times 2 \times 2 = 8$ possible outcomes for the experiment. These outcomes can be seen in Figure 11.2 as we traverse the different paths along the branches of the tree.

The sample space for this experiment is, therefore,

$$S = \{HHH, HHT, HTH, HTT, THH, THT, TTH, TTT\}.$$

These two examples suggest a general principle of counting that is fundamental to the counting techniques we shall discuss shortly.

THE FUNDAMENTAL COUNTING PRINCIPLE If a selection consists of two separate steps, of which the first can be made in m ways and the second in n ways, then the whole selection can be made in $m \times n$ ways.

Example 1 If a girl has 2 sweaters and 3 skirts, then she has 2×3, or 6, different outfits she can wear.

Stated in terms of sample spaces, the Fundamental Counting Principle asserts that if event A has m distinct outcomes and event B has n distinct outcomes, and if A and B are independent of each other, then the sample space involving A and B has $m \times n$ possible outcomes.

The Fundamental Counting Principle also holds for more than two steps, and can be generalized as follows:

> If a selection consists of r different steps of which the first can be made in n_1 ways, the second in n_2 ways, ..., and the rth in n_r ways, then the whole selection can be made in $n_1 \times n_2 \times \ldots \times n_r$ ways.

Example 2 A restaurant offers a choice of 2 appetizers, 4 entrees, and 5 desserts. How many different meals can be made of these choices?

Solution Applying the Fundamental Counting Principle, we have $2 \times 4 \times 5 = 40$ different meals.

Example 3 A class consists of 22 students. In how many ways can the class select a president, vice-president, and secretary, if no student can hold more than one office?

Solution The selection of the officers consists of three steps:

First step Select a president. (22 choices)
Second step Select a vice-president. (21 choices)
Third step Select a secretary. (20 choices)

Note that after each step there is one less choice to choose from for the next step. Applying the Fundamental Counting Principle, we have

$$22 \times 21 \times 20 = 9240$$

ways of selecting the three officers.

11.7 Permutations

How many two-digit numbers can be formed from the three digits 2, 4, and 6? The question asks us, in effect, to determine the number of different arrangements we can make of the three given digits if we take two at a time. We can make the following six arrangements: 24, 26, 42, 46, 62, and 64. Each of these arrangements is called a *permutation* of the three given digits taken two at a time. The symbol for "the number of permutations of 3 things taken 2 at a time" is "${}_3P_2$".

Generally, if r different elements are selected from a set of n elements, then any arrangement of these r elements is called a *permutation.* We now wish to find an efficient method for determining the number of possible permutations we can obtain from n elements taken r at a time, without the need to write out each permutation. We can do this by using the Fundamental Counting Principle in the same way that we applied it to the problem of selecting the class officers as shown in Table 11.1.

TABLE 11.1

Selection	1	2	3	4	...	r
Number of Choices	n	$n-1$	$n-2$	$n-3$	...	$n-(r-1)$ or $n-r+1$

Note that in Table 11.1 the first selection is made from the whole set of n elements; the second selection is made from the $n - 1$ elements, which remain after the first selection has been made; the third selection is made from the $n - 2$ elements, which remain after the first two selections have been made; the fourth selection is made from the $n - 3$ elements, which remain after the first three selections have been made; ...; and the rth selection is made from the $n - (r - 1)$ or $n - r + 1$ elements, which remain after the first $r - 1$ selections have been made.

By the Fundamental Counting Principle, the product of these r factors represents the number of permutations of n things taken r at a time. That is, *the number of permutations of n things taken r at a time is*

$$_nP_r = n(n-1)(n-2)(n-3)\cdots(n-r+1)$$

where r is equal to or less than n.

Example 1 How many different arrangements can be made of the set of vowels {a, e, i, o, u}, if they are taken three at a time?

Solution We are asked to find the number of permutations of 5 things taken 3 at a time:

$$_5P_3 = 5 \times 4 \times 3 = 60.$$

Example 2 In how many ways can four people be arranged in a straight line?

Solution Here, we are required to find the number of permutations of 4 things taken 4 at a time:

$$_4P_4 = 4 \times 3 \times 2 \times 1 = 24.$$

Because the products of consecutive integers, such as $4 \times 3 \times 2 \times 1$ often appear in permutation problems, a simplified notation has been adopted for writing them. The product $4 \times 3 \times 2 \times 1$ is written "4!" and is read "4 factorial." Similarly,

$$3! = 3 \times 2 \times 1, \quad 7! = 7 \times 6 \times 5 \times 4 \times 3 \times 2 \times 1, \quad \text{and} \quad 1! = 1.$$

In the case of zero factorial, it is agreed that $0! = 1$. More generally,

$$n! = n(n-1)(n-2)(n-3)\cdots(3)(2)(1)$$

for any positive integer n.

Example 2 illustrates the fact that *the number of permutations of n things taken all at a time is n!; that is,*

$$_nP_n = n!.$$

Comments

1 *The formula for $_nP_n$ is the special case of the formula for $_nP_r$ where $r = n$ and where the last factor is $n - r + 1 = n - n + 1 = 1$.*

2 *Note that*

$${}_5P_3 = 5 \times 4 \times 3 = \frac{5 \times 4 \times 3 \times 2 \times 1}{2 \times 1} = \frac{5!}{(5-3)!}, \quad \text{and}$$

$${}_7P_2 = 7 \times 6 = \frac{7 \times 6 \times 5 \times 4 \times 3 \times 2 \times 1}{5 \times 4 \times 3 \times 2 \times 1} = \frac{7!}{(7-2)!},$$

and more generally,

$${}_nP_r = \frac{n!}{(n-r)!}.$$

This formula is equivalent to the formula given on page 379 and provides another way of evaluating ${}_nP_r$.

Example 3 How many three-digit numbers can be formed from the digits 2, 4, 6, 8, 9?

Solution 1 ${}_5P_3 = 5 \times 4 \times 3 = 60.$

Solution 2 $${}_5P_3 = \frac{5!}{(5-3)!} = \frac{5 \times 4 \times 3 \times 2 \times 1}{2 \times 1} = 60.$$

EXERCISE SET 11.6, 11.7

1 How many different outfits, each consisting of one skirt and one sweater, can be made out of 3 skirts and 4 sweaters?

2 How many different outfits can be made out of four pairs of slacks, two jackets, and three shirts if each outfit consists of a pair of slacks, a jacket, and a shirt?

3 How many numbers of three or more digits, with no digit repeating, can be formed from the digits 1, 2, 3, 4, 5, 6?

4 A committee consists of four men and six women. How many subcommittees of one man and one woman can be formed?

5 A baseball team consists of nine players. If the catcher and pitcher must bat eighth and ninth, in that order, how many batting orders are possible?

6 In how many different ways can six people be seated in a row of six chairs? In a row of eight chairs?

7 How many different arrangements can be made of the elements of the set $\{\rightarrow, \cap, ?, >, *\}$ if they are taken:
a) One at a time? b) Three at a time? c) Five at a time?

8 In how many ways can ten candidates be listed on a ballot?

9 How many different seating arrangements can a teacher have for ten students in fifteen chairs?

10 In how many different ways can the master of ceremonies of a five-act show arrange his acts?

11 Find the value of each of the following.

a) 4! b) 8! c) $\frac{9!}{3!}$ d) ${}_5P_1$

e) ${}_6P_5$ f) ${}_8P_8$ g) ${}_9P_0$

12 Solve for n.

a) ${}_nP_1 = 24$ b) ${}_nP_2 = 56$ c) ${}_nP_3 = 24$

13 In the decimal numeration system:
a) How many two-digit numbers are possible?
b) How many three-digit numbers are possible?

14 How many outcomes are possible in the experiment of tossing a coin six times?

15 How many numbered tickets, each containing a three-digit number, can be printed from the set of digits $\{0, 1, 2, 3, 4, 5, 6, 7, 8, 9\}$?

16 How many seven-digit telephone numbers are possible if the first digit cannot be zero?

17 How many *distinguishable* arrangements are there of the letters in each of the following words?

a) pop b) area c) radar

[*Note* Repetition of letters in a word alters the number of *distinguishable* arrangements we can make of the letters in the word. For instance, the number of *distinguishable* arrangements of the letters in the word *book* is not ${}_4P_4 = 24$, but $\frac{{}_4P_4}{2!}$, or 12:

book boko bkoo koob kobo kboo
ookb oobk okob okbo obok obko

Justify the rule that to find the number of distinguishable arrangements of n letters (1) when there are r repetitions of the same letter, we divide by $r!$. (2) when there are r repetitions of one letter and s repetitions of another letter, we divide by $r!$ and by $s!$.]

18 If a test consists of 10 true-false questions:
a) In how many different ways is it possible to answer all the questions?
b) In how many ways can a student get 8 or more correct answers?

19 Four couples have theatre tickets for eight seats in a row. In how many different ways can they be seated:
a) So that a man alternates with a woman?
b) So that husband and wife alternate?
c) So that men sit together and women sit together?

11.8 *Combinations*

Imagine a contest in which there are five finalists, A, B, C, D, E, and the judges are asked to select a winner and a runner-up. The number of ways in which the winner and the runner-up can be selected is

$${}_5P_2 = 5 \times 4 = 20: \quad \begin{matrix} AB & AC & AD & AE & BC & BD & BE & CD & CE & DE \\ BA & CA & DA & EA & CB & DB & EB & DC & EC & ED \end{matrix} .$$

In each permutation, the *order* in which the two contestants are named is important: the first name is that of the winner and the second name is that of the runner-up. Obviously, the selections *AB* and *BA* are different since in *AB*, *A* is the winner and *B* is the runner-up, while in *BA*, *B* is the winner and *A* is the runner-up.

Now suppose that the judges are required to select *two equal winners*, not a winner and a runner up. In this case, the order in which the names are listed does not matter since, say, the selections *AB* and *BA* represent the same two winners. A selection of elements from a set where the order in which the elements are listed is not important is called a *combination*. Whereas a *permutation* is an *ordered* arrangement of elements of a set, a *combination* is any selection of elements of a set where *the order of their arrangement is disregarded*. To be clear whether a problem involves permutations or combinations, the controlling consideration is whether or not *order* is important.

How many possible combinations are there in the case of the two equal winners? One way to answer the question is to list all the combinations:

$$AB \quad AC \quad AD \quad AE \quad BC \quad BD \quad BE \quad CD \quad CE \quad DE.$$

In this case it is easy to list all the combinations because there are only ten. In symbols, we write ${}_5C_2 = 10$ to mean that there are 10 combinations of 2 winners out of 10 contestants.

A more efficient way to find the answer is to look back to the case of the contest where we said that the number of permutations is ${}_5P_2 = 20$. But since in the present case we disregard order, *AB* represents the same choices as *BA*, *AC* is the same as *CA*, etc. Thus, every combination gives rise to two permutations and, therefore, there are only *half* as many combinations as there are permutations. So, in the present case, to find the number of combinations, we must divide the number of permutations by 2:

$${}_5C_2 = \frac{{}_5P_2}{2} = \frac{5 \times 4}{2} = 10.$$

Now consider the situation where the judges are required to select *three* equal winners out of the five contestants. How many combinations are now possible? Here, again, we shall compare ${}_5C_3$ with ${}_5P_3$.

One possible combination is *ABC*. This combination gives rise to 3!, or 6, permutations (reminder: ${}_3P_3 = 3!$):

$$ABC \quad ACB \quad BAC \quad BCA \quad CAB \quad CBA.$$

In fact, *every* combination gives rise to 3!, or 6, permutations. Therefore, there are only $\frac{1}{6}$ as many combinations as there are permutations. So to find the number of combinations, we must divide the number of permutations by 3!, or 6:

$$_5C_3 = \frac{_5P_3}{3!} = \frac{5 \times 4 \times 3}{3 \times 2 \times 1} = 10.$$

In general, to obtain a formula for the number of combinations of n objects taken r at a time, note that any r objects have $r!$ permutations. This means that every combination of r objects gives rise to $r!$ permutations, and consequently, $_nP_r$ contains each combination $r!$ times. Therefore, to obtain the number of combinations, we divide the number of permutations by $r!$. That is, *the number of combinations of n things taken r at a time is*

$$_nC_r = \frac{_nP_r}{r!} = \frac{n(n-1)(n-2)(n-3)\cdots(n-r+1)}{r!}.$$

Example 1

A book club offers you nine titles from which you are to select four books. In how many ways can you select the four books?

Solution

$$_9C_4 = \frac{9 \times 8 \times 7 \times 6}{4 \times 3 \times 2 \times 1} = 126.$$

Example 2

How many different 5-card hands can be dealt from a 52-card deck?

Solution

$$_{52}C_5 = \frac{52 \times 51 \times 50 \times 49 \times 48}{5 \times 4 \times 3 \times 2 \times 1} = 2{,}598{,}960.$$

Example 3

How many committees of three members can be chosen from five people:

a) So as to include one particular person?

b) So as to exclude one particular person?

Solution

a) $_4C_2 = \dfrac{4 \times 3}{2 \times 1} = 6.$

b) $_4C_3 = \dfrac{4 \times 3 \times 2}{3 \times 2 \times 1} = 4.$

Comments

1 *Note that $_5C_3 = 10$ and $_5C_2 = 10$; $_6C_2 = 15$ and $_6C_4 = 15$; and, generally, $_nC_r = {_nC_{n-r}}$. This conclusion means: every time we select a set of r objects from a set of n objects, we automatically select a set of n − r objects that are left behind.*

2 *Since $_nC_r = \dfrac{_nP_r}{r!}$ and $_nP_r = \dfrac{n!}{(n-r)!}$, it follows that another formula for $_nC_r$ is $_nC_r = \dfrac{n!}{r!\,(n-r)!}$.*

3 *Another symbol for $_nC_r$ is $\dbinom{n}{r}$.*

EXERCISE SET 11.8

1. Find the value of each of the following.

   a) $\dfrac{5!}{3!\,2!}$ b) ${}_6C_3$ c) ${}_{12}C_8$

   d) ${}_{10}C_1$ e) ${}_{500}C_{498}$ f) ${}_rC_{r-1}$

2. Decide whether permutations or combinations, or neither, are involved in each of the following situations.
   a) The number of 5-letter "words" that can be formed from all the letters of the alphabet, no letter to be repeated.
   b) A combination lock.
   c) Selection of a basketball team of five players from a squad of 30 players.
   d) The number of ways in which customers can line up at a supermarket checkout counter.
   e) The number of ways in which a class can be divided into two teams.

3. How many different juries of 12 people can be selected from a panel of 50 people?

4. A class consists of 5 men and 20 women. How many different committees of three students can be selected from the class if each committee is to consist of:

   a) One man and two women? b) Two men and one woman?

5. How many committees of 3 Democrats and 4 Republicans can be selected from a legislative body consisting of 15 Democrats and 20 Republicans?

6. a) If ${}_nP_3 = 2730$, find ${}_nC_3$.
   b) If ${}_nC_4 = 70$, find ${}_nP_4$.

7. Which permutations of: (a) two of the elements of the set {△, □, ○} represent the same combinations? (b) three of the elements of the set {△, □, ○}?

8. What does ${}_nC_0$ mean? What is its value for any positive integer n?

9. Write a set containing six elements. Then use this set to illustrate the fact that ${}_6C_2 = {}_6C_4$.

10. In how many ways can the first three places of a horse race be decided if there are 8 horses running?

11. How many lines are determined by ten points if no three of the points are collinear?

12. What is the maximum number of points in which 5 lines can intersect each other?

13. In how many ways can 10 people be seated around a circular dinner table if the host occupies a fixed place?

14. a) How many diagonals can be drawn in an octagon?
    b) How many triangles are determined by its vertices?

15. If a positive two-digit number is chosen at random, what is the probability that:

    a) It will be even? b) It will be divisible by 3?

16. a) Show that the value of ${}_2C_0 + {}_2C_1 + {}_2C_2$ is 2^2.
    b) Show that the value of ${}_3C_0 + {}_3C_1 + {}_3C_2 + {}_3C_3$ is 2^3.
    c) Show that the value of ${}_4C_0 + {}_4C_1 + {}_4C_2 + {}_4C_3 + {}_4C_4$ is 2^4.

d) What do you expect the value of ${}_5C_0 + {}_5C_1 + \cdots + {}_5C_5$ to be?
e) What do you expect the value of ${}_9C_0 + {}_9C_1 + \cdots + {}_9C_9$ to be?
f) What do you expect the value of ${}_nC_0 + {}_nC_1 + {}_nC_2 + \cdots + {}_nC_n$ to be?
g) Modify the formula for part f for the case where at least 1 thing is selected.

17 Use your answer to Exercise 16g to find the number of tips it is possible to give if you use one or more of the following coins: a nickel, a dime, a quarter, a half-dollar, and a dollar.

18 A surgeon has six instruments at his disposal.
a) How many different selections of instruments can he make by using none, some, or all of them?
b) What is the probability that he will use exactly two instruments?

11.9 Properties of Probability

So far we have discussed the meaning of probability, how to calculate the probabilities of simple events, and some counting techniques to assist us in these calculations. We shall now see how the probabilities of simple events can help us calculate the probabilities of more complicated events.

When dealing with more than a single event, it is important to consider whether the events are *mutually exclusive.* For instance, in rolling a die, there are six possible outcomes, each equally likely to occur. Since one and only one of these events can occur with a single roll of the die, we say that these six events are *mutually exclusive.* If A and B are mutually exclusive events, then $P(A \cap B) = 0$ since A and B have no outcomes in common.

ADDITION PROPERTY OF MUTUALLY EXCLUSIVE EVENTS

If we roll a die, the probability of obtaining a 2 is $\frac{1}{6}$, and the probability of obtaining a 5 is also $\frac{1}{6}$. But the probability of drawing *either* a 2 *or* a 5 is

$$\frac{1}{6} + \frac{1}{6} = \frac{1}{3}.$$

This result can be verified by the fact that, in the sample space, there are 2 favorable outcomes out of 6 possible outcomes; that is $\frac{2}{6}$ or $\frac{1}{3}$.

If we draw a marble from a bag containing 3 red, 5 blue, and 9 green marbles, the probability of drawing a red marble is $\frac{3}{17}$, and the probability of drawing a blue marble is $\frac{5}{17}$. But the probability of drawing *either* a red marble *or* a blue marble is

$$\frac{3}{17} + \frac{5}{17} = \frac{8}{17}.$$

This result is also verifiable by the fact that we have 3 + 5, or 8, favorable outcomes out of 17 possible outcomes.

These two examples suggest the general *addition property for two mutually exclusive events:*

PROPERTY 1 If A and B are mutually exclusive events, then the probability that A *or* B will occur is the sum of their probabilities. That is,

$$P(A \cup B) = P(A) + P(B).$$

This property can be extended to more than two mutually exclusive events $E_1, E_2, ..., E_n$:

$$P(E_1 \cup E_2 \cup \cdots \cup E_n) = P(E_1) + P(E_2) + \cdots + P(E_n).$$

ADDITION PROPERTY FOR ANY TWO EVENTS

A travel club consists of eight members, $U = \{1, 2, 3, 4, 5, 6, 7, 8\}$. If three members, $A = \{2, 4, 5\}$, have visited Rome, and four members, $B = \{1, 3, 4, 8\}$, have visited London (Figure 11.3), what is the probability that a randomly selected member visited *either* Rome or London?

FIGURE 11.3

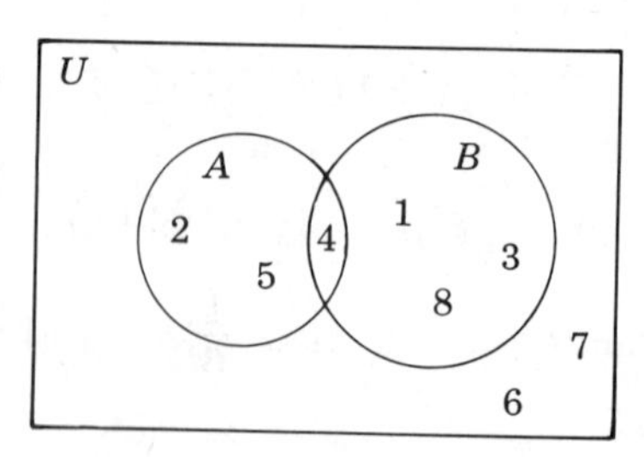

The probability of selecting a member who visited Rome is $\frac{3}{8}$, and the probability of selecting a member who visited London is $\frac{4}{8}$. But adding the two probabilities will *not* give the probability that the member *either* visited Rome *or* London. This is because A and B are not mutually exclusive since they contain member #4 in common; i.e., member #4 has visited Rome *and* London. The error we make by simply adding the two probabilities is that we are counting member #4 twice—once as part of the Rome group and once as part of the London group. We must, therefore, subtract $P(A \cap B)$ from this sum to obtain the correct answer. Since $P(A \cap B)$, the probability of selecting a member who visited both Rome and London, is $\frac{1}{8}$, our answer should be

$$\begin{aligned} P(A \cup B) &= P(A) + P(B) - P(A \cap B) \\ &= \frac{3}{8} + \frac{4}{8} - \frac{1}{8} = \frac{6}{8} = \frac{3}{4}. \end{aligned}$$

This example suggests the more general *addition property for two events that are not mutually exclusive:*

PROPERTY 2 If A and B are *not* mutually exclusive events, then the probability that A *or* B will occur is

$$P(A \cup B) = P(A) + P(B) - P(A \cap B).$$

Example 1 If we draw a card from a 52-card deck, what is the probability of drawing either a spade or a king?

Solution $P(A)$, drawing a spade, is $\frac{1}{4}$.

$P(B)$, drawing a king, is $\frac{1}{13}$.

$P(A \cap B)$, drawing a king of spade, is $\frac{1}{52}$.

$$\begin{aligned} P(A \cup B) &= P(A) + P(B) - P(A \cap B) \\ &= \frac{1}{4} + \frac{1}{13} - \frac{1}{52} = \frac{16}{52} = \frac{4}{13}. \end{aligned}$$

Comment *If A and B are mutually exclusive, then $P(A \cap B) = 0$ and the formula*

$$P(A \cup B) = P(A) + P(B) - P(A \cap B) \qquad (1)$$

becomes

$$P(A \cup B) = P(A) + P(B). \qquad (2)$$

Formula 2 is, therefore, a special case of Formula 1.

MULTIPLICATION PROPERTY FOR INDEPENDENT EVENTS

When two events are related in such a way that the occurrence or nonoccurrence of one of them does not affect the probability of the occurrence or nonoccurrence of the other, we say that the events are *independent.* For example, when we toss two coins the outcome of the second coin is *independent* of the outcome of the first coin. Two events that are not independent are said to be *dependent*.

If we toss a coin and draw a card from a deck of 52 cards, what is the probability of getting a heads *and* an ace?

The probability of getting a heads is $\frac{1}{2}$, and the probability of getting an ace is $\frac{1}{13}$; so we say that the probability of getting *both* a heads and an ace is

$$\frac{1}{2} \times \frac{1}{13} = \frac{1}{26}.$$

We can see the reason for this answer when we recall that the $\frac{1}{2}$ means that, in many trials, we can expect a heads about $\frac{1}{2}$ of the time; the $\frac{1}{13}$ means that, in many trials, we can expect an ace about $\frac{1}{13}$ of the time. Since the two events are independent, we can expect that, in many trials, both events will occur about $\frac{1}{2}$ of $\frac{1}{13}$ of the time, or

$$\frac{1}{2} \times \frac{1}{13}.$$

In other words, the probability that both events will occur is the product of their probabilities.

This discussion suggests the more general *multiplication property for two (or more) independent events:*

PROPERTY 3 If A and B are independent events, then the probability that *both* A and B will occur is the product of the probability that A occurs and the probability that B occurs. That is,

$$P(A \cap B) = P(A) \cdot P(B).$$

Example 2 A man tosses a coin and rolls a die. What is the probability that he will get a tail *and* a 2?

Solution $P(A)$, getting a tail, is $\frac{1}{2}$.

$P(B)$, getting a 2, is $\frac{1}{6}$.

$$\begin{aligned} P(A \cap B) &= P(A) \cdot P(B) \\ &= \frac{1}{2} \times \frac{1}{6} = \frac{1}{12}. \end{aligned}$$

Example 3

What is the probability of obtaining two heads with two tosses of a coin?

Solution

The probability $P(A)$ of getting a head with the first toss is $\frac{1}{2}$; the probability $P(B)$ of getting a head with the second toss is also $\frac{1}{2}$. Therefore,

$$P(A \cap B) = P(A) \cdot P(B)$$
$$= \frac{1}{2} \times \frac{1}{2} = \frac{1}{4}.$$

This result can be verified by an inspection of the sample space of the experiment: $S = \{HH, HT, TH, TT\}$, where the probability of getting HH is $\frac{1}{4}$.

11.10 Conditional Probability

Up to now we were concerned with ways to calculate the probability that certain events will occur. Now we shall concern ourselves with finding the probability that an event will occur under the *condition* that we know something about the event. Such probabilities are called *conditional probabilities.*

Let us suppose that Mean Mary Jean ran for President of her class and that Table 11.2 shows the results of the election. The table contains several sample spaces: students in the class (S), men in the class (M), women in the class (W), students who voted *for* Mary (F), students who voted *against* Mary (A).

TABLE 11.2

	Voted For	*Voted Against*	
Men	20	5	25
Women	8	7	15
	28	12	40

There are some interesting questions that we can ask about this data. For instance, if we draw a name from a box containing the names of all the students in the class, what is the probability of selecting a man if we know that the person selected voted for Mary?

This question involves the concept of *conditional probability*—the probability that a *man* will be selected with the condition that he voted *for* Mary. If M denotes the selection of a man, and F denotes the selection of a student who voted *for* Mary, then we write the symbol $P(M|F)$ to mean *the probability of a man being selected on the condition that he voted for Mary.* More generally, the conditional probability

$$P(A|B)$$

means *the probability that A will occur given that event B has occurred.*

It follows from Table 11.2 that $P(M) = \frac{25}{40} = \frac{5}{8}$; $P(F) = \frac{28}{40} = \frac{7}{10}$. The probability that the student selected is a man who voted *for* Mary is

$$P(M \cap F) = \frac{20}{40} = \frac{1}{2}.$$

Now suppose that we are interested in selecting only a student who voted for Mary. This means that our interest has shifted to another sample space.

The probability $P(M \mid F)$ of selecting a man on the condition that he voted for Mary is

$$P(M \mid F) = \frac{20}{28} = \frac{5}{7}.$$

The result $\frac{20}{28}$ is actually the *ratio* of the number of students who are men and who voted for Mary to the number of students who voted for Mary. Since $\frac{20}{28}$ can be written

$$\frac{\frac{20}{40}}{\frac{28}{40}},$$

we can write

$$P(M \mid F) = \frac{\frac{20}{40}}{\frac{28}{40}} = \frac{P(M \cap F)}{P(F)}.$$

The preceding discussion suggests the following formula which applies to any two events A and B:

FORMULA 1 If $P(B)$ is not equal to zero, then the conditional probability that A will occur once B has occurred is

$$P(A \mid B) = \frac{P(A \cap B)}{P(B)}.$$

Example In Table 11.2, what is the probability of selecting a *woman* who voted *against* Mary?

Solution Let W denote the selection of a woman, and A the selection of a student who voted against Mary. The problem asks us to find the conditional probability that our selection will be a woman, given that she voted against Mary:

$$P(W \mid A) = \frac{P(W \cap A)}{P(A)}.$$

Since $P(W \cap A) = \frac{7}{40}$, and $P(A) = \frac{12}{40}$,

$$P(W \mid A) = \frac{\frac{7}{40}}{\frac{12}{40}} = \frac{7}{12}.$$

Note The answer $\frac{7}{12}$ could have been obtained directly from table 11.2 where we see that 7 of the 12 students who voted against Mary are women.

EXERCISE SET 11.9, 11.10

1 In a single throw of a die, what is the probability of obtaining:
a) A 3 or a 4?
b) Three or a number greater than 4?
c) An odd number or a number greater than 6?
d) An even number or a number less than 6?

2 If two dice are rolled, what is the probability that:
a) The sum of the numbers is 5?
b) One or both of the numbers is 6?
c) The sum of the numbers is 10 or more?
d) The sum of the numbers is 7 or less?

3 If two coins are tossed, find the probability of getting:
a) Two heads or two tails.
b) Two tails or a head only on the first coin.

4 If one card is drawn from a 52-card deck, what is the probability that the card selected is:
a) A jack or a king?
b) A diamond or a spade?
c) An ace or a club?
d) A king or queen or a spade?

5 A box contains 2 red, 4 blue, and 5 black marbles. A second box contains 3 red, 2 blue, and 3 black marbles. If one marble is withdrawn from each box, what is the probability that either both or none is blue?

6 If a four-volume set of books is randomly placed on a shelf, find the probability that they will be arranged in either proper or reverse order?

7 If six people buy theatre tickets for six seats in a row, what is the probability that a certain two of these people will occupy adjacent seats?

8 Which of the following sets of events are independent?
a) Getting a 2 and then a 5 on two successive rolls of a die.
b) Passing a history course and passing a mathematics course.

c) Two dice are rolled once: getting a 4 on at least one die and a sum of 7 on both.
d) Two cards removed from a deck, one at a time, without replacement.
e) Two cards removed from a deck, one at a time, with replacement.
f) Any two mutually exclusive events, given that each can occur.

9 If a die is rolled twice, what is the probability of obtaining:
a) A 2 and then a 3?
b) Two 5's?
c) A 6 and then a number less than 6?

10 A pair of dice is thrown twice. What is the probability that both throws are:
a) 7s? b) 11s?

11 What is the probability that an 11 appears at least once in two rolls of a pair of dice?

12 A die is thrown three times. Find the probability that exactly two of the numbers are the same.

13 A coin is tossed and then a die is rolled.
a) List a sample space for this experiment.
b) Find the probability of getting:
1) A tail and a 4.
2) A head or an odd number.
3) A head and an even number.
4) A head and a number greater than 4.

14 A box contains 1 half-dollar, 2 quarters, 5 dimes, and 3 nickels.
a) In how many ways can three coins be selected?
b) In how many ways can the three coins amount to 60¢.
c) What is the probability of getting exactly 60¢ with three coins?
d) What is the probability of getting exactly 70¢ if four coins are withdrawn from the box?

15 Three coins are tossed.
a) List a sample space for this experiment.
b) Find the probability that:
1) At least one toss is a head.
2) One toss is a head.

16 Three cards are drawn from a deck of 52 cards. Find the probability that:
a) All three cards are aces.
b) Two cards are kings and three are spades.

17 Three dice are rolled simultaneously. What is the probability that:
a) Exactly one 2 will appear?
b) At least one 2 will appear?

18 Two cards are drawn in succession from a 52-card deck without replacing the first card. What is the probability that:
a) Both cards are aces?
b) Both cards are clubs?
c) Both cards are of the same suit?
d) The first card is a king and the second a queen of spades?
e) Neither card is a club?

19 If the first card is replaced in Exercise 18, find the probabilities in parts a through e.

20 A coin is tossed six times. Find the probability that:
a) All tosses are tails.
b) No tosses are tails.
c) At least one toss is a head.

21 A student takes a true-false test containing ten questions. A score of 9 or 10 is graded an A, and a score of 8 is graded a B. Assuming that the student guesses at all the answers, what is the probability that he would be graded (a) an A, (b) a B, (c) less than a B?

22 The probability that a tennis player will win a men's match is 0.65, and the probability that his girlfriend will win the women's match is 0.5. Find the probability that:
a) Both will win.
b) At least one will win.
c) Only the man will win.
d) Only the woman will win.
e) Neither will win.

23 A dealer receives a shipment of 85 television sets of which 6 are defective. If each of three customers buys a set, what is the probability that:
a) All three customers bought defective sets?
b) All three customers bought nondefective sets?
c) Two customers bought nondefective sets and one bought a defective set?

24 A test consists of 10 multiple-choice questions, each of which has five choices. What is the probability of getting:
a) At least 8 correct answers by sheer guessing?
b) Exactly 6 correct answers by sheer guessing?

25 A and B are working on a problem independently. If the probability that A will solve the problem is 0.7 and the probability that B will solve the problem is 0.6, what is the probability that the problem will be solved?

26 Of ten people at a party:
a) What is the probability that at least one of them was born in December (disregarding the differences in the number of days of the months)?
b) What is the probability that exactly one of them was born in December?

27 In Exercise 26, find the probability that at least two of the people at the party have the same birthday.

28 Find the probability of drawing a hand consisting of a pair of aces and three kings in two successive deals.

29 If A and B are events in a sample space such that $P(A) = 0.3$, $P(B) = 0.4$, and $P(A \cap B) = 0.1$, compute each of the following.

a) $P(\overline{A})$ b) $P(\overline{B})$ c) $P(A \cup B)$ d) $P(A|B)$ e) $P(B|A)$

30 If A is the event that a man gets an advancement and B is the event that he is a good bridge player, describe the probability represented by each of the following:

a) $P(A|B)$ b) $P(B|A)$ c) $P(A|\overline{B})$ d) $P(\overline{A}|B)$

31 A card is drawn from a 52-card deck. Given that the card is an ace, what is the probability that it is also a spade?

32 An even number is obtained in a single roll of a die. What is the probability that the number is greater than 2?

33 In a certain college, 35% of all the students have cars, 20% have their own apartments, and 10% have cars and their own apartments.

a) What is the probability that a student has a car *or* an apartment?
b) If a student has a car, what is the probability that he also has his own apartment?
c) If a student has his own apartment, what is the probability that he has a car?

34 If G is the event that a person is a good teacher and K is the event that he knows his subject, express the following probabilities in symbolic form.

a) The probability that a good teacher knows his subject.
b) The probability that one who knows his subject is also a good teacher.
c) The probability that a good teacher does not know his subject.
d) The probability that one who does not know his subject is not a good teacher.

35 A national poll taken of voter opinion of a Presidential candidate shows the following results:

	Favor Candidate	*Do Not Favor Candidate*	*No Opinion*	
Democrats	500	150	50	700
Republicans	100	350	30	480
Independents	200	100	20	320
	800	600	100	1500

On the basis of the data shown in the table, find:

a) The probability that if a Democrat is randomly selected, then he is likely to favor the candidate.
b) The probability that a person who does not favor the candidate is an Independent.
c) The probability that a Republican expressed no opinion of the candidate.
d) The probability that a person who expressed no opinion is an Independent.

36 If we interchange A and B in the formula

$$P(A|B) = \frac{P(A \cap B)}{P(B)},$$

does the relationship still hold? Why?

PUZZLES

1 In counting the ballots in an election contest between two candidates, what is the probability that the eventual winner will always lead his opponent?

2 a) What is the probability of throwing a 12 in n throws of two dice?
b) If you wanted to bet even money on throwing a 12 with two dice, in *how many throws* would you expect a 12?

BASICS REVISITED

1 Add $\frac{5}{8}, \frac{7}{12}, \frac{3}{4}$, and $\frac{7}{42}$.

2 $9\frac{1}{8} - 2\frac{7}{32} = ?$

3 Subtract 11.723 from 14.1.

4 $5\frac{3}{8} \times \left(16 \div \frac{4}{5}\right) = ?$

5 Estimate the product of 3.9 and 251.5.

6 Divide 4.23 by .0082.

7 Find $1\frac{1}{4}$% of 256.

8 6 is what percent of 2?

9 Four pieces of wood, each the same width, have the following lengths: 8 inches, 7 inches, $10\frac{1}{2}$ inches, and $8\frac{1}{2}$ inches.
a) What is the total length of all the four pieces?
b) What is the average length of the four pieces?

10 Find the product of 2.823 and .0087.

11 Divide 10.9 by .23. (Round to 2 places).

12 Write .0775 as a percent.

13 Subtract $29.97 from $95.82.

14 Find the quotient correct to the nearest tenth: $59\overline{)780372.}$

15 Write .01625 as a percent and as a common fraction.

16 14 sq. ft. = ____ sq. in.

17 What percent of 76 is .49?

18 A bank offers $5\frac{3}{4}$% interest annually, compounded monthly. What will a deposit of $4,800 amount to at the end of two months?

19 A person is paid the overtime rate of $1\frac{1}{2}$ times the regular rate for working time over 40 hours a week. Find the weekly earnings of a person who works $47\frac{1}{2}$ hours at $6.43 per hour.

20 The price of a skirt is $23. How much did the skirt cost the storekeeper if he marked up his cost by 40%?

21 Using the short method, multiply 0.025 by (a) 10, (b) 100, (c) 2000.

22 Illustrate, on the number line, the distributive property of multiplication over addition of whole numbers.

23 George used to pay $3 for a haircut, which now costs $3.75. What is the percent increase in the cost of a haircut?

24 If the student drop-out rate at a certain college is 9.8% per year, how many students can be expected to drop out in one year if the college enrollment is 6759 students?

25 A bottle and a cork cost $1.50 together. The bottle costs one dollar more than the cork. How much does each cost?

EXTENDED STUDY

11.11 Probability and Heredity

One of the most interesting applications of the concept of probability is in the science of heredity. Gregor Mendel (1822-1884), an Austrian monk, founded the science of heredity with his experiments on hybrid peas. The main outcome of his research was the discovery that certain parental characteristics are transmitted unchanged, without dilution or blending, because they are carried by distinctive "factors" which we now call *genes.* These genes persist as distinct units from generation to generation and, in each new generation, provide possibilities for wholly new combinations of characteristics.

According to Mendel, some characteristics are *dominant* while others are *recessive.* The dominant characteristics appear in the offspring, while the recessive characteristics do not. But the recessive characteristics do not disappear.

Mendel started his experiments with two pure strains of peas, one yellow and one green. After cross-fertilization, *all* the peas of this first generation turned out to be yellow. When the peas of the first generation were cross-fertilized with each other, the resulting second generation always contained *three times as many yellow peas as green peas;* that is, $\frac{3}{4}$ of all the peas were yellow and $\frac{1}{4}$ were green. These surprising proportions, which baffled other biologists of his day, were explained by Mendel in terms of the concept of probability.

If we relate the results of cross-fertilization to the results of tossing a coin, we can see clearly why Mendel got the results that he did. Let us start with two pure strains of peas: yellow (Y) and green (G). When these are cross-fertilized, the resulting first-generation offspring will be a hybrid pea containing a Y gene and a G gene. This hybrid corresponds to a coin with a "heads" (H) and a "tails" (T):

1st *Generation Pea*	→	*1 Coin*
YG	→	HT

When we cross-fertilize two first generation seeds, we can expect the same four possible outcomes as when we toss two coins:

2nd *Generation Pea*	→	*2 Coins*
YY		*HH*
YG	→	*HT*
GY		*TH*
GG		*TT*

Probability (at least 1 Y) = $\frac{3}{4}$.

Probability (at least 1 H) = $\frac{3}{4}$.

All seeds which contain a yellow gene will give rise to yellow peas because yellow is the dominant color. Since all first generation seeds contain a yellow gene, all first generation peas will be yellow. In second generation seeds, the probability of a seed with at least one yellow gene is $\frac{3}{4}$. This means that a large number of second generation seeds will produce three times as many yellow peas as green peas.

By using the theory of probability, we can predict the proportion of yellow peas in the third and later generations. We can also predict which characteristics will result, and in what proportions, if we cross-fertilize peas with several different pairs of characteristics, such as yellow and green or tall and short. This knowledge is now used to create new fruits and flowers, breed more productive cows, and produce turkeys with large amounts of white meat.

11.12 Binomial Probabilities

Experiments in which there are exactly two possible outcomes in any one trial are called *binomial experiments.* Examples of such experiments are getting heads or tails with the toss of a coin, giving birth to a boy or a girl, "success" or "failure." The probabilities related to such experiments are called *binomial probabilities.* Binomial experiments are sometimes called *Bernoulli experiments,* named after Jacques Bernoulli (1654-1705) who worked extensively with such experiments.

We shall assume that in these experiments, there are a fixed number of trials; the probability of a "success" is the same for each trial; and the trials are independent.

Let us now consider a common type of binomial experiment: the probability of getting heads when we toss one or more coins. We know that the number of ways in which we can obtain a given number of heads when we toss $n = 1, 2, 3$, or 4 coins are as follows:

For 1 *Coin:*

TABLE 11.3

			Sum
Number of Heads	0	1	
Number of Ways	1	1	2
Probability	$\frac{1}{2}$	$\frac{1}{2}$	1

For 2 Coins:

TABLE 11.4

				Sum
Number of Heads	0	1	2	
Number of Ways	1	2	1	4
Probability	$\frac{1}{4}$	$\frac{2}{4}$	$\frac{1}{4}$	1

For 3 Coins:

TABLE 11.5

					Sum
Number of Heads	0	1	2	3	
Number of Ways	1	3	3	1	8
Probability	$\frac{1}{8}$	$\frac{3}{8}$	$\frac{3}{8}$	$\frac{1}{8}$	1

For 4 Coins:

TABLE 11.6

						Sum
Number of Heads	0	1	2	3	4	
Number of Ways	1	4	6	4	1	16
Probability	$\frac{1}{16}$	$\frac{4}{16}$	$\frac{6}{16}$	$\frac{4}{16}$	$\frac{1}{16}$	1

Table 11.7 shows the number of ways in which we can obtain a given number of heads when we toss $n = 1, 2, 3, 4, 5$, or 6 coins. Observe that:

1 The distribution of the numbers in each row is symmetrical: e.g., 1, 2, 1; 1, 3, 3, 1; etc.

2 If n is the number of coins tossed, then there are $(n + 1)$ possible categories of heads to consider. For example, for 2 coins, there are 3 possible categories: 0 *H*, 1 *H*, 2 *H*; for 3 coins there are 4 possible categories: 0 *H*, 1 *H*, 2 *H*, 3 *H*.

3 If we toss n coins, the sum of the outcomes for all categories is 2^n. For example, when we toss 4 coins, there are 2^4 or 16 possible outcomes.

4 The sum of all the probabilities is 1. (See Tables 11.3-11.6.)

TABLE 11.7

n	0 *H*	1 *H*	2 *H*	3 *H*	4 *H*	5 *H*	6 *H*	*Sum*
1	1	1						$2 = 2^1$
2	1	2	1					$4 = 2^2$
3	1	3	3	1				$8 = 2^3$
4	1	4	6	4	1			$16 = 2^4$
5	1	5	10	10	5	1		$32 = 2^5$
6	1	6	15	20	15	6	1	$64 = 2^6$

BINOMIAL EXPANSION

Let us write the expansions of the binomial $(a + b)^n$ for successive values of n.

$$\begin{aligned}
(a+b)^1 &= a + b \\
(a+b)^2 &= a^2 + 2ab + b^2 \\
(a+b)^3 &= a^3 + 3a^2b + 3ab^2 + b^3 \\
(a+b)^4 &= a^4 + 4a^3b + 6a^2b^2 + 4ab^3 + b^4 \\
(a+b)^5 &= a^5 + 5a^4b + 10a^3b^2 + 10a^2b^3 + 5ab^4 + b^5 \\
(a+b)^6 &= a^6 + 6a^5b + 15a^4b^2 + 20a^3b^3 + 15a^2b^4 + 6ab^5 + \\
&\quad b^6.
\end{aligned}$$

If you look only at the *coefficients* and compare them, row by row, with the numbers in Table 11.7, you will find that they are identical. This observation suggests an interesting connection between a binomial experiment and a binomial expansion: the coefficients of the expansion of, say, $(a + b)^4$ represent the number of ways of obtaining 0 H, 1 H, 2 H, 3 H, and 4 H, respectively, when we toss 4 coins. Similarly, the coefficients of the expansion of $(a + b)^n$ represent the number of ways of getting 0 H, 1 H, 2 H, ..., n H, respectively, when we toss n coins. The sum of the coefficients, 2^n, is the total number of possible outcomes, and the ratios of the coefficients to 2^n represent the probabilities for 0, 1, 2, ..., $(n - 2)$, $(n - 1)$, n heads. The exponent n represents the number of coins tossed.

You may recall that the *general* binomial expansion is

$$\begin{aligned}
(a+b)^n = a^n + na^{n-1}b + \frac{n(n-1)}{2 \times 1} a^{n-2}b^2 \\
+ \frac{n(n-1)(n-2)}{3 \times 2 \times 1} a^{n-3}b^3 \\
+ \frac{n(n-1)(n-2)(n-3)}{4 \times 3 \times 2 \times 1} a^{n-4}b^4 + \cdots + b^n.
\end{aligned}$$

The coefficients in this expansion are

$$1, n, \frac{n(n-1)}{2 \times 1}, \frac{n(n-1)(n-2)}{3 \times 2 \times 1}, \frac{n(n-1)(n-2)(n-3)}{4 \times 3 \times 2 \times 1}, \cdots, 1.$$

Note that these coefficients are exactly the numbers

$${}_nC_0, {}_nC_1, {}_nC_2, {}_nC_3, {}_nC_4, \cdots, {}_nC_n.$$

Thus, if $n = 3$, i.e., if we toss 3 coins, the number of ways in which we can obtain 0, 1, 2, and 3 heads are, respectively, ${}_3C_0$, ${}_3C_1$, ${}_3C_2$, ${}_3C_3$, which equal 1, 3, 3, 1, respectively.

Table 11.7 can, therefore be written as shown in Table 11.8.

TABLE 11.8

n	$0H$	$1H$	$2H$	$3H$	$4H$	$5H$	$6H$	$\cdots$	nH	Sum
1	${}_1C_0$	${}_1C_1$								$2 = 2^1$
2	${}_2C_0$	${}_2C_1$	${}_2C_2$							$4 = 2^2$
3	${}_3C_0$	${}_3C_1$	${}_3C_2$	${}_3C_3$						$8 = 2^3$
4	${}_4C_0$	${}_4C_1$	${}_4C_2$	${}_4C_3$	${}_4C_4$					$16 = 2^4$
5	${}_5C_0$	${}_5C_1$	${}_5C_2$	${}_5C_3$	${}_5C_4$	${}_5C_5$				$32 = 2^5$
6	${}_6C_0$	${}_6C_1$	${}_6C_2$	${}_6C_3$	${}_6C_4$	${}_6C_5$	${}_6C_6$			$64 = 2^6$
⋮		⋮		⋮		⋮				⋮ ⋮
n	${}_nC_0$	${}_nC_1$	${}_nC_2$	${}_nC_3$	${}_nC_4$	${}_nC_5$	${}_nC_6$	$\cdots$	${}_nC_n$	2^n

The expansions of $(a + b)^n$ can be used to find *probabilities* directly if we let a represent the probability of "success" and b the probability of "failure" in a single trial, and if n is the number of trials. For instance, if $a = \frac{1}{2}$ is the probability of getting a head in a single toss of a coin, and $b = \frac{1}{2}$ is the probability of *not* getting a head in a single toss of a coin, and we toss a coin three times, then we have, by the expansion formula,

$$(a + b)^n = \left(\frac{1}{2} + \frac{1}{2}\right)^3 = \left(\frac{1}{2}\right)^3 + 3\left(\frac{1}{2}\right)^2\left(\frac{1}{2}\right) + 3\left(\frac{1}{2}\right)\left(\frac{1}{2}\right)^2 + \left(\frac{1}{2}\right)^3$$
$$= \frac{1}{8} + \frac{3}{8} + \frac{3}{8} + \frac{1}{8}.$$

You will recognize these fractions as the *probabilities* of obtaining 0 heads, 1 head, 2 heads, and 3 heads, respectively, if we toss a coin three times.

We can also use the binomial expansion to find probabilities in situations where the probability of success in a single trial is not equal to $\frac{1}{2}$. For instance, consider the probability of rolling fours with two dice. Here, the probability of "success" with one die is $\frac{1}{6}$, the probability of "failure" is $\frac{5}{6}$, and $n = 2$. So we have

$$(a + b)^n = \left(\frac{1}{6} + \frac{5}{6}\right)^2 = \left(\frac{1}{6}\right)^2 + 2\left(\frac{1}{6}\right)\left(\frac{5}{6}\right) + \left(\frac{5}{6}\right)^2$$
$$= \frac{1}{36} + \frac{10}{36} + \frac{25}{36}.$$

These fractions are the probabilities of obtaining 2, 1, and 0 fours, respectively.

If we roll 3 dice, then we have

$$(a + b)^3 = \left(\frac{1}{6} + \frac{5}{6}\right)^3 = \left(\frac{1}{6}\right)^3 + 3\left(\frac{1}{6}\right)^2\left(\frac{5}{6}\right) + 3\left(\frac{1}{6}\right)\left(\frac{5}{6}\right)^2 + \left(\frac{5}{6}\right)^3$$
$$= \frac{1}{216} + \frac{15}{216} + \frac{75}{216} + \frac{125}{216}.$$

These are the probabilities for 3, 2, 1, and 0 fours, respectively. Note that these probabilities are skewed, *not* symmetrical, since a and b have different values.

11.13 The Pascal Triangle

The rows of numbers in Table 11.7 can be shifted to the left in such a way as to produce the triangular arrangement in Table 11.9. This triangle of numbers is known as *Pascal's Triangle*, named after the French mathematician Blaise Pascal (1623-1662). The Chinese, it is believed, were familiar with this triangle as early as the 14th century. Since Table 11.7 and Table 11.8 represent the same array of numbers, Pascal's triangle may be written in the form shown in Table 11.10.

TABLE 11.9

1 1
1 2 1
1 3 3 1
1 4 6 4 1
1 5 10 10 5 1
1 6 15 20 15 6 1

TABLE 11.10

$${}_1C_0 \quad {}_1C_1$$
$${}_2C_0 \quad {}_2C_1 \quad {}_2C_2$$
$${}_3C_0 \quad {}_3C_1 \quad {}_3C_2 \quad {}_3C_3$$
$${}_4C_0 \quad {}_4C_1 \quad {}_4C_2 \quad {}_4C_3 \quad {}_4C_4$$
$${}_5C_0 \quad {}_5C_1 \quad {}_5C_2 \quad {}_5C_3 \quad {}_5C_4 \quad {}_5C_5$$
$${}_6C_0 \quad {}_6C_1 \quad {}_6C_2 \quad {}_6C_3 \quad {}_6C_4 \quad {}_6C_5 \quad {}_6C_6$$

The triangle can be enlarged to as many rows as we wish by writing the expansions of $(a + b)^n$ for as many values of n as we wish. Table 11.11 shows Pascal's triangle for cases up to $n = 10$.

TABLE 11.11

n		*Sum*
1	1 1	2
2	1 2 1	4
3	1 3 3 1	8
4	1 4 6 4 1	16
5	1 5 10 10 5 1	32
6	1 6 15 20 15 6 1	64
7	1 7 21 35 35 21 7 1	128
8	1 8 28 56 70 56 28 8 1	256
9	1 9 36 84 126 126 84 36 9 1	512
10	1 10 45 120 210 252 210 120 45 10 1	1,024

There is an easy way to generate the Pascal triangle: the entries at the ends of any row are always 1. Every other number in the row can be found by adding the pair of numbers above it at the left and right. For example, in the 4th row,

$$4 = 1 + 3, \quad 6 = 3 + 3, \quad 4 = 3 + 1 \qquad \text{(Table 11.12)}.$$

TABLE 11.12

```
    1 1
   1 2 1
  1 3 3 1
 1 4 6 4 1
```

We shall now translate the 8th row of Pascal's triangle (Table 11.13) into probabilities of getting heads by tossing a coin 8 times:

TABLE 11.13

Number of Heads	0	1	2	3	4	5	6	7	8	*Sum*
Number of Ways	1	8	28	56	70	56	28	8	1	256
Probability	$\frac{1}{256}$	$\frac{8}{256}$	$\frac{28}{256}$	$\frac{56}{256}$	$\frac{70}{256}$	$\frac{56}{256}$	$\frac{28}{256}$	$\frac{8}{256}$	$\frac{1}{256}$	1

EXERCISE SET 11.11, 11.12

1. Explain the reason for the connection between the expansions of $(a + b)^n$ and Table 11.7.
2. How many different patterns of numbers can you find in Pascal's triangle?
3. Write the entries in Pascal's triangle for $n = 11; 12; 13; 17$.
4. Is it correct to say that any entry ${}_nC_r$ in Pascal's triangle is equal to the sum of ${}_{n-1}C_{r-1}$ and ${}_{n-1}C_r$? Why?
5. If you wish to complete the shape of the triangle by adding a 1 to the apex, what would this 1 represent?

```
   1
  1 1
 1 2 1
```

6. Use the tenth row of Pascal's triangle to find the probability of getting 7 heads in ten tosses of a coin.
7. A family has seven children. Use the Pascal triangle to find the probability that the children are five boys and two girls.

Chapter 12

STATISTICS

Objectives

Upon completion of this chapter, you should be able to:

1. State what is meant by a *frequency distribution.*
2. Prepare a frequency distribution from a given set of data.
3. Draw a *histogram* and a *frequency polygon* of a frequency distribution.
4. Define *mean*, *median*, and *mode* and illustrate situations where each measure would be more appropriate than the others as an indicator of central tendency.
5. Calculate the mean, median, and mode for both ungrouped and grouped data.
6. Define *range* and *standard deviation.*

7 Calculate the *range* and *standard deviation* for both ungrouped and grouped data.

8 Transform raw scores into z scores.

9 Use *standard normal distribution tables* to solve problems such as Exercises 1–6 on pages 426-428.

Prerequisites You should be familiar with bar and line graphs.

INTRODUCTION

Before Walt Disney spent $17,000,000 to build Disneyland, he knew that he needed to know where to locate it, how big to make it, what amusements to provide, what admission to charge, what hours to stay open, and the answers to many other questions. He decided to consult a group of statisticians at Stanford University. The statisticians proceeded to gather data about people's income, travel habits, amusement preferences, number of children, etc. From this information they calculated the probabilities of certain numbers of people coming to a certain location and paying a certain admission charge. Armed with this knowledge, Disney proceeded to build his successful Disneyland in Southern California.

Recently, a large contingent of U. S. Naval vessels and Air Force planes gathered in the Caribbean to test whether seeding clouds with silver iodide causes them to produce rain. Earlier experiments had been criticized because it was suspected that, in those trials, the clouds that had been selected for seeding might have produced rain even without the seeding. To correct this defect in the experiment, a statistician constructed the following randomized design: first a cloud would be chosen for observation, then a sealed envelope would be opened to learn whether that particular cloud was to be seeded or not. The design was arranged so that about two out of every three clouds were seeded. The conclusion reached after analysis of the data was that seeded clouds do produce more rainfall.

An investigator wishes to find out which of two teaching methods produces higher student achievement. He tests both methods on two groups of students of comparable ability and then, on the basis of information yielded by this small sample of students, he draws an inference about the relative merits of each method when used with very large groups of students.

These three examples illustrate how statistics is used to solve problems. In each case a sample of the population was selected, data about the sample was gathered, and conclusions were drawn about the entire population on the basis of the sample data.

The part of statistics that is concerned with collecting and summarizing of data is called *descriptive statistics.* The part that is concerned with drawing inferences from a sample about the entire population is called *inferential statistics.* We should note that when we talk of a *population* in statistics we do not refer, necessarily, to people, but to the universe of elements under consideration. A population in this sense can refer to such things as a carload of light bulbs, the heights of trees in a forest, and the speeds of cars. The *sample* is a specified number of elements *randomly* selected from the population and, presumably, *representative* of the population.

The nature and function of statistics have grown dramatically in the last fifty years. Statistics has moved from merely gathering data, arranging it in tables and graphs, and summarizing it by a few descriptive numbers like averages to the *science of drawing inferences*, of making predictions about entire populations on the basis of only partial or incomplete information. What has made this exciting forward leap possible is the application of the theory of probability to the collection of data and to their analysis. By selecting a sample and gathering data in accordance with the requirements of the laws of probability, we can employ these laws to compare actual results with expected results, decide whether discrepancies can reasonably be attributed to chance, determine whether statistical signficance can be attached to the results obtained, and deduce, within certain limits of reliability, the properties of the population.

Since the Second World War, the importance of statistics has grown enormously because of the immense amount of data that is collected, processed, and distributed to the public, and because of the increasingly *quantitative* approach being used in all the sciences, especially the social sciences, as well as in business and industry.

Statistics is used in business to assess the market for new products, to design and conduct consumer surveys, and to estimate the volume of sales. In industrial quality control, statistics is used to determine techniques for evaluating quality through adequate sampling. In biology, it is used to create theoretical models of the nervous system and to study genetical evolution.

In psychology, it is used to measure learning ability, intelligence, and personality characteristics and to study "normal" and "abnormal" behavior. In sociology, it is used to test theories about social systems, to study social attitudes, cross-cultural differences, and the growth of human populations. In health, statistics is used to test new drugs; in engineering, to develop safer systems of flight control for airports; in meteorology, to forecast the weather. Statistics is used to forecast elections and to evaluate the popularity of public figures and television shows. The list goes on endlessly.

In this chapter only a few of the very basic ideas used in statistics will be presented. For those who may be interested in pursuing the subject further, there are an adequate number of good statistics books available. Different books stress different applications such as educational statistics, business statistics, psychological statistics, and medical statistics. But most statistical methods are the same regardless of the field of application. By familiarizing yourself with basic statistical ideas, you will be better prepared to read not only the professional literature but also other materials that are so often replete with statistical symbols, concepts, and ideas.

12.1 Frequency Distributions

The results on a mathematics test taken by a class of 50 college freshmen are shown in Table 12.1.

TABLE 12.1

87	71	74	55	83	86	63	59	66	77
71	88	63	44	83	79	86	77	52	57
91	91	71	80	83	81	60	78	89	68
69	83	71	76	73	85	85	52	80	61
89	91	61	90	48	72	53	68	73	81

The students wanted to know more than just their own marks. They wanted to know where they stood in relation to other students, the spread of ability as revealed by the test, the average score of the class, and whether the scores are equally scattered over the entire range of marks or whether they are bunched up at either end or at the middle.

We can readily see that the answers to these questions cannot be found by a mere inspection of Table 12.1. The scores will have to be presented in a more orderly form before we can make sense out of them. A good way to do this is to rearrange the scores in order of size, and then, next to each score, write the number of students who received that score. When we deal with a large number of scores, it is even more helpful to group them, by numerical size, into a small number of categories or *classes* and display them in tabular form as in Table 12.2. We call such a table a *frequency distribution.*

TABLE 12.2

Score	*Tally*	*Frequency, f*
90–94	////	4
85–89	~~////~~ ///	8
80–84	~~////~~ ///	8
75–79	~~////~~	5
70–74	~~////~~ ///	8
65–69	////	4
60–64	~~////~~	5
55–59	///	3
50–54	///	3
45–49	/	1
40–44	/	1
		$\Sigma f = 50 = N$

To construct a frequency distribution we must first decide upon (1) the number of classes into which to divide the data, and (2) the *class interval*, that is, the range of scores that each class is to include. Although both of these decisions are largely arbitrary, the following practices are customary.

1 Usually, 10–20 classes are used. The advantage of a small number of classes is convenience; the advantage of a larger number lies in its greater accuracy.

2 Certain class intervals are preferred. These intervals are 1, 2, 3, 5, 10, and 20.

Let us now apply these ideas to the scores in Table 12.1. The highest score is 91 and the lowest is 44, giving us a spread of 91–44, or 47 points. If we wish each class to cover a range of 5 points, then we divide 47 by 5 and find that we will need 11 classes to accommodate all the data.

To decide with which score to start each class, we usually let the lowest score in each class be a multiple of the class size. For instance, the lowest scores in each class in Table 12.2 are 40, 45, 50, etc.—each a multiple of 5 since we decided to let each class cover a range of 5 points.

CLASS LIMITS

In the class 40–44, the bottom and top scores, 40 and 44, are called the *score limits* of the interval. The *lower limits* of the eleven classes in Table 12.2 are 40, 45, 50, ..., 90; and the *upper limits* are 44, 49, 54, ..., 94.

Note that a score of 40 actually means a score that ranges from 39.5 to 40.5, and the class 40–44 actually extends from 39.5 to 44.5 in the measurement scale (Figure 12.1).

FIGURE 12.1

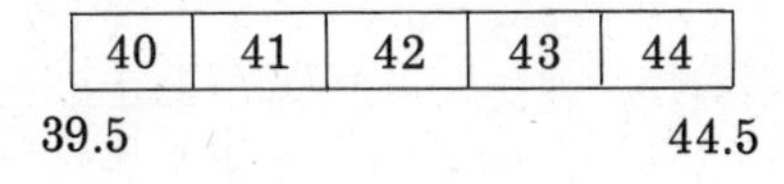

We call 39.5 and 44.5 the *exact limits* of the class interval.

Having decided upon the number of classes to use, the size of each class, and the scores with which to start each class, we complete the first column as in Table 12.2. We then tally the scores that we read from Table 12.1 in the second column, count the number of tally marks, and record the *frequencies* in the third column.

Next we find the sum of the frequencies which is symbolized by Σf. The symbol Σ is the Greek letter sigma and Σf is read *the sum of the frequencies.* The total number of scores is represented by N which, if our tally is accurate, is the same as Σf.

Comment

A frequency distribution presents data in a very usable form, but results in a certain loss of information. For instance, we cannot tell from Table 12.2 how many students got exactly a score of 75, *which is information that we could get from the original, or raw data.*

PICTURING FREQUENCY DISTRIBUTIONS

The information contained in a frequency distribution can be pictured by a graph called a *histogram* which is a bar graph without spaces between the bars as shown in Figure 12.2. Histograms are constructed by representing the scores, or quantities which are grouped, on the horizontal axis. The frequencies are represented on the vertical axis. We then draw rectangles with bases equal to the class intervals and with heights equal to the class frequencies. The markings on the horizontal scale can be the class limits or the exact class limits. Figure 12.2 is a histogram for the distribution of the scores in Table 12.2.

FIGURE 12.2
Histogram of test scores.

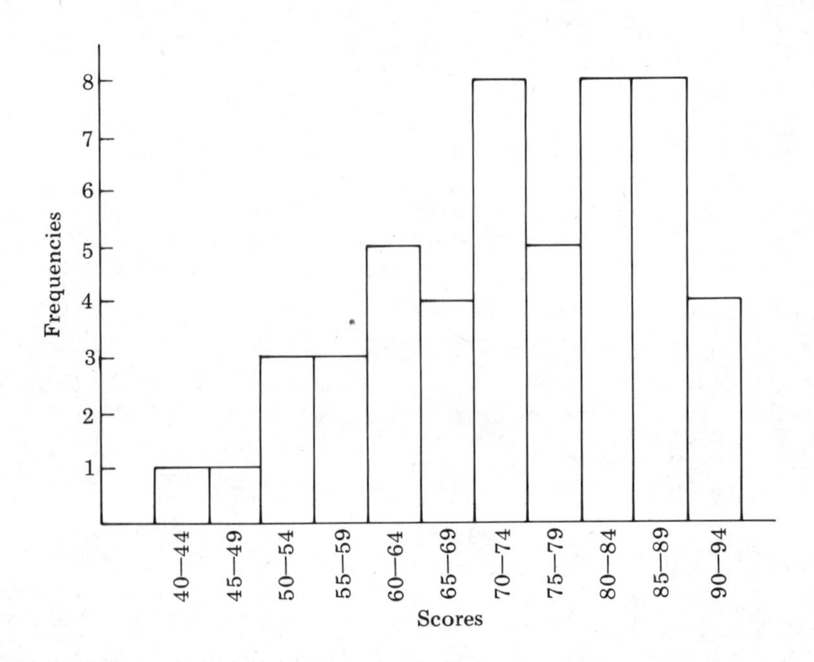

Another way of picturing a frequency distribution is by a *frequency polygon* (Figure 12.3). Here the frequencies are plotted at the *midpoints* of the classes, and the successive points are joined by line segments. The *midpoint* is exactly midway between the *exact* lower and upper limits of the interval. An easy way to find the midpoint is to average either exact or score limits of the interval. For example, the midpoint of the interval 40–44 is 42, and the midpoints of the other class intervals are 47, 52, 57, 62, 67, 72, 77, 82, 87, 92.

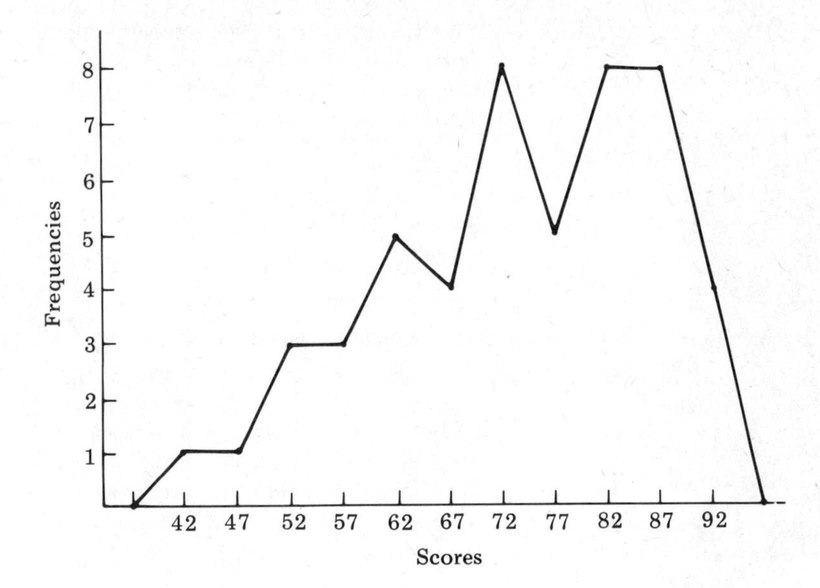

FIGURE 12.3 Frequency polygon of test scores.

Notice that in drawing the frequency polygon we allowed for one more class interval at each end of the scale to "tie down" the graph to the horizontal scale.

EXERCISE SET 12.1

1. What is the difference between descriptive statistics and inferential statistics?
2. How is descriptive statistics used in inferential statistics?
3. What is the difference between statistical inference and prophecy?
4. For each of the ranges of measurements given below, state:
   a) What would be a good size for a class interval.
   b) The score limits of the highest class interval.
   c) The exact limits of the highest class interval.
   d) The midpoint of that interval: (1) 72 to 189; (2) 1 to 37; (3) 25 to 35; (4) 0.2 to 4.9.
5. If the rentals paid by the residents of a city during a given month varied from $50 to $475, indicate:
   a) The number of classes into which these amounts might be grouped.
   b) The score limits of these classes.
   c) The exact limits of these classes.
   d) The class interval of this distribution.
   e) The midpoints of each interval.
6. When we speak of the age group from 2 to 4, we mean all those who have passed their second birthday but have not yet reached their fifth. With this definition in mind, state (1) the score limits, (2) the exact limits, and (3) the midpoints of the following age groups:
   a) 5 to 9. b) 10 to 14. c) 15 to 19.

7 The following is a distribution of the grades on a social studies test obtained by 50 students.

Grades	*Frequencies*
80–99	3
60–79	8
40–59	19
20–39	14
0–19	6

a) Construct a histogram of this distribution.
b) Construct a frequency polygon of this distribution.

8 State whether each of the following is true or false.
a) When we construct a histogram, we are assuming that scores are evenly distributed throughout each class interval.
b) When we construct a frequency polygon, we assume that all cases within each interval are concentrated at the midpoint.

9 The following are the numbers of tires sold by a dealer on 60 consecutive business days.

50 76 63 46 81 52 55 79 45 67
59 51 58 55 68 57 55 53 53 54
56 63 57 82 78 53 52 55 55 54
52 57 59 60 46 48 61 58 62 56
52 75 54 64 73 53 71 67 56 58
51 62 48 59 57 59 55 66 68 65

a) Make a frequency distribution of this data.
b) Construct a histogram of the distribution.
c) Construct a frequency polygon of the distribution.

10 The following are the grocery expenditures (in dollars) of 40 families for the week of October 20–26, 1974.

49.63 38.21 54.30 42.80 41.45 27.16 25.45 33.95 48.57 23.75
29.75 42.87 61.13 38.68 76.88 69.50 54.55 36.55 37.45 43.67
45.58 52.82 26.45 65.75 58.41 28.65 30.60 43.76 41.00 31.27
29.65 35.24 50.14 46.88 25.11 39.69 36.22 42.51 35.56 46.35

a) Make a frequency distribution of this data.
b) Present the data in a histogram.
c) Present the data in a frequency polygon.

11 Obtain the heights (in centimeters) of the members of your class.
a) Make a frequency distribution of the data.
b) Construct a frequency polygon of the data.

12 Show how each of the following statistical statements can be misinterpreted.
a) Eighty percent of all TV servicemen questioned recommend brand A.
b) More people are killed in automobile accidents than were killed in Vietnam. Therefore, it is safer to be in Vietnam than in the U. S.
c) Fifty-four percent of those who listened to the President's TV address disagreed with what he said. Therefore, the majority of Americans disagree with the President.

d) More people are smoking cigarettes today than ever before.
e) The cost of living index has risen by 2%.
f) Research shows that medicine A is twice as effective as medicine B.

13 Given: X takes on the values 1, 2, 3, 4, 5, 6, 7, 8, 9, 10. Calculate each of the following values.

a) ΣX b) $\frac{\Sigma X}{N}$ c) ΣX^2 d) $(\Sigma X)^2$

12.2 Measures of Central Tendency

We often describe a set of data by means of what is loosely called an *average;* that is, by a single number that is indicative of the "center," "middle," or most "typical" of the data. This number is usually near the center of the distribution, and for this reason is called a *measure of central tendency.* There are several such measures of central tendency of which three will be considered here: the *arithmetic mean*, usually called the *mean*; the *median*; and the *mode.* Each of these measures is obtained by a different method, and the results are usually not the same. In different situations, one of these measures may be more appropriate than the other two.

THE MEAN

The most familiar measure of central tendency is the *mean.* It is the measure that the layman calls the *average.* The mean of a set of N numbers is *their sum divided by N.* For instance, if your scores on 5 test are 72, 90, 87, 75, and 68, then your average, or *mean*, is

$$\frac{72 + 90 + 87 + 75 + 68}{5} = \frac{392}{5} = 78.4.$$

More generally, if $X_1, X_2, X_3, \ldots, X_n$ are N measurements, then their *mean* $\overline{X}$ (read "X bar") is

$$\overline{X} = \frac{X_1 + X_2 + X_3 + \cdots + X_n}{N}.$$

Using the summation symbol Σ where ΣX represents the sum of the X's, we can write the formula for *the mean of ungrouped data* as

$$\overline{X} = \frac{\Sigma X}{N} \tag{1}$$

If the data is *grouped*, as in Table 12.2, we must modify Formula 1 in order to get the mean. A good approximation can be obtained by assuming that the *midpoint* of an interval represents all the scores in the interval. We therefore treat each value in the interval as if it equals the midpoint of the interval. Thus, we treat the four values in the interval 90–94 in Table 12.2 as if each is equal to 92.

Therefore, to obtain a formula for the mean of a *grouped* frequency distribution with N scores and k class intervals, let us represent the successive class *midpoints* as $x_1, x_2, ..., x_k$, and the corresponding frequencies as $f_1, f_2, ..., f_k$. Then $x_1 f_1$, the product of the midpoint of the first interval and the number of scores in that interval, represents the approximate sum of all the scores in the first class interval; $x_2 f_2$ is the approximate sum of all the scores in the second class interval, etc. The sum

$$x_1 f_1 + x_2 f_2 + \cdots + x_k f_k \qquad \text{or} \qquad \Sigma x f$$

is the approximate sum of all the scores in the distribution.

Therefore, if N is the number of scores, we can obtain *the mean of grouped data* by the formula

$$\overline{X} = \frac{\Sigma x f}{N} \tag{2}$$

Let us now use Formula 2 to calculate the mean of the grouped data in Table 12.2. Writing the midpoints in the second column and the products xf in the fourth column, we get Table 12.3.

TABLE 12.3

Scores	*Midpoint x*	*Frequency f*	*xf*
90–94	92	4	368
85–89	87	8	696
80–84	82	8	656
75–79	77	5	385
70–74	72	8	576
65–69	67	4	268
60–64	62	5	310
55–59	57	3	171
50–54	52	3	156
45–49	47	1	47
40–44	42	1	42
		$N = 50$	$\Sigma xf = 3675$

The mean of the distribution is

$$\overline{X} = \frac{\Sigma x f}{N} = \frac{3675}{50} = 73.5.$$

The mean has certain characteristics that distinguish it from the other measures of central tendency. For instance, because the mean takes each individual measurement into account, it is sensitive to extreme measurements. This property is not always desirable. The data in a small sample can contain a single extreme value that can throw off the mean to a point where it is no longer "representative" of the entire data. For example, if five employees of a company earn weekly salaries of \$200, \$250, \$225, \$250, and \$950, the mean of these salaries, \$375, is hardly representative of the five salaries.

The mean is also the most stable of the measures of central tendency in the sense that it fluctuates less than the others from sample to sample taken from the same population. Also, the mean is better suited as a basis for further statistical computations, which we shall discuss later.

Finally, the mean may be described as that point in a distribution of scores at which *the algebraic sum of the deviations from it is zero.* The *deviation* of a score from the mean is found by subtracting the mean from the score; that is, a deviation from the mean is equal to $X - \overline{X}$. For example, consider the five scores, 11, 17, 4, 7, 6, whose mean is 9. When we calculate the *algebraic* sum of the deviations of each score from 9, we obtain 0 (Table 12.4). In fact, the sum of the deviations from any mean will always be zero. The mean can therefore be regarded as the "point of balance" or the "center of gravity" of the distribution.

TABLE 12.4

Score	*Deviation from 9*
17	$^{+}8$
11	$^{+}2$
7	$^{-}2$
6	$^{-}3$
4	$^{-}5$
	0

THE MEDIAN

The *median* (Mdn) is defined as that point on the scale of measurement which has exactly half the measurements above it and the other half below it.

We obtain *the median of a set of ungrouped measurements* by first arranging the measurements according to size and then choosing the one in the middle, or the average of the two that are nearest to the middle. When an *odd* number of measurements are arranged according to size, there is always a middle measurement whose value is the median. For instance, if five classes have enrollments of 22, 15, 23, 27, and 35, then by rearranging the numbers according to size we get

15, 22, (23), 27, 35,

and the middle number, or *median* is 23. The median of the scores 5, 9, 4, 3, 5, 8, 2 is 5.

When an *even* number of measurements is arranged according to size, the median is the average of the values of the two measurements that are nearest to the middle. For example, the median weight (in kilograms) of eight packages weighing 5, 2, 9, 6, 4, 7, 12, 3 is

$$\frac{5+6}{2} = 5.5,$$

which is halfway between the two weights 5 and 6 that are nearest the middle. Note that the median is a number or point on the measurement scale and not necessarily an actual measurement in the distribution.

Example 1

To determine *the median of a grouped frequency distribution*, we must again bear in mind that we are looking for that point on the measuring scale which has half the measurements above and half below it. We shall now locate that point in the grouped distribution shown in Table 12.5, which is the distribution of scores by 64 candidates in a civil service examination.

TABLE 12.5

Scores	*f*
55–59	5
50–54	10
45–49	11
40–44	12
35–39	15
30–34	5
25–29	3
20–24	3
	$N = 64$

Here there are 64 cases, so we are looking for that point on the measuring scale above which there are 32 cases and below which there are 32 cases.

Counting the frequencies from the bottom upward, we find that $3 + 3 + 5 + 15 = 26$ cases, 6 cases short of the 32 we need. To make 32 cases exactly, we need 6 of the 12 in the next higher interval 40–44. The median, then, lies somewhere within the interval 40–44 whose exact limits are 39.5 and 44.5. We shall assume that the 12 cases within this interval are spread evenly over the distance from 39.5 to 44.5.

Since we need 6 of the 12 cases to reach the median of the distribution, we must go $\frac{6}{12}$ of the distance from 39.5 to 44.5.

Inasmuch as the distance from 39.5 to 44.5 is 5 units, we must go $\frac{6}{12}$ of 5, or 2.5 units, to reach the median. Adding this 2.5 to the exact lower limit of the class interval, 39.5, we get

$$39.5 + 2.5 = 42.0$$

as the median.

Example 2

Let us now find the median of the distribution in Table 12.6, which shows the distribution of scores by a sophomore class of 245 students on a calculus test.

TABLE 12.6

Scores	*f*
90–94	4
85–89	10
80–84	14
75–79	19
70–74	32
65–69	31
60–64	40
55–59	28
50–54	29
45–49	21
40–44	17
	$N = 245$

Here, half of the cases is $\frac{245}{2}$, or 122.5. Counting upward from the bottom, we get $17 + 21 + 29 + 28 = 95$, which is 27.5 cases short of the 122.5 we need. So we must get the 27.5 cases from the 40 cases in the next higher interval 60–64. Since the exact lower limit of this interval is 59.5 and the size of the interval is 5, we must add

$$\frac{27.5}{40} \times 5$$

to the exact lower limit; that is,

$$\text{Mdn} = 59.5 + \frac{27.5}{40} \times 5 = 59.5 + 3.4 = 62.9.$$

THE MODE

The *mode* of a distribution is defined as the point on the scale of measurement with the greatest frequency. In a grouped distribution, the mode is the midpoint of the class interval having the greatest frequency. In other words, the mode of a set of data is *the value which occurs most frequently.* If more men buy size 15 shirts in a particular store than any other size, then the *modal size* of shirts for that store is 15. In the set of numbers, 7, 2, 9, 5, 9, 13, 9, 6, 8, 2, the mode is 9. For the data presented in Table 12.5, the mode is 37, the midpoint of the class interval 35–39.

Sometimes, a distribution has more than one point of maximum frequency. If a distribution has two such points, we call it a *bimodal* distribution and both values are regarded as modes.

APPROPRIATE USE OF THE MEAN, MEDIAN, AND MODE

Although the question of which measure of central tendency should be used in a given situation cannot always be answered with certainty, there are several guidelines that should be borne in mind.

Use the *mean* when (1) the greatest sampling stability is wanted, (2) other statistical computations are to follow, and (3) the distribution is symmetrical about the center.

Use the *median* when (1) the distribution is markedly skewed, (2) we are interested in whether cases fall within the upper or lower halves of the distribution, and (3) an incomplete distribution is given.

Use the *mode* when (1) the quickest estimate of central tendency is wanted, (2) a very rough estimate of central tendency is adequate, and (3) we are interested in knowing the most "popular" or most "typical" case.

EXERCISE SET 12.2

1 According to the U. S. Bureau of the Census, poverty in the U. S. for the years 1965–1969 was as follows (numbers in millions): 1965-33.2, 1966-28.5, 1967-27.8, 1968-25.4, 1969-24.3.

a) Find the average number of poor people in the U. S. during the years 1965–1969.

b) Find the average *decrease* in poverty per year for these five years.

2 A basketball player scores the following numbers of points in ten games: 20, 22, 15, 23, 27, 35, 18, 24, 19, 30. How many points does he have to score in the eleventh game to give him an average score of 25 points per game?

3 If the final examination in a course counts three times as much as each hour examination, what is the average grade of a student who received grades of 65, 82, and 85 in three one-hour examinations and 75 in the final examination?

4 A man buys 200 shares of stock at $25\frac{1}{2}$ dollars per share. Several months later he buys 100 shares of the same stock at 18, and a year later he buys 300 shares of the same stock at $6\frac{1}{2}$. Find the average price per share of stock.

5 Why are different procedures required for ungrouped and grouped data when calculating the mean?

6 How can you prove the correctness of a mean of a distribution?

7 In what sense is the median an *average*?

8 What is the essential difference between a median and an arithmetic mean?

9 Find (1) the mean, (2) the media, and (3) the mode for each set of data.
a) 10, 21, 18, 15, 15, 12, 11, 16, 17.
b) 20, 15, 12, 15, 16, 13, 11, 15, 15, 13, 16, 12.
c) 13, 14, 13, 12, 25, 13, 14, 30, 14, 14, 13, 20, 13.

10 In Exercise 9,
a) What score do half of the cases in distribution (a) exceed?
b) What is the most probable score of an individual in distribution (c)?
c) Which of the three distributions is most nearly symmetrical?

11 Find the mean, median, and mode for the following 25 distances (in kilometers) from Boston: 67, 74, 60, 84, 77, 72, 75, 66, 76, 59, 61, 68, 73, 66, 62, 78, 79, 70, 71, 65, 73, 72, 68, 79, 71.

12 Find the mean, median, and mode of the grade distribution in Exercise 7 of Exercise Set 12.1 (page 410).

13 Find the mean, median, and mode of the groceries expenditures in Exercise 10 of Exercise Set 12.1.

14 In each of the following distributions find the mean, median, and mode. State which of the measures of central tendency is most appropriate for each distribution and give a reason for your choice.
a) 10, 17, 10, 11, 11, 27, 11, 10, 22, 9, 10, 11, 10.
b) Scores in an English test:

Scores	*f*
95–99	6
90–94	24
85–89	22
80–84	14
75–79	13
70–74	7
65–69	8
60–64	5
55–59	0
50–54	1
	$N = 100$

c) Strength (in kilograms) in hand-grip for women:

Measurements	*f*
33–35	6
30–32	15
27–29	29
24–26	44
21–23	37
18–20	25
15–17	11
12–14	4
9–11	1
	$N = 172$

15 Under what circumstances are the values of the mean, median, and mode of a distribution the same?

16 Suppose the mean score on a test is 72, the median is 70, and the mode is 69. What would be the effect on the mean, median, and mode:
a) If we added 5 points to each person's score?
b) If we doubled each person's score?
c) If we halved each person's score?

17 What is the relationship among the values of the mean, median, and mode of a distribution which is skewed in the direction of the higher measurements? Of the lower measurements? (*Note* When a distribution is not symmetrical with respect to its mean, it is said to be *skewed.*)

18 Words in the English language vary in length from one letter to as many as thirty or more letters. We now wish to find the "average" length, measured in number of letters, of the words in the following quotation from Aristotle (384–322 B.C.):

"What makes men good is held by some to be nature, by others habit or training, by others instruction. As for the goodness that comes by nature, this is plainly not within our control, but is bestowed by some divine agency on certain people who truly deserve to be called fortunate."

a) Make a frequency distribution of the data.
b) Find the mean, median, and mode of the distribution.
c) Which "average" do you consider most appropriate for indicating the length of a typical word in this paragraph?

19 Which measure of central tendency seems most appropriate to represent:

a) The average shirt size worn by men.
b) The average salary of teachers in a given school district.
c) The average loss of weight in a weight-watchers class.

20 The mean height of five people is 64 inches. The mean height of eight other people is 70 inches. What is the mean height of all thirteen people?

12.3 Measures of Variability

Table 12.7 shows the IQ distribution of two classes. Let us assume that IQ scores are a good measure of mathematical aptitude. Then, since the mean IQ of both classes is 104.5, we may conclude that both classes, on the whole, have equal mathematical ability. Yet if we examine the distribution more carefully, we note a major difference between the classes: the *spread* of ability is decidedly different in the two classes. Whereas the spread in class 1 is very wide, ranging from 75 to 134, the spread in class 2 is much narrower, ranging from 95 to 114. For the mathematics teacher, this difference in degree of spread of mathematical ability presents different teaching situations in each class. The large spread in class 1 probably makes this class much more difficult to teach than class 2.

TABLE 12.7

IQ	*Class 1* f	*Class 2* f
125–134	3	0
115–124	5	0
105–114	10	18
95–104	10	18
85–94	5	0
75–84	3	0
N	36	36
$\overline{X}$	104.5	104.5

The way we describe this difference between the two classes is to say that class 1, compared to class 2, has a greater *spread*, or a greater *scatter*, or a greater *dispersion*, or a greater *variability*. To get a more useful description of a distribution, we must know more about it than the mean of its scores. We must also know whether the scores are close together or far apart. That is, we must also know the degree of dispersion, or variability of the scores.

THE RANGE

The easiest and most quickly ascertained indicator of variability is the *range*. The *range* is the *difference between the highest score and the lowest score.* The range of class 1 in Table 12.7 is 134–75, or 59 points; the range of class 2 is 114–95, or 19 points.

A shortcoming of the range is that its value depends entirely upon the two extreme scores. If one of these scores should be changed or eliminated, the range can have a completely different value. Because of this instability, we think of the range as only a rough index of variability much as we think of the mode as only a rough index of central tendency.

THE STANDARD DEVIATION

The most dependable and most commonly used measure of variability is the *standard deviation.* Let us recall that by *deviation* (x) we mean the difference between a raw score X and the mean $\overline{X}$ of all the scores; that is,

$$x = X - \overline{X}.$$

Roughly, we can think of the standard deviation of a distribution as a kind of average of all the deviations about the mean.

Consider the data in Table 12.8 giving the scores of five students. To obtain an average of the deviations, we would expect to divide the sum of the deviations Σx by the number of deviations N. But since, as we have seen, the algebraic sum of the deviations is always zero, the *average* of the deviations $\overline{x}$ will also always be zero. Such an answer tells us nothing useful about the spread of scores in the distribution.

TABLE 12.8

Student	*Score* X	*Deviation* x
A	17	$^{+}7$
B	13	$^{+}3$
C	9	$^{-}1$
D	6	$^{-}4$
E	5	$^{-}5$
$N = 5$	$\Sigma X = 50$ $\overline{X} = 10$	$\Sigma x = 0$ $\overline{x} = 0$

The way we get around this difficulty is to use the *squares* of the deviations, and thereby obtain only *positive* numbers (Table 12.9). If we divide the sum of the squared deviations Σx^2 by N we get a measure called the *variance*. The symbol for the variance of a distribution is σ^2, where "σ" is the Greek letter *sigma*. That is,

$$\sigma^2 = \frac{\Sigma x^2}{N}. \tag{1}$$

The square root of the variance is the *standard deviation*. That is,

$$\sigma = \sqrt{\frac{\Sigma x^2}{N}}. \tag{2}$$

For the data shown in Table 12.9, the *variance* is

$$\sigma^2 = \frac{\Sigma x^2}{N} = \frac{100}{5} = 20,$$

and the *standard deviation* is

$$\sigma = \sqrt{\frac{\Sigma x^2}{N}} = \sqrt{\frac{100}{5}} = \sqrt{20} = 4.47.$$

TABLE 12.9

Student	*Score* X	*Deviation* x	*Deviation Squared* x^2
A	17	$^+7$	49
B	13	$^+3$	9
C	9	$^-1$	1
D	6	$^-4$	16
E	5	$^-5$	25
$N = 5$	$\Sigma X = 50$ $\overline{X} = 10$	$\Sigma x = 0$	$\Sigma x^2 = 100$

To find the standard deviation for the *grouped* data in Table 12.10, we square each deviation, as we did before, and then multiply it by its frequency. The sum of these products Σfx^2 will then be the sum of all the squared deviations in the distribution. So for grouped data, such as in Table 12.10, the formula for the *variance* becomes

$$\sigma^2 = \frac{\Sigma fx^2}{N}, \tag{3}$$

and the formula for the *standard deviation* becomes

$$\sigma = \sqrt{\frac{\Sigma fx^2}{N}}. \tag{4}$$

TABLE 12.10

Score	x	x^2	f	fx^2
20	$^+10$	100	2	200
14	$^+4$	16	3	48
9	$^-1$	1	8	8
4	$^-6$	36	4	144
$\overline{X} = 10$			$N = 17$	$\Sigma fx^2 = 400$

The *variance* for the data in Table 12.10 is

$$\sigma^2 = \frac{\Sigma fx^2}{N} = \frac{400}{17} = 23.53.$$

The *standard deviation* for the data in Table 12.10 is

$$\sigma = \sqrt{\frac{\Sigma fx^2}{N}} = \sqrt{\frac{400}{17}} = \sqrt{23.53} = 4.85.$$

If the grouped data is presented in class intervals, as in Table 12.5, we let the midpoint (m) of each interval represent all the cases within the interval, and the deviation is then

$$x = m - \overline{X}.$$

The rest of the procedure is the same as in Table 12.10. There are more efficient methods of finding the standard deviation from grouped data, but space does not permit us to explain them here.

EXERCISE SET 12.3

1. If the highest score in a distribution is 70 and the lowest is 29, what is the range of the distribution?
2. Compute the standard deviation for the following set of measurements: 5, 7, 11, 13, 16.
3. Why would the range be a misleading indicator of variability for the following set of data: 25, 23, 24, 5, 21, 22?
4. Compute the standard deviation for the distribution in Exercise Set 12.2, Exercise 9 (page 417).
5. Describe the dispersion of the following sets of scores:
   a) 59, 58, 57, 56, 55.
   b) 95, 75, 70, 25, 6, 2.
6. Two normal curves are shown in the figure.
   a) Which curve shows the greater mean?
   b) Which shows the greater standard deviation?
   c) Which shows the greater variance?
   d) Which shows the greater median?

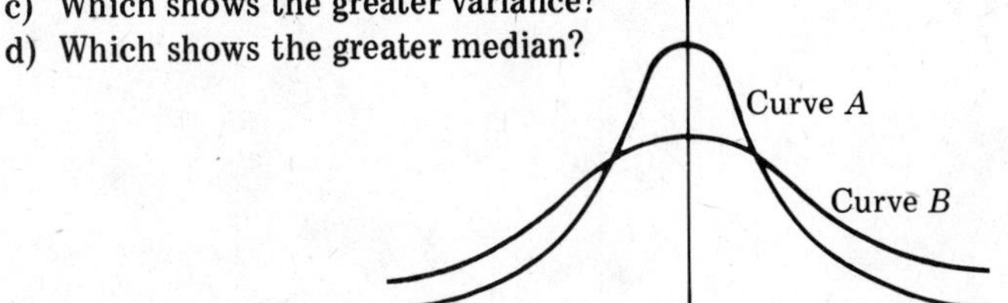

7 Given the following set of measurements: 1, 2, 3, 4, 5, 6, 7, 8, 9, 10.
a) Find the standard deviation.
b) How many of the measurements are withing $\pm 1\sigma$ of the mean?
c) How many are within $\pm 2\sigma$?
d) How many are within $\pm 3\sigma$?

8 Is it possible for all of the measurements in a distribution to be less than one σ from the mean? Why?

9 Compute the standard deviation for the distribution in Exercise Set 12.1, Exercise 7 (page 410).

10 Compute the standard deviation for the data in Table 12.5 (page 414).

11 Compute the standard deviation for the distribution in Table 12.3 (page 412).

12 What does each of the following indicate about a distribution?
a) A small standard deviation.
b) A large standard deviation.
c) The standard deviation equals zero.

12.4 The Normal Distribution

Suppose the mean of a distribution of 10,000 scores is 50 and the standard deviation is 10, what does the number 10 tell us? To answer this question, we shall first consider a special distribution called a *normal distribution*, or a *normal curve*, which is one of the most basic tools in all of statistics. A normal distribution is not a distribution that is obtained by actually measuring something; it is a *theoretical* distribution which assumes that the number of cases N is infinite. The reason that this distribution is so important is that many distributions that are obtained by actually measuring something are distributed like the normal distribution, especially if N is large. Many characteristics in education, psychology, and in the biological and social sciences, when measured, are found to be approximately normally distributed.

A normal curve is shown in Figure 12.4 where μ (the Greek letter mu) indicates the mean of the distribution. Note that the vertical axis indicates frequency and the horizontal axis indicates standard deviation units.

FIGURE 12.4

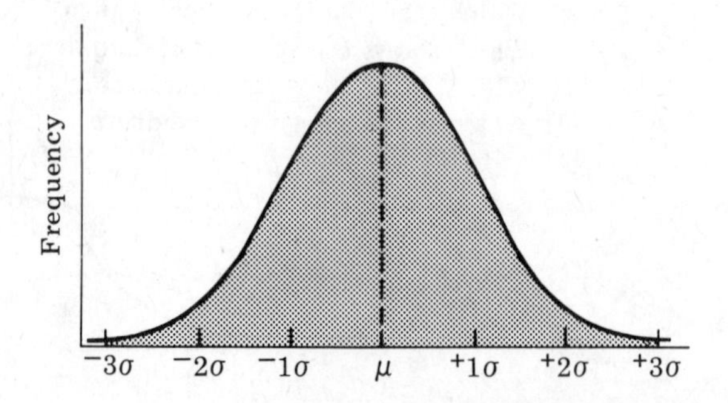

The most obvious characteristic of the normal curve is that it is shaped like a *bell*. It rises to a rounded peak in the middle and tapers off symmetrically at both tails. The tails extend indefinitely in both directions, approaching closer and closer to the horizontal axis without ever reaching it. The shape of the curve tells us that in a normal distribution *both* low scores and high scores are infrequent, and that scores close to the mean occur frequently. Thus, if we assume that intelligence is distributed normally, then there are more people with average intelligence than people with low or high intelligence, and the further we move from the mean IQ, the fewer the number of people who have that score.

AREAS UNDER THE NORMAL CURVE

Much of the use of the normal curve in statistics is concerned with proportions, or percents, of the area under the curve in relation to a given distance on the horizontal axis, or base line. The total area under the curve, that is, the area between the curve and the base line, is taken as 100%. If we draw two lines from the base line up to the curve (Figure 12.5), one line at μ and the other at $^{+}1\sigma$, the area under that portion of the curve will always be .3413, or 34.13% of the total area, *if the curve is normal.* Since the percent area is the same as the percent frequency, we can conclude that in a *normal distribution* 34.13%, or a little more than one-third of all the cases, are included between the mean and 1σ from the mean.

FIGURE 12.5

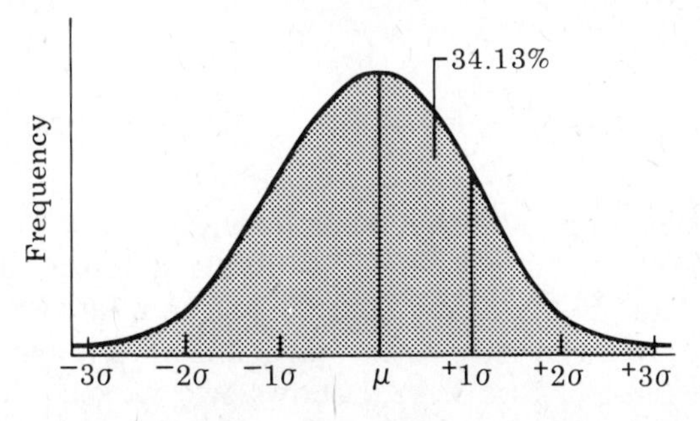

Since the normal curve is symmetrical, another 34.13% of the cases lie between μ and $^{-}1\sigma$. Therefore, 68.26% of the cases fall between $^{-}1\sigma$ and $^{+}1\sigma$ (Figure 12.6). Between μ and $^{+}2\sigma$ lie 47.72% of the cases; so 95.44% of all the cases in a normal distribution fall between $^{-}2\sigma$ and $^{+}2\sigma$. Three σ's above and below μ, that is, six σ's altogether, include 99.74%, or almost all, of the total number of cases.

FIGURE 12.6

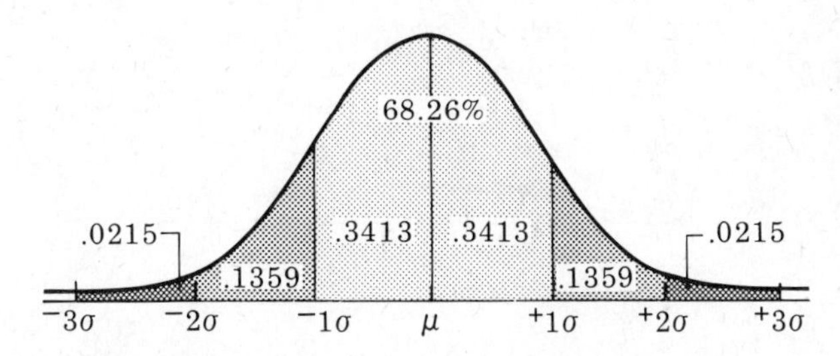

The meaning of these figures can be seen if we assume that the weight distribution for a sample of 10,000 women was exactly normal. In this case, 34.13% of 10,000, or 3413 women would weigh between the mean weight of the distribution and the weight that is one σ above the mean. Likewise, the weight of 9974 women (99.74% of 10,000) would fall between three σ's below the mean and three σ's above the mean.

We can now answer the question asked earlier. Suppose the mean of a distribution of 10,000 scores is 50 and the standard deviation is 10, what does the number 10 tell us? Here,

$$N = 10{,}000, \quad \mu = 50, \quad \text{and} \quad \sigma = 10.$$

A score of $^{+}1\sigma$ would be 50 + 10, or 60; a score of $^{+}2\sigma$ would be 50 + 20, or 70; a score of $^{-}1\sigma$ would be 50 − 10, or 40.

If the distribution of these scores were exactly normal, then 3413 students would score between μ and $^{+}1\sigma$, or between 50 and 60; 6826 students, or better than $\frac{2}{3}$ of all the students, would score between $^{-}1\sigma$ and $^{+}1\sigma$, or between 40 and 60; the scores of nearly all the students, 9974 students, would fall between $^{-}3\sigma$ and $^{+}3\sigma$, or between 20 and 80.

THE STANDARD NORMAL CURVE

Whenever a distribution is approximately normal, we can use the normal curve to help us obtain much useful information about the data. But a slight complication can arise from the fact that normal curves can have different bell shapes, depending on the μ and σ. For example, Figure 12.7 shows two normal curves: one has $\mu = 10$ and $\sigma = 5$; the other has $\mu = 20$ and $\sigma = 10$. So the question is, which normal curve do we use as a standard?

FIGURE 12.7

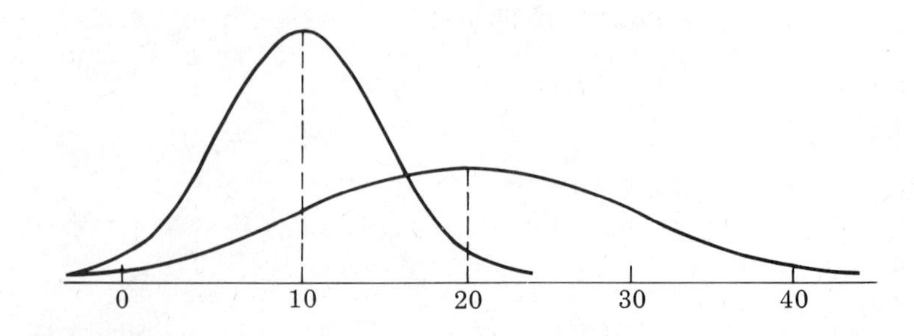

Fortunately, although normal curves can have many different shapes, there is only *one* normal curve with a given μ and a given σ. The unique normal curve with

$$N = 1, \mu = 0, \quad \text{and} \quad \sigma = 1$$

is the *standard normal curve* on which statistical tables are based. Areas under any other normal curve can then be obtained by converting the raw scores into *standard scores*, called *z values*, by means of the formula

$$z = \frac{X - \mu}{\sigma}. \tag{1}$$

A closer look at this formula shows that a standard score z is nothing more than a deviation from the mean in terms of the standard deviation. For example, suppose a normal distribution has a mean of 6 and a standard deviation of 1.5. Then the z corresponding to a raw score of 9 is

$$z = \frac{X - \mu}{\sigma} = \frac{9 - 6}{1.5} = \frac{3}{1.5} = {}^{+}2.0.$$

A z of $^{+}2.0$ tells us that the score of 9 is 2 σ's *above* the mean of 6. The z corresponding to $X = 4$ is

$$z = \frac{X - \mu}{\sigma} = \frac{4 - 6}{1.5} = \frac{{}^{-}2}{1.5} = {}^{-}1.33.$$

A z of $^{-}1.33$ tells us that the score of 4 is 1.33 σ's *below* the mean of 6.

We use such z values in Table 12.11 on page 430 to obtain various areas under the standard normal curve.

THE STANDARD NORMAL DISTRIBUTION TABLE

Table 12.11 on page 430 gives the areas under the standard normal curve between the mean ($\mu = 0$) and z = 0.00, 0.01, 0.02, ..., 3.08, 3.09. To find the area between μ and, say, $z = {}^{+}1.96$, we first locate 1.9 in the z column, then go along that row to the column headed .06 and read off the area, which is .4750. This answer means that 47.50% of the total area under the curve lies between μ and $^{+}1.96\sigma$; that is, 47.50% of the scores in the distribution lie between the mean and 1.96 σ's.

Note that the table has no entries for *negative* values of z since these are not needed. The area between μ and $^{+}1.25\sigma$ is the same as the area between μ and $^{-}1.25\sigma$ because of the symmetry of the curve.

USING TABLE 12.11 TO SOLVE PROBLEMS

Example 1 If $N = 1{,}000$, $\mu = 26$, $X = 35$, and $\sigma = 6$, how many scores will fall between the mean and the score of 35?

Solution $z = \frac{35 - 26}{6} = \frac{9}{6} = {}^{+}1.5$ (Figure 12.8).

FIGURE 12.8

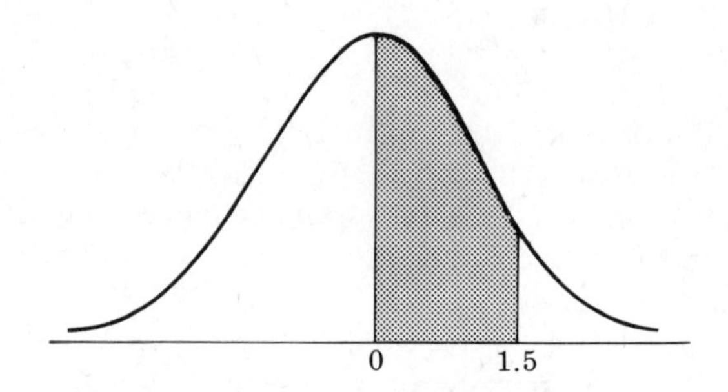

From Table 12.11, the area for $z = {}^{+}1.5$ is .4332; that is, 43.32% of the scores in a normal distribution would be found between the mean and a score of 35. If $N = 1{,}000$, there would be about 433 such scores.

Example 2 Find the area between $z = {}^{+}1.8$ and $z = {}^{+}2.3$.

Solution We are looking for area A in Figure 12.9. We cannot find this area directly from the table. Instead, we look up the area between $z = 0$ and $z = 2.3$, the area between $z = 0$ and $z = 1.8$, and then find the *difference* between the two.

Area between $z = 0$ and $z = 2.3$ is .4893
Area between $z = 0$ and $z = 1.8$ is .4641
Difference = .0252

FIGURE 12.9

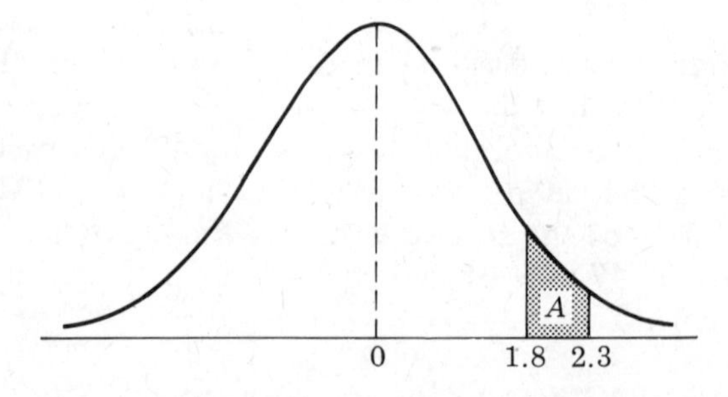

Example 3 Find the area between $z = {}^{-}1.5$ and $z = {}^{-}0.83$.

Solution The area will be the same as the area between $z = {}^{+}1.5$ and $z = {}^{+}0.83$; that is, $.4332 - .2967 = .1365$.

Example 4 Find the area between $z = {}^{-}0.95$ and $z = {}^{+}1.23$.

Solution From Figure 12.10 we see that the area we are looking for is the *sum* of areas A and B:

Area A = .3289
Area B = .3907
Sum = .7196

FIGURE 12.10

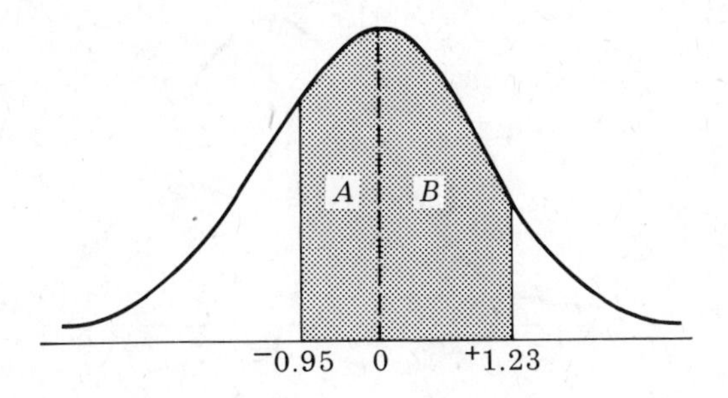

Example 5 For the distribution in Example 1, how many cases can be expected to fall *below* a score of 15?

Solution

1 We are given $N = 1{,}000$, $\mu = 26$, $X = 15$, $\sigma = 6$. Therefore,

$$z = \frac{15 - 26}{6} = \frac{{}^{-}11}{6} = {}^{-}1.83.$$

We now wish to find area A in Figure 12.11.

FIGURE 12.11

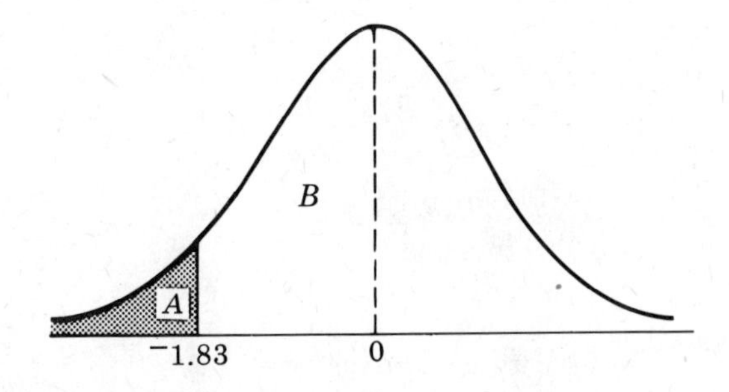

2 From Table 12.11 we note that for $z = {}^{-}1.83$, area B is .4664.

3 Since the sum of area A and area B is .5000, area $A = .5000 - .4664$, or .0336.

4 We can, therefore, expect 3.36% of all the scores to fall below 15. Since $N = 1{,}000$, we can expect 33.6, or about 34 scores to fall below 15.

Example 6 Above what point in the normal curve do the *highest* 3% of the scores fall?

Solution In Examples 1–5 we were given, or were able to find, z scores and were asked to find the corresponding areas. In this problem we are given an area and are asked to find the corresponding z score.

1. In Figure 12.12, the sum of areas A and B is half the total area under the curve; that is, $A + B = .5000$.

FIGURE 12.12

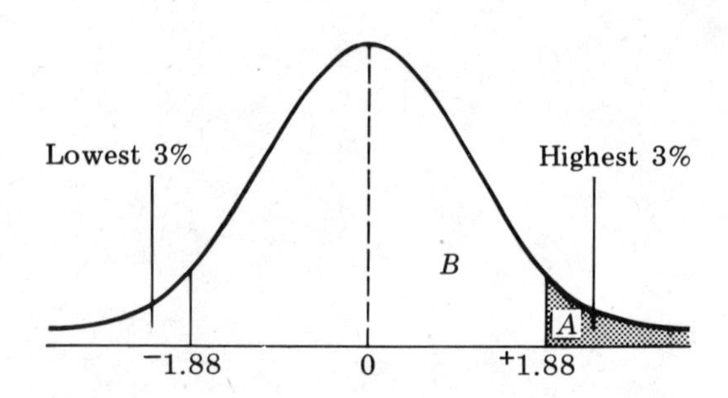

2. If we want area A to be .0300, then $B = .4700$. We now use Table 12.11 to find the z value for B.
3. We note that .4700 is closest to the entry of .4699. The z that corresponds to .4699 is 1.88.
4. Therefore, the highest 3% of the scores fall above the point $z = {}^+1.88$. Note that the z is positive since it is to the right of $z = 0$. Had we asked for the *lowest* 3% of the scores, the answer would have been the point $z = {}^-1.88$.

EXERCISE SET 12.4

1. Find the area under the standard normal curve which lies:
   a) Between $z = 0$ and $z = 0.91$.
   b) Between $z = 0$ and $z = {}^-1.2$.
   c) To the left of $z = 0.87$.
   d) To the left of $z = {}^-1.38$.
   e) To the right of $z = 0.65$.
   f) To the right of $z = {}^-1.05$.
   g) Between $z = 0.98$ and $z = 0.42$.
   h) Between $z = {}^-1.78$ and $z = {}^-0.54$.
   i) Between $z = {}^-0.65$ and $z = 2.40$.
   j) Between $z = 1.89$ and $z = {}^-2.76$.
2. A normal distribution of scores has a mean of 62.2 and a standard deviation of 5.3. Convert the following raw scores to standard scores.
   a) 42 b) 82 c) 53 d) 74

3. Complete the following statements as they relate to a normal distribution.
   a) Between the mean and a point 1σ above it, the percentage of cases is ____.
   b) Between the mean and a point $^{-}1.4\sigma$, the percentage of cases is ____.
   c) The percentage of cases between $^{-}1.3\sigma$ and 0.85σ is ____.
   d) The percentage of cases below $^{-}2.31\sigma$ is ____.
   e) The percentage of cases above $^{-}1.87\sigma$ is ____.
   f) The middle 70% of the cases lie between ____σ and ____σ.
   g) The highest 1% of the cases lies above the point ____.
   h) Almost ____ percent of the scores lie within three standard deviations from the mean.
   i) About ____ percent of the scores are within three standard deviations to the right of the mean.
4. Find z if the normal curve area
   a) Between 0 and z is 0.3708.
   b) To the right of z is 0.7454.
   c) To the left of z is 0.6879.
   d) To the right of z is 0.2373.
   e) Between ^{-}z and z is 0.7416.
5. A student receives a score of 700 on a college entrance test for which the mean is 500 and the standard deviation is 100.
   a) How many standard deviations above or below the mean is this student's score?
   b) What percent of the scores will be above 750?
   c) What percent of the scores will be below 350?
   d) What percent of the scores will fall between 400 and 700?
6. The heights of 10,000 employees of a company are approximately normally distributed with a mean of 68 inches and a standard deviation of 9 inches. If a certain job requires individuals whose height is at least 65 inches but not over 72 inches, how many of the employees should be suitable for the job on the basis of height alone?
7. The mean of an approximately normal distribution of reading scores is 50 and the standard deviation is 10. The mean of an arithmetic test is 48 with a standard deviation of 8. If a student scores 52 on the reading test and 50 on the arithmetic test, has he done better in reading or in arithmetic? Why?
8. A thousand students take a test on which student A receives a standard score of 1.87. If the results are normally distributed:
   a) How many students received a higher score than student A?
   b) What is the probability that a randomly selected student received a lower score than student A?
9. In a given distribution, the mean is 90 and the standard deviation is 7.8. If the measurements are distributed normally:
   a) What percent of the cases would you expect to fall between 80 and 88?
   b) What is the probability that a score, chosen at random, will fall between 80 and 88?
   c) What is the probability that a randomly selected score will be 75 or more?

TABLE 12.11 The Standard Normal Distribution

z	*.00*	*.01*	*.02*	*.03*	*.04*	*.05*	*.06*	*.07*	*.08*	*.09*
0.0	.0000	.0040	.0080	.0120	.0160	.0199	.0239	.0279	.0319	.0359
0.1	.0398	.0438	.0478	.0517	.0557	.0596	.0636	.0675	.0714	.0753
0.2	.0793	.0832	.0871	.0910	.0948	.0987	.1026	.1064	.1103	.1141
0.3	.1179	.1217	.1255	.1293	.1331	.1368	.1406	.1443	.1480	.1517
0.4	.1554	.1591	.1628	.1664	.1700	.1736	.1772	.1808	.1844	.1879
0.5	.1915	.1950	.1985	.2019	.2054	.2088	.2123	.2157	.2190	.2224
0.6	.2257	.2291	.2324	.2357	.2389	.2422	.2454	.2486	.2517	.2549
0.7	.2580	.2611	.2642	.2673	.2704	.2734	.2764	.2794	.2823	.2852
0.8	.2881	.2910	.2939	.2967	.2995	.3023	.3051	.3078	.3106	.3133
0.9	.3159	.3186	.3212	.3238	.3264	.3289	.3315	.3340	.3365	.3389
1.0	.3413	.3438	.3461	.3485	.3508	.3531	.3554	.3577	.3599	.3621
1.1	.3643	.3665	.3686	.3708	.3729	.3749	.3770	.3790	.3810	.3830
1.2	.3849	.3869	.3888	.3907	.3925	.3944	.3962	.3980	.3997	.4015
1.3	.4032	.4049	.4066	.4082	.4099	.4115	.4131	.4147	.4162	.4177
1.4	.4192	.4207	.4222	.4236	.4251	.4265	.4279	.4292	.4306	.4319
1.5	.4332	.4345	.4357	.4370	.4382	.4394	.4406	.4418	.4429	.4441
1.6	.4452	.4463	.4474	.4484	.4495	.4505	.4515	.4525	.4535	.4545
1.7	.4554	.4564	.4573	.4582	.4591	.4599	.4608	.4616	.4625	.4633
1.8	.4641	.4649	.4656	.4664	.4671	.4678	.4686	.4693	.4699	.4706
1.9	.4713	.4719	.4726	.4732	.4738	.4744	.4750	.4756	.4761	.4767
2.0	.4772	.4778	.4783	.4788	.4793	.4798	.4803	.4808	.4812	.4817
2.1	.4821	.4826	.4830	.4834	.4838	.4842	.4846	.4850	.4854	.4857
2.2	.4861	.4864	.4868	.4871	.4875	.4878	.4881	.4884	.4887	.4890
2.3	.4893	.4896	.4898	.4901	.4904	.4906	.4909	.4911	.4913	.4916
2.4	.4918	.4920	.4922	.4925	.4927	.4929	.4931	.4932	.4934	.4936
2.5	.4938	.4940	.4941	.4943	.4945	.4946	.4948	.4949	.4951	.4952
2.6	.4953	.4955	.4956	.4957	.4959	.4960	.4961	.4962	.4963	.4946
2.7	.4965	.4966	.4967	.4968	.4969	.4970	.4971	.4972	.4973	.4974
2.8	.4974	.4975	.4976	.4977	.4977	.4978	.4979	.4979	.4980	.4981
2.9	.4981	.4982	.4982	.4983	.4984	.4984	.4985	.4985	.4986	.4986
3.0	.4987	.4987	.4987	.4988	.4988	.4989	.4989	.4989	.4990	.4990

PUZZLES

1 State police are checking the speed of passing cars at Checkpoint Jimmy where the legal speed limit is 55 miles per hour. The mean speed of passing cars is 49.1 miles per hour with a standard deviation of 5.3 miles. If we assume that the speeds are normally distributed and that the police will ticket any car traveling faster than 55 miles per hour, how many speeding tickets can the police expect to hand out for the first 200 cars that pass Checkpoint Jimmy?

2 Suppose that the actual amount of potato chips that a filling machine puts into a 12-ounce bag varies from bag to bag and is distributed normally with a standard deviation of 0.16 ounces. If only 2% of the bags are to contain less than 12 ounces of potato chips, what must be the average amount of potato chips which the machine puts into the bags?

BASICS REVISITED

1 Add: 56 + 89627 + 3981 + 593047 + 295.

2 Subtract 13587 from 74000.

3 $\left(\frac{3}{5} \times 4\right) \div \left(\frac{7}{8} - \frac{2}{11}\right) = ?$

4 What percent of 270 is .09?

5 Divide 365 by .54. (Round to 2 places.)

6 $30 - 12\frac{3}{8} = ?$

7 Round off to the nearest tenth: $\frac{8 + 5 \times 3}{45 + 17}$.

8 Find .75% of 38.

9 If 3% of a number is 2.91, what is the number?

10 Estimate the product of 2.103 and 3.94.

11 A toy is purchased for $3.20 at a discount of 25%. What is the regular price of the toy?

12 How much is 200% of $89.25?

13 What percent of an inch is a foot?

14 Estimate the quotient of $\frac{56.02}{1.123}$.

15 $\dfrac{\frac{3}{10} \times \frac{7}{8}}{\frac{5}{6} \div \frac{3}{8}}$

16 A road on a street map measures $2\frac{1}{8}$ inches. If the map is drawn to a scale of $\frac{1}{8}$ in. = 2 miles, how long is the road?

17 Separate 100 into three parts such that the second is four times as large as the first, and the third is five times as large as the first.

18 Mr. Jones is 45 years old and his nephew is 9 years old. In how many years will Mr. Jones be three times as old as his nephew?

19 There is a number which, when multiplied by 3 and then has 27 taken away from it, gives 33. Find the number.

20 A man sold his house for $21,500 at a profit of 10% on the selling price. How much did the house cost him?

Complete the following table.

	Common Fraction	Decimal Fraction	Percent
21	?	.625	?
22	$\frac{2}{3}$	?	?
23	?	?	$83\frac{1}{3}$
24	?	?	$87\frac{1}{2}$
25	?	.125	?

EXTENDED STUDY

12.5 Sampling in Statistics

How is it possible for a public opinion poll to predict how 76 million people will vote in a Presidential election from the opinions of only 1500 people? The secret lies in the concept of *sampling*, which is one of the most important ideas in statistics.

A *sample*, as we have already seen, is a relatively small group of elements chosen to represent a larger group of elements called a *population.* For a sample to represent the population accurately, it must be sufficiently large and it must be selected randomly; that is, every element in the population must have the same chance to be selected as a member of the sample as every other element. Also, the selection of one element must in no way affect the selection of another; that is, the selections must be independent of each other. By drawing our sample randomly, we hope to produce a miniature picture of the population.

The sample must be sufficiently large because both experience and theory tell us that the larger the sample the less is the variation from sample to sample in the statistics we seek. This means, for example, that the error in predicting the outcome of an election on the basis of a small sample is likely to be more serious than the error from a large sample.

RANDOMNESS

The selection of a sample must be based on randomness if we are to be able to use the theory of probability to draw inferences about the population from the sample. When the method of selecting a sample does not satisfy this condition, we say that the sample is *biased.*

A famous example of biased sampling occurred in 1936 when the *Literary Digest* failed to correctly forecast the results of the Landon-Roosevelt Presidential election. The magazine's survey chose its sample from automobile registration lists and from telephone directories. The trouble with this procedure was that during the depression of the 1930s large numbers of voters did not own cars and were not telephone subscribers, and consequently were not represented in the sample. The *Literary Digest* went out of business shortly thereafter.

Randomization is usually accomplished in one of two ways: by the use of a lottery technique or by the use of a *table of random numbers.* A lottery technique was used by the Selective Service to select men for induction into the armed services at the beginning of World War II. Each of about 800 numbers was printed on uniform slips of paper which were then rolled and placed in uniform capsules. The capsules were placed in a large bowl, thoroughly mixed, and then drawn one at a time, with thorough mixing between each drawing.

A similar technique can be used to obtain information of, say, a class of 29 children as to the number of hours of television they watch each week. We can draw a sample in the following way. Assign to each child a number from 1 to 29. Place in one bowl three uniform cards numbered 0, 1, 2, and in another bowl place ten uniform cards numbered 0, 1, 2, 3, 4, 5, 6, 7, 8, 9. By drawing a card from each bowl, any number from 0 to 29 can appear. After each drawing, replace the cards, mix all the cards thoroughly, and draw again. If we wish a random sample of five students, we draw five numbers, ignoring the 00 and any repeat number. We can follow this procedure for any size sample we wish.

A better way of assuring randomness is to use a *table of random numbers.* Such a table consists of a rectangular array of numbers which have been placed in sequence by some kind of lottery procedure and electronic method. If individuals in a population are numbered in sequence and identified by their number, then random selections of these individuals can be made by following the random numbers.

STRATIFIED SAMPLING

The sampling procedures just described are called *simple random sampling.* A more sophisticated method of sampling is *stratified sampling*, which is the method commonly used in public opinion polls. A *stratified sample* is obtained by dividing the population into subgroups and then selecting from each subgroup a simple random sample. The stratified sample is the composite of all the subgroup samples. It has been found that a stratified sample often predicts the population characteristics more accurately than a simple random sample.

To poll voter preference from a list of candidates for the Presidency of the U. S., it is likely that variables like party affiliation, socio-economic status, education, age, the section of the country where the voter resides, etc. will influence the voter's choice. Having decided which variables must be reflected in the sample, the pollster then studies the entire population to see what percent falls into each category; that is, what percent are Democrats or Republicans, urban or rural, over age 30 or below, in each socio-economic group. The stratified sample is then selected to reflect the same percentages of the variables as those that exist in the total population. Obviously, the sample will depend on those variables that the pollster decides are important. Judging by the accuracy with which American election results have been forecast in recent years, it is apparent that the pollsters have been selecting the right variables.

SAMPLE SIZE

We said that for a sample to represent the population accurately, the sample must be sufficiently large. How large is *large*? The answer depends on the circumstances and purpose of the study.

From the point of view of probability theory and the behavior of sample statistics, if N is less than 25 or 30, then the sample is considered small. Samples whose N is 30 or more are considered large. In a large sample, it does not matter whether N is 30 or 2,000 since the shape of the distribution will always be approximately the same: about 34% of the area under the curve is between $^{-}1\sigma$ and 1σ, and the curve, generally, approximates the normal curve. But if N is, say, 2, then only about 25% of the area lies between $^{-}1\sigma$ and 1σ, and the curve is much more peaked than the normal curve.

The size of the sample is also related to the type of study being pursued. A sample of 10 test items may be sufficient to distinguish the feeble-minded from the intelligent, but we may need 100 items to distinguish the intelligent from the very intelligent. Public opinion pollsters found that in 1972 a sample of about 1500 people was sufficient for predicting how 76 million Americans would vote in the Presidential election.

Other factors in sample size are the precision requirements of the study as well as the cost of sampling.

12.6 Skewed Distributions

From a frequency distribution, we can obtain important information about the data. We can obtain a measure of its central tendency and a measure of its variability, and we can determine the degree to which the distribution is symmetric. We have already discussed measures of central tendency and measures of variability. Now we will discuss symmetry or lack of symmetry in a distribution.

Frequency distributions, and consequently their graphs, can have an unlimited number of shapes. Some are *symmetrical*; that is, if the graph is folded in half, along a line perpendicular to the base line, the two halves would coincide. Examples of *symmetrical distributions* are shown in Figure 12.13.

FIGURE 12.13

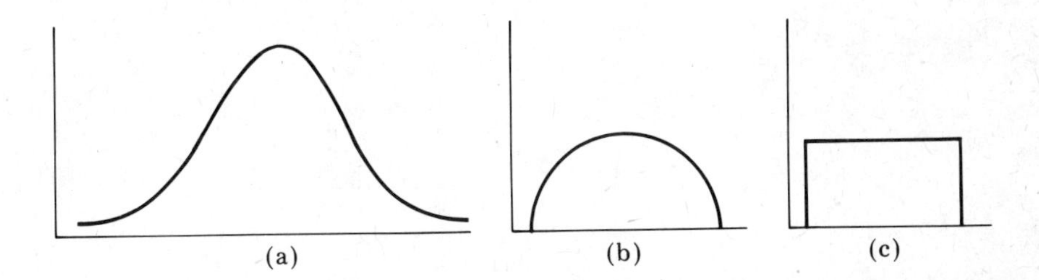

A distribution that is not symmetrical is said to be *skewed.* Several examples of skewed distributions are shown in Figure 12.14. Although all these distributions are skewed, they differ in degree and direction of skewness. By *degree of skewness*, we mean the amount by which the curve deviates from being symmetrical. Thus, in figure 12.14, distribution (a) is more skewed than distribution (b).

FIGURE 12.14

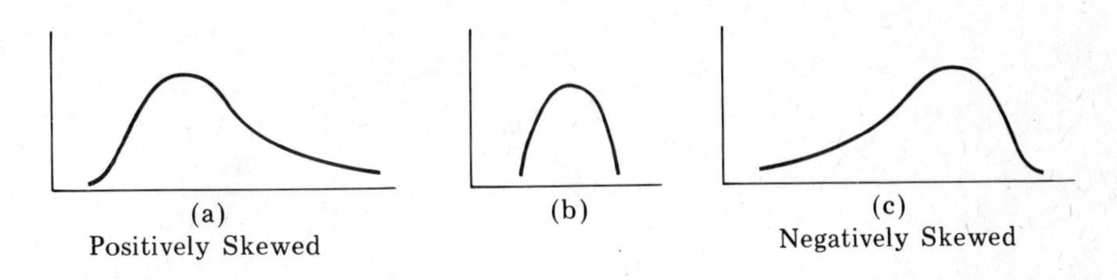

Sometimes the measures of a distribution tend to pile up on the left-hand side of the graph with a tail extending to the right, as shown in Figure 12.14(a). Since the tail extends to the *right*, we say that the curve is *positively skewed.* In Figure 12.14(c), the measures tend to pile up on the right-hand side of the graph with a tail extending to the *left*; we describe this curve as *negatively skewed.* The direction toward which the tail extends gives us the *direction of skewness.*

The degree of skewness can be described roughly as the difference between the mean and median divided by the standard deviation. The *Pearsonian coefficient of skewness* uses the formula

$$\text{Skewness} = \frac{3\,(\overline{X} - \text{Mdn})}{\sigma}.$$

For a symmetrical distribution, where $\overline{X}$ and Mdn are the same, the skewness will be zero.

In a skewed distribution, the mean and the median are pulled away from the mode toward the tail of the distribution with the mean being displaced farther from the mode than the median. The relative positions of the mean, median, and mode for a *negatively skewed distribution* are shown in Figure 12.15(a). A *positively skewed distribution* is shown in Figure 12.15(b).

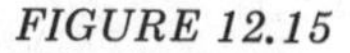

FIGURE 12.15

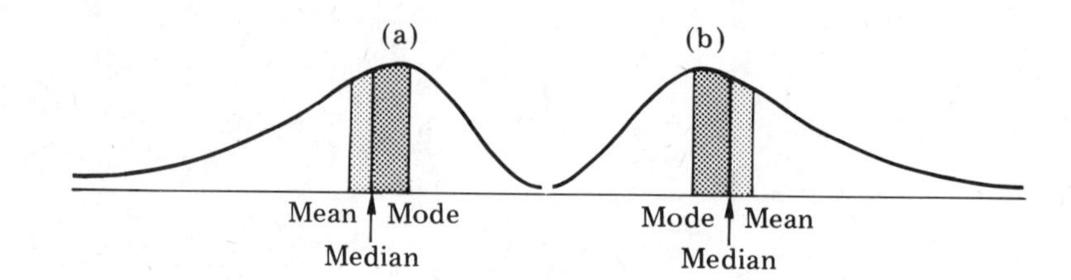

12.7 Correlation

The distributions we have considered so far involved a single variable such as test scores or heights for a group of individuals. There are other distributions, however, that are based on the relatedness of *two* variables, not on a single variable. For example, the data in Table 12.12 shows eight people who have been on a reducing diet for a varying number of months and how much weight each person lost.

TABLE 12.12

Person	*Number of Months*	*Weight Loss in Pounds*
A	2	10
B	3	12
C	4	16
D	5	27
E	6	22
F	9	43
G	12	45
H	15	64

The question we wish to consider now is whether or not there is a relationship between the two variables—between the number of months a person was on the diet and weight loss. If such a relationship does exist, then we would like to have a measure of its strength and direction.

An obvious conclusion we can draw from the given data is that a high number of months on the diet is associated with a high weight loss, and that a low number of months on the diet is associated with a low weight loss. But we still do not know how strong this relationship is. To determine whether or not a relationship exists between two variables and to determine its precise strength and direction, we use a statistical technique called *correlation.*

A *coefficient of correlation* is a number that tells us to what extent two variables are related; that is, to what extent variations in one go with variations in the other, and the direction in which this relatedness goes. We should note that the mere existence of this relatedness between two variables does *not* necessarily imply that one variable *causes* or *influences* the other. Sometimes two variables are related because they are both caused by a third variable. For instance, high math scores may go with high reading scores, not because math ability causes reading ability, but because both may be influenced by the same third variable—intelligence. An important use of a coefficient of correlation is to enable us to predict one variable from another variable. For example, if there is a high correlation between an industrial test score and success in a certain job, then we can predict which applicants are likely to be successful in their work by merely looking at their test scores.

THE PEARSON CORRELATION COEFFICIENT

Statisticians use several different correlation coefficients, depending on the situation under study. But we shall use only the *Pearson product-moment correlation coefficient* (r), named after Karl Pearson (1857–1936), one of the important pioneers in statistics. The size of r varies from $^{+}1.00$ to $^{-}1.00$ and tells us two things about a relation between two variables: its *strength* and its *direction.* A correlation of $^{+}.85$ and a correlation of $^{-}.85$ represent the same strength but opposite directions. When two variables have a correlation of $^{+}.85$, we say that they are *positively* related; that is, as one variable increases the other also increases. The variables in Table 12.12 have a positive correlation since as the number of months on the diet increases, the weight loss also increases. However, an r of $^{-}.85$ indicates a *negative*, or *inverse* relationship; that is, as one variable increases the other *decreases.* For example, the age of a car and its trade-in value have a negative correlation since the greater the age of the car, the smaller the trade-in value.

The *absence* of a relationship is denoted by a correlation coefficient of .00. For example, if the number of cigarettes a man smokes a day is the same regardless of the number of hours he works a day, then we say that there is a *zero correlation* between the number of cigarettes he smokes and the number of hours he works. An r of $^{+}1.00$ is a perfect positive correlation; an r of $^{-}1.00$ is a perfect negative correlation; and an r of .00 indicates complete independence between the variables.

The graphs in Figure 12.16, 12.17, 12.18, and 12.19 illustrate several kinds of correlations. These graphs are called *scatter plots* because they show how the points scatter over the range of possible scores. Every point in the scatter plot represents two values: an individual's score on one variable X, and the same individual's score on a second variable Y.

Figure 12.16 shows a perfect positive correlation between the variables X and Y. Notice that every value of Y is two units more than the corresponding value of X. When all the points are plotted, they fall along a straight line. Because the line runs from the lower left of the scatter plot to the upper right, it represents a perfect positive relationship whose correlation coefficient is $^{+}1.00$. This number signifies that the individual with the highest score on X also has the highest score on Y; the second highest scorer on X is second highest on Y; this pattern continues and the lowest scorer on X is the lowest scorer on Y.

FIGURE 12.16

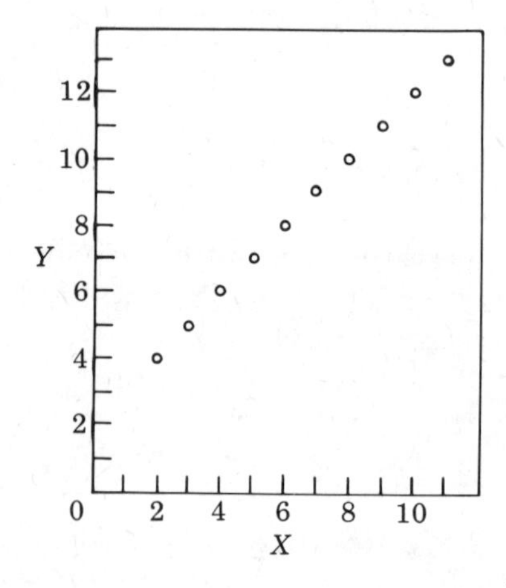

X	2	3	4	5	6	7	8	9	10	11
Y	4	5	6	7	8	9	10	11	12	13

Figure 12.17 illustrates a perfect negative correlation. Notice that for every *increase* of two units in X, there is a corresponding *decrease* of one unit in Y. The points again fall along a straight line, but the line runs from the upper left of the scatter plot to the lower right. Such a graph illustrates a perfect negative relationship whose correlation

coefficient is $^{-}1.00$. This number tells us that the individual with the highest score on X has the lowest score on Y; the second highest scorer on X is second lowest scorer on Y; this pattern continues and the lowest scorer on X is the highest scorer on Y.

FIGURE 12.17

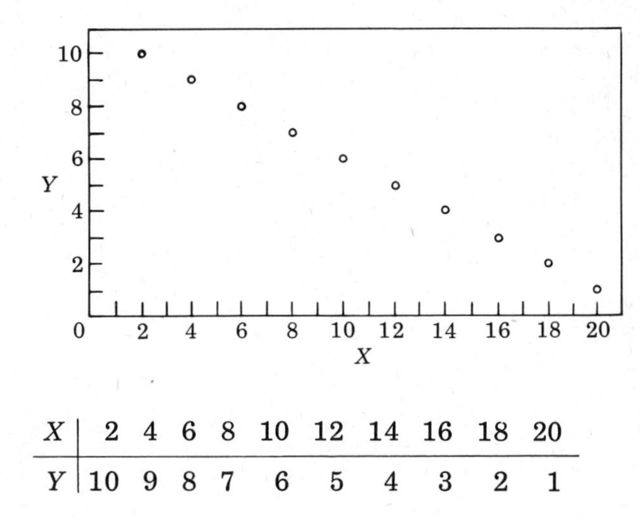

X	2	4	6	8	10	12	14	16	18	20
Y	10	9	8	7	6	5	4	3	2	1

In real life we seldom encounter perfect positive or negative relationships. Figure 12.18 shows a scatter plot of high positive relationship. Note that while the points do not fall along a perfect straight line, they approximate a line.

FIGURE 12.18

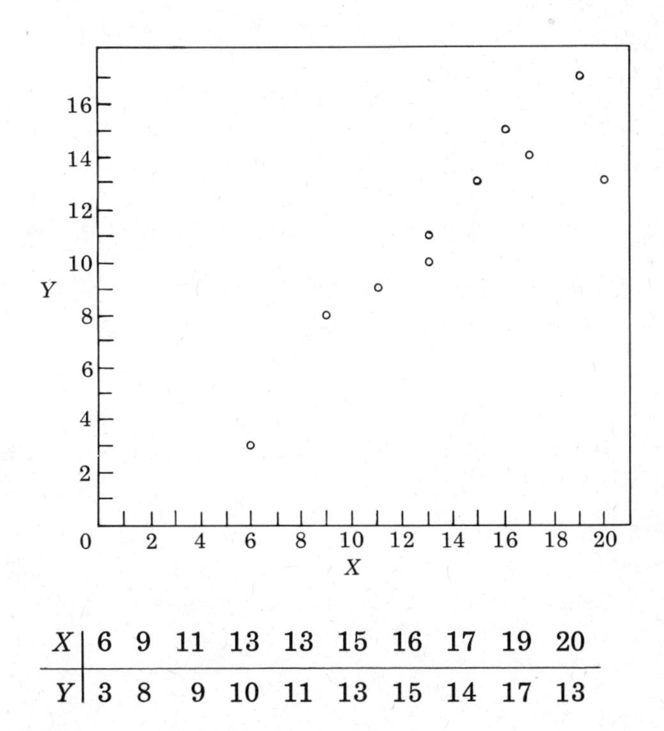

X	6	9	11	13	13	15	16	17	19	20
Y	3	8	9	10	11	13	15	14	17	13

Figure 12.19 shows a scatter plot with zero correlation. This means that no relationship exists between an X score and a Y score. A high score in X is equally likely to be associated with a high, medium, or low score in Y. Here, the coefficient of correlation is .00.

FIGURE 12.19

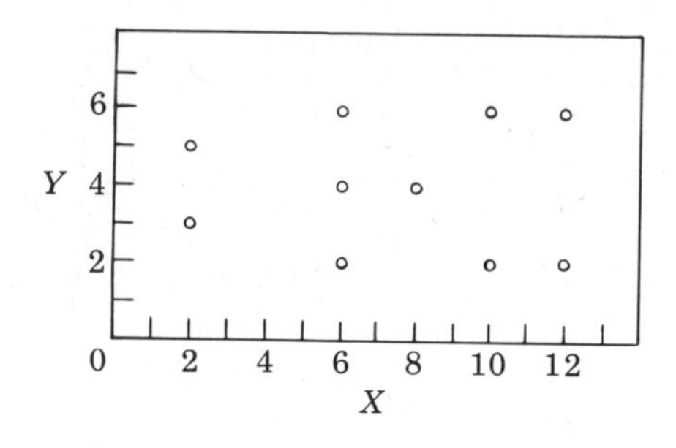

X	2	2	6	6	6	8	10	10	12	12
Y	3	5	2	4	6	4	2	6	2	6

HOW TO COMPUTE THE PEARSON COEFFICIENT OF CORRELATION

One way to compute r is to use the formula

$$r_{XY} = \frac{\Sigma xy}{\sqrt{(\Sigma x^2)(\Sigma y^2)}} \qquad (1)$$

where r_{XY} = the correlation between X and Y,
$x = X - \overline{X}$
$y = Y - \overline{Y}$
Σxy = sum of all the products xy.

Applying Formula 1 to the data in Table 12.13, we compute the Pearson coefficient of correlation between X and Y as shown in Table 12.14.

TABLE 12.13

X	14	13	11	11	9	6	6	5	3	2
Y	10	13	10	6	8	10	2	6	4	1

TABLE 12.14

X	Y	x	y	x^2	y^2	xy
14	10	6	3	36	9	18
13	13	5	6	25	36	30
11	10	3	3	9	9	9
11	6	3	$^-1$	9	1	$^-3$
9	8	1	1	1	1	1
6	10	$^-2$	3	4	9	$^-6$
6	2	$^-2$	$^-5$	4	25	10
5	6	$^-3$	$^-1$	9	1	3
3	4	$^-5$	$^-3$	25	9	15
2	1	$^-6$	$^-6$	36	36	36
80	70	0	0	158	136	113
				↑	↑	↑
$\overline{X} = 8$	$\overline{Y} = 7$			Σx^2	Σy^2	Σxy

$$r_{XY} = \frac{\Sigma xy}{\sqrt{(\Sigma x^2)(\Sigma y^2)}} = \frac{113}{\sqrt{(158)(136)}} = \frac{113}{\sqrt{21488}}$$

$$= \frac{113}{146.58} = {}^+.77$$

INTERPRETATIONS OF r

It is important to note that a coefficient of correlation does not give a percentage relationship. For instance, an r of .80 is *not* twice as strong a relationship as an r of .40, and an increase in r from .30 to .50 is *not* equivalent to an increase from .70 to .90. The coefficient of correlation is an index number, not a measurement on a linear scale of equal units. The correlation coefficient, together with the mean and standard deviation of each of the two variables, enable us to write an equation, called a *regression equation*, by which we can predict either variable from the other.

The question of how to interpret a given size of r depends on the situation. For example, if scores on an aptitude test are known to have a correlation of .45 with measures of vocational success, then the aptitude test is considered a very useful index for predicting vocational success. On the other hand, r should be at least .90 before we can say that two forms of a given test have an acceptable correlation. So the significance of a given size of r depends on the situation for which it was obtained and should not be interpreted in an absolute sense.

An important restriction on the use of the Pearson coefficient of correlation is that the relationship between the variables must be approximately *linear.* One way of telling whether the relationship is linear is from an inspection of the scatter plot. For relationships that are not linear, we use different correlation coefficients.

EXERCISE SET 12.5–12.7

1 Use Formula 1 on page 440 to show that r for the data in Figure 12.16 is $^{+}1.00$.

2 Use Formula 1 to show that r for the data in Figure 12.17 is $^{-}1.00$.

3 Find r for the data in Figure 12.18.

Appendix A

LOGIC

Objectives

At the completion of this chapter, you should be able to:

1 Form the *conjunction*, *disjunction*, *negation*, and *implication* from given simple statements.

2 Form the *converse*, *inverse*, and *contrapositive* of a given implication.

3 Use truth tables to determine whether two statements are *logically equivalent* and whether a compound statement is a *tautology*.

4 Test the validity of an argument by using Venn diagrams, truth tables, or tautologies.

Prerequisites None.

INTRODUCTION

Imagine that you are a member of a jury listening to a criminal case against Sebastian Quattlebaum, and the following facts have been established. (1) If Quattlebaum was in the U. S. on the day of the crime, then he was in New York City on that day. (2) If Quattlebaum did not commit the crime, then he was not in New York City on that day. (3) Quattlebaum was in the U. S. on the day of the crime. What is your verdict, innocent or guilty?

If the facts are as given, then the verdict must be "guilty" whether or not you ever had a formal course in logic. When you reach the end of this chapter you will be able to prove that "guilty" is the only possible conclusion.

Logic is concerned with judging the validity of conclusions such as the one we just drew. It helps us clarify our patterns of thought and guides us to correct reasoning processes. We do not, of course, need a course in formal logic in order to reason correctly any more than a child needs to learn grammar to speak grammatically correct sentences. Indeed, common sense with its intuitive insights and its reliance on both knowledge and experience may be better adapted to everyday affairs than formal logic, which is more suitable to precisely-defined situations. But familiarity with the formal structure of logic may enable us to employ our common sense more systematically and more analytically than we otherwise might.

The invention of *formal logic* goes back more than 2300 years ago to Aristotle who based his system on "syllogisms", the most famous of which is: "All men are mortal; all heroes are men; therefore all heroes are mortal." But Aristotle never got beyond a very small part of the subject because he derived his logical conclusions from ordinary language which contains many words and phrases whose meanings are vague and open to different interpretations.

This limitation in Aristotle's logic was removed in the nineteenth century by George Boole (1815–1864), the English mathematician and logician, who replaced Aristotle's verbalisms with symbols such as x and y and derived his conclusions by algebraic operations. The substitution of symbols for words in logical reasoning makes it possible to reduce complicated logical problems to clear, concise, and precise *symbolic* statements. These statements can more easily reveal such things as whether one statement is equivalent to another or whether two statements are inconsistent. The replacement of words with symbols in Aristotle's logic produced the subject known as *symbolic logic*, which, today, has practical uses in designing parts of telephone circuits and electronic computers.

Boole was followed by other mathematicians who wanted to use symbolic logic to solve logical paradoxes and other fundamental problems of mathematical thinking. By 1913, Alfred North Whitehead (1861–1947) and Bertrand Russell (1872–1970), using a system of symbols invented by the Italian mathematician Giuseppe Peano (1858–1932), developed a formal mathematical logic, which they presented in their famous work, *Principia Mathematica.*

Although there is a vast difference between the formal logic of Aristotle and the symbolic logic of Whitehead and Russell, both systems are mostly concerned with judging the validity of arguments; that is, checking to see whether a certain conclusion necessarily follows from a given set of facts.

This chapter will present an introduction to the elementary concepts of logic. An understanding of these concepts is not only useful in our daily lives and extremely valuable in the study of mathematics but is also interesting in its own right.

A.1 Statements in Logic

Just as numbers are the basic elements of arithmetic, so are simple declarative sentences the building blocks of logic. We start with simple sentences and use these to form more complicated ones. Any sentences considered in logic must be either true or false but not both. Examples of sentences that may be considered in logic are: (1) Mary has a civil service job; (2) The table is green; (3) 10 is a prime number; (4) Chicago is the capitol of the U. S.; (5) $9 - 6 = 5$. Each of these sentences is either true or false.

However, sentences such as (1) She is the prettiest girl in the room, (2) The house has ten rooms, (3) It is an even number, (4) $x + 2 = 9$, (5) $y + 1 > 3$ are not considered in logic because we cannot tell whether they are true or false. These sentences could be considered if in sentence (1) "she" were replaced with a definite person like "Trudy Evans"; if in (2) "house" were replaced by a specific house; if in (3) "it" were replaced by a specific number like "5" or "18"; and if the x and y in (4) and (5) were replaced by specific numbers.

In logic we are not concerned with the *criteria* by which we decide whether a sentence is true or false. It is only required that the criteria be *clear* in every instance as to whether a given sentence is either true or false and that the sentence not be true and false at the same time.

DEFINITION 1 A *statement* is a declarative sentence that is either true or false but not both.

A.2 Compound Statements

Starting with simple statements, we can combine any two or more of them to form a compound statement by the use of certain words called *connectives.* We shall form compound statements from simple statements by using the connectives *and*, *or*, *not*, and *implies*.

DEFINITION 2 A *compound statement* is a statement that contains a connective.

Comments

1 *Since the logic we are discussing here is one in which a sentence is either true or false, it is called a* two-valued logic. *For a discussion of a* three-valued logic *see page 470.*

2 *We call the word "not" a* connective *although, unlike the other* connectives, *it does not "connect" two or more statements. As we shall see later, "not" modifies only one statement at a time.*

CONJUNCTION

DEFINITION 3 When we combine two given statements by placing an "and" between them we form a *conjunction.*

For example, the *conjunction* of the two statements (1) Tim is a barber and (2) I own a camera is: (3) Tim is a barber *and* I own a camera.

We find it useful and convenient to express this conjunction in symbols. We do this by letting a letter such as p represent statement (1), a letter such as q represent statement (2), and letting the symbol $\wedge$ represent the connective "and." Then the conjunction of (1) and (2) becomes

$$p \wedge q,$$

which translates into the compound statement (3) above.

Comments

1 *Sometimes a connective may not appear explicitly, such as when another word appears in its place or when the connective is implied. For example, in the conjunction "Ron is kind as well as rich," the words "as well as" stand for the connective "and"; and in the conjunction "Joan types but does not swim," the word "but" stands for the connective "and."*

2 *$p \wedge q$ may be translated as:*

a) p and q *b*) p as well as q *c*) p but q
d) p also q *e*) p moreover q *f*) p however q.

TRUTH TABLE

Under what conditions will a conjunction be true and under what conditions will it be false? By agreement, a conjunction will be true only if both of its components are true. If either component (or both) is false, the conjunction is considered false.

A good way to show these truth values of a conjunction is to construct a *truth table.* Table A.1 is such a table and exhibits all the information we need to know about $p \wedge q$. In fact, we can define a conjunction or any other compound statement by its truth table. Table A.1 shows the four possibilities for the combined truth values of p and q and the corresponding truth values of their conjunction.

TABLE A.1

p	q	$p \wedge q$
T	T	T
T	F	F
F	T	F
F	F	F

The first row indicates that if both p and q are true, then their conjunction is true; the second row indicates that if p is true and q is false, then their conjunction is false; the third row shows that if p is false and q is true, then their conjunction is false; and the fourth row shows that when both p and q are false, then their conjunction is false.

Examples

Give the truth value of each of the following conjunctions.

1 Two is an even number and A is the first letter in the alphabet.

True because both components are true.

2 Canada is north of the U. S. and cream is meat.

False because the second component is false.

3 A triangle is a square and water is wet.

False because the first component is false.

4 Rome is the capitol of the U. S. and ice is hot.

False because both components are false.

Comment

We said that there are four possible combinations of truth values of p and q and that the truth table for their conjunction (or any other compound statement) will then contain four rows.

Suppose we are interested, not in two simple statements p and q, but in three simple statements p, q, and r. How many possible combinations of truth values do we then have? There is a formula which will give us the number of possible combinations for any number of simple statements.

If we have one simple statement p, then there are two possible truth values for p: p is either true or false (Table A.2). If we have two simple statements, p and q, then there are four possible combinations (Table A.3). Notice that the q column starts with T and then alternates its truth values in each succeeding row, while the p column contains two T's and two F's in that order. If we have three simple statements, p, q, and r, then we have eight possible combinations (Table A.4).

TABLE A.2

p
T
F

TABLE A.3

p	q
T	T
T	F
F	T
F	F

TABLE A.4

p	q	r
T	T	T
T	T	F
T	F	T
T	F	F
F	T	T
F	T	F
F	F	T
F	F	F

Notice that in Table A.4:

a) *Each column starts with a T.*

b) *The r column alternates* one *T with* one *F in each succeeding row.*

c) *The q column alternates* two *T's with* two *F's in each succeeding row.*

d) *The p column alternates* four *T's with* four *F's in each succeeding column.*

The pattern that emerges is shown in Table A.5.

TABLE A.5

n Number of Simple Statements	Number of Possible Combinations
1	2 or 2^1
2	4 or 2^2
3	8 or 2^3
4	16 or 2^4
•	•
•	•
•	•
n	2^n

That is, the number of possible combinations of truth values for n simple statements is 2^n. 2^n is, therefore, the number of rows contained in a truth table involving n simple statements. This is derived from the Fundamental Counting Principle (page 377).

DISJUNCTION

DEFINITION 4 When we combine two given statements by placing an "or" between them, we form a *disjunction.* The symbol for a disjunction is $\vee$.

For example, if p represents the statement, "Tim is a barber," and if q represents the statement, "I own a camera," then the disjunction of p and q, written

$$p \vee q,$$

is the compound statement, "Tim is a barber or I own a camera."

We define the truth values of the disjunction $p \vee q$ by Table A.6. Note that a disjunction of two statements is always true except when both statements are false.

TABLE A.6

p	q	$p \vee q$
T	T	T
T	F	T
F	T	T
F	F	F

Examples Give the truth value of each of the following disjunctions.

1 Two is an even number or A is the first letter in the alphabet.
True because both components are true.

2 Canada is north of the U. S. or cream is meat.
True because the first component is true.

3 A triangle is a square or water is wet.
True because the second component is true.

4 Rome is the capitol of the U. S. or ice is hot.
False because both components are false.

Comment *Note that "or" is defined here to mean p or q, or, both p and q; that is, the disjunction of p and q is true if* (1) either *p or q is true, or* (2) *if* both *p and q are true. We say that we have here defined "or" in the* inclusive *sense.*

In everyday speech we often use the connective "or" in a different sense than the one we used here. For instance, we say, "I shall go to the movies (p) or I shall go to the beach (q)", meaning I shall go to the movies or to the beach, but not to both. In this case, we use "or" in the exclusive *sense.*

In our definition of the truth values of a disjunction, we could have agreed to call a disjunction true only if either *p or q is true, but false when p and q are* both *true. We then would have used "or" in the* exclusive *sense. However, we chose to define "or" in the* inclusive *sense and will be using it only in this sense in the rest of the book.*

IMPLICATION

DEFINITION 5 An *implication* is a compound statement of the form "if p then q," where p and q are any two statements. It is denoted by

$$p \rightarrow q.$$

p is sometimes called the *hypothesis* and q, the *conclusion*.

An example of an implication is "If you fall out of the window, then you will get hurt."

We define the truth values of the implication $p \rightarrow q$ by Table A.7. Note that an implication is always true except when the hypothesis is true and the conclusion false.

TABLE A.7

p	q	$p \rightarrow q$
T	T	T
T	F	F
F	T	T
F	F	T

Comments 1 *The first two rows in Table A.7 are intuitively easy to accept;*

the third and fourth rows are more difficult to accept, but they are part of the definition *of the truth values of an implication. The rationale for the third and fourth rows can be seen if we think of an implication as a promise or guarantee that something will happen if something else happens.*

Consider, for example, the implication, "If it rains then the streets are wet." This implication guarantees that the streets will be wet if it rains. *It makes no promise about what will happen* if it doesn't rain.

The first row of the truth table says that if it rains and if the streets are wet then the implication is true since its promise is kept. The second row says that if it rains but the streets are not wet then the implication is false since the promise is broken.

The third and fourth rows assert that the implication is to be considered to be true if it doesn't rain *since the implication made no promises about this situation and, therefore, no promises were broken.*

2 *In everyday conversation, it is customary for the hypothesis and the conclusion of an implication to be related in a cause and effect sense, as in the implication, "If it rains then the streets are wet." But in mathematical logic, we are not restricted this way, and unrelated statements like "If* 2 *and* 3 *are* 6, *then John will win the game" are perfectly acceptable implications.*

Examples Give the truth values of these implications.

1 If heat melts ice, then $3 + 5 = 8$.

True because hypothesis and conclusion are true.

2 If autos are plants, then plants grow in the earth.

True because the hypothesis is false.

3 If males and females are the same, then all females grow beards.

True because the hypothesis is false.

4 If a square is a quadrilateral, then a circle is a square.

False because the hypothesis is true and the conclusion is false.

An implication can be expressed in different ways. For example, if p represents "You work hard" and q represents "You get tired," then the implication $p \rightarrow q$ can be expressed in these ways:

1 *If p then q.* (If you work hard then you get tired.)

2 p *implies* q. (Your working hard implies your getting tired.)

3 p *only if* q. (You work hard only if you get tired.)

4 q is *necessary* for p. (Your getting tired is necessary for your working hard.)

5 p is *sufficient* for q. (Your working hard is sufficient for you to get tired.)

6 q *if* p. (You get tired if you work hard.)

Other ways of expressing $p \rightarrow q$ are:

7 q follows from p.

8 q is deducible from p.

9 q is implied by p.

NEGATION

DEFINITION 6 The *negation* of a statement p is the statement "It is not true that p," denoted by

$$\sim p.$$

For example, if p is the statement "Howard is six feet tall," then $\sim p$ asserts that "It is not true that Howard is six feet tall." We can improve the sentence structure by saying "Howard is not six feet tall."

We define the truth values of the negation $\sim p$ by Table A.8. That is, if a statement is true, then its negation is false; and if a statement is false, then its negation is true. Both conditions must hold for a statement to be the negation of another statement. Thus, the negation of "The dress is blue" is "The dress is not blue."

TABLE A.8

p	$\sim p$
T	F
F	T

Comments

1 *Note that a statement like "The dress is green" is not the negation of "The dress is blue" because the second row of Table A.8 does not hold in this case. That is, if "The dress is blue" is false, then it is not necessarily true that "The dress is green." The statement "The dress is green" is called a* contradiction *of the statement "The dress is blue."*

2 *What is the negation of a negation? That is, for any given statement p, what are the truth values of $\sim(\sim p)$? From Table*

A.9, it is clear that the negation of a negation has exactly the same truth values as the original statement p.

TABLE A.9

p	$\sim p$	$\sim(\sim p)$
T	F	T
F	T	F

3 *It should be clear by now that in logic we are interested in the truth values of statements rather than in their meanings. The truth values of a compound statement are determined, not by the* meanings *of its components p and q, but by the* truth values *of p and q and by the* connectives *tying them together. If we change the sentences that are represented by p and q, then the meaning of, say, $p \vee q$ will be different although its truth values will remain the same. So that the statement* forms, *$p \wedge q$, $p \vee q$, $p \rightarrow q$, $\sim p$, together with the truth values of p and q are of importance in logic rather than the statement* meanings.

TAUTOLOGY

DEFINITION 7 A *tautology* is a compound statement that is true regardless of the truth or falsity of its components.

For example, consider the statement $p \vee (\sim p)$. Let us construct a truth table (Table A.10) to see under what conditions this statement is true.

TABLE A.10

(a)	(b)	(c)
p	$\sim p$	$p \vee (\sim p)$
T	F	T
F	T	T

Column (a) lists all the possible truth values of p. Column (b) is obtained from column (a) by negating each value in column (a). Column (c) is obtained from columns (a) and (b) by forming the disjunction for each pair of values.

Since all entries in column (c) are true, we conclude that the statement $p \vee (\sim p)$ is always true regardless of the truth or falsity of p. Therefore,

$$p \vee (\sim p)$$

is a *tautology*.

The tautology $p \vee (\sim p)$, known as the Law of the Excluded Middle, says that a statement or its negation must be true. For instance, if p is the statement "Ed is in London," the disjunction "Ed is in London or Ed is not in London" is always true regardless of whether p is true or false.

Comment

A statement which is false in all cases is a self-contradiction. *For instance, the statement* $p \wedge (\sim p)$ *whose truth values are shown in Table A.11 is a* self-contradiction *since it is false regardless of the truth or falsity of p. It says that any statement* and *its negation cannot both be true.*

TABLE A.11

p	$\sim p$	$p \wedge (\sim p)$
T	F	F
F	T	F

Examples of Tautologies

1 $p \vee (\sim p)$ Law of the Excluded Middle (see Table A.10).

2 $\sim[p \wedge (\sim p)]$ Law of Contradiction

3 $[(p \to q) \wedge p] \to q$ Rule of Detachment or Modus Ponens

The Rule of Detachment states that if $p \to q$ is true and p is true, then q must be true. For example, suppose p stands for "You are 18 years old," and q stands for "You may vote." If the implication "If you are 18 years old, then you may vote" is true and if you are 18 years old, then it is also true that you may vote.

4 $[(p \to q) \wedge (q \to r)] \to (p \to r)$ Chain Rule

The Chain Rule says that if the implications $p \to q$ and $q \to r$ are true then the implication $p \to r$ is true. Successive applications of the Chain Rule permit a chain of implications of any desired length. For example, if $p \to q$, $q \to r$, $r \to s$, and $s \to t$, then $p \to t$.

Each of these tautologies can be established by the use of truth tables.

EXERCISE SET A.1, A.2

1 Which of the following are statements? Give the truth value of each statement.

a) August is a summer month in North America.
b) Texas is in South America.
c) Let's have fun.
d) $2 \times 3 = 1 \times 0$.
e) Today is Tuesday.
f) All rectangles are squares.
g) Why did he do it?
h) 12 is divisible by 5.
i) The man is a pilot.
j) A number whose last digit is 0 is divisible by 2.
k) Some women have red hair.
l) $2x + 7 = 15$.
m) $a + 2 > 11$.
n) $x + y = y + x$.

2 Form (1) the conjunction, and (2) the disjunction of the following pairs of statements.
a) The United States has space ships. Men visited the moon.
b) Jack is a machinist. Mary went swimming.
c) $2 + 3 = 5$. Mars is made of plastic.
d) Triangles are not squares. March has 39 days.
e) Today is Tuesday. Tomorrow is Wednesday.

3 Form implications of the pairs of statements in Exercise 2. Use the first statement as the hypothesis and the second statement as the conclusion.

4 In each of the following statements select the hypothesis and the conclusion. If necessary, first rewrite the statement in "if-then" form.
a) If $2 + 3 = 5$, then $5 - 3 = 2$.
b) With practice he will play well.
c) You will succeed if you work hard.
d) Watching too much television makes him sad.
e) Admitting that he was there is a confession of guilt.
f) A triangle is a polygon.
g) A rolling stone gathers no moss.
h) Two lines intersect in only one point.

5 Form the negation of the following statements.
a) The sun is shining.
b) Even numbers are divisible by 2.
c) The sweater is blue.
d) The car rides well.
e) Every triangle contains three angles.
f) $2 < 3$.
g) $10 \leq 9$.
h) $5 \geq 4$.

6 Given:

p: Mat was elected President.
q: Mat had the support of his party.

Translate the following symbolic expressions into words.
a) $p \wedge q$ b) $p \vee q$ c) $p \rightarrow q$ d) $q \rightarrow p$
e) $\sim p \wedge \sim q$ f) $\sim (p \wedge q)$ g) $\sim q \rightarrow \sim p$ h) $\sim [p \wedge (\sim q)]$

7 Given:

p: Anne loves Carl. q: Carl loves Anne.

Express the following statements in symbolic notation.

a) Anne loves Carl but Carl does not love Anne.
b) Anne and Carl do not love each other.
c) If Carl loves Anne, then Anne loves Carl.
d) If Carl does not love Anne, then Anne does not love Carl.

8 Construct truth tables for each of the following.

a) $p \to (p \wedge q)$
b) $\sim(\sim p \vee q)$
c) $\sim p \to \sim q$
d) $(p \wedge q) \to (p \vee q)$

9 Using truth tables, decide whether or not the following are tautologies.

a) $p \to (p \vee q)$
b) $(p \wedge q) \to p$
c) $[(p \to q) \wedge p] \to q$
d) $[(\sim p \vee q) \wedge p] \to q$

10 The sentence "Schools will be closed in the event that we have a blizzard" may be rewritten as

a) If the schools will be closed, then we will have a blizzard.
b) If we have a blizzard, then the schools will be closed.
c) Schools will be closed, and we will have a blizzard.
d) None of the above.

11 Which of the following is false?

a) The conjunction of two statements is false if either statement is false.
b) The disjunction of two statements is true if either statement is true.
c) An implication is true whenever the conclusion is true.
d) A statement of the form $\sim p$ is false.

12 Which of the following is an implication?

a) The table is blue.
b) A resident of Philadelphia is an American.
c) Frank is an American.
d) None of the above.

13 Which of the following statements is not a contradiction of the statement "Some buildings are 100 feet high"?

a) All buildings are not 100 feet high.
b) All buildings are 50 feet high.
c) Every building is 100 feet high.
d) None of the above.

A.3 Derived Implications

From the implication $p \to q$, we can form several related implications some of which may or may not be true if the given implication is true. The most important are the *converse*, the *inverse*, and the *contrapositive*.

DEFINITION 8 The *converse* of the implication $p \to q$ is $q \to p$, That is, the converse of the implication is formed by interchanging the p and the q.

Example 1

Implication If I live in Philadelphia, then I live in the U. S.

Converse If I live in the U. S., then I live in Philadelphia.

Example 2

Implication If today is Tuesday, then tomorrow is Wednesday.

Converse If tomorrow is Wednesday, then today is Tuesday.

Note that if an implication is true, then its converse is not necessarily true. In Example 1, the converse is not necessarily true. In Example 2, the converse is necessarily true.

Comment

Table A.12 discloses under which conditions an implication and its converse are both true.

TABLE A.12

p	q	$p \to q$	$q \to p$
T	T	T	T
T	F	F	T
F	T	T	F
F	F	T	T

Only if both p *and* q *are true or if both* p *and* q *are false will an implication* and *its converse be true.*

DEFINITION 9

The *inverse* of the implication $p \to q$ is $\sim p \to \sim q$. That is, the inverse of an implication is formed by replacing p and q by their negations.

Example 3

Implication If a figure is a square, then it contains four sides.

Inverse If a figure is not a square, then it does not contain four sides.

Example 4

Implication If the three angles of a triangle are congruent, then the triangle is equilateral.

Inverse If the three angles of a triangle are not congruent, then the triangle is not equilateral.

Note that the truth of an implication does not guarantee the truth of its inverse. In Examples 3 and 4 the implications are true, but only in Example 4 is the inverse also true.

Comment

Table A.13 discloses under which conditions are an implication and its inverse both true.

TABLE A.13

p	q	$\sim p$	$\sim q$	$p \to q$	$(\sim p) \to (\sim q)$
T	T	F	F	T	T
T	F	F	T	F	T
F	T	T	F	T	F
F	F	T	T	T	T

An implication and its inverse will both be true only if p and q are both *true or* both *false.*

DEFINITION 10 The *contrapositive* of the implication $p \to q$ is $\sim q \to \sim p$. That is, the contrapositive of an implication is formed by interchanging the p and the q and then negating them.

Example 5 *Implication* If Kathy was at the picnic, then she ate hamburgers.

Contrapositive If Kathy did not eat hamburgers, then she was not at the picnic.

Example 6 *Implication* If a figure has three sides, then it is a triangle.

Contrapositive If a figure is not a triangle, then it does not have three sides.

Note that in Examples 5 and 6 the contrapositive is true if the implication is true. In fact, we shall soon prove that whenever an implication is true its contrapositive is true, and whenever an implication is false its contrapositive is false.

IF AND ONLY IF

Sometimes an implication and its converse are both true; that is $p \to q$ *and* $q \to p$. A shorter way to represent this conjunction is to write $p \longleftrightarrow q$. For example, let p represent "A number is divisible by 2," and let q represent "A number is even." Then

$p \to q$ represents

"If a number is divisible by 2, then it is even."

$q \to p$ represents

"If a number is even, then it is divisible by 2."

and $p \longleftrightarrow q$ represents

"If a number is divisible by 2, then it is even; *and* if a number is even, then it is divisible by 2."

Another way of stating $p \longleftrightarrow q$ is to say p *if and only if* q; that is, "A number is divisible by 2 *if and only if* it is even."

Comments

1 *"p if and only if q" really says*

"p if q and *p only if q."*
$(q \to p) \quad \wedge \quad (p \to q)$

2 *Table A.14 discloses under what conditions* $p \longleftrightarrow q$ *is true.*

TABLE A.14

p	q	$p \to q$	$q \to p$	$(p \to q) \wedge (q \to p)$, or $p \longleftrightarrow q$
T	T	T	T	T
T	F	F	T	F
F	T	T	F	F
F	F	T	T	T

$p \longleftrightarrow q$ *is true if and only if p and q are both true or both false; otherwise the statement is false.*

3 $p \to q$ *is sometimes called a* conditional statement, *while* $p \longleftrightarrow q$ *is called a* biconditional statement *since it represents two conditional statements.*

NECESSARY AND SUFFICIENT

Another way of stating $p \longleftrightarrow q$ is to say that p is *necessary and sufficient* for q. In the example used above, we would say "For a number to be even it is *necessary and sufficient* for it to be divisible by 2."

Comment

"P is necessary and sufficient for q" really says

"p is necessary for q and *p is sufficient for q."*
$(q \to p) \qquad \wedge \qquad (p \to q)$

EXERCISE SET A.3

1 Write (1) the converse, (2) the inverse, and (3) the contrapositive of each of the following implications.
a) If this is March, then three months from now will be June.
b) If two angles are right angles, then they are congruent.
c) The opposite sides of a square are parallel.
d) Every athlete has good coordination.
e) Every prime, except 2, is an odd number.

2 Use the contrapositive to prove that:
a) If $a + b \neq a + c$, then $b \neq c$.
b) If two angles are not congruent, then both are not right angles.

3 Which of the following is true? The converse of an implication is:
a) True when the hypothesis is true.
b) False when the hypothesis is false.
c) Always true if the implication is true.
d) None of the above.

4 Which of the following statements is false?
a) Every implication has a converse.
b) The negation of a false statement is always false.
c) The converse of the converse is the original implication.
d) If circles are triangles, then $1 + 2 = 5$.

5 Let p represent "A number is even," and let q represent "A number is divisible by 2." To say that "A necessary and sufficient condition for a number to be even is for it to be divisible by 2" is the same as saying:
a) p only if q.
b) If p, then q; and if q, then p.
c) q if p.
d) None of the above.

6 Let p represent "There is a heavy snowstorm," and let q represent "The schools are closed." Express the following statements in symbolic form.
a) The schools are closed if there is a heavy snowstorm.
b) Only if there is a heavy snowstorm are the schools closed.
c) The schools are closed if and only if there is a heavy snowstorm.
d) A necessary condition for the schools to be closed is that there be a heavy snowstorm.
e) A heavy snowstorm is a sufficient condition for the schools to be closed.
f) A necessary and sufficient condition for the schools to be closed is that there be a heavy snowstorm.

A.4 Logical Equivalence

DEFINITION 11 Two statements are *logically equivalent* if and only if they have the same truth tables.

For example, compare the truth tables of an implication and its contrapositive as shown in Table A.15.

TABLE A.15

p	q	$\sim p$	$\sim q$	$p \to q$	$\sim q \to \sim p$
T	T	F	F	T	T
T	F	F	T	F	F
F	T	T	F	T	T
F	F	T	T	T	T

Note that for all possible truth values of p and q, the truth values of the implication and its contrapositive are identical. Therefore, an implication and its contrapositive are *logically equivalent*.

Example "If there is a blizzard, then the game will be cancelled," is logically equivalent to "If the game is not cancelled, then there is no blizzard."

Comments

1 *We can say that two statements are logically equivalent if each implies the other.*

2 *Since an implication and its contrapositive are logically equivalent statements:*

a) If either is true, the other is true; if either is false, the other is false.

b) Proving or disproving the contrapositive is equivalent to proving or disproving the implication.

3 *The Law of Substitution, a law used often in logical reasoning, states that we may substitute at any point in our reasoning one statement for an equivalent statement. We may, therefore, substitute at any point in an argument the contrapositive for an implication.*

4 *We defined $p \rightarrow q$ by its truth table. Another, and interesting, definition of $p \rightarrow q$ is the equivalent statement, $\sim(p \wedge \sim q)$. (Use truth tables to show that the two statements are equivalent.)*

5 *Table A.16 shows the relationships between an implication, its converse, its inverse, and its contrapositive.*

TABLE A.16

p	q	$\sim p$	$\sim q$	Implication $p \rightarrow q$	Converse $q \rightarrow p$	Inverse $(\sim p) \rightarrow (\sim q)$	Contrapositive $(\sim q) \rightarrow (\sim p)$
T	T	F	F	T	T	T	T
T	F	F	T	F	T	T	F
F	T	T	F	T	F	F	T
F	F	T	T	T	T	T	T

Note that an implication is equivalent to its contrapositive, while the converse is equivalent to the inverse.

6 *By the truth table, we can show that $\sim(p \rightarrow q)$ is logically equivalent to $(p \wedge \sim q)$. This equivalence gives us a rule for negating $p \rightarrow q$: To negate $p \rightarrow q$, form the conjunction $p \wedge \sim q$. (Note that the negation of an implication is not another implication, but a conjunction.)*

7 *By truth tables, we can also establish the equivalence of*

$$\sim(p \longleftrightarrow q) \quad \text{and} \quad (p \wedge \sim q) \vee (q \wedge \sim p).$$

This equivalence gives us a rule for negating $p \longleftrightarrow q$: To negate $p \longleftrightarrow q$, form the disjunction of the two statements (a) the negation of $p \rightarrow q$, and (b) the negation of $q \rightarrow p$.

EXERCISE SET A.4

1. Prove your answer to each of the following questions.
   a) Is $p \to q$ equivalent to $\sim p \vee q$?
   b) Is $\sim(p \vee q)$ equivalent to $\sim p \wedge \sim q$?
   c) Is $\sim(p \to q)$ equivalent to $p \wedge \sim q$?
2. Which of the following is equivalent to the statement "All Americans are Englishmen"?
   a) If a man is not an American, then he is not an Englishman.
   b) All Englishmen are Americans.
   c) If a man is not an Englishman, then he is not an American.
   d) If a man is an Englishman, then he is an American.
3. For each pair of statements below (1) decide whether they are logically equivalent, (2) restate them, wherever possible, in the *necessary and sufficient* form, and (3) restate them, wherever possible, in the *if and only if* form.
   a) Today is Sunday. Yesterday was Saturday.
   b) A number is composite. A number is not prime.
   c) It is raining. The streets are wet.
   d) If I play, I win. If I don't win, I don't play.

A.5 Valid Arguments

An *argument* is an assertion that a certain statement (the *conclusion*) follows from other statements (the *premises*). How can we tell whether an argument is valid, that is, that the conclusion follows from the premises? Three ways of checking the validity of an argument are through Venn diagrams, truth tables, and tautologies.

USING VENN DIAGRAMS

A convenient way to check the validity of some arguments is through the use of Venn diagrams that help to clarify the relationships between the parts of the argument. These diagrams consist of circles (or other closed figures) and can be used to illustrate the relationships described by the words "some," "all," and "no."

Consider the statement "All residents of San Francisco are residents of California." This relationship between the residents of San Francisco and the residents of California can be represented by Figure A.1. Note that each and every San Franciscan can be associated with a point inside or on the smaller circle, while every Californian who is not a San Franciscan resident is represented by a point outside this circle but inside or on the larger circle.

FIGURE A.1

From Figure A.1 we see that (1) *all* San Franciscans are Californians because every point in or on the smaller circle is contained in the larger circle; (2) not all Californians are San Franciscans because there are points in the larger circle that are not contained in the smaller circle.

Now consider the following argument:

> All San Franciscans (S) are Californians (C).
> All Californians are Americans (A).
> Therefore, all San Franciscans are Americans.

Since circle S in Figure A.2 lies wholly within circle A, it is valid to conclude that all San Franciscans are Americans.

FIGURE A.2

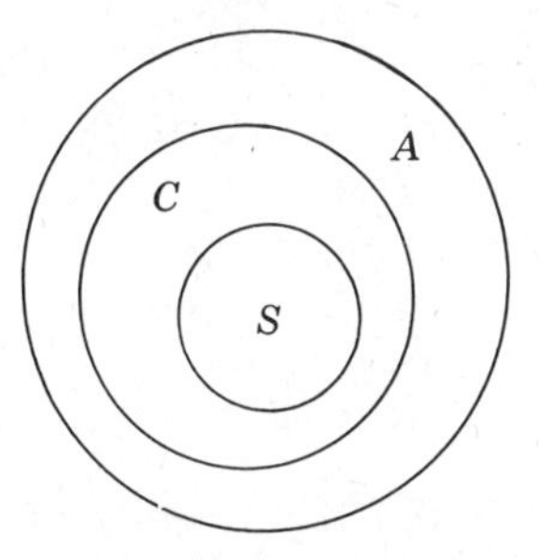

Figure A.3 illustrates the statement "No car (C) is an airplane (A)." We use two circles which have no points in common because no car is also an airplane.

FIGURE A.3

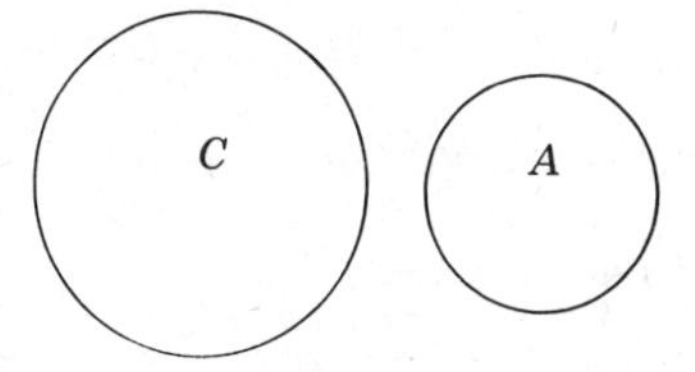

Figure A.4 illustrates the statement "Some students (S) play tennis (T)." The shaded region represents those individuals who are students and who also play tennis.

FIGURE A.4

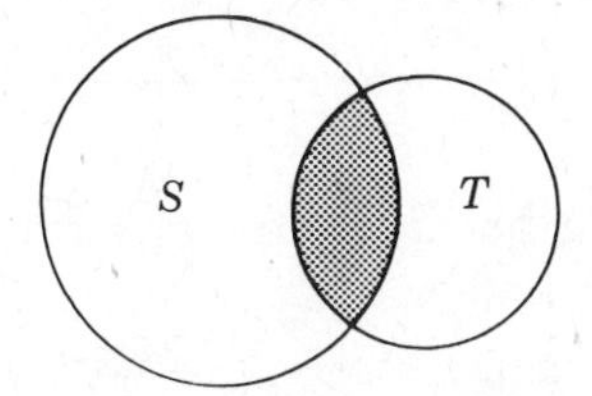

Figure A.1 and Figure A.4 both illustrate the notion of "some." It is important to note that in mathematics "some" can mean just one, several, or all.

Figure A.5 represents the implication $p \rightarrow q$ since it can be interpreted as "all p's are q's." This figure also shows why an implication is equivalent to its contrapositive; that is,

$$(p \rightarrow q) \longleftrightarrow (\sim q \rightarrow \sim p),$$

since the contrapositive can be interpreted as "If an element is not in q, then it is not in p." Figure A.5 shows this to be true.

FIGURE A.5

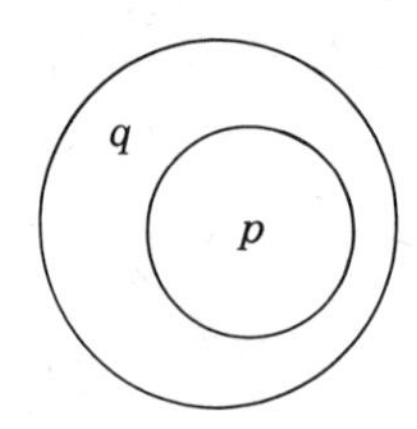

USING TRUTH TABLES

Another way to determine the validity of an argument is to examine the truth table of the premises and the conclusion. If the argument is valid, then the conclusion must be true whenever all the premises are true. In an invalid argument, the conclusion is false in at least one case in which all the premises are true.

Example 1 Consider the argument: (1) If you sell your stock then you will buy a car; (2) You sell your stock; (3) Therefore, you will buy a car. This argument can be expressed in the following way:

(1) $p \rightarrow q$
(2) p
(3) Therefore, q.

The truth table for this argument is shown in Table A.17.

TABLE A.17

Premise 2	*Conclusion*	*Premise 1*
p	q	$p \rightarrow q$
(T)	(T)	(T)
T	F	F
F	T	T
F	F	T

The argument is valid because whenever premise 1 and premise 2 are true the conclusion is true.

Example 2 Consider the argument: (1) If you sell your stock then you will buy a car; (2) You do not sell your stock; (3) Therefore, you do not buy a car. In symbols, the argument becomes

(1) $p \to q$
(2) $\sim p$
(3) Therefore, $\sim q$.

The truth table for this argument is shown in Table A.18.

TABLE A.18

		Premise 2	*Conclusion*	*Premise 1*
p	q	$\sim p$	$\sim q$	$p \to q$
T	T	F	F	T
T	F	F	T	F
F	T	(T)	(F)	(T)
F	F	(T)	(T)	(T)

In the third row of the table, both premises are true but the conclusion is false. Therefore, the argument is not valid.

USING TAUTOLOGIES

Another way to verify the validity of an argument is to reduce it to a tautology. For example, let p represent "The game is postponed," and let q represent "I shall go home." Now consider the argument

(1) If the game is postponed, then I shall go home.
(2) The game is postponed.
(3) Therefore, I shall go home.

Note that (1) and (2) are the premises and (3) is the conclusion. We wish to check the validity of the argument.

Let us first express the argument symbolically:

(1) $p \to q$
(2) p
(3) Therefore, q.

If the argument is valid then the conjunction of premises 1 and 2 implies the conclusion. That is, $[(p \to q) \wedge p] \to q$.

From Table A.19 we see that the argument is a tautology and is, therefore, valid. You may recognize this argument as an illustration of the Rule of Detachment.

TABLE A.19

p	q	$p \to q$	$(p \to q) \wedge p$	$[(p \to q) \wedge p] \to q$
T	T	T	T	T
T	F	F	F	T
F	T	T	F	T
F	F	T	F	T

Comments

1 *It should be noted that the* truth *of the conclusion has no bearing on the* validity *of the argument. For example, consider the argument*

If 2 + 3 = 5, *then the moon is made of orange juice.*
2 + 3 = 5.
Therefore, the moon is made of orange juice.

Note that even though the conclusion is false, the argument is still valid since the conclusion follows from the premises as proven in Table A.19.

Now consider the argument

If Abe Lincoln was President of the U. S. then he was an American citizen.
Abe Lincoln was an American citizen.
Therefore, Abe Lincoln was President of the U. S.

Here, the conclusion is true but the argument is not valid because the conclusion does not follow from the premises as seen in Table A.20 in the next section.

2 *The* conclusion *in an argument is not to be confused with the* conclusion *in an implication. An important difference is that the truth of the implication* $p \to q$ *does not necessarily mean that either* p *or* q *is true, whereas the truth of the premises in an argument guarantees the truth of the conclusion.*

FALLACY

An argument that is not valid is called a *fallacy*. For example, the argument

$p \to q$
q
Therefore, p.

is a fallacy since, as is shown in Table A.20, the conjunction of the two premises does not necessarily imply the conclusion.

TABLE A.20

p	q	$p \to q$	$(p \to q) \wedge q$	$[(p \to q) \wedge q] \to p$
T	T	T	T	T
T	F	F	F	T
F	T	T	T	F
F	F	T	F	T

EXERCISE SET A.5

1 Test the validity of the following arguments.

a) When the program is over, I must leave.
The program is over.
Therefore, I must leave.

b) All squares are quadrilaterals.
All quadrilaterals are polygons.
Therefore, all squares are polygons.

c) If it rained yesterday, then I went to the movies.
Yesterday I went to the movies.
Therefore, it rained yesterday.

d) If you like tennis, then you will like ping-pong.
You do not like tennis.
Therefore, you do not like ping-pong.

e) All desks are green.
Some desks have telephones.
Therefore, some desks with telephones are green.

f) $p \to q$
$q \to r$
Therefore, $r \to p$.

g) $p \vee q$
$r \wedge \sim q$
Therefore, p.

h) $p \to q$
$q \to \sim p$
Therefore, $\sim p$.

i) If Al doesn't play Ted, then he plays Frank.
Whenever Al plays Frank, Ted is disappointed.
Ted is not disappointed.
Therefore, Al plays Ted.

2 Draw valid conclusions, where possible, from the following sets of statements.

a) All A's are B's.
Some C's are A's.
Some C's are not B's.

b) If I can't see you today, I will see you tomorrow.
I can see you today.

c) She will go to town and not go shopping.
If she goes shopping, she will buy a pair of shoes.

d) If Tom works hard, he will get an A.
If Mark does not get an A, Tom will not get an A.
Neither Tom nor Mark fails to work hard.

e) $p \vee q,\ \ q \wedge r,\ \ r \to t.$

3 Write in symbolic form and then test the validity of the argument relating to Sebastian Quattlebaum given at the beginning of this chapter.

PUZZLES

1 Consider these two statements:
a) If it is raining, then I will not go swimming.
b) If I do not mow the lawn, then I will go swimming.

Can you conclude from these statements that if it is raining then I will mow the lawn? Why?

2 An explorer finds himself in a region inhabited by two tribes: the members of one tribe always lie; the members of the other tribe always tell the truth. He meets two natives. "Are you the truth teller?" he asks the tall one. "Goom," the native replies. "He says 'yes'," says the short native who speaks English, "but him big liar." Which tribe did each native belong to?

3 Two boys weighing about 100 pounds each and a man weighing 200 pounds wish to cross a river. Their boat will hold only 200 pounds safely. How can they cross the river in the boat?

4 Sisters and brothers have I none,
But that man's father is my father's son.

Who is *that man*?

5 Two brothers look exactly alike and are exactly the same age, but they are not twins. How is this possible?

EXTENDED STUDY

A.6 More on Negations

NEGATION OF "SOME" AND "ALL"

The words "some," "all," and "no" are called *quantifiers.* They are used to indicate whether a given statement is sometimes true, always true, or never true. In mathematics, "some," called the *existential quantifier*, means *at least one and perhaps all.* "All," called the *universal quantifier*, means *every one.* "No" means *not one.*

The negation of

(1) "Some people are Americans" is

(2) "No people are Americans"

because, as our definition of *negation* requires: when (1) is true, (2) is false; and when (1) is false, (2) is true.

In general, to negate a statement of the form "some (statement)," we use a "no" statement. Similarly, to negate a "no" statement we use a "some" statement.

Example 1 *Statement* Some sweaters are blue.

Negation No sweaters are blue.

Similarly, to negate a statement of the form "all (statement)," we change "all" to "some" and negate the statement.

Example 2

Statement All men are six feet tall.

Negation Some men are not six feet tall.

Again, if we test Example 2 against the definition of negation we find: when the statement is true, the negation is false; and when the statement is false, the negation is true.

Consider again the statement (1) "All men are six feet tall." Does the statement (2) "All men are five feet tall" negate statement 1? The answer is "no" because when we apply the negation test we find: if (1) is true, (2) is false; but if (1) is false, (2) is not necessarily true since if all men are not six feet tall, they are not necessarily five feet tall. They might be some other height.

We call statement 2 a *contradiction* of statement 1, but not its negation. Thus, a contradiction meets only *one* of the two requirements of a negation. Every negation is a contradiction, but not every contradiction is a negation.

NEGATION OF A CONJUNCTION AND A DISJUNCTION

How do we negate the conjunction $p \wedge q$? For example, how do we negate the conjunction "It is raining and Ed is a drummer"? We can prove by Table A.21 that *the negation of the conjunction of two statements is formed by negating each of the original statements and then forming their disjunction*; that is,

$$\sim(p \wedge q) = (\sim p) \vee (\sim q).$$

TABLE A.21

(1) p	(2) q	(3) $p \wedge q$	(4) $\sim(p \wedge q)$	(5) $\sim p$	(6) $\sim q$	(7) $(\sim p) \vee (\sim q)$
T	T	T	F	F	F	F
T	F	F	T	F	T	T
F	T	F	T	T	F	T
F	F	F	T	T	T	T

Note that when column 3 is true, column 7 is false; when column 3 is false, column 7 is true. That is, the negation of $p \wedge q$ is $(\sim p) \vee (\sim q)$, and the negation of the statement "It is raining and Ed is a drummer" is "It is not raining *or* Ed is not a drummer."

We can show in a similar way that *the negation of the disjunction of two statements is formed by negating each of the original statements and then forming their conjunction*; that is,

$$\sim(p \vee q) = (\sim p) \wedge (\sim q).$$

For example, the negation of the disjunction "The team is first or the table is green" is the conjunction "The team is not first *and* the table is not green."

EXERCISE SET A.6

1 Negate each of the following statements.
   a) Some quadrilaterals are squares.
   b) All men are strong.
   c) All numbers ending in zero are divisible by five.
   d) Some natural numbers are odd.
   e) $a = 2$ and $c = 9$.
   f) $x \neq 5$ or $y = 3$.
   g) m is at most 5, and n is at least 10.
   h) $x \geq 4$ or $y < 8$.

2 Which of the following contradicts the statement "The number x is three more than the number y"?
   a) $x = 2$ and $y = 5$.
   b) $x - 3 < y$.
   c) $x - y$ is a positive number.
   d) $2y < 2x$.
   e) $x \geq y + 3$.

3 To say that it is not true that all angles are right angles and all numbers are even is the same as saying that:
   a) All angles are not right angles and all numbers are not even.
   b) All angles are not right angles or all numbers are even.
   c) Neither all angles are right angles nor are all numbers even.
   d) None of the above.

A.7 Three-Valued Logic

The sentences we have been dealing with in our logic so far were required to have one of two possible values; they had to be either true or false. Such a logic is called a *two-valued logic.* But the requirement that sentences be absolutely true or absolutely false often does not correspond to reality. For example, the sentence "John loves dogs" might not necessarily be absolutely true or absolutely false, but might be something in between, such as *maybe.* These considerations led the Polish logician J. Lukasiewicz to suggest *a three-valued logic* in which sentences are designated as having one of *three* possible values: true (T), false (F), or maybe (M).

The three-valued truth tables are then defined as shown in Tables A.22–A.29.

Negation

TABLE A.22 Two-Valued Table

p	$\sim p$
T	F
F	T

TABLE A.23 Three-Valued Table

p	$\sim p$
T	F
M	M
F	T

Conjunction

TABLE A.24 Two-Valued Table

p	q	$p \wedge q$
T	T	T
T	F	F
F	T	F
F	F	F

TABLE A.25 Three-Valued Table

p	q	$p \wedge q$
T	T	T
T	M	M
T	F	F
M	T	M
M	M	M
M	F	F
F	T	F
F	M	F
F	F	F

Disjunction

TABLE A.26 Two-Valued Table

p	q	$p \vee q$
T	T	T
T	F	T
F	T	T
F	F	F

TABLE A.27 Three-Valued Table

p	p	$p \vee q$
T	T	T
T	M	T
T	F	T
M	T	T
M	M	M
M	F	M
F	T	T
F	M	M
F	F	F

Implication

TABLE A.28 Two-Valued Table

p	q	$p \rightarrow q$
T	T	T
T	F	F
F	T	T
F	F	T

TABLE A.29 Three-Valued Table

p	q	$p \rightarrow q$
T	T	T
T	M	M
T	F	F
M	T	T
M	M	T
M	F	M
F	T	T
F	M	T
F	F	T

Since, in every case, the two-valued table is incorporated in the three-valued table, every three-valued tautology is also a two-valued tautology. But not all two-valued tautologies are necessarily three-valued tautologies.

If we change our notation from T, M, F to 1, $\frac{1}{2}$, 0, we can more easily see how it is possible to extend our two-valued logic to n-valued logics. For example, if we wish a 5-valued logic, we could use truth values 1, $\frac{3}{4}$, $\frac{2}{4}$, $\frac{1}{4}$, and 0.

A.8 Proof

We have already noted that the main function of logic is to judge the validity of arguments. The process by which we establish the validity of an argment in mathematics is called a *proof.* Note that in a proof we are testing the *validity* of the conclusion and not its *truth.* We are judging whether our reasoning is correct, not whether the hypothesis is actually true. In testing the validity of an argument we are asking: *if* the hypothesis is true, must the conclusion necessarily be true? The actual truth of the conclusion is neither necessary nor sufficient for the validity of the argument.

DIRECT AND INDIRECT PROOFS

There are two kinds of proofs: *direct* and *indirect.* In a *direct* proof, we start with an hypothesis and then establish a chain of implications, which leads to the desired conclusion. We shall illustrate this method of proof by applying it to the question posed at the beginning of this chapter concerning the guilt or innocence of Sebastian Quattlebaum.

Example If p denotes "Q was in the U. S. on the day of the crime," q denotes "Q was in N. Y. City on the day of the crime," and r denotes "Q committed the crime," then the argument becomes:

$p \rightarrow q$
$\sim r \rightarrow \sim q$
p
Therefore, r.

Assuming the hypotheses are true, we now wish to prove that the conclusion is true.

Direct Proof

1	$p \to q$	(Hypothesis)
2	$\sim r \to \sim q$	(Hypothesis)
3	$q \to r$	(An implication and its contrapositive are logically equivalent.)
4	$p \to r$	(Chain Rule)
5	p	(Hypothesis)
6	Therefore, r.	(Definition of an implication)

In an *indirect* proof, we first assume that what we want to prove is false and then show that this assumption leads to a contradiction. For example, suppose we want to prove that if two distinct lines intersect, then their intersection contains exactly one point (Figure A.6). We can use here the indirect method of proof.

FIGURE A.6

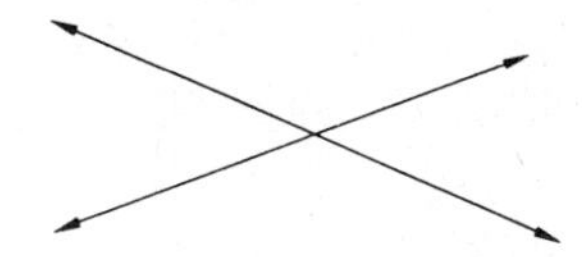

Indirect Proof

1. Assume the two distinct lines intersect not in one point but, say, in two distinct points. This means that two distinct lines will pass through the same two distinct points.
2. But this assumption contradicts the postulate which says that there is exactly *one* line containing two distinct points.
3. Therefore, our assumption is false and the intersection contains exactly one piont.

The indirect method of proof (sometimes called *proof by contradiction*) relies on the fact that if $\sim p$ (the assumption) is false, then p is true. In the above example, we showed $\sim p$ to be false because it contradicts a postulate that we accepted earlier as true.

In general, to use an *indirect* method of proof:

1. Let p represent the conclusion you want to establish. Then assume $\sim p$ is true.
2. Show that if your assumption is true, then it leads to a contradiction of another statement known to be true.
3. Therefore, p must be true.

Appendix B

COMPUTATION

B.1 Percent

For Figure B.1, there are three ways to express what part of the entire rectangular region the shaded region is. One way is to write the common fraction $\frac{1}{2}$; another way is to write the decimal fraction .5; a third way is to use the percent symbol, %, which means "per hundred," and write 50%. All three forms say the same thing in a different way.

FIGURE B.1

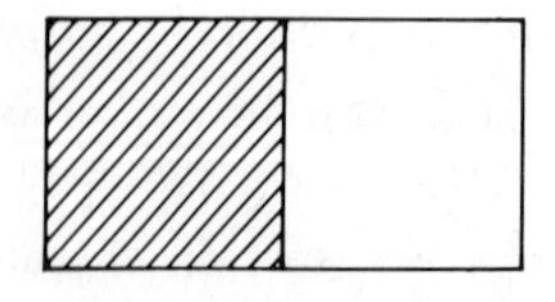

Since the symbol % means *per hundred* or *hundredths*, 50% means $50 \times .01$, or $50 \times \frac{1}{100}$; 7% means $7 \times .01$, or $7 \times \frac{1}{100}$; and .25% means $.25 \times .01$, or $.25 \times \frac{1}{100}$. More generally, $n\%$ means

$$n \times .01, \quad \text{or} \quad n \times \frac{1}{100}.$$

Because it may be more convenient to use one of the three forms rather than the others in a particular situation, it is useful to know how to convert from any one of these forms to the other two.

CONVERTING A PERCENT TO A DECIMAL

Example 1 Find the decimal that corresponds to 17.3%.

$$17.3\% = 17.3 \times .01 \quad \text{(By definition of \%)}$$
$$= .173$$

Since the conversion of a percent to a decimal is achieved by multiplying the percent by .01, what we are really doing is *moving the decimal point two places to the left and removing the % symbol*:

$$17.3\% = .173; \quad 25\% = .25; \quad \text{and} \quad .3\% = .003.$$

CONVERTING A PERCENT TO A COMMON FRACTION

Example 2 Find the common fraction that corresponds to 35%.

$$35\% = 35 \times \frac{1}{100} \quad \text{(By definition of \%)}$$
$$= \frac{35}{100} \text{ or } \frac{7}{20}.$$

Note that to change 35% to a common fraction, we, in effect, take the 35%, remove the % symbol, and then multiply the 35 by $\frac{1}{100}$: Thus,

$$35\% = 35 \cdot \frac{1}{100}$$

Similarly,

$$125\% = 125 \cdot \frac{1}{100} = \frac{125}{100} = \frac{5}{4},$$

and $$12\frac{1}{2}\% = 12\frac{1}{2} \cdot \frac{1}{100} = \frac{25}{2} \cdot \frac{1}{100} = \frac{1}{8}.$$

Example 3 Convert 4.95% to a fraction.

$$4.95\% = 4.95 \times \frac{1}{100} \quad \text{(By definition of \%)}$$
$$= \frac{4.95}{100} = \frac{495}{10000} \text{ or } \frac{99}{2000}.$$

CONVERTING A DECIMAL TO A PERCENT

To change a decimal to a percent, we reverse the steps followed when changing a percent to a decimal.

To change a percent to a decimal is equivalent to moving the decimal point two places to the left and removing the % symbol. So to change a decimal to a percent, *we move the decimal point two places to the right and add the % symbol.*

Example 4 Change .153 to a percent.

$$.153 = 15.3\%$$

Example 5 Change 1.75 to a percent.

$$1.75 = 175\%$$

CONVERTING A FRACTION TO A PERCENT

To convert a percent to a fraction, we remove the % symbol and multiply by $\frac{1}{100}$; then to change a fraction to a percent, we reverse the process. *We divide the fraction by* $\frac{1}{100}$ *and add the % symbol.*

Example 6 Express $\frac{3}{4}$ as a percent.

$$\frac{3}{4} = \left(\frac{3}{4} \div \frac{1}{100}\right)\%$$
$$= \left(\frac{3}{4} \times 100\right)\% = \frac{300}{4}\% = 75\%$$

Example 7 Express $\frac{7}{8}$ as a percent.

$$\frac{7}{8} = \left(\frac{7}{8} \div \frac{1}{100}\right)\%$$
$$= \left(\frac{7}{8} \times 100\right)\% = \frac{700}{8}\% = 87\frac{1}{2}\%$$

SOLVING PROBLEMS INVOLVING PERCENTS

Problems involving percents are easily solved by translating the problem situation to an equation (or number sentence). Then simply solve the equation. The unknown will be denoted by y; the word "is" will be denoted by = (equals); and the word "of" will be denoted by $\times$ (times).

Example 8 What is 15% of 89?

Solution *Translate* What is 15% of 89?

$$\begin{array}{ccccc} \text{What} & \text{is} & 15\% & \text{of} & 89? \\ \downarrow & \downarrow & \downarrow & \downarrow & \downarrow \\ y & = & 15\% & \times & 89 \end{array}$$

Solve

$$\begin{aligned} y &= 15 \times .01 \times 89 \\ y &= .15 \times 89 \\ y &= 13.35 \end{aligned}$$

That is, 15% of 89 is 13.35.

Example 9 3 is what percent of 18?

Solution *Translate*

$$\begin{array}{cccccc} 3 & \text{is} & \text{what} & \text{percent} & \text{of} & 18? \\ \downarrow & \downarrow & \downarrow & \downarrow & \downarrow & \downarrow \\ 3 & = & y & \% & \times & 18 \end{array}$$

Solve

$$\begin{aligned} 3 &= y \times .01 \times 18 \\ 3 &= y \times .18 \\ y &= \frac{3}{.18} \\ y &= 16\frac{2}{3} \end{aligned}$$

That is, 3 is $16\frac{2}{3}$% of 18.

Example 10 9 is $12\frac{1}{2}$% of what number?

Solution *Translate*

$$\begin{array}{ccccc} 9 & \text{is} & 12\frac{1}{2}\% & \text{of} & \text{what number?} \\ \downarrow & \downarrow & \downarrow & \downarrow & \downarrow \\ 9 & = & 12.5\% & \times & y \end{array}$$

Solve

$$\begin{aligned} 9 &= 12.5 \times .01 \times y \\ 9 &= .125y \\ y &= \frac{9}{.125} \\ y &= 72 \end{aligned}$$

That is, 9 is $12\frac{1}{2}$% of 72.

Example 11 A student's salary is increased from \$50 to \$65 a week. What is the percent increase over the original salary?

The question here is: \$15 (increase) is what percent of \$50 (original salary)?

Solution

Translate \$15 is what percent of \$50?

$$\begin{array}{cccccc} \$15 & \text{is} & \text{what} & \text{percent} & \text{of} & \$50? \\ \downarrow & \downarrow & \downarrow & \downarrow & \downarrow & \downarrow \\ 15 & = & y & \% & \times & 50 \end{array}$$

Solve

$$\begin{aligned} 15 &= y \times .01 \times 50 \\ 15 &= y \times .5 \\ y &= \frac{15}{.5} \\ y &= 30 \end{aligned}$$

That is, the student's percent of increase over his original salary is 30%.

Example 12

A store's profit mark-up is 18% of the selling price. What must a coat sell for if the store whishes to make a profit of \$22.50?

Here, the question is: 18% of what amount is \$22.50?

Solution

Translate 18% of what amount is 22.50?

$$\begin{array}{ccccc} 18\% & \text{of} & \text{what amount} & \text{is} & 22.50? \\ \downarrow & \downarrow & \downarrow & \downarrow & \downarrow \\ 18\% & \times & y & = & 22.50 \end{array}$$

Solve

$$\begin{aligned} 18 \times .01 \times y &= 22.50 \\ .18\,y &= 22.50 \\ y &= \frac{22.50}{.18} \\ y &= \$125 \end{aligned}$$

That is, the store must sell the coat for \$125 since 18% of \$125 is \$22.50.

B.2 Rounding Off Numbers

If the distance from your home to school is 18 miles, you can say that you live *approximately* 20 miles from school. If the actual distance is 13 miles, you can say that you live *about* 10 miles from school. In both cases, you *approximated* the distance *to the nearest ten miles.* The number line in Figure B.2 shows how you arrived at these approximations: 18 miles is closer to 20 than to 10, while 13 miles is closer to 10 than to 20. We describe this process of approximating a number as *rounding off the number.* We *rounded off* 18 to 20; and we *rounded off* 13 to 10. The 18 was rounded *up*; the 13 was rounded *down*.

FIGURE B.2

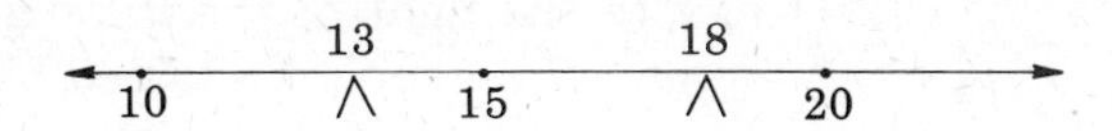

Rounding off numbers is convenient when a precise number is not needed. If we only wish an *estimate* of the cost of $2\frac{7}{8}$ yards of cloth at $7.25 a yard, we can round off $2\frac{7}{8}$ yards to the nearest yard and $7.25 to the nearest dollar. We then obtain, as the approximate cost, 3 × $7, or $21.

Large numbers are often rounded off to the nearest ten, hundred, thousand, or million. Decimals are often rounded off to the nearest tenth, hundredth, or thousandth. Before rounding off a number, we must first decide what is the last digit in the number that we wish to retain. Thus, if we wish to round off 342 *the the nearest ten*, the last digit retained will be the *tens digit*, 4: 340. If we wish to round off 342 *to the nearest hundred*, the last digit retained will be the *hundreds digit*, 3: 300.

After we decide which digit of a number is to be the last one retained, we generally observe the following rules for rounding off the number:

1 If the digit to the *right* of the last digit retained is *less* than 5, leave the last digit unchanged.

Example 1 Round off 742 *to the nearest ten.*

The last digit to be retained is 4, the tens digit: 742. Since 2, the digit to the right of 4, is *less* than 5, we leave the 4 unchanged. We say that 742 rounded off to the nearest ten is 740 (Figure B.3).

FIGURE B.3

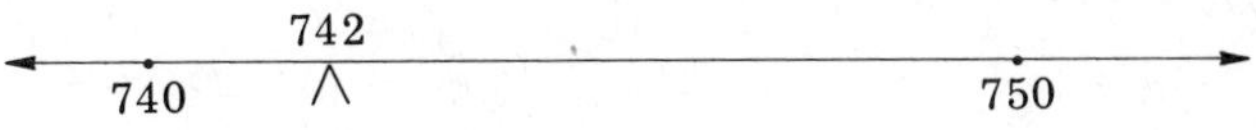

742

740 750

Example 2 Round off 1.564 *to the nearest hundredth.*

The last digit to be retained is 6, the hundredths digit: 1.564. Since 4, the digit to the right of 6, is *less* than 5, we leave the 6 unchanged and say that 1.564 rounded off to the nearest hundredth is 1.56 (Figure B.4).

FIGURE B.4

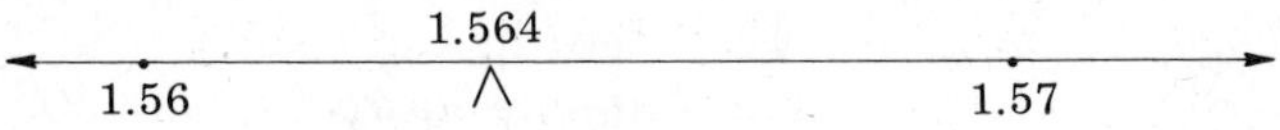

1.564

1.56 1.57

2 If the digit to the right of the last digit to be retained is *5 or more*, increase the last digit retained by 1.

Example 3 Round off 2873 *to the nearest hundred.*

The last digit to be retained is 8, the hundreds digit: 2873. Since 7, the digit to the right of 8, is *more than 5*, we add 1 to the 8. Therefore, 2873 rounded off to the nearest hundred is 2900 (Figure B.5)

FIGURE B.5

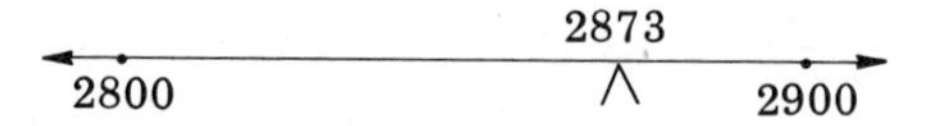

Example 4 Round off .54273 *to the nearest thousandth.*

The last digit to be retained is 2, the thousandths digit: .54273. Since 7, the digit to the right of 2 is *more than 5*, we add 1 to the 2 and say that .54273 rounded off to the nearest thousandth is .543 (Figure B.6).

FIGURE B.6

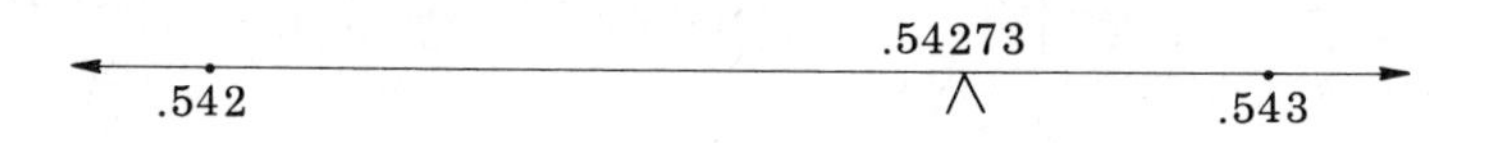

Example 5 Round off 375 *to the nearest ten.*

The last digit to be retained is 7, the tens digit: 375. Since 5 is the digit to the right of 7, we add 1 to the 7 and say that 375 rounded off to the nearest ten is 380 (Figure B.7).

FIGURE B.7

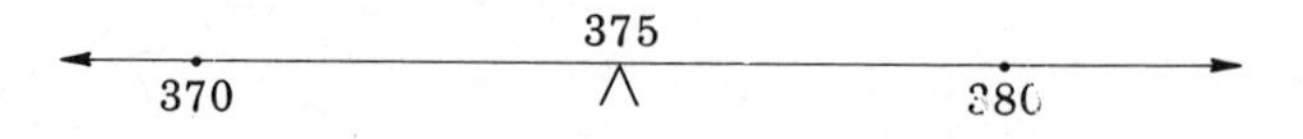

Comment *If the digit to the right of the last digit to be retained is* 5, *the usual practice, as just indicated, is to increase the last digit to be retained by* 1. *For example,* 45 to the nearest ten *is* 50, *and* 55 to the nearest ten *is* 60.

In statistical work, however, this procedure is not followed because it introduces a bias into the data. If, for instance, we are finding the sum of a set of numbers and always round off all "half-way" numbers up, our answer would probably be too large. What is done, therefore, is to round off the number so that the digit preceding the 5 is always even. *So,* 65 *rounded to the nearest ten is* 60, *while* 75 *rounded to the nearest ten is* 80. *By following this "even-handed" procedure, we can expect that about half of the time we will round* down *and about half of the time we will round* up.

*B.3 Computation: Historical Notes**

CALCULATION

The word *calculation* is derived from the Latin word *calculus*, which means a small stone or pebble. Pebbles were used by ancient peoples to perform calculations. They put pebbles in grooves in the sand and let each pebble in the first groove represent one unit; each pebble in the second groove represented ten units; and so on.

FIRST COMPUTER

The earliest known device for performing the basic arithmetic operations is the *abacus.* The word *abacus* derives from the Greek word *abax*, which meant a board strewn with sand used for computing. Eventually, the pebbles and the sand board were replaced by the *abacus*, a table with ruled lines to guide counters that represented numbers. The modern abacus, which is essentially the same as the one used by the Romans, is a portable frame with counters or beads that are moved along wires or wooden rods. The abacus can be used to calculate rapidly and efficiently and is still widely used in India, China, Japan, and Russia.

In 1946, in a contest between a Japanese abacus expert and an American expert with an electrical desk calculator, the abacus was faster in addition, subtraction, and division. In multiplication, the desk calculator was faster only when both factors contained more than ten digits each. In a problem involving all the operations, the abacus won decisively.

A drawback in using the abacus is that each step in a calculation erases the preceding step. Therefore, it is not possible to check the answer except by doing the problem over.

*A number of these notes are based on *Historical Topics for the Mathematics Classroom* (Thirty-First Yearbook), National Council of Teachers of Mathematics, Washington D.C., 1969.

FINGER COMPUTATION

Most ancient peoples, including the Greeks, Romans, Arabs, and Hindus, developed some system for representing counting numbers by means of various positions of the fingers and hands. An example is shown in Figure B.8. In Europe, during the Middle Ages, finger numbers were used for bargaining at international fairs because language was sometimes a barrier, and in the Orient today, finger numbers are still used for the same reason

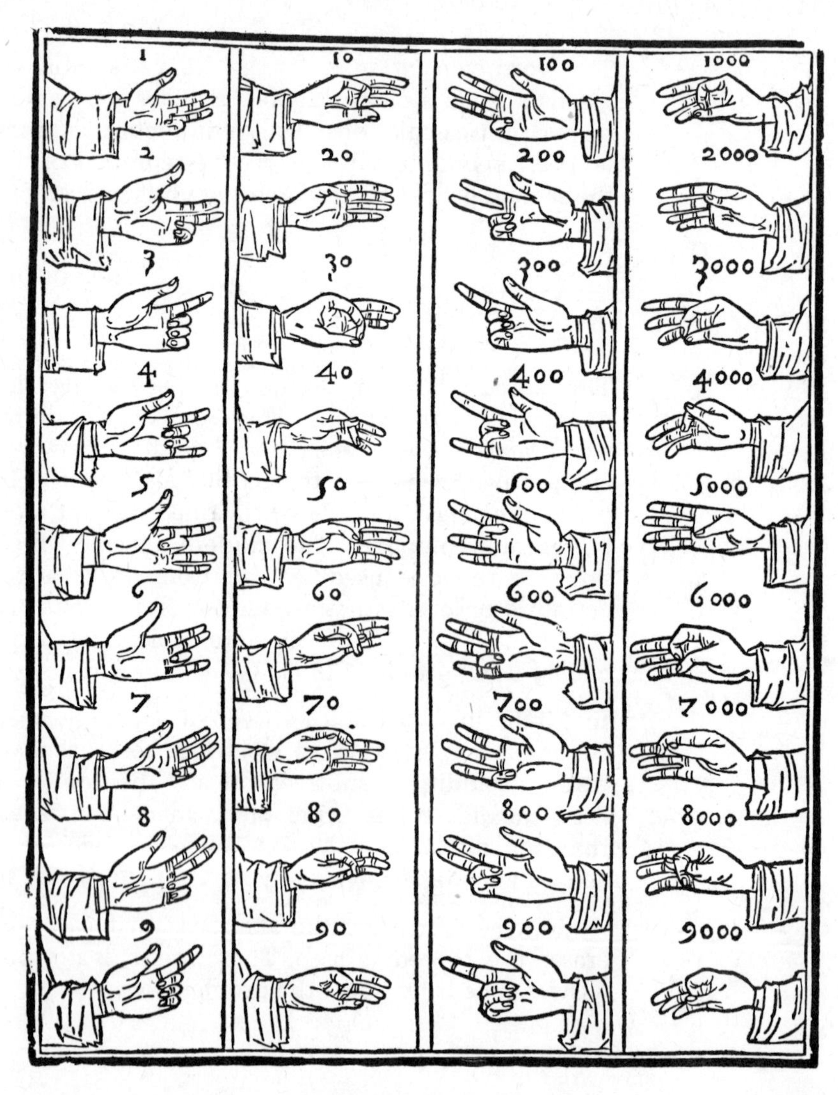

FIGURE B.8
Finger Numbers.
(From Luca Pacioli, *Summa de arithmetica, geometrica, proportioni et proportionalita,* Tusculeno, 1523. Courtesy of the David Eugene Smith Collection, Columbia University Libraries.)

Finger representation of numbers led to computation by fingers. Suppose you wish to multiply 7 by 9. Spread your hands on the table and *bend the seventh finger from the left* (Figure B.9). You now have your answer: *six* fingers to the *left* of the bent finger, and *three* fingers to the *right* of the bent finger, or 63.

FIGURE B.9

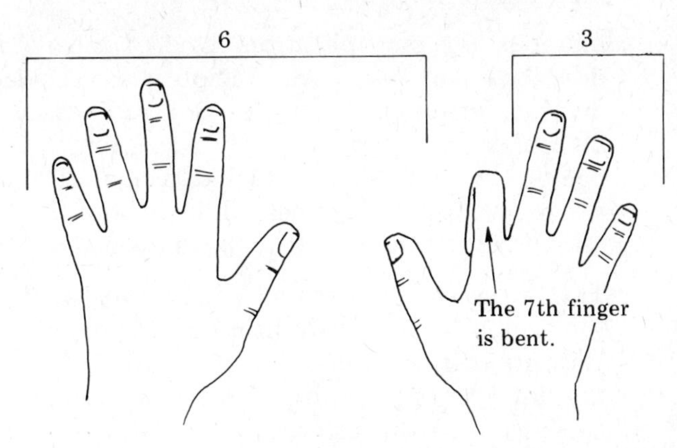

To multiply 38 by 9, start with the 8 of "38" by bending the *eighth* finger. Then count off the 3 of "38" by putting the first three fingers together as shown in Figure B.10. Your fingers are now arranged in three groups—*three* together at the *far left*, *four* to the *left* of the bent finger, and *two* to the *right* of the bent finger—giving a product of 342.

FIGURE B.10

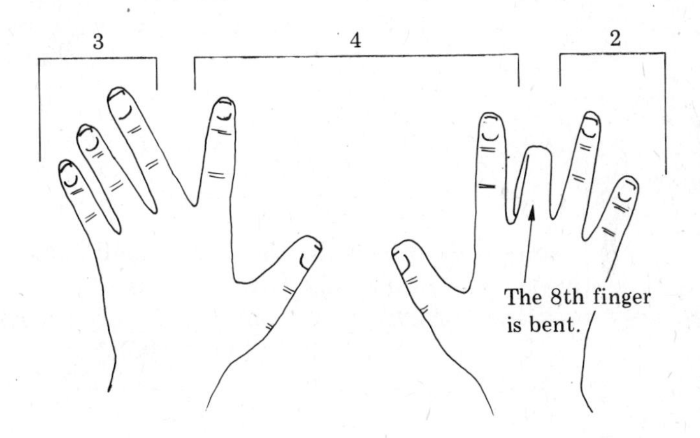

ADDITION AND SUBTRACTION

The symbols + and − first appeared in a book in 1498. Many historians believe that these symbols were first used in commerce where the + indicated an *excess* and the − indicated a *deficiency*.

MULTIPLICATION AND DIVISION

The first rules for multiplication and division appeared at the end of the fifteenth century, although a 10×10 multiplication table appeared in a book about 400 years earlier. In the fifteenth century, the Italian mathematician, Luca Pacioli, suggested eight different methods for multiplication. One of them was the *lattice method*, described on page 96. Pacioli also showed four methods for division, including the *galley* or *scratch* method that was used on a sand abacus. The name *galley* was given to the method because the completed problem looked like a boat. The division algorithm that we use today first appeared in a book in 1491 and became well established by the end of the seventeenth century.

The English mathematician William Oughtred (1574–1660) first used the symbol × for multiplication in 1631. Sixty-seven years later, the German mathematician Leibniz (1646–1716) proposed the use of the dot · to show multiplication because he thought that × could be confused with the letter x.

The horizontal bar in the fractional symbol for division was invented in 1200 A.D.; the symbol $\overline{)\quad}$ was introduced in 1544; and the symbol ÷ was invented in 1659.

The decimal fraction first appeared in 1585, but historians do not agree when and by whom the decimal point was introduced. John Napier (1550–1617), the inventor of logarithms, recommended the use of "a period or a comma" to separate units and tenths. Today, in many European countries, the comma is still used instead of the dot for the decimal point.

FRACTIONS

Fractions date back to the oldest mathematical records found, but the ancients used only special kinds of fractions. The Babylonians used fractions whose denominators were successive powers of sixty; the Egyptians used the *unit fraction*, that is, only fractions with numerators of 1; while the Romans usually kept the denominator a constant 12.

ZERO

The word *zero* probably comes from the Latin *zephirum*, which came from the Arabic *sifr*, which, in turn, is a translation of the Hindu *sunya*, meaning *void* or *empty. Sifr* was introduced into Germany in the thirteenth century as *cifra* from which the word *cipher* is derived. No one knows when or where the symbol "0" for zero was first introduced although it is generally attributed to the Hindus.

The inability for a long time to distinguish *zero* from *nothing* probably explains why zero was so long in coming. With the appearance of the symbol, 0, mathematicians were finally able to develop our decimal numeration system.

CALCULATING PRODIGIES

Although very few people are able to perform mentally, and quickly, long and intricate calculations such as multiplying two ten-digit numbers, those who have shown this rare ability have almost never shown any other mathematical talent. This fact was revealed in a study of the outstanding calculating prodigies of the last two hundred fifty years. Some of these prodigies were self-taught, while some never learned to read or write.

In 1751, Jedediah Buxton calculated mentally the number of cubic inches in a block of stone 23,145,789 yards long, 5,642,732 yards wide, and 54,965 yards thick. Thomas Fuller, who was brought to Virginia as a slave in 1724, could multiply mentally two nine-digit numbers, compute the number of seconds in a given period of time, and calculate the number of grams of corn in a given mass. Neither Buxton nor Fuller ever learned to read or write.

About 160 years ago, Zerah Colburn was able to compute in seconds the value of 8^{16}. He also gave the cube root of 268,336,125 instantaneously. In 1818, a self-taught English boy named George Parker Bidder calculated compound interest on £11,111 for 11,111 days at 5%. Bidder later pursued formal studies and became a distinguished civil engineer.

The most recently (1976) reported calculating prodigy is Shakuntala Devi from India. Although she forgot that she had visited the United States in 1952 and has difficulty remembering her birth date, she can do square and cube roots and logarithms in her head and recall the days of the week of any given dates in the last century.

The following are some of the problems that Mrs. Devi answered in 20 seconds or less during her visit to the United States in 1976:

1 Add:

$$\begin{array}{r} 25\,842\,278 \\ 111\,201\,721 \\ 370\,247\,830 \\ 55\,511\,315 \\ \hline \end{array}$$

Multiply result by 9878.

2 $\sqrt[3]{188{,}132{,}517}$

3 On what days of the week did the 14th of each month occur in 1935?

COMPUTATION TODAY

John Napier (1550–1617), a Scotch nobleman who turned to mathematics for amusement and relaxation, contributed greatly to the increase of efficiency in computational work. First, he devised a simple and ingenious device for multiplying and dividing known as Napier's Bones (see page 97). Then, by studying the relationships between points moving on two separate straight lines with different velocities, he invented *logarithms.* By means of logarithms, every multiplication or division example can be reduced to an addition or subtraction example.

The next advance in computational efficiency came in the mid-seventeenth century with Blaise Pascal who, at the age of 19, invented an adding machine. The principle of this adding machine is now used in the construction of cash registers, the odometer in an automobile, and the automatic recorder on a Geiger counter.

About ten years later, G. W. Leibniz improved Pascal's machine by enabling it to multiply by repeated additions and to divide by repeated subtractions. In 1875, the American Frank Stephen Baldwin (1838–1925) invented the first practical calculating machine that could perform the four fundamental arithmetic operations without the need to reset the machine.

Then came the electronic computers which increased the speed of what could be done with Pascal's machine more than a million times. The digital computers not only perform arithmetic computations, but they also store information, make comparisons, select numbers on the basis of such comparisons, and follow a sequence of instructions. Although these computers can now be programmed to "think" logically, they still cannot exhibit insights, intuition, and imagination that only man possesses.

ANSWERS

Answers to Odd-Numbered Exercises

EXERCISE SET 2.1, 2.2, PAGE 16

1 We mean that there must be a reliable way of telling whether or not a given element belongs to the set.

3 a) $\{x | x \text{ is a whole number and } 0 < x < 9\}$
b) $\{x | x \text{ is odd}\}$ c) $A \subset B$ d) $S \subseteq S$ e) $\phi \subset G$ f) $a \in \{a\}$

5 a) Yes b) No c) Yes

7 a) Equal; equal and equivalent b) Neither equal nor equivalent c) Equivalent

9 The set of all girls in the class

11 "Is an element of" refers to exactly one member of the set; "is a subset of" refers to a set of elements contained in the set.

13 a) $D = \{18, 19, 20, 21, 22, 23, 24, 25, 26, 27, 28\}$
b)

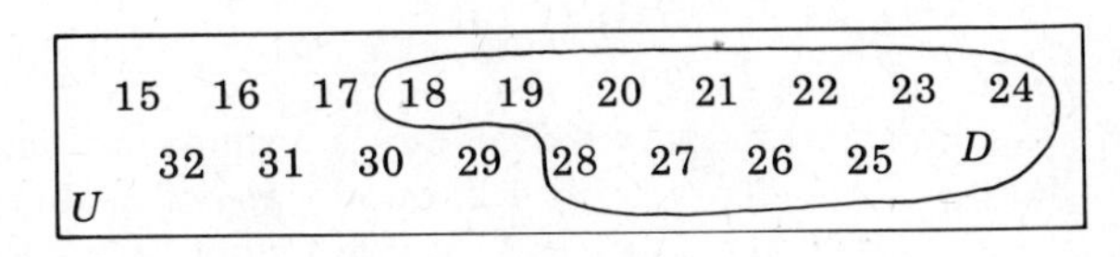

14 Set up a one-to-one correspondence by getting men to dance with women. Then see whether any men or any women are left.

EXERCISE SET 2.3, PAGE 22

1 a) $\{1, 2, 3, 4, 5, 6, 8\}$ b) $\{0, 1, 2, 3, 4, 5, 6, 8, 10, 12, ...\}$
c) $\{1,3,5,7, . . .\}$ d) $\{2, 4, 6, 8\}$ e) $\{1, 3, 5\}$ f) ϕ
g) $\{2, 4, 6, 8\}$ h) ϕ

3

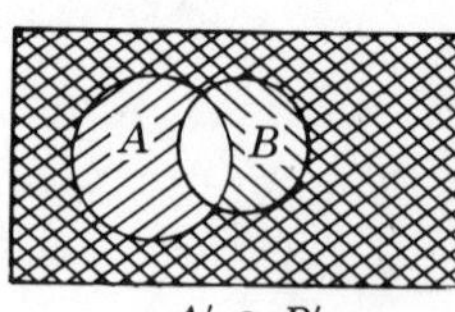

$A' \cap B'$
Cross-hatched Region

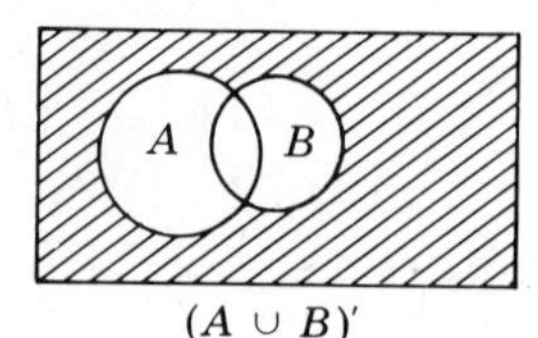

$(A \cup B)'$
Hatched Region

5 4 people

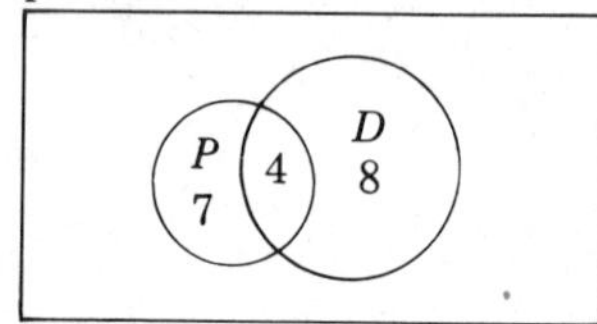

EXERCISE SET 2.4 a, PAGE 24

1 a) $\phi; D = \phi \quad R = \phi$ b) $\{(1, 3)\}; D = \{1\}, R = \{3\}$
c) $\{(2, 4)\}; D = \{2\}, R = \{4\}$

3 $2^6 - 1$, or 63

EXERCISE SET 2.4 b, PAGE 26

1

	a)	b)	c)	d)	e)	f)	g)	h)	i)	j)	k)	l)
Reflexive	No	Yes	Yes	Yes	No	Yes	No	No	Yes	No	No	Yes
Symmetric	No	Yes	No	Yes	No	Yes	Yes	No	Yes	No	Yes	No
Transitive	Yes	Yes	Yes	Yes	No	Yes	No	Yes	No	Yes	Yes	Yes

3 C_1 = {Ann, Laura, Eve}, C_2 = {John, Ben}

5 a) $\{(1, 1), (1, 2), (2, 2)\}$ b) $\{(1, 3), (2, 4), (4, 2), (3, 1)\}$
c) $\{(1, 2), (2, 4), (1, 4)\}$
d) $\{(1, 2), (1, 3), (2, 3), (2, 1), (3, 1), (3, 2), (1, 1), (2, 2), (3, 3)\}$
e) $\{(1, 2), (2, 3), (1, 3)\}$

7 (1) a) $\{(x, y)|x = y\}$ b) $\{(x, y)|x + y < 5\}$ c) $\{(x, y)|x = y^2\}$
(2) a) $\{(x, y)|y < x$; x and y are whole numbers between 3 and 6$\}$
b) $\{(x, y)|x = y + 2$; x and y are whole numbers between 1 and 4$\}$
c) $\{(x, y)|y - x < 3$; y is a whole number less than 5 and x is a whole number less than 7$\}$

EXERCISE SET 2.5, PAGE 29

1 a) Yes b) Yes c) No d) Yes

3 a) $R^{-1} = \{(2, 1), (3, 2), (4, 3)\}$; this is a function.
b) $S^{-1} = \{(1, 1), (2, 2), (3, 3), (4, 4)\}$; this is a function.
c) $A^{-1} = \{(3, 1), (4, 1), (5, 2), (7, 3)\}$; this is a function.
d) $C^{-1} = \{(5, 1), (7, 2), (5, 3), (9, 4)\}$; this is not a function.

5 a) $y = 2x + 1$
b)

x	1	2	3	4	5	6	7	8	9	10
y	3	5	7	9	11	13	15	17	19	21

c) $\{(1, 3), (2, 5), (3, 7), (4, 9), (5, 11), (6, 13), (7, 15), (8, 17), (9, 19), (10, 21)\}$

7 a) $\{(1, 4), (2, 8), (3, 12), (4, 16), (5, 20)\}$
b) $\{(0, 7), (1, 8), (2, 9), (3, 10), (4, 11)\}$
c) $\{(4, 1), (5, 2), (6, 3), (7, 4), (8, 5)\}$
d) $\{(1, 1), (2, 3), (3, 5), (4, 7), (5, 9)\}$

9 a)

t	0	1	2	3	4
s	0	16	64	144	256

b) $\{(0, 0), (1, 16), (2, 64), (3, 144), (4, 256)\}$
c) $s = 16(2.5)^2 = 100$ ft
d) $150 = 16t^2$; $t = 3.06$ sec

EXERCISE SET 3.1, 3.2, PAGE 50

1 a) 3 b) 24 c) 212

3 a) 17 b) 49 c) 120 d) 154 e) 678

5 a) True b) True c) False

7 a) True b) True
c) False (The symbol for zero was first introduced into the Chinese system about 800 years ago.)

9 The Hindu-Arabic system is based on the place-value principle and uses a symbol for zero both as a number and as a place holder.

11 a) XXVII b) CIX c) CCCLXIX d) LXVIII e) DCCL
f) MCMLXXVI

13 a) 896 b) 3204 c) 58010

15 38

EXERCISE SET 3.3, PAGE 59

1 a) 102_{five} b) 11011_{two} c) 36_{seven} d) 23_{twelve}

3 a) 16_{ten} b) 2210_{three} c) 11101_{two} d) 1100_{five} e) 101_{twelve}
f) 31_{six} g) 102_{five} h) 36_{seven} i) 101110_{two} j) 127_{eight}

5 a) Base 8 b) Base 6

7 a) Base 5 b) Base 9

9 a) 343_{eight} b) 30312_{five} c) 245_{six} d) 100_{two}

11

+	0	1	2	3
0	0	1	2	3
1	1	2	3	10
2	2	3	10	11
3	3	10	11	12

×	0	1	2	3
0	0	0	0	0
1	0	1	2	3
2	0	2	10	12
3	0	3	12	21

13 a) Base 3

+	0	1	2
0	0	1	2
1	1	2	10
2	2	10	11

×	0	1	2
0	0	0	0
1	0	1	2
2	0	2	11

b) Base 8

+	0	1	2	3	4	5	6	7
0	0	1	2	3	4	5	6	7
1	1	2	3	4	5	6	7	10
2	2	3	4	5	6	7	10	11
3	3	4	5	6	7	10	11	12
4	4	5	6	7	10	11	12	13
5	5	6	7	10	11	12	13	14
6	6	7	10	11	12	13	14	15
7	7	10	11	12	13	14	15	16

×	0	1	2	3	4	5	6	7
0	0	0	0	0	0	0	0	0
1	0	1	2	3	4	5	6	7
2	0	2	4	6	10	12	14	16
3	0	3	6	11	14	17	22	25
4	0	4	10	14	20	24	30	34
5	0	5	12	17	24	31	36	43
6	0	6	14	22	30	36	44	52
7	0	7	16	25	34	43	52	61

c) Base 12

+	0	1	2	3	4	5	6	7	8	9	*t*	*e*
0	0	1	2	3	4	5	6	7	8	9	*t*	*e*
1	1	2	3	4	5	6	7	8	9	*t*	*e*	10
2	2	3	4	5	6	7	8	9	*t*	*e*	10	11
3	3	4	5	6	7	8	9	*t*	*e*	10	11	12
4	4	5	6	7	8	9	*t*	*e*	10	11	12	13
5	5	6	7	8	9	*t*	*e*	10	11	12	13	14
6	6	7	8	9	*t*	*e*	10	11	12	13	14	15
7	7	8	9	*t*	*e*	10	11	12	13	14	15	16
8	8	9	*t*	*e*	10	11	12	13	14	15	16	17
9	9	*t*	*e*	10	11	12	13	14	15	16	17	18
t	*t*	*e*	10	11	12	13	14	15	16	17	18	19
e	*e*	10	11	12	13	14	15	16	17	18	19	1*t*

×	0	1	2	3	4	5	6	7	8	9	*t*	*e*
0	0	0	0	0	0	0	0	0	0	0	0	0
1	0	1	2	3	4	5	6	7	8	9	*t*	*e*
2	0	2	4	6	8	*t*	10	12	14	16	18	1*t*
3	0	3	6	9	10	13	16	19	20	23	26	29
4	0	4	8	10	14	18	20	24	28	30	34	38
5	0	5	*t*	13	18	21	26	2*e*	34	39	42	47
6	0	6	10	16	20	26	30	36	40	46	50	56
7	0	7	12	19	24	2*e*	36	41	48	53	5*t*	65
8	0	8	14	20	28	34	40	48	54	60	68	74
9	0	9	16	23	30	39	48	53	60	69	76	83
t	0	*t*	18	26	34	42	50	5*t*	68	76	84	92
e	0	*e*	1*t*	29	38	47	56	65	74	83	92	*t*1

15 a)

0	1	2	3	4	5	6	7	8	9	10	11	12	13	14	15
0	*	?	[	*0	**	*?	*[	?0	?*	??	?[	[0	[*	[?	[[

16	17	18	19	20	21	22	23	24	25
*00	*0*	*0?	*0[	**0	***	**?	**[	*?0	*?*

b) 2(1024) + (256) + 3(64) + 2 = 2498

17 Base 8

19 $(\rightarrow \times \text{pati}^3) + (* \times \text{pati}^2) + (? \times \text{pati}) + >$

21 A zero is added at the end of the numeral.

23 a) 33_{eight} b) 377_{eight} c) 207_{eight}

25 a) 1 b) 4 c) $b - 1$

27 False; 10^3 means 10 × 10 × 10.

29 False. For instance, base 2 has 4 basic addition facts; base 5 has 25 basic addition facts; base 10 has 100 basic addition facts.

31 False. A numeration system may be based on as many basic symbols (> 1) as we wish. The most familiar, however, are those systems that use some or all of the symbols 0, 1, 2, 3, 4, 5, 6, 7, 8, 9, *t*, *e*—always including 0.

33 False. In base 2, 1 + 1 = 10, not 2.

35 False. (e.g. 14 expressed in base 5 is 24_{five}; 14 expressed in base 12 is 12_{twelve}. Note that each numeral contains two digits.)

37 a) Base 7.
b) ⬠, ⬡, |*, ||, |∧, |△, |□, |⬠, |⬡, ∧*, ∧|, ∧∧, ∧△, △□, ∧⬠,
∧⬡, △*, △|, △∧, △△, △□, △⬠, △⬡, □*, □|, □∧, □△, □□,
□⬠

39 15_{seven}

EXERCISE SET 4.1–4.3, PAGE 92

1 0; there is no largest whole number.

3 a) Yes b) No. 2 × 5 ≠ 5 × 2 since (2 + 3) × 5 ≠ (5 + 3) × 2.

5 a) Transitive b) Addition c) Multiplication

7 a) Multiplication property of 0 b) Commutative property of addition
c) Commutative property of multiplication d) Closure for addition

9 Commutative *or* associative

11 b, c

13 a) $(5 + 9)12 = 14 \times 12 = 168$; $(5 + 9)12 = (5 \times 12) + (9 \times 12) = 60 + 108 = 168$

b) $17(3 + 8) = 17 \times 11 = 187$; $17(3 + 8) = (17 \times 3) + (17 \times 8) = 51 + 136 = 187$

15 a) $12 - 5 = 7$; $12 - 7 = 5$ b) $t - r = m$; $t - m = r$

17 a) $27 \div 9 = 3$; $27 \div 3 = 9$ b) $h \div f = g$; $h \div g = f$

19 a) $a + c = b$ b) $<$

21 $a < b$

23 $(24 \div 4) \div 2 = 6 \div 2 = 3$; $24 \div (4 \div 2) = 24 \div 2 = 12$

25 a) False. $0 \times 5 = 0$ b) True. There is no counting number less than 1.
c) False. For *any* whole number a, $a + 1$ is also a whole number.
d) False. There is none between 1 and 2.
e) True. If 0 is subtracted from a whole number, the result is the same whole number.
f) False. This definition is true only if A and B are disjoint sets.
g) False. The statement given illustrates the commutative property.
h) False. The given statement illustrates the zero property of multiplication.
i) False. $\frac{a}{0}$ is undefined.
j) False. The conclusion here is justified by the transitive property.
k) False. What is stated here is the associative property for multiplication.
l) True, by definition.

EXERCISE SET 5.1–5.3, PAGE 112

1 a) There are 35 such primes: 2, 3, 5, 7, 11, 13, 17, 19, 23, 29, 31, 37, 41, 43, 47, 53, 59, 61, 67, 71, 73, 79, 83, 89, 97, 101, 103, 107, 109, 113, 127, 131, 137, 139, 149.
b) There are 21 such primes: 101, 103, 107, 109, 113, 127, 131, 137, 139, 149, 151, 157, 163, 167, 173, 179, 181, 191, 193, 197, 199.

3 a) $3 + 5$ b) $3 + 11$; $7 + 7$ c) $5 + 19$; $7 + 17$; etc.
d) $5 + 31$; $7 + 29$; etc.

5 a) $2 \times 3 \times 5$ b) 2×3^3 c) $3^2 \times 17$ d) $2^2 \times 5 \times 31$

7 a) Yes b) Yes c) No. $837 = 27 \times 31$ d) Yes

9 $37 = 1 + 6^2$; $101 = 1 + 10^2$; $257 = 1 + 16^2$

11 2, 3. Any number which differs by 1 from a prime greater than 3 must be even and, therefore, composite.

EXERCISE SET 5.4, PAGE 114

1 Let $2k + 1$ and $2p + 1$ (where k and p are natural numbers) be the two odd numbers. Then the product of the two numbers is

$$(2k + 1)(2p + 1) = 4kp + 2p + 2k + 1 = 2(2kp + p + k) + 1,$$

which is odd.

3 Let $2p + 1$ be an odd number, and let $2n$ be an even number (where p and n are natural numbers). Then their sum is

$$(2p + 1) + 2n = 2p + 2n + 1 = 2(p + n) + 1,$$

which is odd.

5 a) Even b) Even c) Odd

7 a) Even b) Either c) Odd d) Either e) Odd f) Odd

9 An even number. Since the product of an even number and any number of numbers will contain 2 as a factor, the result must be even.

11 Let p and q be any two natural number. If p and q are *both* odd, then their product, $p \times q$, would be odd. (See Exercise 1.)

EXERCISE SET 5.5, PAGE 119

1 a) 2, 3, 4, 8, 9 b) None c) 3 d) 2, 4, 5, 10 e) 2, 4, 8

3 a) Yes. Since the number has 2 and 3 as factors, it will have 6 as a factor.
b) Yes. Since the number has 4 and 3 as factors, it will have 12 as a factor.

5 a) 12,045 b) 24,624 c) 143,280

7 N is divisible by 6 if N is divisible by 2 and by 3.

9 N is divisible by 15 if N is divisible by 3 and by 5.

EXERCISE SET 5.6, 5.7, PAGE 125

1 a) 1, 2, 4, 8 b) 1, 2, 3, 6

3 a) 18 b) 1 c) 8 d) 3 e) 14 f) 33 g) n h) 35
i) $11^2 \times 13^2$ j) 6 k) 3

5 The prime factorization of two numbers, a and b, is independent of the order of the numbers.

7 The GCD(a, b) consists of the product of the lowest powers of the prime factors *common* to a and b.

9 a) 6 b) 7 c) 7 d) 6

11 a) {0, 7, 14, 21, 28, ...} b) {0, 12, 24, 36, 48, ...}
c) {0, 15, 30, 45, 60, ...}

13 $2^2 \times 3^2 = 36$

15 $n \times (n + 1)$

17 If the GCD(a, b) = 1, then the prime factorizations of a and b have no factors in common. Therefore, the LCM(a, b) will include the product of *all* the factors of a and b; that is, the LCM(a, b) = $a \times b$.

If the LCM(a, b) = $a \times b$ and the GCD(a, b) $\neq$ 1, then the LCM(a, b) would contain *fewer* factors than would be found in $a \times b$. Therefore, the GCD(a, b) = 1.

EXERCISE SET 6.1–6.4, PAGE 146

Answers to Exercises 1 and 3 are arranged in the following format:

I *Expanded Notation*, II *Partial Sums*, III *Algorithm.*

1 a) I

$36 = (3 \times 10) + (6 \times 1)$ (Expanded notation)
$\underline{+42 = (4 \times 10) + (2 \times 1)}$
$= [(3 + 4) \times 10] + [(6 + 2) \times 1]$ (Distrib. prop.)
$= (7 \times 10) + (8 \times 1)$ (Basic add. facts)
$= 78$ (Place-value notation)

II
36
$\underline{+42}$
8 (2 ones + 6 ones)
$\underline{70}$ (4 tens + 3 tens)
78

III
36
$\underline{+42}$
78

b) I

$27 = (2 \times 10) + (7 \times 1)$ (Expanded notation)
$\underline{+98 = (9 \times 10) + (8 \times 1)}$
$= [(2 + 9) \times 10] + [(7 + 8) \times 1]$ (Distrib. prop.)
$= (11 \times 10) + (15 \times 1)$ (Basic add. facts)
$= [(10 + 1) \times 10] + [(10 + 5) \times 1]$ (Decimal notation)
$= (10 \times 10) + (10 \times 1) + (1 \times 10) + (1 \times 5)$ (Distrib. prop.)
$= (1 \times 100) + (1 \times 10) + (1 \times 10) + (5 \times 1)$ (Place value)
$= (1 \times 100) + (2 \times 10) + (5 \times 1)$ (Associative, distrib., basic add. facts)
$= 125$ (Place-value notation)

II
27
$\underline{+98}$
15 (8 ones + 7 ones)
$\underline{110}$ (9 tens + 2 tens)
125

III
27
$\underline{+98}$
115

c) I

$145 = (1 \times 100) + (4 \times 10) + (5 \times 1)$ (Expanded notation)
$\underline{+339 = (3 \times 100) + (3 \times 10) + (9 \times 1)}$
$= (4 \times 100) + (7 \times 10) + (14 \times 1)$ (Distrib., basic add. facts)
$= (4 \times 100) + (7 \times 10) + (1 \times 10) + (4 \times 1)$
(Decimal notation distrib.)
$= (4 \times 100) + (8 \times 10) + (4 \times 1)$ (Distrib., basic add. facts)
$= 484$ (Place-value notation)

II
145
$\underline{+339}$
14 (9 ones + 5 ones)
70 (3 tens + 4 tens)
$\underline{400}$ (3 hund. + 1 hund.)
484

III
145
$\underline{+339}$
484

d) I

$267 = (2 \times 100) + (6 \times 10) + (7 \times 1)$ (Expanded notation)
$\underline{+558 = (5 \times 100) + (5 \times 10) + (8 \times 1)}$
$= (7 \times 100) + (11 \times 10) + (15 \times 1)$ (Distrib., basic add. facts)
$= (7 \times 100) + [(10 \times 10) + (1 \times 10)] + [(10 \times 1) + (5 \times 1)]$
(Decimal notation, distrib.)
$= (7 \times 100) + (1 \times 100) + (2 \times 10) + (5 \times 1)$ (Place-value, distrib., basic add. facts)
$= (8 \times 100) + (2 \times 10) + (5 \times 1)$ (Assoc., distrib., basic add. facts)
$= 825$ (Place-value notation)

II

$$\begin{array}{rl} 267 & \\ +558 & \\ \hline 15 & \text{(8 ones + 7 ones)} \\ 110 & \text{(5 tens + 6 tens)} \\ 700 & \text{(5 hund. + 2 hund.)} \\ \hline 825 & \end{array}$$

III

$$\begin{array}{r} 267 \\ +558 \\ \hline 825 \end{array}$$

e) I

$$\begin{array}{rl} 145 = & (1 \times 100) + (4 \times 10) + (5 \times 1) \quad \text{(Expanded notation)} \\ +\ 67 = & \qquad\qquad\quad (6 \times 10) + (7 \times 1) \\ 237 = & 2(100) \quad\ + (3 \times 10) + (7 \times 1) \\ \hline \end{array}$$

$= (3 \times 100) + (13 \times 10) + (19 \times 1)$ (Distrib., basic add. facts)

$= (3 \times 100) + (10 \times 10) + (3 \times 10) + (10 \times 1) + (9 \times 1)$ (Decimal notation, distrib.)

$= (3 \times 100) + (1 \times 100) + (3 \times 10) + (1 \times 10) + (9 \times 1)$ (Place-value)

$= (4 \times 100) + (4 \times 10) + (9 \times 1)$ (Distrib., basic add. facts)

$= 449$ (Place-value notation)

II

$$\begin{array}{rl} 145 & \\ +\ 67 & \\ 237 & \\ \hline 19 & \text{(7 ones + 7 ones + 5 ones)} \\ 130 & \text{(3 tens + 6 tens + 4 tens)} \\ 300 & \text{(2 hund. + 1 hund.)} \\ \hline 449 & \end{array}$$

III

$$\begin{array}{r} 145 \\ +\ 67 \\ 237 \\ \hline 449 \end{array}$$

3 a) I $83 \times 9 = (80 + 3)9$ (renaming)

$= (80 \times 9) + (3 \times 9)$ (Distrib. property)

$= 720 + 27$ (Basic mult. facts)

$= 747$ (Addition algorithm)

II

$$\begin{array}{rl} 83 & \\ \times\ 9 & \\ \hline 27 & (9 \times 3) \\ 720 & (9 \times 80) \\ \hline 747 & \end{array}$$

III

$$\begin{array}{r} 83 \\ \times\ 9 \\ \hline 747 \end{array}$$

b) I $62 \times 75 = (60 + 2)(70 + 5)$ (Renaming)

$= 60(70 + 5) + 2(70 + 5)$ (Distrib. property)

$= (60 \times 70) + (60 \times 5) + (2 \times 70) + (2 \times 5)$ (Distrib. prop.)

$= 4200 + 300 + 140 + 10$ (Basic mult. facts)

$= 4650$ (Addition algorithm)

II

$$\begin{array}{rl} 62 & \\ \times 75 & \\ \hline 10 & (5 \times 2) \\ 300 & (5 \times 60) \\ 140 & (70 \times 2) \\ 4200 & (70 \times 60) \\ \hline 4650 & \end{array}$$

III

$$\begin{array}{r} 62 \\ \times 75 \\ \hline 310 \\ 434\ \\ \hline 4650 \end{array}$$

c) I $384 \times 56 = (300 + 80 + 4)(50 + 6)$ (Renaming)

$= 300(50 + 6) + 80(50 + 6) + 4(50 + 6)$ (Distrib. prop.)

$= 15000 + 1800 + 4000 + 480 + 200 + 24$ (Distrib. prop.)

$= 21504$ (Addition algorithm)

II

384	
$\times$ 56	
24	(6×4)
480	(6×80)
1800	(6×300)
200	(50×4)
4000	(50×80)
15000	(50×300)
21504	

III

384
$\times$ 56
2304
1920
21504

5 *Repeated Subtraction* / *Algorithm*

a) Repeated Subtraction

$3 \overline{)59}$	
30	(10×3)
29	
27	(9×3)
2	(19×3)

$59 = (19 \times 3) + 2$

Algorithm

19
$3 \overline{)59}$
3
29
27
2

b) Repeated Subtraction

$12 \overline{)654}$	
600	(50×12)
54	
48	(4×12)
6	(54×12)

$654 = (54 \times 12) + 6$

Algorithm

54
$12 \overline{)654}$
60
54
48
6

c) Repeated Subtraction

$9 \overline{)32957}$	
27000	(3000×9)
5957	
5400	(600×9)
557	
540	(60×9)
17	
9	(1×9)
8	(3661×9)

$32957 = (3661 \times 9) + 8$

Algorithm

3661
$9 \overline{)32957}$
27
59
54
55
54
17
9
8

EXERCISE SET 7.1–7.3, PAGE 158

1 a) 5 b) $^-12$ c) ^-b d) 0 e) $a + b$ f) $^-(x + y)$

3 ϕ

5 a) $x = {}^-3, {}^-4, {}^-5, \ldots$ b) $x = 0, 1, 2, 3, \ldots$ c) $x = 3, 2, 1, 0, {}^-1, {}^-2, \ldots$
d) $x = 11, 12, 13, 14, \ldots$

7 a) 9 b) 28 c) 5 d) 47 e) 0 f) 129

9 a) $<$ b) $<$ c) $<$ d) $<$ e) $=$ f) $>$ g) $<$

11 a)

2 3 4 5 6

$|5 - 3| = 2$

b) 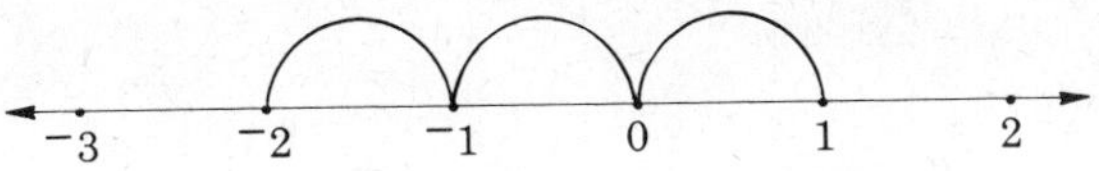

$|1 - {}^{-}2| = |3| = 3$

c) 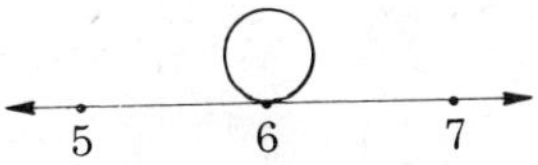

$|6 - 6| = 0$

d) 

$|f - c|$

e) 

$|x - {}^{-}2| = |x + 2|$

13 a) False. $^{-}1$ is not a whole number.
b) True. The counting numbers correspond to the positive integers.
c) False. The opposite of $^{-}1$ is 1.
d) True. The absolute value of an integer is always a nonnegative integer, that is, a whole number.
e) True. Every whole number is an integer, but not every integer is a whole number.
f) True. Since $^{-}(^{-}a) = a$, $^{-}(^{-}a) + b = 0$ means that $a + b = 0$. Therefore, a and b are additive inverses.
g) True. Every natural number is an integer.

EXERCISE SET 7.4a, PAGE 163

1 a) 11 b) 2 c) $^{-}16$ d) $^{-}13$ e) $^{-}11$ f) 3 g) $^{-}9$ h) $^{-}32$

3 a) 18 b) $^{-}12$ c) 10

5 a) Positive b) Negative

EXERCISE SET 7.4b, PAGE 167

1 a) 20 b) $^{-}63$ c) $^{-}120$ d) 66 e) $^{-}665$ f) $^{-}432$ g) 80 h) 204

3 a) 14 b) $^{-}32$ c) 36 d) $^{-}80$

5 a) $^{-}6$ b) $^{-}6$ c) 8 d) $^{-}27$ e) 3

7 a) c^2 is positive; c^3 is positive. b) c^2 is positive; c^3 is negative.

9 No. *Counterexample* $12 \div (4 + 2) \stackrel{?}{=} (12 \div 4) + (12 \div 2)$

$$12 \div 6 \stackrel{?}{=} 3 + 6$$

$$2 \neq 9$$

11 a) 0° b) 37.8° c) 30° d) $^{-}40°$

EXERCISE SET 8.1–8.5, PAGE 182

1 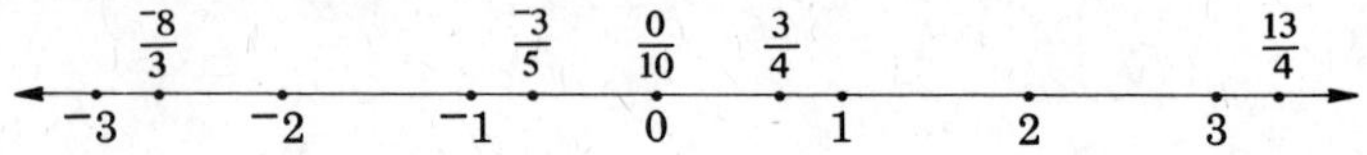

3 a) $\frac{{}^-41}{35}$ b) $\frac{23}{24}$ c) $\frac{{}^-1}{6}$ d) $\frac{{}^-53}{20}$ e) $-\frac{cn + dm}{dn}$ f) $\frac{{}^-91}{60}$

g) $\frac{7}{5a}$

5 a) $\frac{{}^-20}{21}$ b) $\frac{3}{20}$ c) $\frac{{}^-5}{4}$ d) $\frac{ab}{st}$ e) ab^2 f) $\frac{5}{3}$

7 a) $\frac{8}{3}$ b) $\frac{41}{12}$ c) $\frac{6}{5}$ d) $\frac{20}{9}$ e) $\frac{5}{11}$ f) $\frac{bdx}{(ad + bc)y}$

9 a) $\frac{{}^-1}{16}$ b) $\frac{{}^-1}{15}$ c) $-\frac{cn + dm}{2dn}$

11 Positive

13 If $a > 0$ and $b > 0$, then $\frac{1}{a} + \frac{1}{b} > \frac{1}{a + b}$.

Proof $\frac{1}{a} + \frac{1}{b} = \frac{a + b}{ab}$.

$\frac{a + b}{ab} > \frac{1}{a + b}$ because $(a + b)(a + b) > ab$;

that is, $a^2 + 2ab + b^2 > ab$.

Therefore, $\frac{1}{a} + \frac{1}{b} > \frac{1}{a + b}$.

15 $\frac{ad - bc}{bd} + \frac{c}{d} = \frac{ad - bc + bc}{bd} = \frac{ad}{bd} = \frac{a}{b}$.

Therefore, $\frac{ad - bc}{bd} = \frac{a}{b} - \frac{c}{d}$.

17 a) Negative b) Positive c) Zero d) Zero e) Positive
f) Negatives or opposites

19 a) Nonpositive rational b) $\frac{{}^-1}{3}$ c) Not defined

21 If $2d < n$ and $0 < d$, then $2 < \frac{n}{d}$

EXERCISE SET 8.6, PAGE 187

1 a) A number y such that $x + y = 0$.
b) A number z such that $xz = 1$.

3 a) $\frac{1}{6}$ b) There is no multiplicative inverse of zero. c) $\frac{{}^-3}{2}$

d) $\frac{t}{r}$ if $r \neq 0$ e) $\frac{{}^-d}{c}$ if $c \neq 0$ f) There is no multiplicative inverse of zero.

5 a) $3 - 1 \neq 1 - 3$, and $(3 - 1) - 1 \neq 3 - (1 - 1)$.

b) $\frac{3}{2} \neq \frac{2}{3}$, and $\frac{\frac{3}{2}}{3} \neq \frac{3}{\frac{2}{3}}$.

7 Every rational number, except 0, has a multiplicative inverse; and the set of rational numbers is dense. Neither of these properties holds for the integers.

9 a) The counting numbers have all the properties listed except *identity* +, *inverse* +, and *inverse* X.
b) The whole numbers have all the properties except *inverse* + and *inverse* X.
c) The integers have all the properties except *inverse* X.
d) The rationals have all the listed properties.

11 a) Let a be any rational number, and suppose that *both* e and f are additive identities such that $a + e = e + a = a$ and $a + f = f + a = a$. Then $f = e + f = e$ since the identity properties hold when $a = f$ and $a = e$. Therefore, e and f are the same element.
b) If $ae = ea = a$ and $af = fa = a$ for all rational numbers a, then $f = fe = e$ by letting $a = f$. Then $a = e$. Therefore, e and f are the same element.

EXERCISE SET 8.7, PAGE 191

1 For any two rational numbers $\frac{a}{b}$ and $\frac{c}{d}$, $\frac{a}{b} < \frac{c}{d}$ if $ad < bc$ for $bd > 0$; and $ad > bc$ if $bd < 0$.

3 a) $\frac{7}{5} < \frac{3}{2}$ b) $\frac{^-2}{5} < \frac{^-1}{3}$ c) $\frac{^-54}{^-12} = \frac{9}{2} < \frac{26}{5}$ d) $\frac{^-3}{37} < \frac{^-6}{75}$

5 $\frac{a}{x} < \frac{a}{y}$ if $x > y$.

7 a) $\frac{n+1}{n} > 1$ b) $\frac{n-1}{n} < 1$

9 Find the average, a_1, of $\frac{1}{1000}$ and $\frac{2}{1000}$; then find the average, a_2, of $\frac{1}{1000}$ and a_1; then find the average, a_3, of $\frac{1}{1000}$ and a_2; etc. Expressed exponentially, we have the following a_1, a_2, a_3, a_4, a_5, and a_6 between $\frac{1}{1000}$ and $\frac{2}{1000}$:

$\frac{1}{10^3}, \frac{15}{10^4}, \frac{125}{10^5}, \frac{1125}{10^6}, \frac{10625}{10^7}, \frac{103125}{10^8}, \frac{1015625}{10^9}, \frac{2}{10^3}$.

11 4 oranges for 53¢ is less expensive.

13 a) $^-1, \frac{^-5}{6}, 0, \frac{2}{3}$ b) $\frac{8}{17}, \frac{9}{19}, \frac{17}{37}, \frac{4}{7}$ c) $\frac{^-8}{29}, \frac{^-3}{13}, \frac{17}{80}, \frac{7}{30}$

EXERCISE SET 8.8, PAGE 195

1 a) .625 b) .325 c) .9375 d) $13.\overline{6}$ e) 2.375

3 a) $\frac{6}{25}$ b) $\frac{17}{1000}$ c) $\frac{5307}{1000}$ d) $\frac{1}{9}$ e) $\frac{5}{99}$ f) $\frac{274}{45}$

g) $\frac{1491}{495}$ h) $\frac{131}{999}$ i) $\frac{851661}{49995}$

5 $r = 2^n \times 5^m$, where n and m are whole numbers.

7 a) 16 b) 32 c) 120 d) 400 e) 2841.6

9 64.5 cm^2

EXERCISE SET 9.1–9.3, PAGE 209

1 c, d, e, f, g, i

3

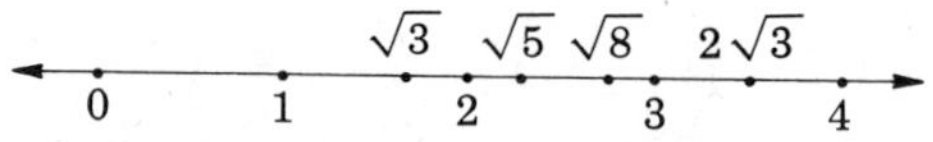

5 a) Yes, because the product of two rational numbers is a rational number.
b) No. $\sqrt{2}$ is irrational, but $(\sqrt{2})^2 = 2$.

7 Let x be an irrational number, and let y be a rational number.
(1) Assume that $x + y = z$, where z is a rational number.
(2) Then $x = z - y$.
(3) Since $z - y$ is a rational number (closure), the equation in (2) says that an irrational number equals a rational number.
(4) Therefore, the assumption in (1) is false, and we conclude that the sum of an irrational number and a rational number is irrational.

9 a) 2.236 b) 4.583 c) .283 d) 5.408

11 (1) Assume that $a\sqrt{2} + b\sqrt{3} = c$, where c is rational.
(2) Then $a\sqrt{2} = c - b\sqrt{3}$.
Squaring both sides of (2), we get:
(3) $2a^2 = c^2 - 2bc\sqrt{3} + 3b^2$, or
(4) $2bc\sqrt{3} = c^2 + 3b^2 - 2a^2$.
(5) Since the left side of (4) is irrational (see Exercise 6) and the right side of (4) is rational, we have a contradiction.
(6) Therefore, assumption (1) is false, and we conclude that $a\sqrt{2} + b\sqrt{3}$ is irrational.

13 a) Counting numbers, whole numbers, integers, rationals, reals b) Reals
c) Counting numbers, whole numbers, integers, rationals, reals
d) Rationals, reals e) Integers, rationals, reals f) Complex

15 a) Counting numbers: closure +, closure ×, order
b) Whole numbers: closure +, closure ×, order
c) Integers: closure +, closure ×, inverse +, order
d) Rationals: closure +, closure ×, inverse +, inverse ×, order, denseness
e) Reals: all the properties listed

EXERCISE SET 10A.1, PAGE 232

1 a) No b) Yes

3 4

5 a) Infinitely many b) One

7 They must be collinear.

9 Yes. Two planes are either disjoint or intersect in a line or coincide. In the first case, the two planes have no common point.

11 Yes. The first statement is the contrapositive (see page 458) of the second statement.

EXERCISE SET 10A.2, PAGE 239

1 Angles BAE, BAD, BAC, CAE, CAD, DAE

3 $m\angle A = 60$; $m\angle B = 30$

5 (Reminder: We have not defined a "0" angle.)
a) True b) False c) False d) False e) False f) False

7 150. The measure of its supplement is 30.

9 m; 180-m; 180-m

11 a) $\overline{ZY}$ b) $\{X, Z\}$ c) ϕ

EXERCISE SET 10A.3, PAGE 247

1 Prove $\triangle ABC \cong \triangle FED$ by SAS.

3 Prove $\triangle ACD \cong \triangle DBA$ by SSS, making $\angle 1 \cong \angle 2$. Therefore $\triangle AED$ is isosceles.

EXERCISE 10A.4, PAGE 256

1

	Rectangle	Parallelogram	Rhombus	Square	Trapezoid
a)	x	x	x	x	
b)	x	x	x	x	
c)	x			x	
d)	x			x	
e)			x	x	
f)	x	x	x	x	x
g)			x	x	
h)	x			x	
i)	x	x	x	x	

3 Rhombus (and, therefore, also a trapezoid).

5 By congruent triangles ABE and CBE,

(1) $AE = EC$

(2) $m\angle 1 = m\angle 2$. Since $\angle 1$ and $\angle 2$ are supplementary, each angle has a measure of 90.

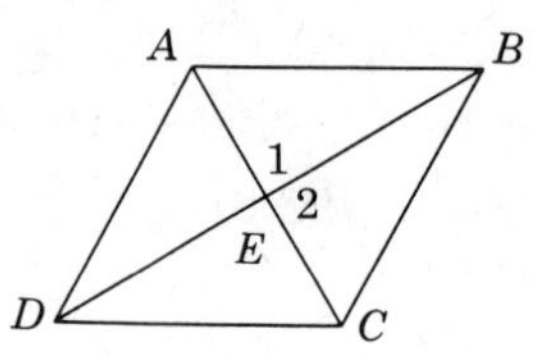

7 (1) Draw line parallel to $\overline{AB}$ through C, meeting $\overline{AD}$ at E. Then $CE = AB = CD$.

(2) Therefore $\angle 1 \cong \angle D$. But $\angle 1 \cong \angle A$.

(3) Therefore $\angle A \cong \angle D$.

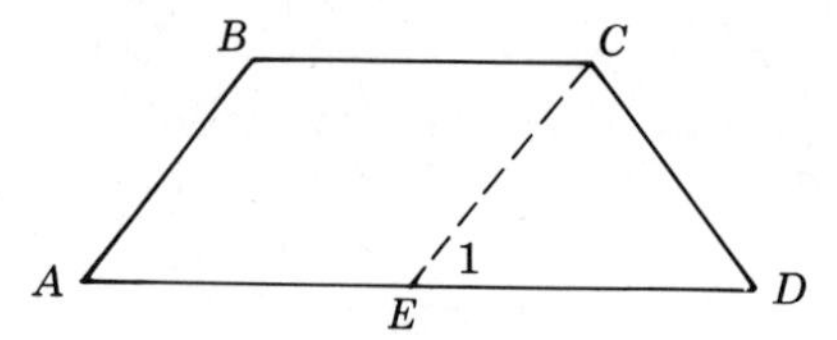

EXERCISE SET 10A.5, PAGE 259

1 a) The square of the length of one side of a triangle is equal to the sum of the squares of the lengths of the other two sides *if and only if* the triangle is a right triangle whose right angle is opposite the longest side.

b) A *necessary and sufficient condition* for the square of the length of one side of a triangle to be equal to the sum of the squares of the lengths of the other two sides is that the triangle be a right triangle whose right angle is opposite the longest side.

3 a) True. $2s^2 = d^2$, $s = \dfrac{d}{\sqrt{2}}$

b) False. To solve for a or b in $d^2 = a^2 + b^2$ we must know *two* values.

c) False. $7^2 \neq 3^2 + 5^2$.

d) False. If $c^2 = a^2 + b^2$, then $9c^2 = 9a^2 + 9b^2$. Therefore, if we triple the lengths of the sides of the triangle, the triangle remains a right triangle by the converse of the Pythagorean Theorem.

e) False. according to historians, Pythagoras (or one of his followers) was the first to discover this theorem.

5 $200\sqrt{2}$

7 $4\sqrt{2}$

9 10×24

11 30 miles

13 a) $\sqrt{2}$ b) $\sqrt{3}$ c) 2 d) $\sqrt{5}$

15 a) $9\sqrt{3}$ b) $25\sqrt{3}$ c) $\frac{a^2}{4}\sqrt{3}$

17 a) Find the area of the triangle by taking $\frac{1}{2}$ the product of the legs. Knowing the area of the triangle, we find the altitude to the hypotenuse by using the fact that the area of the triangle is also equal to $\frac{1}{2}$ the product of the hypotenuse and the altitude to the hypotenuse.

b) Area $\triangle LMN = \frac{1}{2} \times 3 \times 6 = 9$; $LM = 3\sqrt{5}$.

$$\text{Area } \triangle LMN = \frac{1}{2} \times NR \times 3\sqrt{5}$$

$$9 = \frac{1}{2} \times NR \times 3\sqrt{5}$$

$$NR = \frac{6}{\sqrt{5}} \text{ or } \frac{6\sqrt{5}}{5}.$$

EXERCISE SET 10A.6, PAGE 264

1 a) Similar b) Congruent c) Neither d) Neither e) Similar

3 a) True. All corresponding angles are congruent.

b) False. The two quadrilaterals are not necessarily similar, so the ratios of their corresponding sides are not necessarily equal. For example,

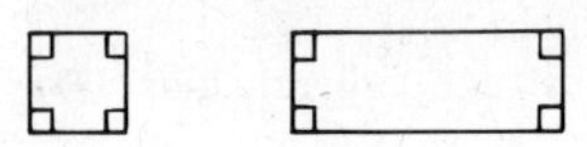

c) False. For example,

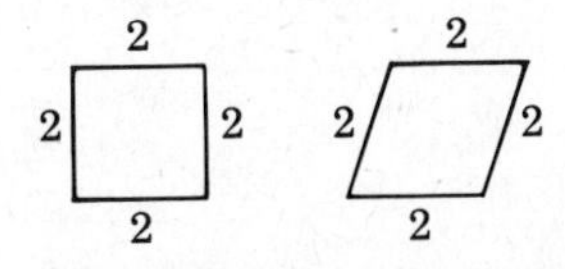

d) True. If the ratios are equal, then the triangles are similar.

e) True. Congruence satisfies the conditions for similarity.

f) True. Congruent figures are similar figures of the same size.

5 175 miles

7 120 feet

9 By definition, all the sides of each regular polygon are congruent and all its angles are congruent. If the two polygons have the same number of sides, their corresponding angles are congruent. Since the ratios of their corresponding sides are equal, the two polygons are similar.

EXERCISE SET 10B.1–10B.6, PAGE 273

1 (1) On $\angle BAC$ lay off equal distances AD and AE.
(2) Construct median $\overline{AF}$.
(3) $\triangle ADF \cong \triangle AEF$ (SAS), making $\angle 1 \cong \angle 2$.

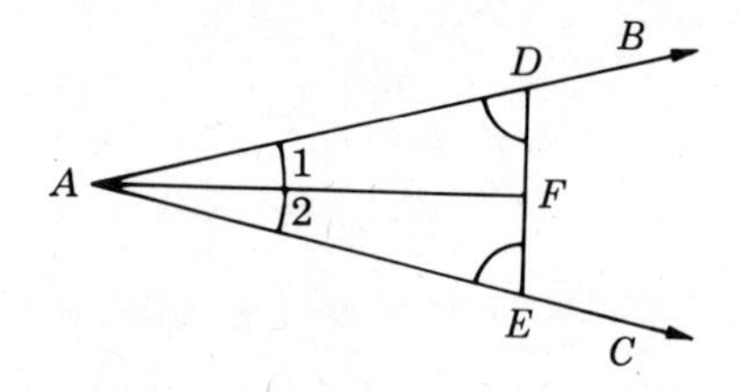

3 At each endpoint of the diagonal construct two 45° angles.

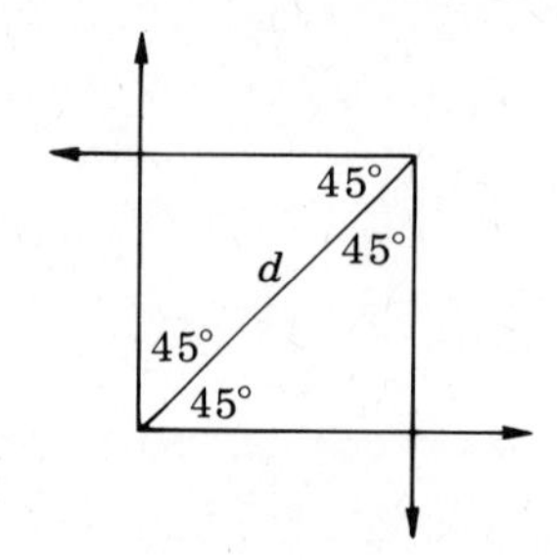

5 On RX, lay off $RS = AB$; on RY lay off $RT = CD$. Draw ST. This construction is based on the SAS postulate.

7 (1) Construct $BC = 1$ inch.
(2) At midpoint D, construct perpendicular $AD = 1$ inch.
(3) Draw $\overline{AB}$ and $\overline{AC}$.

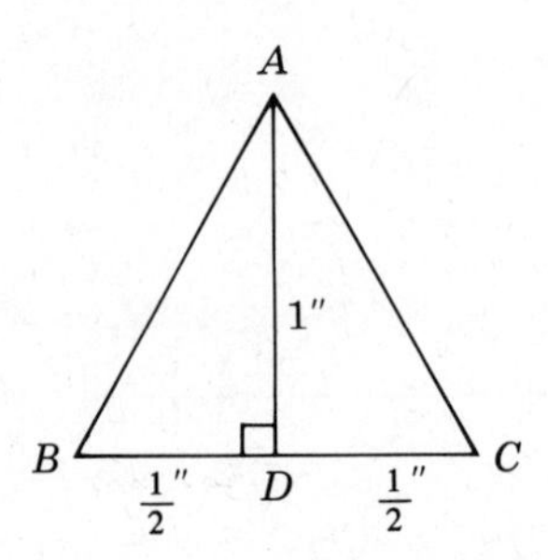

9

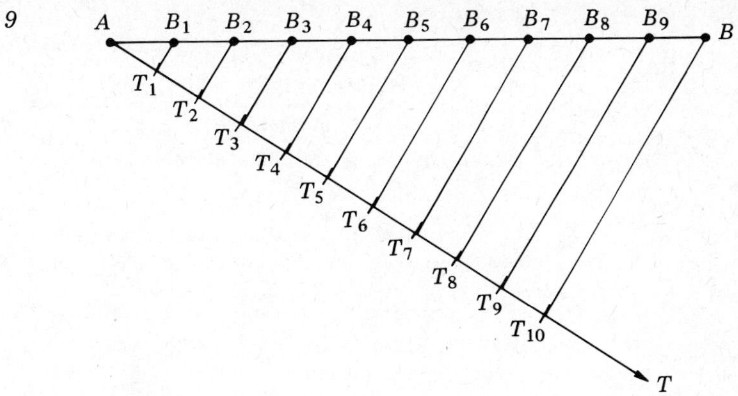

(1) Draw any ray $\overrightarrow{AT}$, not on line $\overleftrightarrow{AB}$.

(2) Choose any radius on your compass and lay off on $\overrightarrow{AT}$, ten congruent segments $\overline{AT_1}$, $\overline{T_1T_2}$, $\overline{T_2T_3}$, etc. end to end.

(3) Join T_{10} to B.

(4) Through the other points T_9, T_8, T_7, ..., T_1 construct segments parallel* to $\overline{T_{10}B}$. These segments intersect $\overline{AB}$ in $B_1, B_2, B_3, ..., B_9$.

(5) The points $B_1, B_2, B_3, ..., B_9$ divide $\overline{AB}$ into ten equal parts.

EXERCISE SET 10C.1, PAGE 279

1

	To Nearest Inch		*To Nearest*	*To Nearest*	*To Nearest*
	Estimate	*Ruler*	$\frac{1}{4}$ *Inch*	$\frac{1}{8}$ *Inch*	$\frac{1}{16}$ *Inch*
1	1	1	1″	1″	$1\frac{1}{16}$
2	0	0	$\frac{2}{4}$	$\frac{4}{8}$	$\frac{7}{16}$
3	1	1	$\frac{3}{4}$	$\frac{7}{8}$	$\frac{14}{16}$
4	1	1	$1\frac{2}{4}$	$1\frac{4}{8}$	$1\frac{7}{16}$
5	3	3	$2\frac{3}{4}$	$2\frac{6}{8}$	$2\frac{13}{16}$

*The method for constructing parallel segments involves concepts not discussed in this book. For an explanation of this method, see *Geometry: Content and Strategy for Teachers* by Sol Weiss, Prindle, Weber and Schmidt, 1972.

EXERCISE SET 10C.2, PAGE 283

1 a) Inch b) Meter c) Mile d) Liter e) Ounce f) Pound g) Kilogram h) .05 ounces

5 a) 10; 100; 1000 b) 60; 600; 6000 c) 90 d) 140 e) .8 f) 108 g) 3000; 3×10^6 h) 2300 i) 10; 100 j) 2.5; .25; .025 k) 3.6; .36 l) 18.5; .000185 m) 3.25

7 a) 3.937 b) 328 c) 21.872 d) 9.936 e) 3.28 f) 3.937 g) 91.44 h) 80.465 i) .4572 j) 68.039 k) 2.839 l) 60.565 m) 85048.5

9 437; 875; 1750; 3500; 6562

11 1620 km

EXERCISE SET 10C.3, PAGE 289

1 The area of a region bounded by a square whose side is 1 meter; etc.

3 a) doubled b) doubled c) quadrupled

5 13 inches

7 2 to 1

9 42 feet

11 A cubic foot is the volume of a cube of which each side is 1 foot; etc.

13 12.8 m^3

15 202.5 cu in.

17 648 cm^3

19 Partition the rock into basic geometric shapes, estimate the volume of each shape, and then find the sum of all the volumes. [By nongeometric means, you can find the volume by immersing the rock in a container of water and then measuring the volume of water it displaced.]

EXERCISE SET 10C.4, PAGE 295

1

	Radius	*Diameter*	*Circumference*	*Area*
a)	5 mm	10	31.4	78.5
b)	4	8 ft	25.12	50.24
c)	1	2	6.28	π sq ft
d)	7.01	14.01	44 yd	154.14
e)	1¼ in.	2.5	7.85	4.91
f)	2.39	4.77	15 m	17.91
g)	1.3	2.6 cm	8.16	530.66
h)	23.89	47.77	150 yd	1791.40

3 $13

5 19.63 inches

7 15,700 inches

9 a) 168 pieces b) 325.08 sq in.

11 a) 2.36 gallons b) $15.34

13 The larger one is 4 times the smaller one.

15 108.18 pounds

17 $\frac{4}{3}\pi r^3 = 4\pi r^2$, $r = 3$; diameter = 6.

EXERCISE SET 10D.1–10D.5, PAGE 309

1 a) ⊥ b) ⊤ c) ⊣ or ⊢

3 a) A b) C c) B d) $\overline{AB}$ e) $\overline{AC}$ f) $\overline{CB}$ g) $\overline{AD}$

5 a) A translation b) None c) None
d) A rotation followed by a translation

7 a) 0 b) 1 (if the rotation is not the identity) c) Infinitely many

9 A B C D E H I M O S T U V W X Y (The answer will vary with the way the letters are formed.)

11 a) 0, 8 b) 0, 8 c) 0, 8

13 a)

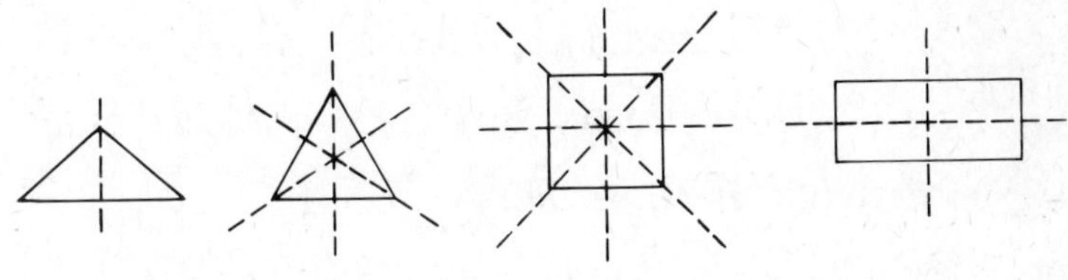

b)

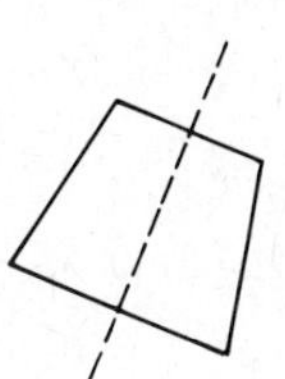

15 a)

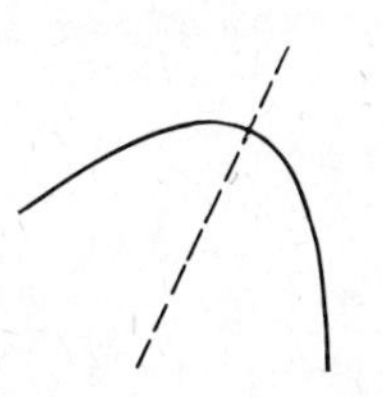

b)

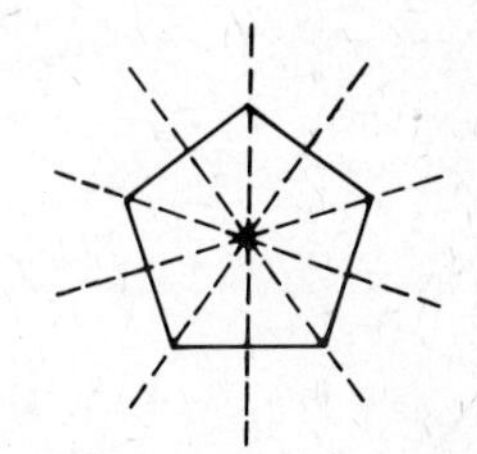

17 $0 + 2n\pi, \left(\frac{1}{2} + 2n\right)\pi, (2n + 1)\pi, \left(2n + \frac{3}{2}\right)\pi$, where $\pi = 180°$ and $n = 0, \pm 1, \pm 2, \ldots$

19 A space figure F is symmetric about a plane α if for every point P of F there exists a corresponding point P' of F (not necessarily distinct) such that α is the perpendicular bisector of $\overline{PP'}$. The plane α is called a *plane of symmetry*.

EXERCISE SET 10E.1–10E.5, PAGE 327

1

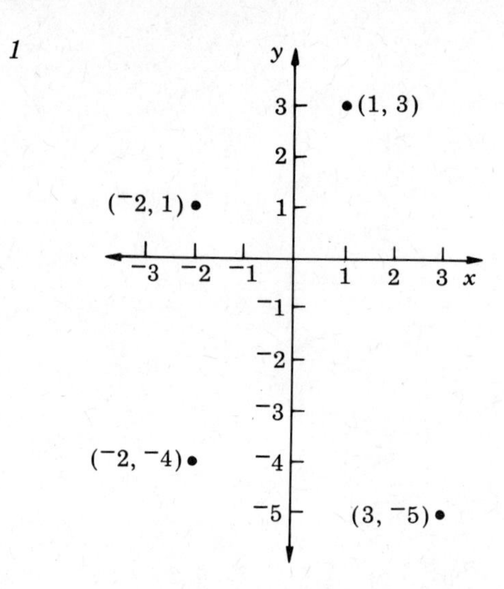

3 a) y-axis b) x-axis c) Vertical line passing through $(^-2, 0)$
d) Horizontal line 3 units above the x-axis.

5 3

7 $\left(1, \frac{1}{2}\right)$

9 a) $^-2$ b) x_2 c) $^-3$ d) $\frac{10}{3}$

11 a) (4, 3.5) b) $(^-1, 3.5)$ c) $(^-2.5, ^-1.5)$

13 a) $\overline{AB}$: $\frac{1}{3}$; $\overline{BC}$: $\frac{4}{3}$; $\overline{AC}$: $\frac{^-2}{3}$
b) to $\overline{AB}$: $^-3$; to $\overline{BC}$: $\frac{^-3}{4}$; to $\overline{AC}$: $\frac{3}{2}$
c) to $\overline{AB}$: 3; to $\overline{BC}$: 4.5; to $\overline{AC}$: $\sqrt{29.25} = 5.4$

15 a) $\frac{^-3}{2}$ b) $\frac{^-5}{2}$ c) No slope d) 0

17 $2x + 3y = 0$

19 They all satisfy the equation of the line: $2x + y + 8 = 0$.

EXERCISE SET 11.1–11.3, PAGE 371

1 If an experiment is repeated many times under identical conditions, then the probability p of an event E occurring is the quotient:

$$p = \frac{\text{number of favorable outcomes}}{\text{number of outcomes in the sample space}}.$$

That is, the probability is what can be expected in the long run.

3 a) $\frac{5}{17}$ b) $\frac{3}{17}$ c) $\frac{9}{17}$ d) $\frac{14}{17}$ e) 0 f) 1

5 a) $\frac{1}{2}$ b) $\frac{1}{2}$ c) $\frac{1}{6}$ d) $\frac{1}{6}$ e) $\frac{1}{6}$ f) $\frac{5}{6}$

g) $\frac{3}{6} = \frac{1}{2}$ (2, 3, and 5 are prime.) h) $\frac{2}{6} = \frac{1}{3}$ (4 and 6 are composites.)

7 $S = \{HHH, HHT, HTH, HTT, THH, THT, TTH, TTT\}$

9 a) $\frac{1}{8}$ b) $\frac{3}{8}$ c) $\frac{3}{8}$ d) $\frac{1}{2}$ e) $\frac{1}{8}$ f) $\frac{1}{8}$

11 a) $\frac{2}{25}$ b) $\frac{1}{25}$ c) $\frac{15}{25} = \frac{3}{5}$ d) $\frac{6}{25}$

EXERCISE SET 11.4, 11.5, PAGE 374

1 a) 1 to 1 b) 1 to 3 c) 1 to 3 d) 1 to 12 e) 2 to 11

3 a) 1 to 1; 1 to 1; 1 to 1; 1 to 2
b) 1 to 5 (The number 1 is neither prime nor composite.)

5 a) 32 to 4, or 8 to 1 b) 28 to 8, or 7 to 2 c) 30 to 6, or 5 to 1
d) 10 to 26, or 5 to 13

7 0

9 $^{-}\$\frac{4}{9}$, or about $^{-}44$ cents

11 $4

EXERCISE SET 11.6, 11.7, PAGE 380

1 12

3 1920 (${}_6P_3 + {}_6P_4 + {}_6P_5 + {}_6P_6 = 120 + 360 + 720 + 720 = 1920$)

5 ${}_7P_7 = 7! = 5040$

7 a) ${}_5P_1 = 5$ b) ${}_5P_3 = 60$ c) ${}_5P_5 = 120$

9 ${}_{15}P_{10} = 15 \times 14 \times 13 \times 12 \times 11 \times 10 \times 9 \times 8 \times 7 \times 6 = 10897286400$

11 a) 24 b) 40320 c) 60480 d) 5 e) 720 f) 40320 g) 1

13 a) 90 b) 900

15 900

17 a) 3 b) 12 c) 30

19 a) 1152 ($4 \times 4 \times 3 \times 3 \times 2 \times 2 \times 1 \times 1 + 4 \times 4 \times 3 \times 3 \times 2 \times 2 \times 1 \times 1 = 1152$)
b) 48 c) 1152 ($4! \times 4! \times 2 = 1152$)

EXERCISE SET 11.8, PAGE 384

1 a) 10 b) 20 c) 495 d) 10 e) 124750

f) $\dfrac{r!}{(r-1)!\,[r-(r-1)]!} = \dfrac{r!}{(r-1)!\,1!} = r$

3 ${}_{50}C_{12} = \dfrac{50!}{12!\,38!}$

5 ${}_{15}C_3 \times {}_{20}C_4 = 455 \times 4845 = 2204475$

7 a) △□ and □△ are the same combination.
△○ and ○△ are the same combination.
□○ and ○□ are the same combination.

b) △□○, △○□, □△○, □○△, ○△□, ○□△ are the same combination.

9 For each subset of 2 picked, there is a subset of 4 remaining. Therefore, there are just as many 2-element subsets as there are 4-element subsets.

11 ${}_{10}C_2 = 45$ (Every pair of points determines a line.)

13 $9! = 362880$

15 a) $\dfrac{45}{90} = \dfrac{1}{2}$ b) $\dfrac{30}{90} = \dfrac{1}{3}$

17 $2^5 - 1 = 31$

EXERCISE SET 11.9, 11.10, PAGE 391

1 a) $\dfrac{2}{6} = \dfrac{1}{3}$ b) $\dfrac{3}{6} = \dfrac{1}{2}$ c) $\dfrac{3}{6} = \dfrac{1}{2}$ d) $\dfrac{3}{6} + \dfrac{5}{6} - \dfrac{2}{6} = 1$

3 a) $\dfrac{2}{4} = \dfrac{1}{2}$ b) $\dfrac{2}{4} = \dfrac{1}{2}$

5 $P(\text{both blue}) + P(\text{no blue}) = \dfrac{4}{11} \times \dfrac{2}{8} + \dfrac{7}{11} \times \dfrac{6}{8} = \dfrac{50}{88} = \dfrac{25}{44}$

7 $\dfrac{240}{720} = \dfrac{1}{3}$ (5 pairs of adjacent seats, 2 orders for each pair, 24 permutations for the remaining 4 seats gives $10 \times 24 = 240$ favorable outcomes of the $6! = 720$ possible outcomes.)

9 a) $\dfrac{1}{36}$ b) $\dfrac{1}{36}$ c) $\dfrac{5}{36}\left(\dfrac{1}{6} \times \dfrac{5}{6}\right)$

11 $\dfrac{35}{324}\left(\dfrac{2}{36} + \dfrac{2}{36} - \dfrac{2}{36} \times \dfrac{2}{36}\right)$

13 a) $S = \{H1, H2, H3, H4, H5, H6, T1, T2, T3, T4, T5, T6\}$

b) (1) $\dfrac{1}{12}$ (2) $\dfrac{3}{4}\left(\dfrac{6}{12} + \dfrac{6}{12} - \dfrac{3}{12}\right)$ (3) $\dfrac{1}{4}$ (4) $\dfrac{1}{6}$

15 a) $S = \{HHH, HHT, HTH, HTT, THH, THT, TTH, TTT\}$

b) (1) $\dfrac{7}{8}$ (2) $\dfrac{3}{8}$

17 a) $\dfrac{25}{72}\left(3 \times \dfrac{1}{6} \times \dfrac{5}{6} \times \dfrac{5}{6}\right)$

b) $\dfrac{91}{216}\left[P(1\text{ two}) = \dfrac{75}{216}, P(2\text{ twos}) = \dfrac{15}{216}, P(3\text{ twos}) = \dfrac{1}{216}\right]$

19 a) $\frac{4}{52} \times \frac{4}{52} = \frac{1}{169}$ b) $\frac{13}{52} \times \frac{13}{52} = \frac{1}{16}$ c) $\frac{4}{16} = \frac{1}{4}$

d) $\frac{4}{52} \times \frac{1}{52} = \frac{1}{676}$ e) $\frac{39}{52} \times \frac{39}{52} = \frac{9}{16}$

21 a) $P(9) + P(10) = \frac{10}{1024} + \frac{1}{1024} = \frac{11}{1024}$ (There are ${}_{10}C_9 = 10$ ways of getting 9 right and 1 wrong; there is 1 way of getting 10 right; and there are altogether $2^{10} = 1024$ possible outcomes.)

b) $P(8) = \frac{45}{1024}$ (There are ${}_{10}C_8 = 45$ ways of getting 8 right and 2 wrong.)

c) $1 - [P(8 \text{ or } 9 \text{ or } 10)] = 1 - \frac{56}{1024} = \frac{968}{1024} = \frac{121}{128}$

23 a) $\frac{6}{85} \times \frac{5}{84} \times \frac{3}{83} = \frac{4}{19754}$ b) $\frac{79}{85} \times \frac{78}{84} \times \frac{77}{83} = \frac{11297}{14110}$

c) $3 \times \frac{79}{85} \times \frac{78}{84} \times \frac{6}{83} = \frac{9243}{49385}$

25 $.7 + .6 - .7 \times .6 = .88$

27 $1 - P(\text{no shared birthdays}) =$

$$1 - \left(\frac{365}{365} \times \frac{364}{365} \times \frac{363}{365} \times \frac{362}{365} \times \frac{361}{365} \times \frac{360}{365} \times \frac{359}{365} \times \frac{358}{365} \times \frac{357}{365} \times \frac{356}{365}\right)$$

29 a) .7 b) .6 c) $.3 + .4 - .1 = .6$

d) $\frac{.1}{.4} = \frac{1}{4} = .25$ $[P(A \text{ and } B) = P(B) \times P(A|B)]$

e) $\frac{.1}{.3} = \frac{1}{3}$ $[P(A \text{ and } B) = P(A) \times P(B|A)]$

31 $\frac{\frac{1}{52}}{\frac{4}{52}} = \frac{1}{4}$ $\left[P(S|A) = \frac{P(S \text{ and } A)}{P(A)}\right]$

33 a) $.35 + .2 - .1 = .45$

b) $\frac{.1}{.35} = \frac{2}{7}$ $\left[P(A|C) = \frac{P(A \text{ and } C)}{P(C)}\right]$

c) $\frac{.1}{.2} = \frac{1}{2}$ $\left[P(C|A) = \frac{P(C \text{ and } A)}{P(A)}\right]$

35 a) $\frac{500}{700} = \frac{5}{7}$ b) $\frac{100}{600} = \frac{1}{6}$ c) $\frac{30}{480} = \frac{1}{16}$ d) $\frac{20}{100} = \frac{1}{5}$

EXERCISE SET 12.1, PAGE 409

1 Descriptive statistics is concerned with collecting and summarizing of data either numerically or graphically. Inferential statistics draws inferences from a sample about the entire population.

3 Statistical inference is based on sampled data and draws conclusions that are based on the laws of probability. Prophecy is not based on such data or on the laws of probability but on faith, intuition, special insights, or extra-sensory perception.

5 a) 9

b)	c)	d) 50	e)
50–99	49.50–99.50		74.50
100–149	99.50–149.50		124.50
150–199	149.50–199.50		174.50
200–249	199.50–249.50		224.50
250–299	249.50–299.50		274.50
300–349	299.50–349.50		324.50
350–399	349.50–399.50		374.50
400–449	399.50–449.50		424.50
450–499	449.50–499.50		474.50

7 a), b)

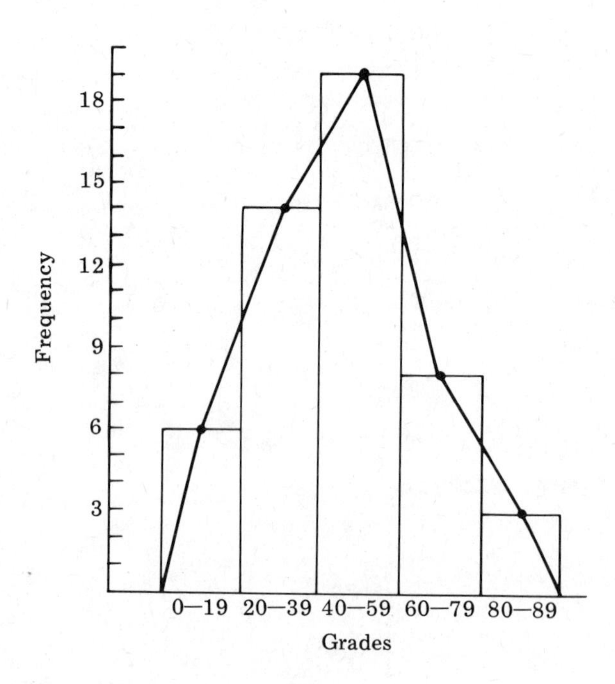

9 a)

Number of Tires	*Midpoint*	*Frequency*
44–50	47	6
51–57	54	26
58–64	61	14
65–71	68	7
72–78	75	4
79–85	82	3
		60

b), c)

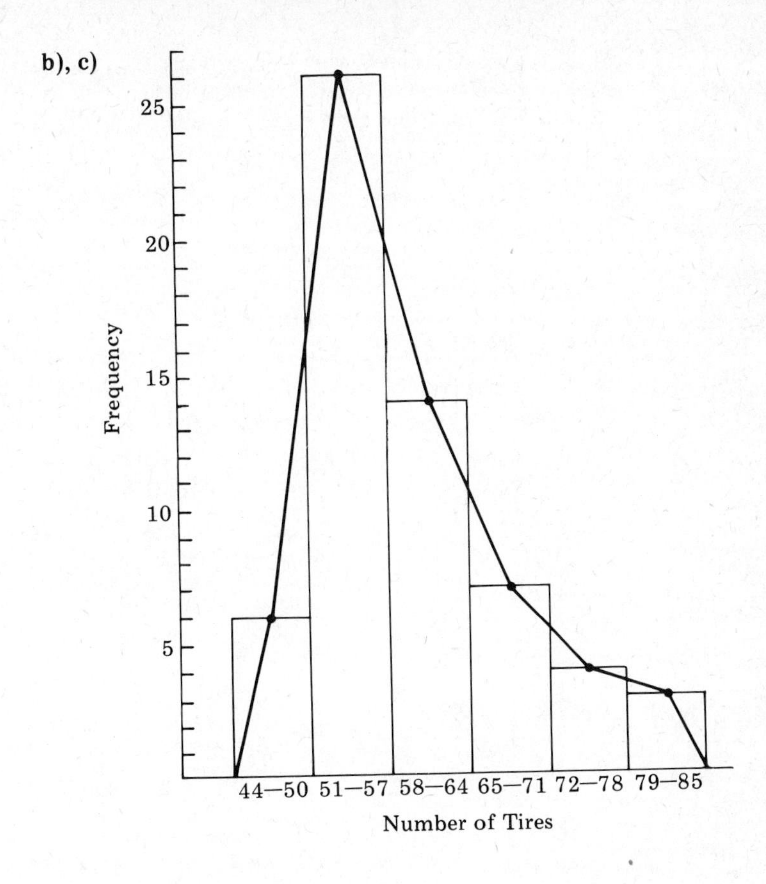

13 a) 55 b) 5.5 c) 385 d) 3025

EXERCISE SET 12.2, PAGE 416

1 a) 27.84 b) 2.23

3 76.17

5 Different procedures are required because, in ungrouped data, we use the actual scores. In grouped data, we use only approximations of these scores.

7 The median is an average in the sense that the median is that point on the scale of measurement above which are exactly half the cases and below which are the other half.

9

Mean	Median	Mode
a) 15	15	15
b) 14.42	15	15
c) 16	14	13

11 Mean: 70.64; median: 71; there are 5 modes: 66, 68, 71, 72, 79.

13 Mean: $42.20; median: $41.22; there is no mode.

15 The values are the same if the distribution is symmetrical about the mean.

17 If the distribution is skewed negatively, i.e., the tail is toward the left, then mean < median < mode. If the distribution is positively skewed, i.e., the tail is toward the right, then mean > median > mode.

19 a) Mode b) Median c) Mean

EXERCISE SET 12.3, PAGE 421

1 41

3 The range would be a misleading indicator of variability for this data because an extreme score (5) varies greatly from the other scores.

5 a) Range: 4; standard deviation: 1.41; the dispersion is very narrow.
b) Range: 93; standard deviation: 36.05; the dispersion is very great.

7 a) 2.87 b) 6 c) 10 d) 10

9 21.00

11 12.82

EXERCISE SET 12.4, PAGE 428

1 a) .3186 b) .3849 c) .8078 d) .0838 e) .2578 f) .8531
g) .1737 h) .2571 i) .7340 j) .9677

3 a) 34.13% b) 41.92% c) 70.55% d) 1.04% e) 96.93%
f) -1.04σ and 1.04σ g) 2.33σ h) 100% i) 50%

5 a) 2 b) .62% c) 6.68% d) 81.85%

7 He has done better in arithmetic. On the reading test, he scored $.20\sigma$ above the mean. On the arithmetic test, he scored $.25\sigma$ above the mean.

9 a) 29.71% b) .2971 c) .9726

EXERCISE SET A.1, A.2, PAGE 454

1 a) Statement. True.
b) Statement. False.
c) Not a statement.
d) Statement. False.
e) Statement. Truth values depends on when statement is made.
f) Statement. False.
g) Not a statement.
h) Statement. False.
i) Not a statement until "the man" is defined.
j) Statement. True.
k) Statement. True.
l) Not a statement because x is not defined.
m) Not a statement because a is not defined.
n) Statement. True for all x and y even though no values are assigned.

3 a) If the U.S. has space ships, then men visited the moon.
b) If Jack is a machinist, then Mary went swimming.
c) If 2 + 3 = 5, then Mars is made of plastic.
d) If triangles are not squares, then March has 39 days.
e) If today is Tuesday, then tomorrow is Wednesday.

5 a) The sun is not shining.
b) Some even numbers are not divisible by 2.
c) The sweater is not blue.
d) The car does not ride well.
e) Some triangles do not contain three angles.
f) $2 \geq 3$.
g) $10 > 9$.
h) $5 < 4$.

7 a) $p \wedge \sim q$
b) $\sim p \wedge \sim q$
c) $q \rightarrow p$
d) $\sim q \rightarrow \sim p$

9 a) Yes.

p	q	$p \vee q$	$p \rightarrow (p \vee q)$
T	T	T	T
T	F	T	T
F	T	T	T
F	F	F	T

b) Yes.

p	q	$p \wedge q$	$(p \wedge q) \rightarrow p$
T	T	T	T
T	F	F	T
F	T	F	T
F	F	F	T

c) Yes.

p	q	$p \rightarrow q$	$(p \rightarrow q) \wedge p$	$[(p \rightarrow q) \wedge p] \rightarrow q$
T	T	T	T	T
T	F	F	F	T
F	T	T	F	T
F	F	T	F	T

d) Yes.

p	q	$\sim p$	$\sim p \vee q$	$(\sim p \vee q) \wedge p$	$[(\sim p \vee q) \wedge p] \rightarrow q$
T	T	F	T	T	T
T	F	F	F	F	T
F	T	T	T	F	T
F	F	T	T	F	T

11 d

13 c

EXERCISE SET A.3, PAGE 459

1 a) (1) If three months from now is June, then this is March.
(2) If this is not March, then three months from now is not June.
(3) If three months from now is not June, then this is not March.

b) (1) If two angles are congruent, then they are right angles.
(2) If two angles are not right angles, then they are not congruent.
(3) If two angles are not congruent, then they are not right angles.

c) (1) If the opposite sides of a figure are parallel, then it is a square.
(2) If a figure is not a square, then its opposite sides are not parallel.
(3) If the opposite sides of a figure are not parallel, then it is not a square.

d) (1) If you have good coordination, then you are an athlete.
(2) If you are not an athlete, then you do not have good coordination.
(3) If you do not have good coordination, then you are not an athlete.

e) (1) If a number is odd, then it is a prime other than 2.
(2) If a number is not prime other than 2, then it is not odd.
(3) If a number is not odd, then it is not a prime other than 2.

3 a

5 b

EXERCISE SET A.4, PAGE 462

1 a) Yes.

p	q	$\sim p$	$p \to q$	$\sim p \vee q$
T	T	F	T	T
T	F	F	F	F
F	T	T	T	T
F	F	T	T	T

b) Yes.

p	q	$\sim p$	$\sim q$	$p \vee q$	$\sim(p \vee q)$	$\sim p \wedge \sim q$
T	T	F	F	T	F	F
T	F	F	T	T	F	F
F	T	T	F	T	F	F
F	F	T	T	F	T	T

c) Yes.

p	q	$\sim q$	$p \to q$	$\sim(p \to q)$	$p \wedge \sim q$
T	T	F	T	F	F
T	F	T	F	T	T
F	T	F	T	F	F
F	F	T	T	F	F

3 a) Equivalent.

Today being Sunday is a necessary and sufficient condition for yesterday having been Saturday.

Today is Sunday if and only if yesterday was Saturday.

b) Not equivalent. (Composites are not primes but nonprimes are not necessarily composites.)

c) Not equivalent.

d) Equivalent.

My playing is a necessary and sufficient condition for me to win.

I play if and only if I win.

EXERCISE SET A.5, PAGE 467

1 a) Valid b) Valid c) Not valid d) Not valid e) Valid
f) Not valid g) Valid h) Not Valid i) Valid

3 Let U = Quattlebaum was in the U.S. on the day of the crime.
N = Quattlebaum was in New York on the day of the crime.
C = Quattlebaum committed the crime.

The hypotheses are (1) $U \rightarrow N$, (2) $\sim C \rightarrow \sim N$, (3) U.

From (1) and the contrapositive of (2), we get $U \rightarrow C$. Then, combining this result with (3), the conclusion is C; that is, that Quattlebaum committed the crime.

Answers to Puzzles

CHAPTER 1, PAGE 5

1 No such barber can exist under the given conditions.

2 This question cannot be answered from the given information.

3 Divide the 8 boxes of candy into two sets of three each and one set of two. Then place the two sets of three on the balance scale.

a) If they balance, then the lighter box must be in the remaining set of two boxes. Weigh the two boxes. The box that is lighter on the balance scale is the one that's underweight.

b) If they do not balance, then the underweight box is among the set of three that is lighter on the balance scale. Weigh any two of these against each other. If they balance, then the remaining box is underweight. If they do not balance, then the underweight box is the one that is lighter on the balance scale.

4 If everybody at the party were born in a different month, then there could be no more than 12 people altogether. Since there are 15 people, more than one person had to be be born in at least one month.

5 Let A, B, and C be the three men. Since all three men knocked when the blindfolds were removed, at least *two* of them had to have ink spots.

Let A be the man who stood up. He reasoned as follows: "Let me assume that I *don't* have an ink spot. Then B and C must have ink spots since at least two people must have them. If B (or C) saw that I did *not* have an ink spot, and knowing that at least two of us must have ink spots, he would have concluded that he, *himself*, must have an ink spot and would have stood up. Since he didn't stand up, I must have an ink spot on my forehead."

CHAPTER 2, PAGE 30

1 There are 105 guests.

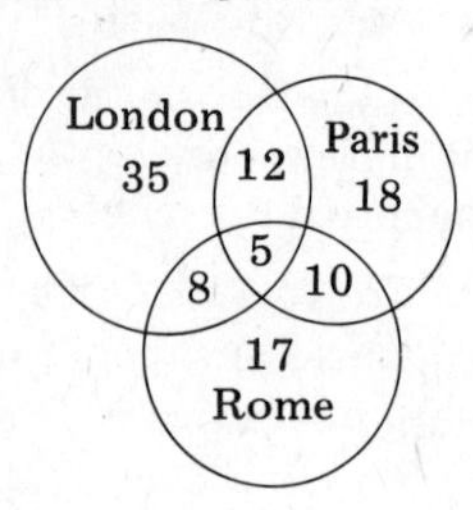

2 The total number of inhabitants under the given conditions except the last must be 14, not 15 as stated in the last condition.

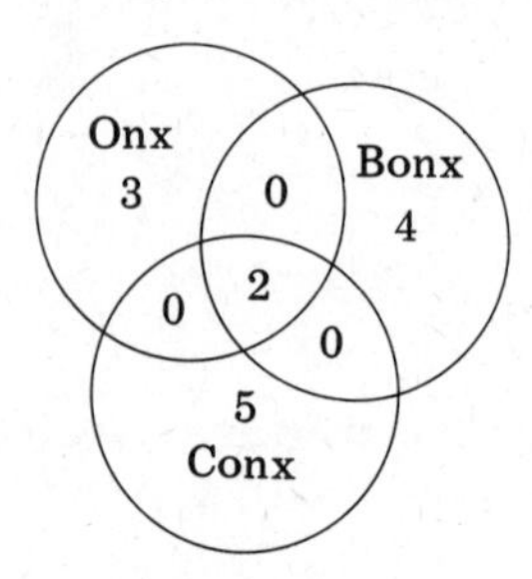

CHAPTER 3, PAGE 61

1 Four weights are needed: 1 lb; 2 lbs; 4 lbs; 8 lbs.

2 $A = 4; E = 3; H = 1; L = 2; T = 0.$

3 The story is written in base-seven: 482 Grant Ave; 23 years old; 488 Grant Ave; $1058; $1433; $375; 29 days; 1529 miles.

CHAPTER 4, PAGE 94

1 $4 = 4 + [(4 - 4) \div 4]$

2 a) One solution.

6	1	8
7	5	3
2	9	4

b) One solution.

1	15	14	4
12	6	7	9
8	10	11	5
13	3	2	16

3 The groceries he sold to the customer and the $15 in genuine bills he gave the customer.

4 a) Here is one solution of 26 possible solutions.

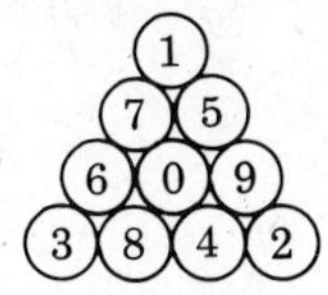

b) Here is one solution of 16 possible solutions.

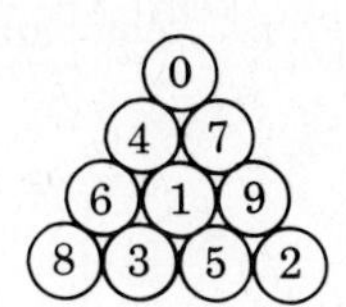

CHAPTER 5, PAGE 126

1 If the jar is filled in an hour, then it must have been half filled one minute before, or in 59 minutes.

2 The first offer will net you about $1,150,000,000,000,000,000 in 5 years. You will receive more than a million dollars by the end of the 20th month.

3 Fill the 3-quart measure and pour it into the 5-quart measure.

Fill the 3-quart measure a second time and pour as much as you can into the 5-quart measure. This will leave 1 quart in the 3-quart measure.

Empty the 5-quart measure and pour the 1 quart into it. Fill the 3-quart measure and pour it into the 5-quart measure. This will make 4 quarts.

CHAPTER 6, PAGE 147

1 a)

$$\begin{array}{r} 96233 \\ +62513 \\ \hline 158746 \end{array}$$

b)

$$\begin{array}{r} 125 \\ \times 125 \\ \hline 625 \\ 250 \\ 125 \\ \hline 15625 \end{array}$$

CHAPTER 7, PAGE 168

1 a) The A, B, and S scales are parallel, and they are perpendicular to the line that includes the zero point of each scale. The A and B scales are the same distance from the S scale. The length of the unit segment on the A scale is the same as on the B scale, but the unit segment on the S scale is *half* as long as the unit segment on the A and B scales. To add any two integers shown on the A and B scales, draw a segment connecting the two points. The point of intersection of the segment with the S scale represents the sum of the two integers. The illustration shows that $^{-}5 + {}^{+}3 = {}^{-}2$.

b) The operation of the nomograph is based on the geometric fact that in the trapezoid $BACD$ with bases a and b, as shown in the figure on page 520, the median s is parallel to the bases and equal to half their sum; that is

$s = \frac{1}{2}(a + b)$. But since the numbers marked on the S scale are doubled compared to the numbers on the A and B scales, $s = a + b$ *not* $\frac{1}{2}(a + b)$.

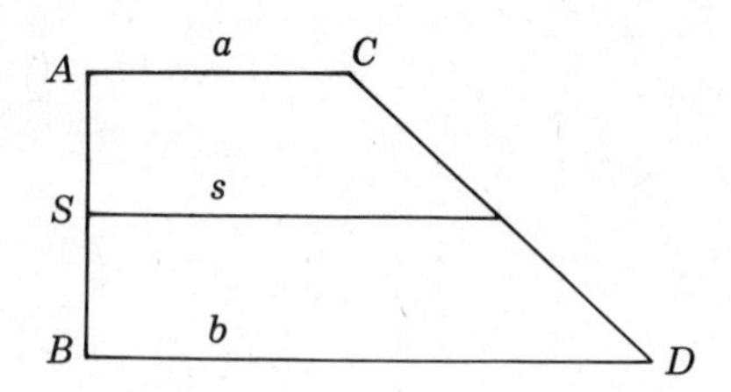

CHAPTER 8, PAGE 196

1 \$54

2 12 yards

3 Let x = the number of shells collected originally. Then,

A left: $\frac{2}{3}(x - 1)$

B left: $\frac{2}{3}\left[\frac{2}{3}(x - 1) - 1\right]$

C left: $\frac{2}{3}\left\{\frac{2}{3}\left[\frac{2}{3}(x - 1) - 1\right] - 1\right\} = y \qquad (1)$

where y is the number of shells left at the end. Equation 1 reduces to

$8x - 27y = 38 \qquad (2)$

Since x and y must be positive whole numbers, and since x must be the smallest positive integer that will permit y to be a positive integer, $x = 25$ and $y = 6$. That is, the least number of shells the boys originally collected was 25, in which case there would have been 6 shells left.

CHAPTER 9, PAGE 210

1 $(\sqrt{.2})^{-2}$

2 There are no two extra dollars. The correct computation of this problem is: (\$60 - \$6) - \$4 = \$50; or, \$50 + \$6 + \$4 = \$60. The incorrect computation suggested in the puzzle is (\$60 - \$6) + \$4.

3 The total of these four numbers must always be *twice the current year.* So if the puzzle is asked in the year 1976, the sum will always be $2 \times 1976 = 3952$. The explanation for this result is as follows:

Let x = person's year of birth,
y = current year, and
a = the number of years between the person's birth and the year the important event took place.

Expressing the four numbers in the puzzle in terms of x, y, and a, we get

a) x b) $x + a$ c) $y - x$ d) $y - (x + a)$

The sum of these numbers is:

$$x + x + a + y - x + y - (x + a) = 2y.$$

CHAPTER 10, PAGE 329

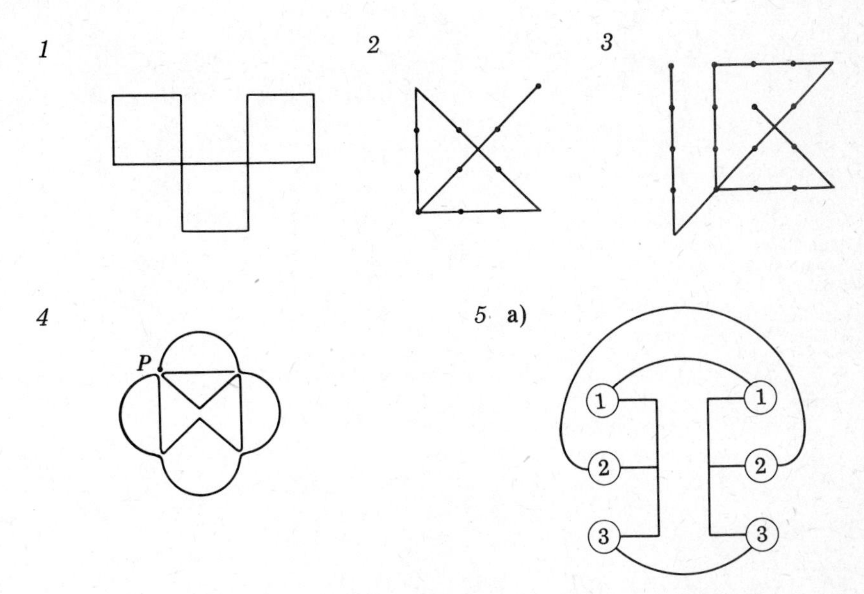

5 b) It is impossible to connect like numerals in this figure without crossing a line. We can see intuitively why this is so. When we connect 1 to 1 and 2 to 2, we obtain a simple closed curve. This leaves one 3 inside and the other 3 outside the simple closed curve, and you cannot get from the inside to the outside of a simple closed curve without intersecting the curve.

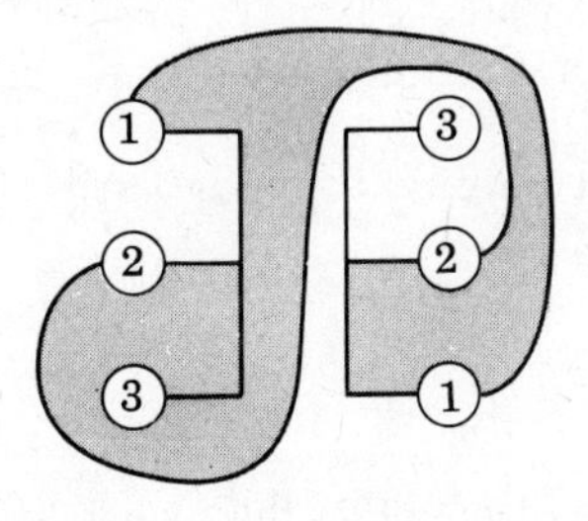

6 35.

7 Arrange the toothpicks in the shape of a tetrahedron.

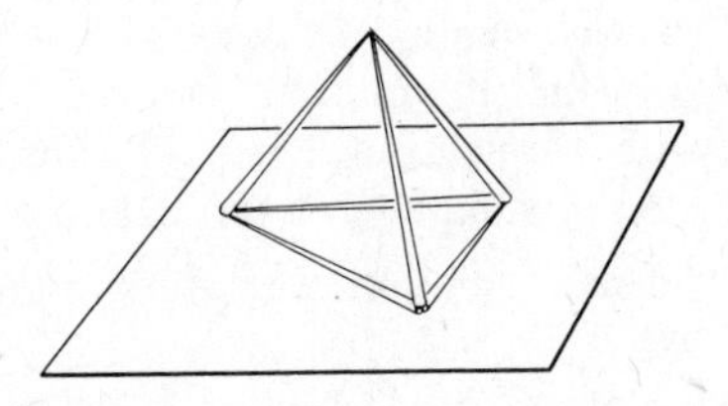

8

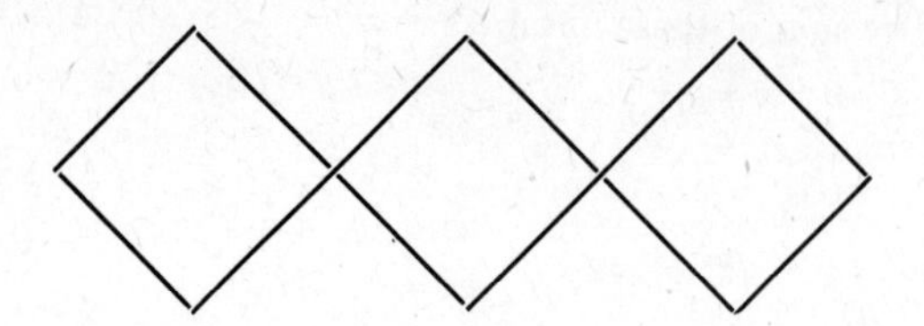

9 The coins cannot be so arranged.

10 Inscribe any triangle in the circle. The intersection of the perpendicular bisectors of any two sides is the center of the circle.

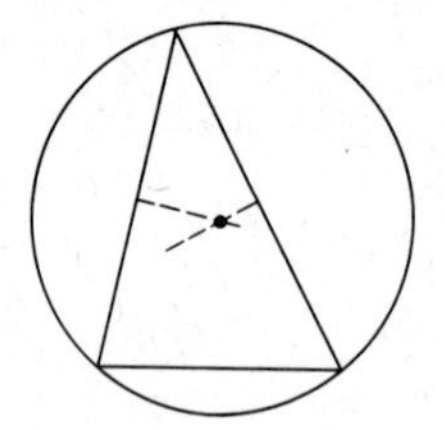

CHAPTER 11, PAGE 394

1 If w votes are cast for the winner, and l votes for the loser, then the probability that the eventual winner will always lead his opponent is $\frac{w-l}{w+l}$.

2 a) The probability of throwing a 12 in one throw of two dice is $\frac{1}{36}$, and the probability of throwing something else is $1-\frac{1}{36}$, or $\frac{35}{36}$. In n throws, the probability of throwing something else is $\left(\frac{35}{36}\right)^n$, and the probability of throwing at least one 12 is

$$1-\left(\frac{35}{36}\right)^n.$$

b) To bet even money means that the probability is $\frac{1}{2}$. The question, then, is equivalent to asking: what is the value of n in the equation

$$\left(\frac{35}{36}\right)^n=\frac{1}{2}?$$

A solution to the equation shows that $n = 24.6$ (approximately). This answer indicates that it would be somewhat to your disadvantage to bet even money on throwing a 12 in *24* throws, but somewhat to your advantage to bet even money on throwing a 12 in *25* throws.

As a check to this conclusion, we note that the probability of throwing a 12 in *24* throws is

$$1 - \left(\frac{35}{36}\right)^{24} = 1 - 0.5086 = 0.4914;$$

whereas the probability of throwing a 12 in *25* throws is

$$1 - \left(\frac{35}{36}\right)^{25} = 1 - 0.4984 = 0.5016.$$

This problem was proposed to Pascal by the Chevalier de Méré.

CHAPTER 12, PAGE 430

1 We are looking for area A as shown in the figure.

$$z = \frac{55 - 49.1}{5.3} = \frac{5.9}{5.3} = 1.11$$

Area B = .3665
Area A = .5000 − .3665 = .1335

Therefore, 13.35%, or about 27 of the 200 motorists can be expected to be ticketed.

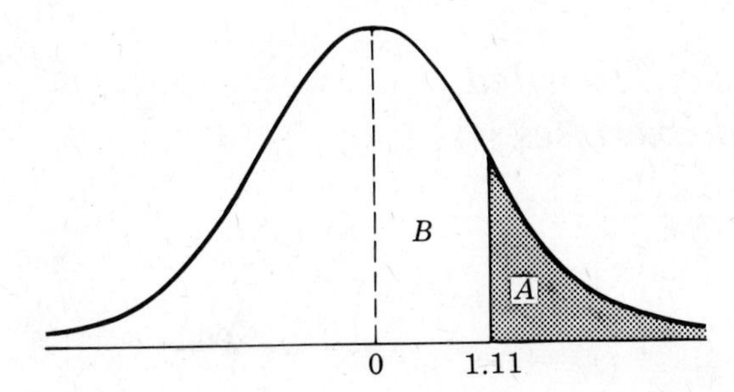

2 We have to find μ in the formula $z = \frac{X - \mu}{\sigma}$. If area A shown in the figure is .0200, then area B = .4800 and z = $^-2.05$ (approximately). To find μ, we have

$$^-2.05 = \frac{12 - \mu}{0.16}$$

$$12 - \mu = {}^-2.05(0.16) = {}^-0.33$$

$$\mu = 12.33 \text{ ounces.}$$

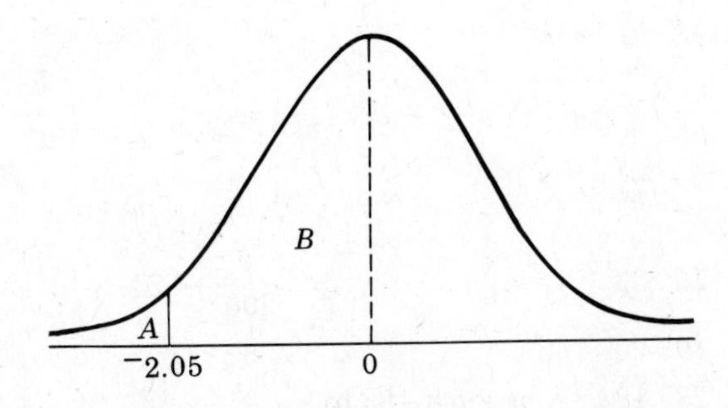

APPENDIX A, PAGE 468

1 This conclusion is valid. For, if

p represents "It is raining;"
q represents "I will go swimming";
r represents "I will mow the lawn",

then statement a can be represented by $p \to \sim q$; statement b can be represented by $\sim r \to q$ or $\sim q \to r$ (contrapositive). Therefore, $p \to r$, which says that "If it is raining, then I will mow the lawn."

2 The tall native must answer "yes" regardless of whether he is a liar or a truth-teller. Since the short native told the truth about the tall native's response, he must be a truth-teller and his companion must be the liar.

3 Two boys cross; one returns. Man crosses; boy returns. Two boys cross.

4 *That man* is the speaker's son.

5 They are part of a triplet.

Answers to Basics Revisited, Odd-Numbered Exercises

CHAPTER 1, PAGE 6

1 a) 6 b) 8

3 399,000

5

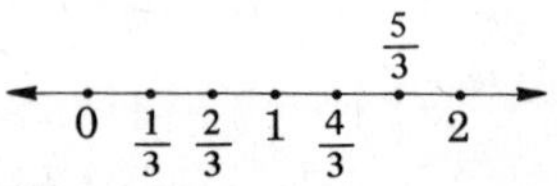

7 1.549

9 .534

11 175.3

13 $\frac{19}{40}$

15 $2\frac{7}{9}$

17 $\frac{2}{3}$

19 11%; .11; $\frac{11}{100}$

21 15

23 150

25 $378.75

CHAPTER 2, PAGE 30

1 6,274

3 1868.18

5 1.977

7 71.78

9 1

11 $6\frac{1}{24}$

13 83

15 $\frac{9}{200}$; 4.5%

17 17^6

19 20

21 2.8

23 182.92, or about 183 bags.

25 $4.50

CHAPTER 3, PAGE 62

1 3,265,004   *3* 400   *5* $\frac{2}{5}$

7 76   *9* (1, 72), (2, 36), (3, 24), (4, 18), (6, 12), (8, 9)

11 a) 228 b) 828   *13* a) $\frac{2}{3} < \frac{3}{4}$ b) $\frac{9}{5} > \frac{11}{7}$

15 $\frac{27}{40}$ *17* $4\frac{1}{7}$ *19* $4\frac{55}{63}$

21 20%   *23* 0, 1, 2, 3, 4, 5, 6, 7, 8, 9

25 64.4 km/hr

CHAPTER 4, PAGE 95

1 385,000,000   *3* $\frac{13}{20}$ *5* $8\frac{8}{9}$

7 145.17   *9* $\frac{8}{28}$ *11* 84

13 .529   *15* a) 3 b) 16   *17* 160

19 The value given to a certain position in a numeral.

21 7808.8 times   *23* 320   *25* 225 lbs

CHAPTER 5, PAGE 126

1 0   *3* $60\frac{5}{8}$ *5* .2286

7 370%   *9* $\frac{27}{100}$ *11* 6.8%

13 Five million, two hundred thousand, nine hundred eight.

15 72   *17* a) 243 b) 1 c) 15

19 37,338 men; 6223 women   *21* 39.62%

23 $250.77   *25* $33.02

CHAPTER 6, PAGE 147

1 1,276,704   *3* a) 29.5 b) 2.95 c) .295

5 a) 9 b) 27 c) $\frac{1}{9}$ d) 3 e) 1   *7* 150

9 Three thousand five hundred sixty-two and two thousandths.

11 3600   *13* 31.06   *15* 17,500

17 $\frac{1}{26}$ *19* $\frac{11}{32}$

21 a) 5 b) .75 c) 0 d) Undefined e) Undefined f) Undefined

23 \$27 *25* 945

CHAPTER 7, PAGE 168

1 4.7489 *3* .025998 *5* 2

7 \$.28 *9* 70% *11* \$14.80

13 $\frac{3}{19}$ *15* \$31 *17* 70

19 a) .002 b) .0002 c) .00002

21

23 True *25* 94

CHAPTER 8, PAGE 197

1

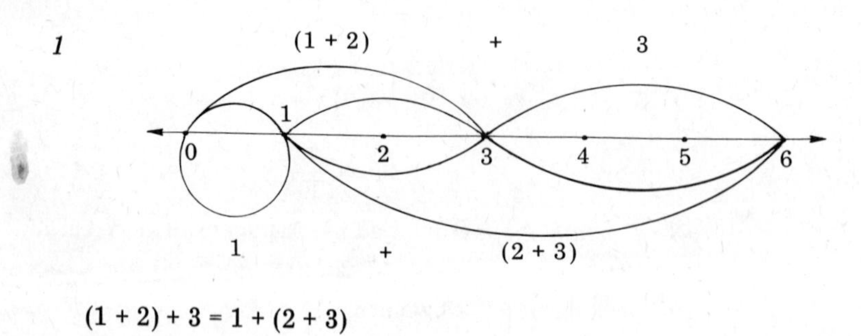

$(1 + 2) + 3 = 1 + (2 + 3)$

3 False *5* $4\frac{4}{11}$ lbs *7* 16.1

9 .0025; $\frac{1}{400}$ *11* 240,000 *13* 14.73258

15 \$140 *17* 8.8% *19* $1\frac{1}{14}$

21 $\frac{24}{67}$ is larger by $\frac{9}{3953}$. *23* $33\frac{1}{3}\%$

25 \$194.80

CHAPTER 9, PAGE 210

1 \$15.30 *3* \$5,645.16 *5* 22

7 5 *9* $86\frac{1}{8}$ *11* 3

13 \$210 *15* $3\frac{21}{40}$ *17* 30

19 23 minutes *21* .47141

23 a) .3049 b) .03049 c) .003049 *25* 12%

CHAPTER 10, PAGE 331

1 2.75 *3* 13 *5* $\frac{21}{64}$

7 3.836 *9* .09, .1, 2.41, 6 *11* 1.425

13 .1 is larger by .01133 *15* 72

17 804 *19* a) 12.03 b) 120.3 c) 1203

21 \$14.59 *23* 71.5 ft *25* \$17,241.38

CHAPTER 11, PAGE 395

1 $2\frac{1}{8}$ *3* 2.377 *5* 1000

7 3.2 *9* a) 34 in. b) $8\frac{1}{2}$ in.

11 47.39 *13* \$65.85 *15* 1.625%; $\frac{13}{800}$

17 .64% *19* \$329.54 *21* a) .25 b) 2.5 c) 50

23 25% *25* Bottle: \$1.25; cork: \$.25.

CHAPTER 12, PAGE 431

1 687,006 *3* $3\frac{141}{305}$ *5* 675.93

7 .4 *9* 97 *11* \$4.27

13 1200% *15* $\frac{189}{1600}$ *17* 10; 40; 50

19 20 *21* $\frac{5}{8}$, .625, $62\frac{1}{2}$% *23* $\frac{5}{6}$, .833, $83\frac{1}{3}$%

25 $\frac{1}{8}$, .125, $12\frac{1}{2}$%

INDEX